4

Date Due

OCT 13

OMEGA

This book is printed on paper containing a substantial amount of recycled fiber. The paper has been supplied by the Wausau Paper Company of Wausau, Wisconsin.

In addition to making considerable use of recycled fiber, Wausau is one of the first major paper mills to develop the high temperature recovery process of paper making chemicals. Their recovery process has contributed greatly to pollution abatement of the Wisconsin river—a problem of grave concern to environmentalists.

Dali, Salvador. *The Persistence of Memory (Persistance de la memoire.)* 1931. Oil canvas, 9 1/2'' X 13''. Collection, The Museum of Modern Art, New York.

OMEGA

Murder of the Ecosystem and Suicide of Man

Paul K. Anderson

University of Calgary
Alberta, Canada

WM. C. BROWN COMPANY PUBLISHERS
Dubuque, Iowa

Consulting Editor
E. Peter Volpe
Tulane University

Library of Congress Catalog Card Number: 70-144389

ISBN 0–697–07575–3

Third Printing, 1972

Printed in the United States of America

to
Kristin and Erik
and
the hope
that
things
aren't
as
bad
as
they
seem

CONTENTS

Preface: A Man Said to the Universe (poem) / Stephen Crane, **xi**
Introduction: An Ecological Jeremiad, **1**

Chapter I GETTING A COMMON VIEW FROM MOUNT OLYMPUS, 13
Man's Use of the Environment—The Need for Ecological Guidelines / R.O. Slayter, **14**
Evolution in Environments / Ronald Goldring, **25**
All About Ecology / William Murdoch and Joseph Connell, **29**

Chapter II THE ECOLOGICAL NICHE: MAN AND ECOSYSTEM, 41
Pleistocene Overkill / Paul S. Martin, **42**
Evolution (poem) / Langdon Smith, **49**
The Agency of Man on the Earth / Carl O. Sauer, **51**
Man Adapting: His Limitations and Potentialities / Rene Dubos, **71**

Chapter III QUANTITY AND QUALITY OF MAN, 81
Degrees of Concentration: A Note on World Population Distribution / David Grigg, **84**
Population Density and the Style of Social Life / Nathan Keyfitz, **87**
Overpopulation and Mental Health / George M. Carstairs, **97**
Casualties of Our Time / Amasa B. Ford, **106**
The Consumer (poem) / Marge Piercy, **120**

Chapter IV METABOLITES, 123
The Smell of Cologne (poem) / Samuel Taylor Coleridge, **128**
Radioactivity and Fallout: The Model Pollution / George M. Woodwell, **128**
Pesticide Pollution / Clarence Cottam, **134**
Some Effects of Air Pollution on Our Environment / Vincent J. Schaefer, **141**
Winds, Pollution, and the Wilderness / Lester Machta, **145**
Environmental Noise Pollution: A New Threat to Sanity / Donald F. Anthrop, **149**
The Myth of the Peaceful Atom / Richard Curtis and Elizabeth Hogan, **156**
Thermal Pollution / LaMont C. Cole, **169**
Effects of Pollution on the Structure and Physiology of Ecosystems / George M. Woodwell, **176**
Darker Woods (poem) / Elijah L. Jacobs, **184**

Chapter V WE HAVE ONLY ONE BIOSPHERE, 187
Impacts of Man on the Biosphere / F. Fraser Darling, **189**
Ecology and Northern Development / Ian McTaggart Cowan, **203**
The California Desert—Problems and Proposals / Jack F. Wilson, **211**
Controlling the Planet's Climate / J.O. Fletcher, **216**

Chapter VI ATTITUDES: THE INTANGIBLE ENVIRONMENT, 233
The Historical Roots of Our Ecologic Crisis / Lynn White, Jr., **236**
The Economics of Pollution / Edwin L. Dale, Jr., **244**
A Corollary to the Dismal Theorem / Conrad Istock, **250**
Can a Progressive Be a Conservationist / Edward Teller, **255**
Confronting the A.E.C. / John W. Gofman, **260**
Clear Air and Future Energy / E.F. Schumacher, **264**
World Power and Population / Colin Clark, **274**
Second Edition / The Politics of Population / Aldous Huxley, **280**
Who is Really Uprooting This Country? / Josephine W. Johnson, **289**

Chapter VII PROSPECTS FOR THE FUTURE: POPULATION, 293
Optimum World Population / H.R. Hulett, **296**
Population Policy: Will Current Programs Succeed? / Kingsley Davis, **299**
Japan: A Crowded Nation Wants to Boost Its Birthrate / Philip M. Boffey, **317**

Chapter VIII PROSPECTS FOR THE FUTURE: RESOURCES, 325
The Future of World Energy / Sir Harold Hartley, **328**
Non-Fuel Mineral Resources in the Next Century / T.S. Lovering, **332**
Dimensions of the World Food Crisis / H.F. Robinson, **348**
The Green Revolution / Anonymous, **357**
Population and Panaceas / A Technological Perspective / Paul R. Ehrlich and John P. Holdren, **360**

Chapter IX RIDERS OF THE APOCALYPSE, 375
What We Must Do / John Platt, **379**
History's Greatest Dead End / Frank Blackaby, **391**
Eco-catastrophe! / Paul R. Ehrlich, **397**

Chapter X ALTERNATIVES, 409
The Unanswerable Questions / W.H. Ferry, **411**
Criteria for an Optimum Human Environment / Hugh H. Iltis, Orie L. Loucks, and Peter Andrews,
   419
Ecumenopolis, World-City of Tomorrow / Constantinos A. Doxiadis, **427**
Four Changes / Anonymous, **440**

IN APPRECIATION

I am grateful to those authors whose works are included in this anthology, and to numerous others whose efforts have influenced my views and my selections, even though their own works have not been included. I am grateful, also, to publishers who have permitted republication of articles.

My wife, Donna, has contributed greatly to the task of scanning the outpouring of current literature. Mrs. Margo Travis helped considerably with the clerical tasks, as did Mrs. Elise Wittig of the University of Calgary's Stenographic Services. I would like to acknowledge also the support received from the Department of Biology of the University of Calgary, and the National Research Council of Canada, during the period when this anthology was being assembled and prepared for publication.

Paul K. Anderson

UPI

PREFACE

A man said to the universe:
"Sir, I exist!"
"However," replied the universe,
"The fact has not created in me
A sense of obligation."

STEPHEN CRANE

ACKNOWLEDGEMENTS

* *The Collected Poems of Stephen Crane,* New York: Alfred A. Knopf, 1930. Reprinted with permission.

1. *Australian Journal of Science* 32(4): 146-153, 1969. Reprinted with permission.
2. *New Scientist* 671: 141-143, 1969. Reprinted with permission.
3. Reprinted, with permission, from the January 1970 issue of *The Center Magazine,* a publication of the Center for the Study of Democratic Institutions in Santa Barbara, California.
4. *Natural History* 76(10): 32-38, 1967. Reprinted with permission.
5. *A Treasury of the Familiar,* edited by R. L. Woods, New York: The Macmillan Co., 1950. Reprinted with permission.
6. *Man's Role in Changing the Fact of the Earth,* edited by William L. Thomas, Jr., Chicago: The University of Chicago Press. Reprinted with permission.
7. *Environment for Man: The Next Fifty Years,* edited by W. R. Ewald, Jr., Bloomington: The Indiana University Press, 1967. Reprinted with permission.
8. *Geography* 54(244): 325-329, 1969. Reprinted with permission.
9. *BioScience* 16(12): 868-873, 1966. Reprinted with permission.
10. *Exploding Humanity,* edited by H. Regier, and J. B. Falls, Toronto: Anansi Press, 1969. Reprinted with permission.
11. *Science* 167(3916): 256-263, 1970. Reprinted with permission.
12. *The Carleton Miscellany* 10(3): 102. Reprinted with permission.
13. no permission necessary
14. *BioScience* 19(10): 884-887, 1969. Reprinted with permission.
15. *National Parks Magazine* 43(266): 4-9, 1969. Reprinted with permission.
16. *The Living Wilderness* 36(103): 3-8, 1969. Reprinted with permission.
17. Reprinted by permission of Science and Public Affairs, the *Bulletin of the Atomic Scientists.* Copyright 1969 by the Educational Foundation for Nuclear Science.
18. *Natural History* 68: 6-16, 71-77, 1969. Reprinted with permission.
19. *BioScience* 19(11): 989-992. Reprinted with permission.
20. *Science* 168(3930): 429-433, 1970. Copyright 1970 by the American Association for the Advancement of Science. Reprinted with permission.
21. *Southwest Review* 55(1): 52, 1970. Reprinted by permission.
22. *Unasylva* 22(2): 3-13. Reprinted with permission.
23. *Arctic* 22(1): 3-12. Reprinted with permission.

24. A paper delivered at a meeting of the Western Society of Naturalists, Los Angeles, California, December 1969.
25. *Impact of Science on Society* 14(2): 151-168. Reprinted with permission. Copyright, UNESCO.
26. *Science* 155(3767): 1203-1207, 1967. Reprinted with permission. Copyright 1967 by the American Association for the Advancement of Science.
27. Reprinted from *New York Times Magazine,* April 19, 1970.
28. *BioScience* 19(12): 1079-1081, 1969. Reprinted with permission.
29. *New Scientist* 45(689): 346-349, 1970. Reprinted with permission.
30. Reprinted, by permission, from the May 1970 issue of the *Center Magazine,* a publication of the Center for the Study of Democratic Institutions, Santa Barbara, California.
31. no permission necessary
32. *National Review* 21(19): 481-484, 1969. Reprinted with permission.
33. *The Center Magazine* 2(2): 13-19, 1969. A publication of the Center for the Study of Democratic Institutions, Santa Barbara, California. Reprinted with permission.
34. *The New York Times,* May 10, 1969. Reprinted with permission.
35. *BioScience* 20(3): 160-161, 1970. Reprinted with permission.
36. *Science* 158(3802): 730-739, 1967. Copyright 1967 by the American Association for the Advancement of Science. Reprinted with permission.
37. *Science* 167(3920): 960-962, 1970. Copyright 1970 by the American Association for the Advancement of Science. Reprinted with permission.
38. This article was first published in *New Scientist,* the weekly news magazine of science and technology, London. *New Scientist,* 13 November 1969.
39. *Texas Quarterly* 11(2): 127-147. Reprinted with permission.
40. *BioScience* 19(1): 24-29, 1969. Reprinted by permission.
41. Reprinted from the UNESCO *Courier,* February, 1970, pp. 31-32.
42. *BioScience* 19(12): 1065-1071, 1969. Reprinted by permission.
43. *Science* 116(3909): 1115-1121. Copyright 1969 by the American Association for the Advancement of Science. Reprinted by permission.
44. *Saturday Review* 53(11): 19-21,46; 1970. Copyright 1970, Saturday Review, Inc. Reprinted by permission of Saturday Review, Inc. and the author.
45. *Ramparts* 8(3): 24-28. Reprinted by permission of the author.
46. Reprinted, by permission, from the July 1969 issue of the *Center Magazine,* a publication of the Center for the Study of Democratic Institutions, Santa Barbara, California.
47. Reprinted by permission of Science and Public Affairs, the *Bulletin of the Atomic Scientists.* Copyright 1970 by the Educational Foundation for Nuclear Science.
48. *Impact of Science on Society* 19(2): 179-193, 1969. Reprinted by permission of UNESCO. Copyright UNESCO.
49. From *The Environmental Handbook* edited by Garrett de Bell. Copyright 1970 by Garrett de Bell. A Ballantine/Friends of the Earth Book. Reprinted with permission.

AN ECOLOGICAL JEREMIAD

The earth is finite and its capacity to support our species, or any other species, without the deterioration of its biosphere is limited. Beyond that bald statement lie the enormous complexities of human culture and the evolving planetary ecosystem, as well as a clash of variables that change in time at vastly different rates. We are inevitably part of an uncontrolled, and non-replicable experiment to determine how large our population can become and how far the ability of a planetary life support system can be extended to accommodate it.

From an ecological point of view, the biosphere can be seen as the product of three and one half billion years of evolution, involving not only the generation of new forms of life, but also their integration as a single functioning entity. Over the last one hundred years we have learned a good deal about evolution, and over the last fifty years we have learned something about ecology, but as yet we can come to no agreement as to the degree of organization which exists in the biosphere as a whole, the process by which this organization was achieved, or the limits of its capacity to adjust and to endure. We are in desperate need of information. Yet, in our everyday culture, ecology and evolution are still regarded as interesting pursuits which are largely irrelevant to practical concerns. Despite this attitude, and our scientific uncertainties, we have suddenly become aware that we are faced with ecological disaster. We are beginning to realize that ecology, and evolution, have immense and immediate practical importance.

We cannot now enumerate all of the variables in our experimental test of the resilience of the life support system of Earth. As the number of human beings increases, the effects are clearly being compounded by per capita increases in consumption of resources and production of wastes. There is much to suggest that the delicate and intricate film of life on earth is tearing, crumbling, rotting, and dying around us. Despite this, neither the avenues of response open to us nor our ability to choose among them are evident.

Consider for a moment the limiting effects and consequences of the differing rates of change by means of which adjustments are possible. Biologically, the nature of individual men changes on the 100,000 year scale of organic evolution. Our reproductive systems are appropriate for the environmental demands of the Old Stone Age.

Culturally, we need one or two hundred years to make major changes. Our mores, the attitudes of our political and military leaders, and our political, economic, and social institutions are appropriate to the Victorian era. In that tradition, our political and economic commitment, in particular, is toward growth in numbers of people, in gross national product, and in consumption of resources.

Technologically, we can produce major revolutions in less than ten years. Technological "fixes" arise which commit us to doing those things which are technologically possible, whether or not they are culturally or environmentally advisable, and whether or not they are attuned to our biological natures. The one factor in which our biology, our culture, and our technology are in agreement is *growth,* which is the least appropriate. Anachronistic though it may be, human biology, culture, and technology are all growth adapted: they may clash with each other in the time scales of their operation, but in their relation to population growth all are in conflict with the asymptotic reality of man's present relation to the carrying capacity of "Spaceship Earth".

This is the essential message of this anthology. Technology, economics, politics, and the intangible world of human ideas all have their significant contributions to make to any solution which may be forthcoming. Still, the problem is fundamentally ecological, for only an ecological perspective can encompass the full scope of our present dilemma, and only an ecological perspective can disclose those paths which will keep open our options and guide us to the best choices.

The relationships which have evolved between living organisms taken collectively, and the physical environment, have shown remarkably stable development over the span of evolutionary time. In terms of human generations change has been imperceptibly slow, generally progressive, and without wild or sudden fluctuations. The evolutionary process has produced, and discarded, numerous kinds of organisms. While this has been going on, the effectively self-regulating total relationship between organisms and environment has also evolved. Life, and its evolution, have a large responsibility for present conditions in the biosphere. Among the critical, and planet-encompassing, consequences of the evolutionary process, we can note the composition of the atmosphere, the distribution and availability of many chemical elements, the nature of the soil, the atmospheric and surface temperatures of the earth, and the patterns of precipitation and erosion. The records of the sedimentary rocks show that the planetary ecosystem has only rarely passed through periods of major crisis and relatively rapid change. Such crises, when they occur, are recorded in the extinction of unusually large numbers of organic forms, and the appearance of unusually large numbers of new species adapted to new conditions, as well as by such physical phenomena as changing rates of sedimentation. At this moment, though we usually do not realize it, we are living in one of these rare and major revolutions in the biosphere. This crisis, unlike most of those in the past, has roots which are fundamentally biological. It is the consequence of the explosive and uncontrolled expansion of the population, and the ecological niche, of our own species.

Seen in the perspective of evolutionary and geological time, the current planetary crisis has developed with lightning speed. Man has only recently become a major factor in the biosphere as a whole. Extremely rapid, and accelerating, growth of human populations dates back only two thousand years, even in technologically advanced areas. At the time it began, the "standing crop" of human beings (the total number alive at any one time) numbered around 350 million. Between 1 A.D. and 1800 this number doubled, on the average, once every 500 to 700 years. About 1830, when man may have existed for 1.5 million years, the standing crop reached one billion. The second billion was added in one hundred years, the third billion in thirty years. It appears certain that the fourth billion will have been added (by 1975) in 15 years, and that an additional three to three and one-half billion will be added in the remaining 25 years of this century.

It is important that we understand the nature and causes of this increase if we are to grasp the impact on the stability of the biosphere, and the possibilities for control of human population growth. Prior to 1900, human population growth, even though it was unprecedented and "abnormal" for a mammal of such body size, was largely the consequence of technological innovations which expanded the resource base, and/or the application of existing technology to new land areas. As long as this was so, population growth was more or less in balance with the resources available and the ability of the existing technology to provide for human needs. From about 1900 onwards, however, death control supplanted increase in available resources as the major stimulant to population growth. Death control is the result of the developing technology of public health. Public health technology can be applied quickly and efficiently by a relatively small staff of trained personnel, with a minimum expenditure of funds and resources. The motivation for such application in the cause of relieving human illness, suffering, and unhappiness, is strong. The consequence is that death rates, particularly among infants, are rapidly reduced.

On the basis of our experience to date, birth rates, in contrast to death rates, cannot be changed so easily. Control of birth rates involves a different set of emotions and mores, and the technology involved has not proven adaptable to mass application in the same way as we apply techniques for controlling malaria, or smallpox. As a consequence birth rates have remained high, or dropped only slowly, in the areas where death control has been dramatically applied.

Sustained high birth rates in conjunction with reduced mortality among infants and children rapidly changes the age composition of the populations in question. A very high proportion then consists of children and young adults who consume food and resources, but are not in a position to contribute significantly to the economy. Furthermore, as these young people reach reproductive age they tend to give a strong additional push to the birth rate, and thus to the rate of increase of the population as a whole.

The "humane" introduction of death rate controls, and the resulting increase in the dependent population, as well as the total population, are relatively swift in comparison with the slow and complex process of developing natural and industrial resources. As a result mankind has been divided into "haves" and "have-nots" and the gap between the two has been widening. The rate of population growth has been maximized among the have-nots and they now constitute 69% of the world's adults and 80% of the world's children. These people are caught in a dilemma. The underdeveloped countries in which they live seem unable to reduce their rates of population growth without industrializing and are unable to industrialize until they have reduced their rates of population growth. This dilemma will somehow be resolved, but in the meantime conditions of life for the human beings trapped in the underdeveloped areas remain static or deteriorate. The situation is exacerbated by the availability of vastly improved means of communication. With access to radio and television, those who starve are no longer consoled by their ignorance of the luxurious life available in the technologically advanced countries. People who are young, and informed, starve less gracefully than those who are old and ignorant. Together, growth in population and in communications technology are rapidly increasing the proportion of people in the underdeveloped countries who are young, angry, and destitute. Their unrest creates chronic political emergency.

Political unrest is not a characteristic of the underdeveloped countries alone. Young adults are particularly sensitive to the tensions induced by overcrowding. The advanced countries are afflicted with accumulating wastes, decreasing privacy, and increasing expectations (generated by growth politics, growth economics, and growth ideologies) as well as by international competition for declining resources. These corporate problems interact with personal frustrations to alienate and radicalize large numbers of people, particularly the young who increase in relative numbers, in awareness, and in impatience.

I have emphasized the climate of political unrest because it is a significant factor in the environmental equation. Political tension distracts governments from urgent environmental problems, diverting needed funds and resources. The air of political emergency encourages risk-taking (e.g., the use of nuclear or chemical weapons, of excess fertilizer, of dangerous biocides, of marginal lands), and vastly increases the probability of major environmental blunders. The risks of politically motivated actions, like their cost-benefit relationships, are not usually assessed in environmental terms, and political instability becomes a major factor in local and worldwide degradation of the environment.

In this situation many long-standing attitudes are dangerously inappropriate. Our cultural tradition regards the environment as an enemy to be subjugated, exploited, and controlled. Cooperative coexistence in a dynamic biosphere comes as a novel, even amusing, concept to most of us. This is particularly true of the "elders" of our society, those who now occupy the decision-making positions. Our recent history of geographical and technological expansion has nurtured and continues to encourage the insane beliefs that our technology is all-powerful, and that our opportunities for growth in population, affluence, consumption of resources, and production of wastes are unlimited. For example, we cling to the belief that all human beings can eventually utilize resources, and pollute their environments, at the rate now enjoyed in the presently "developed" nations. Yet Hulett (page 296) has calculated, on the basis of available information, that the long-term carrying capacity of the planet at the North American level of affluence may be only about one billion people. Some estimates have placed the figure at half that. If these figures are indeed realistic, the disparity between them and our present and anticipated numbers will produce enormous stresses. Such stresses are very likely to lead to the taking of such "emergency-dictated" risks as the production and use of atomic weapons, or the increased use of persistent pesticides, which have been mentioned.

Environmentally, the crisis has clearly arrived. Join me in tracing a few aspects of the current status of man and the planetary ecosystem and consider the risks we are now taking on the basis of partial and clearly inadequate information!

The activities of each individual organism require energy. For the present, at least, the supply of energy for physiological activity of all except a few minor life forms is derived from solar radiation through photosynthesis. The human population which now exists, if provided with an adequate diet, might consume as much as two percent of the annual net plant photosynthesis of the planet as a whole. *Homo sapiens* is but one of perhaps two million heterotrophic species. For one species out of two million to acquire as much as 1/50th of the photosynthetically available energy supply is indeed remarkable, yet at our present rate of growth, and our indulgence in activities which inhibit or prevent photosynthesis over large areas, the human energy requirement might equal the total annual net photosynthesis of the planet in as little as two hundred years. This would mean the elimination of all other species of animals as well as the heterotrophic plants. Such a possibility is an unqualified fantasy, yet it illustrates the ecological irrationality of our current faith in unlimited expansion, and the pressure which the sheer mass of human beings is exerting on other passengers of "Spaceship Earth".

Numerous signs of stress within the human population, and stress on the ecosystem in which it must exist, are already obvious in the light of our current ecological knowledge, which includes some general observations on overcrowding in mammalian populations, and on the stability of ecosystems. Are human populations showing symptoms of intolerable overcrowding? Among vertebrates overcrowded populations are characterized, initially, by pressure on younger individuals and consequent restlessness among them. This normally leads to emigration and colonization of new areas. Many individuals settle in areas which are marginal in quality, where the risks of existence are high. In such areas there may be rapid over-exploitation of the habitat, and conditions for survival may be minimal with a high incidence of disease and generally decreased individual quality and life expectancy. If the mechanisms inducing emigration and its population-regulating consequences are frustrated, psychological and physiological abnormalities appear, including altered social systems, increased aggression, homosexuality, and maternal inadequacy. Man has now filled the earth beyond the possibility of operation of the emigration mechanism. Many of the symptoms of blocked emigration-inducing mechanisms which I have described, only moderately transformed in the matrix of human mentality and culture, may be recognized in both the ghettos and suburbs of human cities.

If the human population has, in fact, exceeded the carrying capacity of its ecosystem, we might anticipate violent fluctuations in numbers. William and Paul Paddock argue, in *Famine 1975!*, that the turning point has come, and that mankind is irrevocably committed to a mass die-off through starvation within five to ten years. Their arguments predicate 1970 as the time of a sharp break in the rate of increase in food production, in conjunction with a further increase in the overall rate of population growth. As a consequence, they anticipate that sometime between 1975 and 1980 the present era of surplus food production in North America will end. At that time most or all of the food produced in the United States and Canada together will be required by the population of the North American continent. Exports from these countries have, in past years, filled the gap between production and consumption in the underdeveloped countries, and the end of these shipments could trigger mass starvation elsewhere. Should major famines occur, "emergency" measures, ranging from cultivation of submarginal lands and unrestrained use of fertilizers and pesticides, up to and including conventional and even nuclear warfare, become highly probable.

The signs of stress in ecosystems include the elimination of the more sensitive species, and the increasing instability of populations of those which remain. We would predict, also, tendencies for accumulation of unused and partially decomposed organic matter as food chains are decimated or interrupted. Changes may also be expected as a consequence of inadequate recycling of inorganic materials.

A famine is an obvious emergency, but the fact that we have already committed ourselves to risks which have grave or unpredictable consequences for our ecosystem is not obvious to the non-ecologist. For example, we meet current food needs only through an exaggerated version of the ecological phenomenon known as competitive exclusion. In our agricultural operations we replace stable natural communities of from 10,000 to 50,000 co-adapted species with monocultures of a single crop plant. We attempt to eliminate all potential competitors from these monocultures. During a recent five-year period in South Africa some four hundred million red-billed finches were killed as "pests" of grain crops. Most insecticide use is in the cause of competitive exclusion, and during the 20 years preceeding 1966 the United States alone produced and released into the planetary ecosystem more than 2.5 billion pounds of DDT. The consequences to the ecosystem, and their gravity, were not assessed in advance and are only now becoming apparent. In central Africa hundreds of thousands of large mammals have been slaughtered in an effort to clear areas of disease-carrying tsetse flies and open such areas for grazing by domestic cattle in order to increase the supply of protein available to human beings. Agriculture, and construction, drastically alter habitats and supplement direct killing in the elimination of other organisms, the decrease of ecosystem complexity, and the increase in instability.

A few activities which represent risks to the survival of the biosphere have been pointed out. Have we, up to the present, thoroughly evaluated the possible consequences of such risks in relation to the desired benefits? The extremely toxic and long-lasting poisons which we have allowed to penetrate every cranny of the biosphere are carried in the bodies of all human beings, and the milk of nursing mothers exceeds the "safe" limit set for dairy products in the United States. The obvious contradiction in that fact alone would suggest that we are taking unacceptable risks, and the level of such contamination in the biosphere as a whole is rapidly and unavoidably rising, with the peak levels consequent on past use alone still to be achieved. There is growing evidence that the accumulation of pesticides in the oceans may be sufficient to seriously impair, or to destroy, the photosynthetic activity of marine plankton. If that happens we shall have lost a major mechanism responsible for maintenance of oxygen and carbon dioxide balances in our atmosphere!

Even excluding pesticides, our present activities involve serious threats to the composition of the planetary atmosphere. We pay a high price in energy consumption in order to maintain today's population and support its further growth. Part of that price is devoted to propping up the increasingly unstable local ecosystems created by agricultural practices necessary for our present (less-than-adequate) production of food. We also find it necessary to pay a higher price in terms of energy as we turn to more refractory mineral resources. While our population is increasing, per capita energy requirements are increasing even more rapidly and the majority of this energy must be derived from sources other than immediate photosynthesis. To meet our needs for transportation, for extraction of raw materials, and for industrial production, we are oxidizing, in a few decades, all the fossil fuels (coal and oil) accumulated beneath the earth over the past 500 million years. The consequent rapid release of vast quantities of carbon dioxide is overloading the balancing mechanisms of the ecosystem and CO_2 levels in the atmosphere are rising. The theoretical consequence is an increase in atmospheric temperature ("greenhouse effect") followed by the melting of the ice caps and a rise of sea level of as much as 300 feet with a consequent loss of much habitable and agriculturally productive land.

The "greenhouse effect" may be supplemented by direct release of waste heat into the atmosphere, or both effects may be counterbalanced by a man-caused increase in the turbidity of the atmosphere, but there are still other consequences

Female Peregrine Falcon Eating It's Own Egg Under Influence of Pesticide Poisoning. Canadian Scientist II (4):17 1968. Reprinted by permission.

of spiralling energy demands. It is generally accepted that an increasing proportion of electrical energy must be produced by nuclear installations. These are being rapidly constructed, despite the unsolved problem of unavoidable environmental contamination during their operation (usually evaluated only in relation to its direct effect on human beings), and the still more serious problem of radioactive waste disposal.

Meeting energy requirements involves production of thermal "waste". The unusable heat produced in the generation of electricity, by any means, must be dissipated. The requirement for cooling generating plants in the U.S. which is forecast for the turn of the century involves the equivalent of all fresh water on the North American continent. Such use of surface waters may produce an average temperature increase of from 15 to 30 degrees Fahrenheit.

The impact of this warming of surface waters on climate, and on local ecosystems, is only partially predictable, but it will certainly aggravate problems which are superficially far removed from those of energy production. For example, the agricultural techniques which we utilize in order to support our population are extractive. Food is produced on land fertilized with nutrients brought from distant areas. These nutrients are extracted, and transported, at a high energy cost. Crops in monoculture also require the use of biocides, which must also be manufactured and transported at an energy price. The crops are cultivated by machines using fossil fuels. Both the machines and their fuels are made available at an energy price. The food that is produced is then transported, re-manufactured and packaged, and transported once again to cities where it is finally cooked (still further energy consumed) and converted into organic wastes. These wastes are either flushed directly into a river or lake, or (in an energy-consuming sewage disposal plant) converted into nutrients which are then released into the river or lake. In either case the result is the same: the original organic waste, or the organic material produced via the nutrients released from the treatment plants, causes an overload on decomposer elements of the aquatic ecosystems. The "waste" heat of the generating plants where much of the energy was produced enters the picture again at this point, since, in thermally stratified water bodies, the effect of the heat injected with "waste" cooling water is likely to be a greater and more rapid conversion of nutrients into organic material in the warmed upper layers of the water. There is unlikely to be any compensatory effect on the rate of decomposition in the cool bottom waters. The net result will be a more rapid depletion of oxygen, with a consequent reduction in the rate of breakdown of organic wastes in the cooler waters.

The sequence is typical. Many steps removed from the original question of the "waste heat" of energy production we find ourselves involved in the eutrophication of lakes. Complex as this may seem, it exemplifies, in what is actually simplified form, the kind of involved relationships to which we have been committed. With the present world population, technological man is a destructive factor in the biosphere. The ecological childhood of man is over, and it has ended without the gift of ecological wisdom. The voices of our technomaniacs urge us to ever greater interferences with, and depredations on, the tottering ecosystem. Our ecologically ignorant politicians and economists preach the necessity, and proclaim the desirability, of interminable growth. Yet it may be, even now, too late for a solution short of total destruction of the living earth.

UPI

Getting a Common View from Mount Olympus

Like the perception of beauty, the perception of meaning lies largely in the mind of the beholder. The intellectual voyage proposed for this anthology has real need of a common departure point in shared perceptions of meaning. What, in the context of both evolutionary and human histories, *do* we mean by an ecological equilibrium? What are "balance" and "instability" in "The Ecological Theater and The Evolutionary Play"?

Even if most readers can arrive at the starting point, or at least leave it, with some common perceptions of meaning, the enterprise to which this anthology presumes will be difficult enough. We must encompass, simultaneously, some Olympian differences of scale. We must consider both the frailty and the resilience of the biosphere in the contexts of human and of cosmic time. We must deal with the complexities of both human societies, and ecosystems, without losing sight of the falling of human sparrows: the beauty, the pathos, and the solitary individuality of human existence.

In all of this, we must gather knowledge and wisdom sufficient to apprehend and consider some extreme viewpoints. That is perhaps the best excuse for an anthology. From the common Olympian ground for which this first chapter strives, our minds must be open to those who see around them an on-coming and unending high summer of evolutionary time. We must also hear those for whom the extinction of falcon and eagle represents the shedding of leaves from the evolutionary tree in the melancholy fall preceding the last of all winters in life's only known habitat in the universe. Is the glow of nuclear fission and fusion a sign of dawn, or of sunset?

Man's Use of the Environment — the Need for Ecological Guidelines

R. O. Slayter

Man is an animal, and is part of the so-called "web of life". There is, however, an important difference between man and all other organisms. To an unprecedented degree, man has been able to manipulate other species and the environment itself. In the process his numbers and needs have increased dramatically, and are still increasing. Yet the capacity of the environment to supply these needs is limited, and man's exploitation of the environment tends to reduce that capacity. Clearly, this constitutes a collision course, vividly reflected in the axiom that man's future existence on earth is not threatened by any species other than himself.

The thesis I wish to expound is that failure to observe some ecological ground rules has put man into this position. Ecology is the study of organisms in relation to their environment; it deals with the environmental requirements of single species and with whole populations or communities, with the way in which organisms influence, and are influenced by, their environment; and with the way in which organisms interact with one another. If man is to persist on this planet indefinitely, I believe he must adopt a new ethic, based on ecological premises, to guide his future activity.

I would like to consider first aspects of the evolution of the global environment itself, and of the way in which organisms live together, finally looking at the effect of man's past activities, and pointers to the future.

Evolution of the Global Environment

The history of the earth as a planet still contains many secrets and considerable controversy surrounds aspects of its formation and early development. As far as man is concerned, it is the evolution of the atmosphere and the hydrosphere which is of particular interest, since both are quite essential to life as we know it. Without these thin surface films, the earth would be exposed to wide temperature extremes, the land would be bombarded by lethal radiation, and it would not contain the water and gases necessary for biological activity.

As a crude analogy, the barren surface of the moon provides a useful illustration of their role in creating a favourable environment for life on earth.

The atmosphere and hydrosphere only represent a tiny fraction—about 0.24 per cent —of the earth's mass. The atmosphere itself represents only 0.37 per cent of the hydrosphere (Mason, 1958). As we all learn in high school physics, its mass would be equivalent to a layer of water covering the earth to a depth of about 33 feet. Even so, the actual quantities involved are very large. The atmosphere, converted to the density of water and heaped on the continent of Australia, would form a layer almost half a mile deep. The oceans would form a layer about 150 miles deep.

It now seems to be generally accepted that the primitive earth had virtually no atmosphere or hydrosphere (at least in the context of their present characteristics), and that both have accumulated gradually during geological time as a result of the escape of water vapour, carbon dioxide, nitrogen and other substances, from intrusive and extrusive rocks within the earth's interior (Rubey, 1963).

Before the evolution of life much of the water vapour so evolved had condensed to form the oceans; many of the other gases dissolved in the water so formed. Carbon dioxide and nitrogen, after water probably the main constituents of the internal emissions, followed different paths. Carbon dioxide, as well as dissolving in water, was continuously removed by reaction with silicate rocks to form, in a series of geochemical reactions, carbonates and silica, the former becoming the first of the great sedimentary rocks of today, the latter the ubiquitous mineral of the contemporary earth's surface. Largely in consequence, only trace amounts of CO_2 remain in the atmosphere although these amounts are of great significance to man. By comparison, nitrogen, relatively insoluble in water, was also relatively inert chemically and tended to accumulate in the free atmosphere.

By comparison with today's atmosphere, this pre-life atmosphere was probably almost oxygen-free, and its carbon dioxide concentration may have been significantly higher. The general temperature range was probably not much different from today, although mean temperatures may have been higher than now because higher concentrations of carbon dioxide would have tended to accentuate the "glass-house" effect of the atmosphere. (Like glass, CO_2 permits short-wave solar radiation to reach the earth's surface, but impedes the escape of long-wave thermal radiation.)

Another important difference, linked closely with the relative absence of oxygen, was the absence of ozone. Ozone is a remarkably effective filter of U-V (ultra-violet) radiation; its absence meant that radiation of this wave-length reached the earth. U-V radiation has a substantially higher energy content than visible radiation, so much so that it can photochemically dissociate many molecules. This meant that the atmosphere and surface layers of the hydrosphere became a vast chemical factory in which this energy enabled many types of organic compounds to be synthesized, including the chemical building blocks required for organic metabolism and self-replication. In consequence, in the period between the earth's formation and the appearance of life, the oceans became what Cole (1966) has termed a 'dilute soup' of organic molecules.

From these building blocks, and in this reducing environment, the first life processes appear to have emerged 3.0 to 3.5 billion years ago (Cloud, 1968), and the first tremendous step towards the evolution of man occurred. It seems likely that the first self-replicating entities appeared in deep and protected waters. Since U-V radiation destroys biologically active molecules and is in fact lethal to life as we know it, life must have emerged in water at depths below the zone of penetration of ultra-violet radiation (about 30 ft.). The waters were almost certainly protected ones, since active stirring of the water by wave or tidal movements would have brought these first lifeforms closer to the surface and into the zone of influence of disintegrating U-V.

It is not appropriate now to detail what is known of the exciting story of organic evolution—the development of the cell membrane, of the single cell and of multicellular organisms, of the various refinements and adaptations, that led to man. Many books have been written, and will probably be rewritten, on this subject. There are, however, some highlights of the evolutionary record which I do wish to mention because of their environmental as well as their biological significance.

The chief one of these concerns photosynthesis, the process by which green plants convert a flow of solar energy into stored chemical energy. This is the process on which all of us, directly or indirectly, are still dependent for our food supplies, our clothing, almost all of our fuel and power, and much of our shelter.

The first forms of life were heterotrophic, dependent on outside food sources—of chemical nutrients—for their survival. Although, when life first appeared, the "dilute soup" was so abundant that ample chemical energy, in the form of nutrients, was available for nutrition, it is clear that life could not persist indefinitely unless it became autotrophic—able to be self-sufficient for its basic energy needs.

The evolution of photosynthesis provided the key for this autotrophism. Photosynthesis utilized carbon dioxide and water, as raw materials, producing carbohydrates and oxygen, while storing solar energy in the intramolecular structure of the carbohydrate molecules. In the presence of adequate carbon dioxide the rate of photosynthesis is closely dependent on light intensity.

The building blocks for photosynthesis were also formed in the "dilute soup;" the first photosynthetic organisms are thought to have appeared about 2.7 billion years ago (Hoering and Abelson, 1961). Like their non-photosynthetic cousins, they were restricted to deep water by the need to avoid U-V damage; at these depths light intensity was also much reduced, so that the rate of photosynthesis was initially very slow.

Until photosynthesis began, it seems that the only oxygen in the atmosphere was that formed by the photodissociation of water molecules by U-V radiation (Berkner and Marshall, 1965). While significant quantities of oxygen were made available by this process, although at a slow rate, and also some ozone, by the photodissociation of the oxygen molecules themselves, it seems likely that both oxygen and ozone were maintained at near zero levels by their rapid involvement in oxidation reactions.

The oxygen and energy produced by photosynthesis is consumed by the metabolism of the chains of organisms which feed on the photosynthetic products. This respiratory metabolism represents, in chemical terms, a net reverse of the overall photosynthetic reaction, energy being released as oxygen is consumed and carbon dioxide is produced. A net yield of oxygen is therefore only obtained, and a net store of energy conserved, when the photosynthetic products themselves, or fossil remnants from other organisms, accumulate at protected sites, such as in the ocean depths.

When photosynthesis first appeared, the oxygen evolved was also rapidly utilized in oxidative reactions both in the water itself, in the atmosphere, and in weathering processes on land. Very slowly, however, the oxygen concentration began to rise. When it reached a level sufficient to maintain a permanent concentration of ozone in the atmosphere, a major step forward in the evolution of life began, because the ozone began to filter out U-V radiation and reduce its intensity at the surface. In consequence its depth of penetration in the oceans began to decline and organisms began to occupy shallower waters, in which visible radiation was higher and photosynthesis could be much more rapid.

Gradually this trend accelerated, oxygen and ozone increasing until ozone levels were high enough to shield the land surfaces of the earth from lethal U-V. This provided tremendous opportunities for biological diversity; the scope for colonization on land can be appreciated when it is realized that the land surfaces were bare at this time. It is not surprising that in a relatively brief span of 30 million years, between the late Silurian period, 420 million years ago, and the early Devonian, virtually all the land surfaces were vegetated.

Since that time continued evolution has brought us the global environment we know today. Without an ozone screen to protect us from U-V, without oxygen for respiratory metabolism, without CO_2 for photosynthesis, life as we know it could not exist.

Many other features of the environment are also vital for our existence. Almost without exception all are in a dynamic state of balance, in which biological and geological factors interact to provide the attributes that now prevail. Thus our present global environment is not a relic from a bygone age, nor is it a resource which can be exploited by man as though it were limitless, or as though change can be made in one aspect without regard to the repercussions of that change to the whole.

In this overall environment biological evolution has led, through genetic innovation and natural selection, to the colonization of almost every environmental niche on earth, with a range of plant and animal species, each competing with, yet often dependent on, others for space and nourishment. In this way a range of biological environments has been developed; collectively they are termed *the biosphere.*

Man, although he frequently attempts to pretend otherwise, is a creature of his heredity and his environment. He has evolved through the mainstream of biological evolution and the biochemical building blocks which constitute his structure and regulate his function reflect the restraints and demands of the environments in which they evolved and to which they became adapted. Man cannot live in a range of climates much different from those on terrestrial earth (I do not regard planetary or deep ocean existences for man as living!); he cannot exist without a range of foodstuffs that contain the chemical nutrients essential for his metabolism. He undoubtedly has other requirements associated with his inherited behavioural patterns. Yet man is capable of changing his environment to such a degree that some of these primary requirements could be destroyed.

Before we turn to examine aspects of man's impact on his environment, let us look a little further at some of the main attributes of life itself.

Ecology and Environment

A primary feature of life on earth is that organisms do not exist in isolation; instead the entire biosphere is composed of a range of ecosystems, each of which contains a number of species and a number of micro-environments, with each species tending to utilize and occupy an environmental niche more effectively than its competitors; the whole assemblage of species tending to cohabit in a manner that provides a high degree of internal self-regulation. A forest, or a lake, provides examples of typical ecosystems, but the scale can vary widely; the entire biosphere constitutes the earth's ecosystem.

A primary feature of an ecosystem is that it tends towards self-regulation. Solar energy is absorbed by the green plants of an ecosystem, to provide, through photosynthesis, the basic energy input. Plants also absorb water and mineral elements from the soil. The plant components thus produced are then passed through a food chain, in which the initial products are eaten by herbivores, herbivores by carnivores, carnivores by other carnivores, and so on until decomposer organisms return the organic wastes and the remnants of the organisms in the food chain to the soil, in a form that enables their reabsorption by the green plants.

In most natural ecosystems, therefore, there tends to be no net production—in the human context of a net harvest of materials. The solar energy absorbed and stored by the green plants is gradually consumed by metabolism through the food chain and dissipated as heat. Thus there is a flow of energy through an ecosystem, starting from solar energy, passing through successive forms of chemical energy —at each stage some energy being lost as heat —until it is all dissipated. Associated with this flow of energy is a cycling of nutrients through food chains so that the ecosystem as a whole tends to be balanced and self-contained.

Within an ecosystem the numbers of any one species change in response to opportunities afforded to it in the way of nutrients and space. An environmental catastrophe, such as fire or drought, generally affects some species more than others; the more successful species increase in number until other restraints develop to prevent their further increase. Similarly, invasion of an ecosystem by a new species better able to compete than some existing ones will be associated with an increase in the number of the better adapted species and a reduction in the number of the less well adapted ones. The weeds in your garden provide examples of this.

An important phenomenon associated with such a disturbance is that, when a new species enters an ecosystem, it first encounters competition from existing species. If it is successful in gaining entry, its numbers increase until, in the absence of other competition, its own increasing numbers and demands for space and nutrients prevent further increase—i.e., it encounters competition from within its own species. A moral for man's explosive population increase can be seen here.

The stability of an ecosystem, its ability to adapt to invasion or catastrophe without major change, is largely a matter of its diversity. In turn, this is largely a matter of the rate of nutrient cycling, or the rate of energy flow. An ecosystem with little diversity is vulnerable to invasion, and especially if energy flow is slow, is often unable to adapt to the change, at least without a period of marked instability. The

successful invasion of Australia by rabbits is a good example of limited species diversity in Australian ecosystems. The ability of rabbits to compete favourably with other animals for forage, and the absence of effective predators, meant that rabbit numbers increased dramatically until in many areas competition for food, between rabbits themselves, was the main factor limiting their numbers. The successful invasion of Queensland pastures by prickly pear is another example of ineffective competition and the absence of suitable animals in the food chain to consume it. Fortunately, the absence of predators for the insect *Cactoblastis,* introduced to control prickly pear, meant that this animal could control it effectively and the existing ecosystems, perhaps enlarged by these two species alone, forming another loop in the food chain, returned to a degree of stability.

The most impressive examples of potentially unstable communities, ripe for invasion by other species, are agricultural crops, where a single species may be grown over thousands of square miles. The opportunities for invasion by "weed," "pest" and "disease" species—all words of modern man's vocabulary—are tremendous.

Management of an ecosystem, in the sense of increasing the numbers of one, or a group of species within it, and perhaps in removing them from the ecosystem so that there is a net yield, need not disturb internal self-regulation. The primary need is to ensure that nutrients continue to be recycled, that other important organisms are not adversely affected, and that there is sufficient diversity in species composition to prevent the community from becoming unstable.

So we see that nature obtains stability by allowing energy to flow smoothly through ecosystems, by retaining and recycling nutrients and by encouraging species diversity. We should not think that natural ecosystems do not change; geological processes and climatic change bring slow changes in the composition and structure of ecosystems, as does the constant geographic movement of species and continued genetic evolution. Abnormal weather, fire and similar phenomena bring rapid changes.

Furthermore, the species themselves change. Apart from the evolution of distinct new species, existing species change in character as environmental change induces natural selection. The degree to which a species can change in a given period of time depends on its intrinsic capacity to change—its inbuilt genetic diversity—and on its generation time. The most resilient and adaptable species tend to be those with short generation times. It goes without saying that these include many of our pest and disease species. The rapid development of immunity of insects to DDT, of disease organisms to penicillin and of rabbits to myxomatosis reflects this property.

Let us now look at man's impact on the environment to see how he has adjusted ecologically to the biosphere in which he evolved. To my mind it is easiest to do this by looking at man in three stages of his cultural evolution —Man the hunter-gatherer, Man the herder-farmer and Man the technologist.

Man's Impact on the Environment

When man first appeared, on the order of a million or so years ago—a very brief period in geological time—the earth contained many of the species of plants and animals which exist today, most of the climates which exist today and many of today's topographic features. Although the distribution of climates and the location of shorelines have changed, the range of ecological situations available for life has not changed to a pronounced degree. In many regions the first men enjoyed the same type of weather, breathed the same kind of air, ate the same kinds of animals as did Neanderthal man only a few tens of thousands of years ago, or today's huntsmen-campers in areas remote from industrial centres.

Just why man appeared when he did, and rapidly demonstrated his mastery over his environment and over other species, is still a subject of some speculation. However, even if it is not surprising that man's superiority rapidly ensured his survival, man's place in nature, until his first deliberate activities in cultivation and animal herding, was virtually the same as that of any other creature. Man the hunter-gatherer preyed on, and was preyed on by, other animals. He gathered plant foods when and where they were available. He was, in all respects, part of the food chain of the ecosystems he lived in; the changes he was able to make to his immediate ecosystems, apart from reduction in the numbers of the animals he consumed, were probably less than those of rabbits. Clearly, he made no changes to the biosphere as a whole.

The Australian aboriginal, in many respects, lived this way prior to European colonization of Australia. If aborigines could not be seen, evidence of their previous presence at any location was meagre and short-lived. In most respects they were as well adjusted to the natural environment as the animals and plants they consumed.

These men really lived their ecological role. Admirable though this was in the sense of permitting nature's overall fulfillment, it clearly left man vulnerable as a species. So it was that man the farmer emerged.

Those of us who admire the noble savage must conclude that life since the emergence of agriculture has been one long downhill slide— from fun to work. Whether or not one agrees with this view, the fact is that when man became a farmer he began a commitment from which there was no escape.

As I mentioned previously the numbers of any species in an ecosystem will tend to rise if it is protected from its competitors. Man the farmer sought to ensure his own survival by protecting his food supplies—the plants and animals he wished to consume—from competition and predation, and by deliberately cultivating them. In ecological terms he sought to maximize the energy flow passing through himself by maximizing the energy flow passing through the species directly ahead of him in the food chain.

Because he was also able to protect himself against predation, his numbers increased, and immediately his dependence on the "managed" ecosystems increased further. There was no going back. Not only did he become dependent on agriculture, but his increasing numbers started a demand-supply spiral which meant that continually he had to attempt to increase both the area under cultivation and the yield of any one area.

Furthermore, the more specialized his agriculture became—the more, if you like, he attempted to reduce the number of species in his managed ecosystem—the more unstable they tended to become. The tendency for undesirable species to invade his farms increased, the removal of existing vegetation exposed his lands to erosion, and this factor combined with the removal of nutrients from his ecosystems without replacement, began to reduce their productive potential.

Thus many of the early agricultural efforts tended to be highly exploitative, and a shifting agriculture arose. When productivity dropped, new farms were developed, and the old farmlands were left to revert to a natural ecosystem. The shifting agriculture still evident in parts of New Guinea is a good example of this.

With time, man learnt to replace nutrients on his fields by using crop residues and human and animal manure; he learnt to prevent erosion by maintaining ground cover and cul-

tivating only level land. He learnt to give his domesticated species an edge over their competitors and predators by farming practice. Crop rotation, sanitation practices, and other procedures all helped in this regard. In consequence, his ecological management practices improved, nutrient cycling was restored, and levels of stability were achieved.

Although some specific ecosystems were badly damaged (for example, those on the fringes of the Sahara where herding of goats caused almost completely new and much less productive ecosystems to be established), man's activities, in most cases, were still of little consequence to the biosphere as a whole. Not only was his capacity for major change limited by his muscle-power and that of his domesticated animals, but his numbers were still so small that there was always more land over the next hill or in the next valley.

Man the farmer was therefore able, if not to live an ecological role as fully as man the hunter, at least to avoid large-scale environmental change, although his low numbers were the main factor here. Even so, he barely managed to match food production against numbers—as periodic famine in peasant agricultural systems, even today, testifies. Thus in ecological terms he was stretching capacity.

The first signs of a new problem in ecological management were also evident whenever man congregated into villages and towns. In these centres there developed great accumulations of organic debris—body wastes, food scraps, the detritus of the range of human activities. At this stage the word "pollution" could well have been introduced to man's vocabulary. Pollution, in ecological terms, generally means the addition of materials to an ecosystem which accumulate because they overload the existing food chain pathways or because, in the modern idiom, they are "non-bio-degradable," that is, they cannot be recycled at all, or fast enough to keep their levels low. In the former case the opportunities for species capable of capitalizing on the extra nutrients are increased considerably; in human terms these species are frequently pathogens or carry pathogens, and epidemic disease is the logical result. In both cases, however, the materials may affect species other than man in quite different ways from those in which man is affected, and pronounced changes in ecosystem characteristics can follow.

The story of man the technologist is, in most respects, simply an extension of man the farmer. However, with the industrial revolution, man's ability to harness power to his needs meant that the impact of his activities on the environment increased tremendously.

To my mind this impact has two main, closely linked, aspects. The first has been the dramatic, and continuing, increase in man's numbers as his capacity to manipulate agricultural ecosystems for food production, and his control over human disease, have increased. The second has been the development of a great diversity of human activities, human demands, and human products. Thus not only has population itself increased rapidly, but the demands of each human being for factors other than basic food needs, have also increased. Man the technologist expects, not merely to survive, but to enjoy a socio-economic infra-structure which provides transportation, education, housing, recreational space and many other cultural facets.

To satisfy these desires and needs, man has affected the environment both directly and indirectly. In a direct sense his mechanical activity—in constructing cities, highways, dams and in soil cultivation and mining—is the most striking and obvious. Indirectly, though, the other products of modern technology are also of great importance as agents of change.

These products, in the main, are those of chemical and engineering technology. They include the wide variety of products needed by modern industry, agriculture and commerce —agricultural and industrial chemicals, for example; as well as the by-products of these industries—manufacturing residues, exhaust fumes, hot water. They also include the debris of the modern consumer society—packaging materials, sewage and food wastes, detergents, worn-out and useless "durable" consumer goods, and so on.

In these activities, man the technologist has attempted to ignore the capacities and characteristics of his ecosystems. The agricultural ecosystems in which he produces his food have been still further removed from stability. He has loaded them with the products of chemical technology thinking only of maximizing food yield from a few species. In the process the nutrient and non-nutrient chemicals (fertilizers, pesticides, weedicides) which he has added have had repercussions far beyond the ecosystems to which they were applied.

Excess nutrition of agricultural fields can cause the enrichment of inland waters into which the nutrients drain. In time this can cause such an increase in aquatic photosynthesis that the aquatic ecosystems can be completely changed. In some cases lakes can fill in with photosynthetic products (Sperry, 1967). With regard to pesticides, there are many examples of target organisms developing immunity to the substances used to kill them, yet of other species remaining susceptible. Furthermore, not only are vulnerable species directly damaged or destroyed, but many of these substances remain biologically active for lengthy periods of time, and may accumulate, particularly in food chains involving sequences of carnivorous animals, until they kill species originally unaffected. Fish and birds are often affected in this way. In consequence agricultural practices have, in some cases, endangered the ecological stability of extensive regions.

In the ecosystems in which man has built his cities, he has also added vast quantities of nutrient and non-nutrient compounds—both domestic and industrial wastes. These too have affected regions far greater than the areas where these materials have been dumped (Beeton, 1965).

The results of these activities are all around us, and I do not need to recount them. A week seldom goes by without the press providing examples. The stark realism of the situation is that, in the case of agriculture, I believe we are now dependent in large measure on these practices; and in the case of our population centres, I think we have almost come to accept them as a fact of life.

The impact of these changes on the environment as a whole is now just beginning to be appreciated. Ever since his appearance on earth, man has attempted to exploit his local environment for his own ends. In a sense, all organisms have done this, whether deliberately or not. As long as man used natural methods for this exploitation, and the power of his own metabolism, and as long as his numbers were low, there was little likelihood of the changes he induced affecting more than his immediate environment—and his early attempts at exploitation were still essentially conservative, in the sense that they preserved the basic diversity and character of the environment. Now, however, the situation has gone full circle; man is so abundant and so powerful that he is changing the properties of the entire biosphere. The rate of change is far greater than occurred even during the great transitions from one geological epoch to another. This is the collision course referred to before; clearly the trend must be reversed.

Man and the Future

The motivation for many of man's actions has been what Hardin (1968) has termed the philosophy of the Commons. He illustrates this philosophy by the following example: "Visualize a pasture open to all, on which a number of herdsmen are free to graze their stock." Each, as an independent entrepreneur, will attempt to keep as many stock as possible on this common ground.

Few problems arise until the total number of stock reaches the carrying capacity of the pasture. Then any one herdsman asks, "What is the value to me of adding one more animal to my herd?"

His profit is related to the number and condition of his stock. He knows that the addition of one animal to the total herd on the commons will reduce the nourishment available to each by a very small fraction indeed. By comparison that animal represents a relatively large increase in the size of his own herd.

As Hardin points out, "The rational herdsman concludes that the only sensible course for him to pursue is to add another animal to his herd. And another, and another. . . . But this is the conclusion reached by each and every rational herdsman sharing a commons. Therein lies the tragedy."

In how many cases can we see aspects of this philosophy being acted out: the chemical manufacturer who decides that rather than clean up the effluent from his factory it is cheaper for him to pollute a stream and pay only a share of the taxes levied by the city council to clean up the pollution so induced; or the farmer who keeps adding agricultural chemicals to increase his yield, without thought to the effect elsewhere; or even the family, in countries with pressing population problems, that adds more and more children to the load society as a whole largely bears.

Clearly, the freedom of the individual in matters affecting other people cannot continue without restraint. The last great "commons" are the air and the sea, yet already changes are occurring there.

Let us look, for example, at changes which are occurring in atmospheric CO_2. The energy that man has used to power his industrial society has come primarily from the combustion of fossil fuels. You will recall that these have been laid down during geological time, and that it was largely their laying down that caused oxygen levels to rise to values which permitted man's evolution. Fortunately, consumption of these fuels is not likely to directly affect oxygen levels, at least to a significant degree, but already it has affected the level of CO_2 in the atmosphere as a whole.

The rate of combustion of these fuels is staggering. Last year the amount which was burnt was equivalent to a bush-fire which would destroy the entire vegetation of Australia. Small wonder that, if present trends continue, the amount of CO_2 added to the atmosphere is likely to be equivalent to the present concentration within 50 years (Revelle and Suess, 1957). So far it has not been possible to calculate the likely climatic effects of this change; suffice it to say that some estimations predict an increase of mean world temperature of several degrees (Plass, 1956). Should this occur, dramatic changes in, amongst other things, cloudiness, rainfall and agricultural productivity may take place.

What are the solutions to this collision course between man's numbers and demands on the one hand, and his environment on the other?

It is clear that the goal of the "greatest good for the greatest number" is impossible to achieve. Despite the bliss that this phrase may conjure up in some people's minds, man cannot maximize for both of these factors at once. Ecologically, it is probable he cannot maximize for either, unless the goal "of the greatest good" is identified with conservation of his own species. But our present numbers, our present technology, and our present attitudes are not consistent with this goal. What are the

requirements if man is to maximize for conservation of his own species?

First, he must regulate his numbers. Unless population stability is achieved, everything else must ultimately fail.

Secondly, he must conserve and recycle the basic materials he uses to the greatest possible degree.

Thirdly, he must ensure that his food supply is adequate for his regulated numbers, and that the means of its production are not in themselves leading to environmental deterioration.

I believe the principle of conservation and recycling should be applied not only to renewable, but also to non-renewable materials. For example, the nutrients in domestic sewage (including food scraps as well as body wastes) at present move resources such as rock phosphate, or atmospheric nitrogen, to factories, to farms, to human mouths, and then through sewage systems to inland waters or to the sea. There is no effective link to recycle these nutrients to the agricultural ecosystems where they could be used again. In consequence, the sewage tends to pollute the ecosystems into which it is discharged, by over-enrichment even if not in a pathological sense.

In much the same way metals for industry are largely obtained by digging fresh ore out of the ground, rather than by re-using it from worn-out commodities. These non-renewable resources are not inexhaustible. Furthermore, their extraction, processing and accumulation frequently pose problems in ecosystem stability. Although short-term "commons" type economics may indicate that present practices are cheaper than those which would involve resource conservation and recycling, I contend that serious attempts to reduce the cost of re-separation of primary substances from manufactured products—minerals from sewage, metals from durable consumer goods, etc.—

may well make such procedures competitive with ore processing.

Similar remarks apply to energy. At present it is cheaper for us to use fossil fuel energy than energy in nature in the form of solar or tidal power. Only hydro-electric power appears to be competitive. Perhaps nuclear energy can be regarded separately, but it is clearly undesirable to continue to exploit fossil fuels purely as a source of energy. Apart from atmospheric contamination, these fuels represent chemical energy which could play an important role in recyclable human food supplies. In energy terms, the fossil fuels being burnt each year would feed the present world population for 20 years. Clearly, intensive research should be directed to the harnessing of natural energy sources.

The challenge to agriculture, as I see it, is to use, as far as possible, ecological principles to nourish crops, to control pests and diseases, and to prevent erosion. Agricultural chemicals should preferably be of a type which can be rapidly utilized in food chains. As far as possible these chemicals should be retained on each farm. In this way agricultural ecosystems would not only gain more internal stability, but the reduced degree of erosion and chemical pollution would help to minimize the risk of contaminating other areas.

Clearly, there are limits to the degree to which ecological principles can be applied, for the reasons given before. To me this suggests that consideration should be given to isolating at least some agricultural ecosystems from natural ecosystems. Already there is a trend in this direction, reflected in the increasing use of greenhouses for horticultural crops, and in the methods used for chicken and egg production. In many cases these practices are more economic than conventional agriculture, and it would seem that much of our vegetable and meat production could take place in this way,

thereby separating part of man's basic food chain from natural ecosystems and enabling foodstuffs to be produced in environments where pests and diseases could be safely and easily controlled without threat to other species.

However, the main use to which agricultural land is now put is to produce food grains —our primary energy foods. The areas used for such production are so vast that controlled environment culture would appear to be impossible, at least with present or foreseeable technology. In such areas agricultural scientists must devise ways of ensuring productivity while maintaining stability. This task should not be too great as long as the pressure to maximize productivity is relaxed and the pressure to maintain ecological stability is increased.

The only alternative to broad-acre agriculture for producing our main energy foods is to produce them synthetically. Instead of solar energy being captured by photosynthesis in field crops and converted into carbohydrates, is it conceivable to manufacture them in chemical factories? To date, chemical synthesis of only a few food compounds has been achieved, but present knowledge and technology would appear to be capable of developing procedures for producing the simple carbohydrates which are the basic energy foods for human and animal nutrition.

If such a development could occur, it would raise many problems associated with the energy needed for synthesis; associated with waste disposal; associated with cost or the integration of such forms of production with existing agriculture. However, it would be a logical extension of present trends in agriculture. We have seen food production change from a way of life to a highly technical business—perhaps we must recognize that the difficulties of conducting such a business out-of-doors, exposed to the vagaries of nature, must lead us to more and more control and, finally, at least for foods where energy rather than flavour is the main product, to chemical synthesis. Such a development would also have the long-term attribute of enabling much of man's food to be produced without a direct impact on the biosphere and on other species, and would enable much crop land to revert to other, or multiple, forms of land use. Again, it seems that much more research should be conducted on this subject.

Regardless of whether chemical food factories are practicable or not, these three main ground rules I have suggested—control of our numbers, control of our food supply in a way that minimizes the degree to which we jeopardize other species, and the recycling and conservation, as far as possible, of the materials which constitute our basic needs—all add up to living ecologically. There is no doubt in my mind that this is possible, but I realize that, in the short term, it probably makes economic nonsense.

What is required is a change in our basic philosophy—from an attitude to our environment which regards it as a resource to be exploited for short-term personal, regional or national gain—the attitude of the "commons"—to an attitude of living ecologically in a way that is essentially conservative of the environment. Perhaps one must ask where is the basic economy of economics if it is not consistent with this latter objective.

I think the biologist has an important role to play in pointing out the ecological rules for these changes, but the task is one for all mankind. If men of all nations can be made to see themselves, not engaged in battle against each other for a share of a "commons" which no longer exists, but rather as fellow members of a species which has passed from the stage of competition against other species to the infinitely more dangerous stage of competition within itself, perhaps we can really set ourselves on a new path of human progress in partnership.

ACKNOWLEDGEMENT

I wish to acknowledge the valuable assistance of Dr. Nita Mortlock in the preparation of the basic information on which this paper is based.

REFERENCES

Beeton, A. M. (1965): *Limnol. Oceanog.,* 10, 240.
Berkner, L. V., and Marshall, L. C. (1965): *J. Atmos. Sci.,* 22, 225.
Cloud, P. E., Jr. (1968): *Science,* 160, 729.
Cole, L. C. (1966): *BioScience,* 16, 243.
Hardin, G. (1968): *Science,* 162, 1243.
Hoering, T. C., and Abelson, P. H. (1961): *Proc. Natl. Acad. Sci. U.S.,* 47, 623.
Mason, B. (1958): *Principles of Geochemistry,* 2nd ed. (New York: John Wiley & Sons, Inc.)
Plass, G. N. (1956): *Tellus,* 8, 140.
Revelle, R., and Suess, H. E. (1957): *Tellus,* 9, 18.
Rubey, W. W. (1963): In *The Origin and Evolution of Atmospheres and Oceans,* P. J. Brancazio and A. G. W. Cameron, eds. (New York: John Wiley & Sons, Inc.)
Sperry, K. (1967): *Science,* 158, 351.

Evolution in Environments
Roland Goldring

The extent to which Man is rapidly changing the natural environments on the surface of the Earth is all too familiar today. Over the regions he has actively colonized by farming, by the destruction of forests, by chemical pollution of the atmosphere and so on, he has greatly modified relatively stable environments that existed formerly. Man's actions over the past few hundred years mark the beginning of a major revolution in the pattern and types of environment present on the Earth's surface. As yet, this revolution is most clearly in evidence in the land; oceanic environments, and those of the larger seas, are so far scarcely affected.

When one looks at the geological record of the past 1000 million years it is interesting to see that this is not the first time that organisms have affected the Earth's surface on so vast a scale. Indeed, it may be argued that, in the past, natural environments have been modified to an even greater degree by bursts of organic activity, though never at quite the pace of the present revolution.

Land plants began to evolve slowly about 400 million years ago, but during the Carboniferous period (some 300 to 350 million years ago) there was a prolific increase in their abundance and diversity, which led to the formation of major coal deposits. Professor Fairbridge, of Columbia University, has argued that the result was an immense removal of carbon dioxide from the atmosphere. The increase in oceanic alkalinity that followed possibly contributed to the unusually great extinctions about 225 million years ago, which were most marked among marine invertebrates.

The effects of organic growth and activity are recorded in many of the sediments that were deposited in the formerly existing environments. Such sediments can be traced back beyond 3000 million years to about the time when life itself began to exercise a slowly increasing control over the natural environments.

Sedimentary rocks cover much of the Earth's surface and it is from these rocks that fossil fuels stem, and in which many natural resources occur—such as iron ore and basic building materials like gravel, clay, and limestone for bricks and cement. An appreciation of the effect of organisms in past and present environments is important for geologists' attempts to interpret fossil sediments in terms of natural environments and in prospecting for these resources. For example, although the oil and gas fields of Alberta are associated with 360-million-year-old Devonian "coral reefs," the overall form of these reefs and the structure of the fossil elements is quite different from any present day coral reef, calling for an evolutionary approach on the part of the oil-field geologist.

The two major controls which geologists generally hold to be responsible for the type and association of fossil sediments are climate, and the structural, tectonic, state of the source and depositional areas. One might contrast the deposits in deep ocean troughs opposite lofty, tropical mountain ranges with those in shallow warm seas opposite dry deserts. Climate and structure are subject to the basic physical constants and though they might combine to give many types of environments, there is as yet no evidence that the constants have changed over the past 3000 million years, though the gravitational force at the surface of the Earth must have decreased, assuming the Earth to have expanded. Thus hot arid deserts must always have looked the same.

Organic control must undoubtedly be recognized as a third major control and, moreover, one which is evolving with geological time. The effects of organisms on sedimentary environments may be grouped under four headings.

First, there is the vast bulk of sediment actively contributed by organisms in the form of macroscopic and microscopic shell and other skeletal materials. Sometimes this sediment will accumulate where the animals actually lived and died as, for instance, the rocks formed by the extensive calcite skeletons of an extinct group of sea-lillies (crinoids) during the Devonian and Carboniferous periods. In a bore into the Carboniferous Limestone at Filton near Bristol a few years ago almost 30 per cent of the 480 feet cored was of crinoidal limestone, the welded, fossilized skeletal debris of the crinoids. The Chalk of southern England and elsewhere in northwest Europe is often 99 per cent calcium carbonate—of organic origin. In the Isle of Wight, the Chalk is 1800 feet thick. The equivalent amount of sediment in Ordovician times, long before the evolution of the minute organisms that make up much of the Chalk, would have been only a few feet of claystone.

Second, organisms disturb the normal distribution of the sediments by diverting and retaining material. Nowhere is this more evident than in the Middle East where many of the oil fields are associated with the reefs and lagoonal sediments which flourished during the Mesozoic and Tertiary periods. Without the organically constructed reefs there would be no lagoons. In an area where the Earth's surface is subsiding the lagoon is a vast sediment trap and a highly favoured niche for organic growth. There were no organically formed lagoons in Pre-Cambrian and early Palaeozoic times. Only relatively ephemeral sand bars thrown up by waves served to break wave attack. For the past 50 million years reefs have been very efficient structures leading to the formation of vast lagoonal areas and also protecting the shores.

In Britain we are familiar with tree-shaded river banks and the main function of the trees in protecting the bank against erosion seems somewhat incidental. The evolution of the prop-rooted, salt-tolerant, mangroves some 65 million years ago probably changed the appearance of all younger tropical deltas and many tropical coastlines. Around our own shores in modern times the effects of the hybrid grass, *Spartina townsendi,* since it ap-

peared some 50 years ago, are obvious in fixing mud, leading to the silting up of many estuaries and creeks.

Third, living organisms inhibit erosion. They achieve this end in several ways, though perhaps the effect of plants slowing down rain run-off is the most important. Though not so strong in the early evolutionary stages of plants during the Devonian and Carboniferous periods when most of the plants were growing in depositional areas (swamps, etc.), it has been especially significant since the beginning of the Tertiary period when grasses colonized much of the land surface. We know the effect of removing the grass cover; the terrible look of the overgrazed badlands of North America and Australia; even the gullying which may begin on patches of cleared waste ground in temperate northern Europe.

Around the periphery of the oceans the sediment is mostly in a relatively unstable position with the surface being aggraded at one time of the year and degraded at another. Various organisms, particularly algae, polychaete worms and mussels, are able to create sediment baffles, leading to deposition and stabilization of sediment which would otherwise be hydrodynamically mobile. Some of the rock-boring organisms, conversely (particularly certain molluscs and the boring sponge *Clione)* are effective agents of erosion.

Fourth, organisms influence the chemistry of the environment. The effects of certain organic processes are so important in this respect as to make Man's efforts appear miserably insignificant (though his potential is now far greater with nuclear power). I have mentioned the possible effect of plant upsurgence during the Carboniferous. More lowly structures form a second example; the planktonic, calcite-secreting organisms, including certain foraminifera, and the minute coccolith algae. These showed a tremendous burst in abundance during the Cretaceous period (about 100 million years ago). Previously, carbonate in the oceans was almost entirely deposited in shallower water, particularly on the continental shelves, by animals and plants favouring such conditions. Over geological time the continental shelves have been frequently uplifted, warped and, in places, folded. When uplift has been sufficient to expose the sediments to the forces of weathering and erosion the carbonates have been returned to the oceans in solution and recycled. However, the trend now is for calcium carbonate to be deposited on the relatively stable floors of the oceans.

Does this mean that the potential of the seas for the production of carbonate rocks is being gradually reduced because of permanent deposition in deeper regions of the ocean? If so this tendency might eventually lead to a decrease in carbon dioxide in the atmosphere and a reduction in plant growth—in spite of the vast amount of carbon dioxide now being added to the atmosphere by the burning of fossil fuels.

I have discussed several of the major organic events in the Earth's history. Professor Cloud (see "The Primitive Earth," *New Scientist,* vol. 43, p. 325) has discussed those in the Pre-Cambrian. Organic evolution has not been as orderly as might be generally thought and there are many perplexing instances where Darwinian factors do not seem to have been dominant. No good explanation is yet forthcoming for many of the major extinctions of animal groups. Often there is no obvious, more efficient group which replaced the declining group; in other instances no group seems to have taken immediate advantage of an apparently available ecological niche. For instance, at the base of the Cambrian period (about 600 million years ago), a group of organisms, known as the Archaeocyathida, with a stout calcareous skeleton and most closely resembling the sponges, evolved very rapidly and their fossils are now classified into eleven sub-

orders (according to *The Fossil Record,* published by the Geological Society of London and the Palaeonotological Association). Their remains account for hundreds of feet of limestone in Siberia and Australia. They had become extinct by the beginning of the Middle Cambrian and the ecological niche they occupied in the warm shallow seas was not as effectively inhabited again for about another 80 million years when the earliest corals established themselves. During that period the environments to which they belonged presumably reverted almost to those existing in Pre-Cambrian times.

Man's actions are really only another stage in the evolution of environments—though, since he now has some measure of control, how he plays it is rather for his own future. This discussion could now lead into science fiction but one might reasonably ask the question: From what it is feasible to discover about the evolution of environments over the past 1000 million years or so, what possible events might be forecast which would significantly affect man or lead to new patterns of natural environments?

Plants, or plants in association with groups of invertebrates, offer both the greatest potential and the greatest potential danger. In a relatively small way it has been possible to record the changing environments in Lake Kariba (see *New Scientist,* vol. 36, p. 750) where, for a time, mats of water fern were a definite danger to the control of the lake. Such effects are minute compared with, say, the extensive colonization of the seas and even oceans by long-stalked sea-weeds with sea-anchor-like roots.

A rather more subtle danger is the plague of starfishes now preying on living coral in Australia and in the Red Sea. Starfish "control" of coral growth could lead to a reduction in the protective function of coral reefs and the destruction of coral islands. However, the lessons for geologists are those that can be learned from the present day and, as the editors of a recent geological text-book remarked, an appreciation of the nature of biological processes is essential in the training of a geologist.

Without this, the history of the Earth's cover is meaningless, because environments are essentially the products of biological activity.

All About Ecology
William Murdoch and Joseph Connell

The public's awakening to the environmental crisis over the past few years has been remarkable. A recent Gallup Poll showed that every other American was concerned about the population problem. A questionnaire sent to about five hundred University of California freshmen asked which of twenty-five topics should be included in a general biology course for non-majors. The top four positions were: Human Population Problems (85%), Pollution (79%), Genetics (71.3%) and Ecology (66%).

The average citizen is at least getting to know the word ecology, even though his basic understanding of it may not be significantly increased. Not more than five years ago, we had to explain at length what an ecologist was. Recently when we have described ourselves as ecologists, we have been met with respectful nods of recognition.

A change has also occurred among ecologists themselves. Until recently the meetings of ecologists we attended were concerned with the esoterica of a "pure science," but now ecologists are haranguing each other on the necessity for ecologists to become involved in the "real world." We can expect that peripatetic "ecological experts" will soon join the ranks of governmental consultants jetting back and forth to the Capitol—thereby adding their quota to the pollution of the atmosphere. However, that will be a small price to pay if they succeed in clearing the air of the political verbiage that still passes for an environmental policy in Washington.

Concern about environment, of course, is not limited to the United States. The ecological crisis, by its nature, is basically an international problem, so it seems likely that the ecologist as "expert" is here to stay. To some extent the present commotion about ecology arises from people climbing on the newest bandwagon. When the limits of ecological expertise become apparent, we must expect to lose a few passengers. But, if only because there is no alternative, the ecologist and the policymakers appear to be stuck with each other for some time to come.

While a growing awareness of the relevance of ecology must be welcomed, there are already misconceptions about it. Further, the traditional role of the expert in Washington predisposes the nation to a misuse of its ecologists. Take an example. A common lament of the socially conscious citizen is that though we have enough science and technology to put a man on the moon we cannot maintain a decent environment in the United States. The implicit premise here seems clear: the solution to our ecological crisis is technological. A logical extension of this argument is that in this particular case, the ecologist is the appropriate "engineer" to resolve the crisis. This reflects the dominant American philosophy (which is sure to come up after every lecture on the environment) that the answer to most of our problems is technology and, in particular, that the answer to the problems raised by technology is more technology. Perhaps the most astounding example of this blind faith is the recent assurance issued by the government that the SST will not fly over the United States until the sonic boom problem is solved. The sonic boom "problem," of course, cannot be "solved." One job of the ecologist is to dispel this faith in technology.

To illustrate the environmental crisis, let us take two examples of how the growth of population, combined with the increasing sophistication of technology, has caused serious problems which planning and foresight could have prevented. Unfortunately, the fact is that no technological solutions applied to problems caused by increased population have ever taken into consideration the consequences to the environment.

The first example is the building of the Aswan High Dam on the upper Nile. Its purposes were laudable—to provide a regular supply of water for irrigation, to prevent disastrous floods, and to provide electrical power for a primitive society. Other effects, however, were simply not taken into account. The annual flood of the Nile had brought a supply of rich nutrients to the eastern Mediterranean Sea, renewing its fertility; fishermen had long depended upon this annual cycle. Since the Aswan Dam put an end to the annual flood with its load of nutrients, the annual bloom of phytoplankton in the eastern Mediterranean no longer occurs. Thus the food chain from phytoplankton to zooplankton to fish has been broken; and the sardine fishery, once producing eighteen thousand tons per year (about half of the total fish catch), has dropped to about five hundred tons per year.

Another ecological effect of the dam has been the replacement of an intermittent flowing stream with a permanent stable lake. This has allowed aquatic snails to maintain large populations, whereas before the dam was built they had been reduced each year during the dry season. Because irrigation supports larger human populations, there are now many more people living close to these stable bodies of water. The problem here is that the snails serve as intermediate hosts of the larvae of a blood fluke. The larvae leave the snail and bore into humans, infecting the liver and other organs. This causes the disease called schistosomiasis. The species of snail which lives in stable water harbors a more virulent species of fluke than that found in another species of snail in running water. Thus the lake behind the Aswan Dam has increased both the incidence and virulence of schistosomiasis among the people of the upper Nile.

A second example we might cite is the effect of DDT on the environment. DDT is only slightly soluble in water, so is carried mainly on particles in the water for short distances until these settle out. But on tiny particles in the atmosphere it is carried great distances; it may even fall out more heavily in distant places than close to where it was sprayed. DDT is not readily broken down by microorganisms; it therefore persists in the environment for many years. It is very soluble in fats so that it is quickly taken up by organisms. Herbivores eat many times their own weight of plants; the DDT is not broken down but is accumulated in their bodies and becomes further concentrated when the herbivores are eaten by the carnivores. The result is that the species at the top of the food chain end up with high doses of it in their tissues. Evidence is beginning to show that certain species of predators, such as ospreys, are being wiped out as a result of physiological debilities which lead to reproductive failure, all caused by accumulations of DDT.

The reproduction of top carnivores such as ospreys and pelicans is being reduced to negligible amounts, which will cause their extinction. No amount of technological ingenuity can reconstruct a species of osprey once it is extinct.

The tendency of DDT to kill both the herbivorous pest as well as its predators has produced some unpredicted consequences. In natural circumstances, herbivores are often kept at rather low numbers by their predators, with occasional "outbreaks" when there is a decrease in these enemies. Once spraying is started, and both the pests and their natural enemies are killed, the surviving pests, which have higher rates of increase than their predators, can then increase explosively between applications.

Before pesticides were applied to North American spruce and balsam forests, pest populations exploded once every thirty years or so, ate all the leaves, and then their numbers plummeted. Since spraying began, the pests, in the absence of a balancing force of predators, are continually able to increase between sprayings. In two instances, in cotton fields in Peru and in cocoa plantations in Malaysia, the situation became so bad that spraying was stopped. The predators returned and the damage by pests was diminished to the former tolerable levels. Another consequence of spraying has been that any member of the pest population which happens to be physiologically resistant to an insecticide survives and leaves offspring; thus resistant strains are evolved. Several hundred of these resistant strains have evolved in the last twenty years.

Because DDT is not present in concentrated form in the environment, it does not represent an energy resource common enough to support microorganisms. None has yet evolved the ability to break it down, even though it has been used as a pesticide for twenty-five years. Chlorinated hydrocarbons may even reduce drastically the plant productivity of the oceans. These plants are not only the base of the ocean food chain but also help maintain the oxygen supply of the atmosphere.

In sum, the indiscriminate use of DDT throughout the world, its dispersal by the atmosphere, its property of killing both pests and their enemies, and the evolution of resistant strains, have combined to create a crisis in the environment. The reaction has been to stop spraying some crops and to ban the use of DDT in some countries. Probably the correct solution, though, is to use pesticides carefully, applying them very locally (by hand if possible) to places where pest outbreaks are threatening, and to introduce or encourage enemies of the pests. This is called "integrated control." It is the hope of the future.

Since this article concerns pure ecology, it is probably worth distinguishing between pure and applied ecology. Applied ecologists are concerned with such problems as controlling pests and maximizing the yield from populations. Pure ecologists study interaction among individuals in a population of organisms, among populations, and between populations and their environments. (A population is a more or less defined group of organisms that belong to the same species.)

A brief indication of how some ecologists spend their time may be in order here. One of us (Connell) became interested in discovering what determines the distribution on the rocky seashore of a species of barnacle. He made frequent visits to the shore, photographed the positions of barnacles, counted their numbers at different levels on the shore at different life stages, noted the density and positions of predators, other barnacle species, and so forth. He developed hypotheses (in one area, that the limit to distribution is set by the presence of another barnacle species; in another, that beyond a certain height on the seashore a snail species eats them all) and tested the ideas by various experiments such as placing cages on the shore to exclude predators or removing the competing species. This work went on for several years and has now firmly established the two hypotheses.

Murdoch spent the past three years in the laboratory examining an idea about predators.

The idea was that predators keep the numbers of their various prey species stable by attacking very heavily whichever species is most abundant. (The idea is a bit more complicated than that, but that is approximately it.) This entailed setting up experiments where different predators were offered different mixtures of two prey species at a variety of densities, and then counting the number eaten of each species. These experiments led to others, in order to test different sub-hypotheses. The conclusion was that predators would "switch" only under very particular conditions.

Other ecologists spend long periods in the field trying to measure what happens to the vegetable material in a field. How much is produced and what percentage goes to rabbits, mice, insects? What percentage of the total weight of mice produced (biomass) is eaten by weasels and how efficient are weasels at converting mouse biomass to weasel biomass? Such work takes a great deal of time, estimates are rough, shaky assumptions have to be made, and in the end we have only approximate answers.

Other ecologists try to build mathematical models which might suggest how a community or some subset of a community comes to have the structure which our rough measurements tell us it may have. In pursuing all these activities they hope to build models of how nature works. The models, while not being copies of nature, should catch the essence of some process in nature and serve as a basis for explaining the phenomena that have been observed. They hope these models will be generally, though not necessarily universally, applicable. They study particular systems in the hope that these systems are not in all respects, or even in their major aspects, unique. Thus the aspirations of ecologists are not different from those of any other scientists.

Ecologists face problems which make their task difficult and at times apparently insurmountable. It is a young science, probably not older than forty years; consequently, much of it is still descriptive. It deals with systems which are depressingly complex, affected by dozens of variables which may all interact in a very large number of ways. Rather than taking a census of them, these systems must be sampled. Ecology is one of the few disciplines in biology in which it is not clear that removing portions of the problem to the laboratory for experimentation is an appropriate technique. It may be that the necessary simplification this involves removes exactly the elements from the system which determine how it functions. Yet field experiments are difficult to do and usually hard to interpret. Ecology, moreover, is the only field of biology which is not simply a matter of applied physics and chemistry. The great advances in molecular biology resulted from physicists looking at biological systems (such as DNA), whose basic configuration is explicable in terms of the positions of atoms. But the individual or the population is the basic unit of ecology. It seems certain, then, that a direct extension of physics and chemistry will not help ecologists.

Finally, there is the problem that each ecological situation is different from every other one, with a history all its own; ecological systems, to use a mathematical analogy, are non-Markovian, which is to say that a knowledge of both the past and the present is necessary in order to predict the future. Unlike a great deal of physics, ecology is not independent of time or place. As a consequence, the discipline does not cast up broad generalizations. All this is not a complete list of the general problems ecologists face, but it may be enough to provide a feeling for the difficulty of the subject.

Ecologists, though, do have something to show for forty years' work. These are some of the general conclusions they have reached. (Not all ecologists, by any means, would agree that they are generally applicable—and those who do agree would admit that exceptions occur—but they are the kind of basic conclusions that many ecologists would hope to be able to establish.)

Populations of most species have negative feedback processes which keep their numbers within relatively narrow limits. If the species itself does not possess such features, or even if it does, the community in which it exists acts to regulate numbers, for example, through the action of predators. (Such a statement obviously is not precise, e.g., how narrow are "relatively narrow limits"? A measure of ecology's success, or lack of it, is that, in forty years, there are no more than a half-dozen populations in which regulation has been adequately demonstrated; and the basis for belief in regulation is either faith or very general observations, such as the fact that most species are not so abundant that they are considered pests.)

The laws of physics lead to derivative statements in ecology. (For example, the law that matter cycles through the ecosystem, to be used again and again. Or the law that energy from the sun is trapped by plants through photosynthesis, moves up the food chain to herbivores and then to carnivores as matter, losing energy at each successive conversion so that there is generally less energy and biomass in higher food levels than in lower ones.) Ecologists have tried to take such truths from physics and construct more truly ecological generalities from them. Thus, to stay with the same example, it appears likely that there are never more than five links in any one chain of conversions from plant to top predator.

It is probably true, on a given piece of the earth and provided that the climate doesn't change, that a "climax" ecosystem will develop which is characteristic of the area's particular features and that places with similar features will develop similar ecosystems if left undisturbed. Characteristically, a "succession" from rather simple and short-lived communities to more complex and more persistent communities will occur, though there may be a reduction in the complexity of the final community. We use "final" to mean that a characteristic community will be found there for many generations. We might go further and say that during the period of development disturbances of the community will result in its complexity being reduced. (Again, such statements will certainly arouse the dissent of some ecologists.)

Finally, most ecologists would agree that complex communities are more stable than simple communities. This statement illustrates the difficulties faced by theoretical ecologists. Take some of its implications: What is complexity and what is stability in an ecological setting? Charles Elton embodied the idea in a simple, practical, and easily understood way. He argued that England should maintain the hedgerows between its fields because these were complex islands in a simple agricultural sea and contained a reservoir of insect and other predators which helped to keep down pest populations. The idea here seems quite clear. Ecologists, though, want a more precise exposition of the implications of the statement. What kind of complexity? What is stability?

Physical complexity, by providing hiding places for prey, may increase stability. Certainly biological complexity in general is thought to lead to stability—more species or more interspecific interactions, more stability. Here a variety of answers is available. It has been suggested that complex communities are stable, i.e., able to resist invasion by species new to the area, by having all the "niches" filled. Thus sheer numbers of kinds of organisms in all food levels were considered the appropriate sort of complexity. To keep the numbers of prey stable, the most likely candidates are predators. Now other questions arise: Do we just want more species of predators? Do we want more species of predators which are very specific in the prey they eat, implying that prey are stabilized by having many species feed on them? Do we want predators which are very general and attack many prey species, so that we still have a large number of interspecific interactions which are made up in a different way? The answer is not obvious, and indeed there is disagreement on it. Furthermore, if one studies the way some predators react to changes in the numbers of their prey, their short-term responses are such as to cause *instability*. Thus only some types of biological complexity may produce stability.

What do we mean by stability? In the examples cited, we have meant numerical constancy through time, but this is by no means the only meaning. It has even been suggested that numerical *in*constancy is a criterion for stability. Stability might also mean that the same species persist in the same area over long periods, showing the same sort of interspecific interactions (community stability). A community or population might be considered stable because it does not change in response to a great deal of environmental pressure, or because it changes but quickly returns to its original state when the disturbing force is removed. It is worth noting that if a population or community is observed merely not to change, we cannot tell whether this is owing to its ability to resist perturbing factors or merely to the absence of such factors. If we want to know about the *mechanisms* which might lead to the truth of our original statement, "complexity leads to stability," all the above points are important.

This general statement about complexity and stability rests upon the kind of observation readily apparent to most intelligent laymen. Thus simple agricultural systems seem to be much more subject to outbreaks of herbivores than the surrounding countryside. Ecosystems in the tropics appear to be more stable than in the simpler temperate zone. In turn the temperate zone seems to be more stable than the Arctic. This seems to be mainly an article of faith. However, even this classic sort of evidence is questioned—for example, small mammals may actually be more unstable numerically in the United States than in the much simpler Arctic environment. Other evidence comes from the laboratory. If one takes small species of prey and predator—for example, two single-celled animals or two small mites— and begins culturing them together, the numbers of prey and predators fluctuate wildly and then both become extinct quickly, for the predators exhaust their food source. "Simple" predator-prey systems tend to be unstable. There is some evidence that if physical complexity is added the system may become more stable. From these examples of the generalizations ecologists have arrived at, an important question emerges. Even if we dispense with the idea that ecologists are some sort of environmental engineers and compare them to the pure physicists who provide scientific rules for engineers, do the tentative understandings we have outlined provide a sound basis for action by those who would manage the environment? It is self-evident that they do not.

This conclusion seems to be implied in a quotation from an article published in *Time* on the environment, which underlines the point that application of the ecologist's work is not the solution to the environmental crisis. According to *Time:* "Crawford S. Holling was once immersed in rather abstract research at the University of British Columbia—mathematical models of the relationship between predators and their prey. 'Three years ago, I got stark terrified at what was going on in the world and gave it up.' Now he heads the university's interdepartmental studies of land and water use, which involve agriculture, economics, forestry, geography, and regional planning. 'What got me started on this,' says Holling, 'were the profound and striking similarities between ecological systems and the activities of man: between predators and land speculators; between animal-population growth and economic growth; between plant dispersal and the diffusion of people, ideas, and money.' "

The "rather abstract research" was ecology. Holling's testimony is that it would not provide a solution. Yet, by and large, ecologists are concerned and probably have the best understanding of the problem.

We submit that ecology as such probably cannot do what many people expect it to do; it cannot provide a set of "rules" of the kind needed to manage the environment. Nevertheless, ecologists have a great responsibility to help solve the crisis; the solution they offer should be founded on a basic "ecological attitude." Ecologists are likely to be aware of the consequences of environmental manipulation; possibly most important, they are ready to deal with the environmental problem since their basic ecological attitude is itself the solution to the problem. Interestingly enough, the supporting data do not generally come from our "abstract research" but from massive uncontrolled "experiments" done in the name of development.

These attitudes and data, plus obvious manifestations of physical laws, determine what the ecologist has to say on the problem and constitute what might be called environmental knowledge. Some examples of this knowledge follow, though this is not to be taken as an encapsulation of the ecologist's wisdom.

Whatever is done to the environment is likely to have repercussions in other places and at other times. Because of the characteristic problems of ecology some of the effects are bound to be unpredictable in practice, if not in principle. Furthermore, because of the characteristic time-dependence problem, the effects may not be measurable for years—possibly not for decades.

If man's actions are massive enough, drastic enough, or of the right sort, they will cause changes which are irreversible since the genetic material of extinct species cannot be reconstituted. Even if species are not driven to extinction, changes may occur in the ecosystem which prevent a recurrence of the events which produced the community. Such irreversible changes will almost always produce a simplification of the environment.

The environment is finite and our nonrenewable resources are finite. When the stocks run out we will have to recycle what we have used.

The capacity of the environment to act as a sink for our total waste, to absorb it and recycle it so that it does not accumulate as pollution, is limited. In many instances, that limit has already been passed. It seems clear that when limits are passed, fairly gross effects occur, some of which are predictable, some of which are not. These effects result in significant alterations in environmental conditions (global weather, ocean productivity). Such changes are almost always bad since organisms have evolved and ecosystems have developed for existing conditions. We impose rates of change on the environment which are too great for biological systems to cope with.

In such a finite world and under present conditions, an increasing population can only worsen matters. For a stationary population, an increase in standard of living can only mean an increase in the use of limited resources, the destruction of the environment, and the choking of the environmental sinks. There are two ways of attacking the environmental crisis.

The first approach is technology; the second is to reverse the trends which got us into the crisis in the first place and to alter the structure of our society so that an equilibrium between human population and the capacities of the environment can be established.

There are three main dangers in a technological approach to the environmental crisis. The first threatens the environment in the short term, the second concerns ecologists themselves, and the third, which concerns the general public attitude, is a threat to the environment in the long term.

Our basic premise is that, by its nature, technology is a system for manufacturing the need for more technology. When this is combined with an economic system whose major goal is growth, the result is a society in which conspicuous production of garbage is the highest social virtue. If our premise is correct, it is unlikely that we can solve our present problems by using technology. As an example, we might consider nuclear power plants as a "clean" alternative to which we can increasingly turn. But nuclear power plants inevitably produce radioactive waste; this problem will grow at an enormous rate, and we are not competent to handle it safely. In addition, a whole new set of problems arises when all these plants produce thermal pollution. Technology merely substitutes one sort of pollution for another.

There is a more subtle danger inherent in the technological approach. The automobile is a blight on Southern California's landscape. It might be thought that ecologists should concern themselves with encouraging the development of technology to cut down the emission of pollutants from the internal combustion engine. Yet that might only serve to give the public the impression that something is being done about the problem and that it can therefore confidently await its solution. Nothing

significant could be accomplished in any case because the increasing number of cars ensures an undiminishing smog problem.

Tinkering with technology is essentially equivalent to oiling its wheels. The very act of making minor alterations, in order to placate the public, actually allows the general development of technology to proceed unhindered, only increasing the environmental problems it causes. This is what sociologists have called a "pseudo-event." That is, activities go on which give the appearance of tackling the problem; they will not, of course, solve it, but only remove public pressure for a solution.

Tinkering also distracts the ecologist from his real job. It is the ecologist's job, as a general rule, to oppose growth and "progress." He cannot set about convincing the public of the correctness of this position if in the meantime he is putting his shoulder behind the wheel of technology. The political power system has a long tradition of buying off its critics, and the ecologist is liable to wind up perennially compromising his position, thereby merely slowing down slightly or redirecting the onslaught of technology.

The pressures on the ecologist to provide "tinkering" solutions will continue to be quite strong. Pleas for a change of values, for a change to a non-growth, equilibrium economy seem naive. The government, expecting sophistication from its "experts," will probably receive such advice cooly. Furthermore, ecologists themselves are painfully aware of how immature their science is and generally take every opportunity to cover up this fact with a cloud of obfuscating pseudo-sophistication. They delight in turning prosaic facts and ideas into esoteric jargon. Where possible, they embroider the structure with mathematics and the language of cybernetics and systems analysis, which is sometimes useful but frequently is merely confusing. Such sophistication is easily come by in suggesting technological solutions.

Finally, there is always the danger that in becoming a governmental consultant, the ecologist will aim his sights at the wrong target. The history of the Washington "expert" is that he is called in to make alterations in the model already decided upon by the policymakers. It would be interesting to know what proportion of scientific advice has ever produced a change in ends rather than in means. We suspect it is minute. But the ecologist ought not to concern himself with less than such a change; he must change the model itself.

We should point out that we are not, for example, against substituting a steam-driven car for a gas-driven car. Our contention is that by changing public attitudes the ecologist can do something much more fundamental. In addition, by changing these attitudes he may even make it easier to force the introduction of "cleaner" technology, since this also is largely a *political* decision. This certainly seems to be so in the example of the steam-driven car. We do not believe that the ecologist has anything really new to say. His task, rather, is to inculcate in the government and the people basic ecological attitudes. The population must come, and very soon, to appreciate certain basic notions. For example: a finite world cannot support or withstand a continually expanding population and technology; there are limits to the capacity of environmental sinks; ecosystems are sets of interacting entities and there is no "treatment" which does not have "side effects" (e.g., the Aswan Dam); we cannot continually simplify systems and expect them to remain stable, and once they do become unstable there is a tendency for instability to increase with time. Each child should grow up knowing and understanding his place in the environment and the possible consequences of his interaction with it.

In short, the ecologist must convince the population that the only solution to the problem of growth is not to grow. This applies to population and, unless the population is declining, to its standard of living. It should be clear by now that "standard of living" is probably beginning to have an inverse relationship to the quality of life. An increase in the gross national product must be construed, from the ecological point of view, as disastrous. (The case of underdeveloped countries, of course, is different.)

We do not minimize the difficulties in changing the main driving force in life. The point of view of the ecologist, however, should be subversive; it has to be subversive or the ecologist will become merely subservient. Such a change in values and structure will have profound consequences. For example, economists, with a few notable exceptions, do not seem to have given any thought to the possibility or desirability of a stationary economy. Businessmen, and most economists, think that growth is good, stagnation or regression is bad. Can an equilibrium be set up with the environment in a system having this philosophy? The problem of converting to non-growth is present in socialist countries too, of course, but we must ask if corporate capitalism, by its nature, can accommodate such a change and still retain its major features. By contrast, if there are any ecological laws at all, we believe the ecologists' notion of the inevitability of an equilibrium between man and the environment is such a law.

We would like to modify some details of this general stand. Especially after the necessary basic changes are put in motion, there *are* things ecologists as "experts" can do: some of them are sophisticated and others, in a very broad sense, may even be technological. Certainly, determining the "optimum" U.S. population will require sophisticated techniques.

Ecologists, willy-nilly, will have to take a central role in advising on the management of the environment. They already are beginning to do this. The characteristics of ecology here determine that this advice, to be good, will be to some extent sophisticated to fit particular cases. Thus, good management will depend on long-term studies of *particular* areas, since ecological situations are both time-dependent and locale-dependent. These two features also ensure that there will be a sizable time-lag between posing the question and receiving the ecological advice, and a major job of the ecologists will be to make the existence of such lags known to policymakers.

Ecologists sometimes will have to apply technology. As one instance, integrated pest control (that is, basically biological control with occasional small-scale use of pesticides) will surely replace chemical control, and integrated pest control can be considered biological technology. In this area there is some promise that sophisticated computer modeling techniques applied to strategies of pest control may help us design better techniques. The banning of DDT, for example, could no doubt be a laudable victory in the war to save the environment, but it would be disastrous to mistake a symbolic victory like this for winning the war itself.

The Ecological Niche: Man and Ecosystem

CHAPTER

II

Somewhere back down the road, perhaps fifteen or twenty thousand years ago, man may have become a "superpredator" capable of eliminating whole species and genera of prey to the last individual without suffering disaster in return. Such extinction of prey, with immunity from serious consequences, represents defiance of a fundamental ecological "rule" about predator-prey relationships.

More recently, man may have achieved even greater ecological status as a "supercompetitor." At that rank he would be granted the possibility of out-competing all other species or, in other words, simultaneously monopolizing all ecological niches. This would denote superiority to a considerable number of "ecological rules." The new ecological equation would be: man + physical environment = ecosystem. Actual, or at least potential, supercompetitor status is an implicit assumption of "The Establishment" of the present day.

We must grant that the ecological niche of pre-technological man was already unusually, even uniquely, broad. Morphologically and physiologically, for example, we are equipped to deal with an enormous range of food sizes, qualities, and sources. It is not surprising that man should have diverse talents. He is the creature of (and still exists in) the Pleistocene, a period of extreme temporal and geographic diversity. It is no accident that a large mammal evolving in a period of recurrent environmental crisis should become an aggressive, highly adaptable, ecological generalist.

Climatologically, the earth is in a not-particularly-warm interglacial episode. Genomes change less rapidly than does climate, and much less rapidly than culture. Biologically, man has retained the innate qualities of a giant, omnivorous, ground dwelling, Pleistocene primate. Changing environments, cultural included, have yet to generate any major alterations in the genetic bases of physique, tool using ability, imagination and creativity, gregariousness, territoriality, aggressiveness or other of man's ice-age characteristics. There is, in fact, considerable evidence of growing dissonance between man and the environment to which

41

headlong cultural and technological innovation are committing him. Even the bounds of Pleistocene adaptability can be exceeded! Ominously, it appears that orthogenesis is possible in cultural evolution, even if it may not occur in the more refined biological process. In terms of needs less obvious than food and shelter man has demonstrated surprising clumsiness in creating new environments truly adequate for man. Subtle needs for space, diversity, identity, and even less subtle needs like sex, seem frequently warped and frustrated. Man makes for himself worlds for which he was never made. Much that recent civilization has created seems to press on the limits of individual tolerance, to evidence a widening gap between the species and its rapidly changing ecological niche.

An evaluation of man's past and present ecological niche is a very pertinent exercise. At the least it will reveal some unrecognized assumptions. It will also remind us that past civilizations have had different ecological bases and effects. Some of these civilizations deteriorated and disappeared, and the concordance of their disappearance with the over-exploitation and degradation of their environments may provide food for thought. The thoroughly man-created environments so far produced (cities, new and old) have generally proven to be rather nasty places for the vast majority of their inhabitants. It is not a record to inspire great confidence. Further, domestication, including the domestication of man, runs counter to both ecological succession and natural selection in many important respects.

Cultural invention has the consequence of rapidly expanding the ecological niche of the species; thus the possibility of supercompetitor status. It is a general principle with radical new designs that the early model may prove unsound and be recalled. The analogy is useful, for if the nature of modern man permits expansion and some degree of self-determination of his ecological niche, there is no guarantee that man will select the best of the available choices. Niches and ecosystems are, and hopefully will always be, dynamic. Within the evolving ecosystem, *what ecologic niche for man?*

Pleistocene Overkill
Paul S. Martin

About ten thousand years ago, as glaciers retreated into Canada and as man moved southward at the end of the last Ice Age, North America suddenly and strangely lost most of its large animals. Native North American mammals exceeding 100 pounds in adult body weight were reduced by roughly 70 per cent. The casualty list includes mammoths, mastodons, many species of horses and camels, four genera of ground sloths, two of peccary, shrub oxen, antelope, two genera of saber-toothed cats, the dire wolf, the giant beaver, tapirs and others totaling over 100 species. Despite this fantastic loss of large animals during the Pleistocene, the most recent geologic epoch, the fossil record shows no loss of

small vertebrates, plants, aquatic organisms, or marine life.

One need not be a Pleistocene geologist to ask the obvious question: What happened? To date there is no obvious answer, certainly not one acceptable to any consensus of scientists interested in the mystery. The question of just what occurred to bring about this unprecedented extinction continues to provoke a storm of controversy.

Extinction, we know, is not an abnormal fate in the life of a species. When all the niches, or "jobs," in a biotic community are filled, extinction must occur as rapidly as the evolution of new species. The fossil record of the last ten million years bears witness to this fact, for it is replete with extinct animals that were sacrificed to make room for new and presumably superior species. But this is a normal state of affairs from a paleontologist's point of view.

However, the extinction that took place at the end of the Pleistocene did not comply with biological rules of survival. Unlike former extinctions, such as occurred in the Miocene, Pliocene, and Early Pleistocene, Late Pleistocene extinction of large mammals far exceeded replacement by new species that could easily have been accommodated by the prevailing habitat. The complete removal of North American horses, for example, represents the loss of a lineage of grass-eaters, without the loss of the grass! It left the horse niche empty for at least eight thousand years, until the Spaniards introduced Old World horses and burros. Some of these then escaped to reoccupy part of their prehistoric range. Today, tens of thousands of wild horses and burros still live along remote parts of the Colorado River and in the wild lands of the West. Certainly nothing happened at the end of the Pleistocene to destroy the horse habitat. What, then, caused these animals, not to mention mammoths, camels, sloths, and others, to become extinct?

Like the horses, camels first evolved in North America many millions of years ago. They then spread into South America and crossed to the Old World by means of the Bering bridge. Crossing in the opposite direction were elephants, which soon prospered in the New World, judging by the abundance of mammoth and mastodon teeth in Pleistocene outcrops. From their evolutionary center in South America came a variety of edentates, including ground sloths and glyptodonts, the former spreading north to Alaska. But by the end of the Pleistocene, the majority of these herbivorous species had completely disappeared. Only relatively small species within these groups survived—the alpaca and llama among the camels of South America, and the relatively small edentates such as anteaters, armadillos, and tree sloths.

There was no obvious ecological substitution by other large herbivores competing for the same resources. Too many large mammals were lost, too few were replaced, and there was too little change among smaller plants and animals to accept this extinction as a normal event in the process of North American mammalian evolution.

One hypothesis commonly proposed for the abrupt and almost simultaneous extinction of large mammals is that of sudden climatic change. We know that climates did change many times in the Pleistocene—a three-million-year period of repeated glacial advance and retreat—so perhaps the great herds were decimated in this way. A sizable group of vertebrate paleontologists believes that that is indeed what happened. They maintain that with the retreat of the glaciers the early postglacial climate grew more continental—summers became hotter, and winters colder and supposedly more severe than they had been during the time of the ice advance. The result was an upset in the breeding season, a lethal cold sterility imposed on species of large mammals adapted for reproduction at what came to be the wrong season. Perhaps also the large Ice Age mammals were confronted for the first time with excessive snow cover and blue northers, which even today can kill thousands of cattle and sheep in the High Plains. Paleozoologist John Guilday at the Carnegie Museum believes that accelerated competition occurred among the large mammals before they could readjust to the change in vegetation and climate. They proceeded to exterminate their food supply and themselves in a morbid togetherness. Then, according to Guilday, the early North American hunters who arrived over the Bering bridge delivered the coup de grace to the few remaining large mammals after the great herds were already sadly depleted.

But we have no evidence that the large mammals were under competitive stress, then or at any other time in the Pleistocene. We know that they had witnessed, and certainly survived, the advance and retreat of earlier glacial ice sheets. And among today's large mammals most are remarkably tolerant of different types of environments. Some large desert mammals can endure months without drinking; others, such as musk ox, live the year round in the high Arctic. Reindeer and wildebeest migrate hundreds of miles to pick their pasture. Why should we believe that the great mammals of the Pleistocene were less adaptable?

Furthermore, while climatic changes had some effect upon existing fauna and its habitat, extinction apparently occurred when range conditions were actually improving for many species. From the fossil pollen record we know that mastodon and woodland musk ox of eastern North America occupied spruce forests ten to twelve thousand years ago, a habitat

then rapidly expanding northward from its constricted position bordering the Wisconsin ice sheet. And the western plains grassland was extensive and spreading at the time of the extinction of grazing horses, mammoths, and antelopes.

Another objection to cold winter climates as an explanation for extinction arises when one looks to the American Tropics. There, far more species became extinct during this period than in the temperate regions. More extinct Pleistocene genera were found in a single fauna in Bolivia than are known in the richest of the fossil faunas of the United States. However, the Tropics never experienced the zero temperatures of North America. This being the case, the climatic change hypothesis cannot account for the large-scale extinction in that part of the world.

Nor can it account for the extinction that occurred on the large islands of the world, such as on Madagascar and New Zealand, which did not take place until less than a thousand years ago. In the case of the giant bird, *Aepyornis,* of Madagascar, and the giant moa from New Zealand, carbon 14 dates indicate that these birds did not perish until long after the time of major world-wide climatic upset.

Without doubt, the climatic disturbances that affected North America during the Pleistocene proved equally disturbing in New Zealand. During the last glaciation one-third of the South Island was ice-covered and the remainder of the island was much colder than today. The subsequent melting of the glaciers brought a worldwide rise in sea level and divided the country in two. During the postglacial period intense volcanic eruptions blanketed the North Island, so that by 2,000 years ago large parts of it were covered by sterile ash supporting only dwarfed vegetation. In fact, sheep raising failed in these areas until cobalt and other trace elements were added to the pastures. Yet, some 27 species of moas apparently survived the natural climatic catastrophes of the Pleistocene and disappeared only after East Polynesian invaders, the predecessors of the Maori, arrived sometime about or before A.D. 900. Thus, any credibility the climatic change hypothesis may have when applied to a single region vanishes when the global pattern is considered. Pleistocene experts generally believe that whatever their magnitude, major climatic changes of the last 50,000 years occurred at approximately the same time throughout the world—the extinctions did not.

My own hypothesis is that man, and man alone, was responsible for the unique wave of Late Pleistocene extinction—a case of overkill rather than "overchill" as implied by the climatic change theory. This view is neither new nor widely held, but when examined on a global basis, in which Africa, North America, Australia, Eurasia, and the islands of the world are considered, the pattern and timing of large-scale extinction corresponds to only one event—the arrival of prehistoric hunters.

Some anthropologists, such as Loren Eiseley of the University of Pennsylvania, have challenged the man-caused extinction theory on the grounds that African megafauna did not suffer the same fate as the large mammals of North America. At first, this appears to be a sound argument since that continent now contains some 40 genera of large mammals that were around during the Pleistocene. Africa's fabulous plains fauna has long been regarded as a picture of what the American Pleistocene was like prior to extinction, at least in terms of size and diversity of the big mammals. During the million years of hominid evolution in Africa, it seems as if man and his predecessors would have had ample time to exterminate its fauna. And if, as I believe, Late Paleolithic hunters in the New World could

have succeeded in destroying more than 100 species of large mammals in a period of only 1,000 years, then African hunters, it seems, should at least have made a dent on that continent's mammals.

It turns out that they did, for today's living megafauna in Africa represents only about 70 per cent of the species that were present during the Late Pleistocene. Thus, while the proportion of African mammals that perished during the Pleistocene was less than that in North America, the loss in number of species was still considerable. In addition to the large mammals that now inhabit the African continent, an imaginary Pleistocene game park would have been stocked with such species as the antlered giraffe, a number of giant pigs, the stylohipparion horse, a great long-horned buffalo, a giant sheep, and an ostrich of larger size than is known at present. In Africa, as in America, the wave of Pleistocene extinction took only the large animals.

The African extinction has also been attributed to climatic and climate-related change. L.S.B. Leakey would explain extinction of the giant African fauna as the result of drought. If so, the drought strangely did not affect nearby Madagascar. On that island, barely 250 miles from the African shore, extinction of giant lemurs, pygmy hippopotamuses, giant birds, and tortoises did not occur until a much later date, in fact not until within the last 1,000 years.

African big game extinction appears to coincide in time with the first record of fire, or at least of charcoal, in archeological sites. In addition, most extinct fauna is last found in many locations associated with the distinctive stone tools of Early Stone Age (Acheulean) hunters. If fire was used in hunting, man-caused extinction becomes easier to understand, because fire drives necessarily involve large amounts of waste—whole herds must be decimated in order to kill the few animals sought for food. Perhaps fire became a major weapon in the hands of the Acheulean big game hunters enabling them to encircle whole herds of animals.

In any event, African extinction ended during the period of the Early Stone Age hunters. This fact raises the possibility that the cultures that succeeded the Acheulean developed more selective methods of hunting and may even have learned to harvest the surviving large mammals on a sustained yield basis. Even during the last 100 years, when modern weapons have reduced the ranges of many species, there has been no loss of whole genera of terrestrial mammals as occurred during the time of the early hunters.

The case of Australia also supports the hypothesis of man-caused extinction. On that continent, no evidence of extinction, without replacement by other species, can be found until after men had inhabited the island, at least 14,000 years ago. About this time, various species of large marsupials perished, including the rhino-sized *Diprotodon* and the giant kangaroo.

About 12,000 years ago, when the Paleo-Indians swept into North America across the Bering bridge, through unglaciated Alaska, and down the melting ice corridor east of the Cordilleras, we can be confident that they were old hands at hunting woolly mammoths and other large Eurasian mammals. In contrast, the New World mammoth and other species of big game had never encountered man, and were unprepared for escaping the strange two-legged creature who used fire and stone-tipped spears to hunt them in communal bands. Probably the New World fauna of the time was no more suspicious of man than are the fearless animals that now live in the Galapagos and other regions uninhabited by man. In any case, radiocarbon dates indicate that North Ameri-

can extinction followed very closely on the heels of the big game hunters. The Paleo-Indians easily found and hunted the gregarious species that ranged over the grasslands, deserts, or other exposed habitat. As the hunters increased in number and spread throughout the continent, large animals whose low rate of reproduction was insufficient to offset the sudden burden of supporting a "super-predator" soon perished.

Early man may not have been able to avoid killing the herd animals in excess. To capture *any* members of a bison or elephant herd, it was necessary to kill them all, for instance, by driving them over a cliff. Even when big game became scarce and small animals became more important in the human diet, the pride and prestige associated with killing an elephant may have continued. This, in fact, seems likely, judging from the prestige associated with the unnecessary killing that persists even today within our society. By virtue of his cultural development, man became a superpredator, less susceptible to the biological checks and balances that apparently prevent such predators as Arctic foxes from annihilating their prey, the Arctic hare.

If the overkill hypothesis is valid, how did *any* large mammals manage to survive? There are several explanations. Some species, such as tapirs, capybaras, deer, white-lipped peccaries, anteaters, and tree sloths, took refuge in the vast forests of tropical America. In temperate regions, solitary moose and bear also found refuge in wooded areas, perhaps with a few small herds of woodland bison, at the time when soon-to-be-extinct species of giant bison were being annihilated by Folsom hunters on the plains. Mountain sheep found protection on the roughest desert ranges while mountain goats escaped only in the northern Rockies.

The musk ox, that conspicuous and easily hunted game animal of the open tundra, was wiped out in Eurasia, but in America it escaped through a piece of paleogeographic good luck. Since parts of the Canadian Arctic Archipelago and Greenland were untouched by glaciation, tundra habitat was available through the Pleistocene for this species and for barren ground caribou. Some of these animals remained in the "safe region" north of the continental ice sheet, a zone unknown to the early hunters. They thus escaped the fate suffered by most species located along the path taken by nomadic hunters as they pushed into Alaska, down western Canada, across the northern United States, and into New England. The Keewatin and Laurentian ice sheets provided a barrier to the early hunters whose distinctive spear points and other artifacts are unknown in the eastern Canadian Arctic. With the final melting of ice, less than 6,000 years ago, the Greenland musk oxen were at last exposed to the New World Indians and Eskimos. But the wandering superpredators— the Paleo-Indians—were no longer present. The Eskimo had the good fortune, or good sense, to harvest musk oxen on a sustained yield basis, and the species was able to spread westward through northern Canada, ultimately recovering part of its Alaskan range. If the woolly mammoth had also occupied the Greenland refuge, it too might have survived the Pleistocene.

Can the overkill hypothesis be disproved by future experiments or discoveries? To discount the hypothesis one need simply identify a major wave of extinction anywhere in the world in the Late Pleistocene prior to man's arrival. To date, such evidence has not been found. Quite the opposite, in fact, since the chronological sequence of extinction follows closely upon man's footsteps—occurring first in Africa and southern Asia, next in Australia, then through northern Eurasia and into North and South America, much later in the West

Indies, and finally, during the last 1,000 years, in Madagascar and New Zealand. The pattern shows that Late Pleistocene extinction did not occur in all locations at the same time, as it would have if there had been a sudden climatic change or perhaps a cataclysmic destruction of the earth's atmosphere with lethal radiation caused by cosmic ray bombardment, another common hypothesis. Since no synchronous destruction of plants or of plant communities is known, the long-held belief that climatic change caused extinction lacks credibility.

I do not pretend that the overkill hypothesis explains how, why or even how many early hunters were involved. It seems reasonable to assume that fire and fire drives were a major weapon; possibly plant poisons were used in the Tropics. To the objection that too few spear points or other Stone Age artifacts have been found in the Americas to prove there was a sizable prehistoric human population, one may assert, tongue in cheek, that too few fossils of Pleistocene ground sloths, mammoths, camels, and saber-toothed cats have been found to prove there was a sizable prehistoric population of them either. The obvious difficulty with the "spear point" argument is that even the best fossil localities, with or without artifacts, do not yield data that can be reliably converted into population estimates. The case for overkill is best presented as a "least improbable hypothesis," and is not based on extensive knowledge of how prehistoric hunters may have carried out their hunting. Nor is there much hope that we will ever learn more of their techniques than the little we now know. The essence of the argument is based upon the simple matter of Late Pleistocene chronology. In no part of the world does massive unbalanced faunal extinction occur without man the hunter on the scene.

To certain comfortable concepts about pristine wilderness and ancient man, the implications of this hypothesis are startling, even revolutionary. For example, that business of the noble savage, a child of nature, living in an unspoiled Garden of Eden until the "discovery" of the New World by Europeans is apparently untrue, since the destruction of fauna, if not of habitat, was far greater before Columbus than at any time since. The subtle lesson of sustained yields, of not killing the goose that lays the golden eggs, may have been learned the hard way, and forgotten, many times before the twentieth century.

A related conceptual mistake, if the hypothesis holds, may be the long-held opposition of range ecologists to the introduction of exotic large mammals in America. Part of the opposition to the introduction of alien species is based on the idea that native North American mammals are already using all the available browsing and grazing space that could or should be occupied in this country. But remembering the numerous species of the Pleistocene, it is difficult to imagine that native mountain sheep, bison, antelope, deer, and elk occupy all available niches in the American ecosystem. The concept of game ranching, of keeping both cattle and native game species on the same range in order to make maximum use of pastures, is catching on in Africa. Since our own ranch industry is essentially a monoculture of either cattle or sheep, perhaps it's time to take a fresh look at the unfilled niches on the American ranges. Domestic livestock, wild game, and the range itself may well benefit from a greater diversity of fauna and a partial restoration of the complex ecosystem that was America for millions of years, until man arrived.

Evolution
Langdon Smith

When you were a tadpole and I was a fish
In the Paleozoic time,
And side by side on the ebbing tide
We sprawled through the ooze and slime,
Or skittered with many a caudal flip
Through the depths of the Cambrian fen,
My heart was rife with the joy of life,
For I loved you even then.

Mindless we lived and mindless we loved
And mindless at last we died;
And deep in the rift of the Caradoc drift
We slumbered side by side.
The world turned on in the lathe of time,
The hot lands heaved amain,
Till we caught our breath from the womb of
death
And crept into light again.

We were amphibians, scaled and tailed,
And drab as a dead man's hand;
We coiled at ease 'neath the dripping trees
Or trailed through the mud and sand.
Croaking and blind, with our three-clawed feet
Writing a language dumb.
With never a spark in the empty dark
To hint at a life to come.

Yet happy we lived and happy we loved,
And happy we died once more;
Our forms were rolled in the clinging mold
Of a Neocomian shore.
The eons came and the eons fled
And the sleep that wrapped us fast
Was riven away in a newer day
And the night of death was past.

Then light and swift through the jungle trees
    We swung in our airy flights,
Or breathed in the balms of the fronded palms
    In the hush of the moonless nights;
And oh! what beautiful years were there
    When our hearts clung each to each;
When life was filled and our senses thrilled
    In the first faint dawn of speech.

Thus life by life and love by love
    We passed through the cycles strange
And breath by breath and death by death
    We followed the chain of change.
Till there came a time in the law of life
    When over the nursing side
The shadows broke and the soul awoke
    In a strange, dim dream of God.

I was thewed like an Auroch bull
    And tusked like the great cave bear;
And you, my sweet, from head to feet
    Were gowned in your glorious hair.
Deep in the gloom of a fireless cave,
    When the night fell o'er the plain
And the moon hung red o'er the river bed
    We mumbled the bones of the slain.

I flaked a flint to a cutting edge
    And shaped it with brutish craft;
I broke a shank from the woodland lank
    And fitted it, head and haft;
Then I hid me close to the reedy tarn,
    Where the mammoth came to drink;
Through the brawn and bone I drove the stone
    And slew him upon the brink.

Loud I howled through the moonlit wastes,
    Loud answered our kith and kin;
From west and east to the crimson feast
    The clan came tramping in.
O'er joint and gristle and padded hoof
    We fought and clawed and tore;
And cheek by jowl with many a growl
    We talked the marvel o'er.

I carved that fight on a reindeer bone
    With rude and hairy hand;
I pictured his fall on the cavern wall
    That men might understand.
For we lived by blood and the right of might
    Ere human laws were drawn,
And the age of sin did not begin
    Till our brutal tush were gone.

And that was a million years ago
    In a time that no man knows;
Yet here tonight in the mellow light
    We sit at Delmonico's.
Your eyes are deep as the Devon springs,
    Your hair is dark as jet,
Your years are few, your life is new,
    Your soul untried, and yet —

Our trail is on the Kimmeridge clay
    And the scarp of the Purbeck flags;
We have left our bones in the Bagshot stones
    And deep in the Coralline crags;
Our love is old, our lives are old,
    And death shall come amain;
Should it come today, what man may say
    We shall not live again?

God wrought our souls from the Tremadoc
    beds
    And furnished them wings to fly;
He sowed our spawn in the world's dim dawn,
    And I know that it shall not die,
Though cities have sprung above the graves
    Where the crook-bone men make war
And the oxwain creaks o'er the buried caves
    Where the mummied mammoths are.

Then as we linger at luncheon here
    O'er many a dainty dish,
Let us drink anew to the time when you
    Were a tadpole and I was a fish.

The Agency of Man on the Earth
Carl O. Sauer

The Theme

As a short title for the present conference we have spoken at times and with hope of a "Marsh Festival," after the statesman-scholar, George Perkins Marsh, who a century ago considered the ways in which the Earth has been modified by human action (Marsh, 1864, 1874). The theme is the capacity of man to alter his natural environment, the manner of his so doing, and the virtue of his actions. It is concerned with historically cumulative effects, with the physical and biologic processes that man sets in motion, inhibits, or deflects, and with the differences in cultural conduct that distinguish one human group from another.

Every human population, at all times, has needed to evaluate the economic potential of its inhabited area, to organize its life about its natural environment in terms of the skills available to it and the values which it accepted. In the cultural *mise en valeur* of the environment, a deformation of the pristine, or prehuman, landscape has been initiated that has increased with length of occupation, growth in population, and addition of skills. Wherever men live, they have operated to alter the aspect of the Earth, both animate and inanimate, be it to their boon or bane.

The general theme may be described, therefore, in its first outline, as an attempt to set forth the geographic effects, that is, the appropriation of habitat by habit, resulting from the spread of differing cultures to all the *oikoumene* throughout all we know of human time. We need to understand better how man has disturbed and displaced more and more of the organic world, has become in more and more regions the ecologic dominant, and has affected the course of organic evolution. Also how he has worked surficial changes as to terrain, soil, and the waters on the land and how he has drawn upon its minerals. Latterly, at least, his urban activities and concentrations have effected local alterations of the atmosphere. We are trying to examine the processes of terrestrial change he has entrained or originated, and we are attempting to ask, from our several interests and experiences, relevant questions as to cultural behaviours and effects. Thus we come properly also to consider the qualities of his actions as they seem to affect his future well-being. In this proper study of mankind, living out the destiny ascribed in Genesis—"to have dominion over all the earth"—the concern is valid as to whether his organized energies (social behavior) have or should have a quality of concern with regard to his posterity.

On the Nature of Man

The primordial condition of man setting our kind apart from other primates involved more than hands, brain, and walking upright. Man owes his success in part to his digestive apparatus, which is equaled by none of his near-kin or by few other similarly omnivorous animals as to the range of potential food which can sustain him on a mixed, vegetarian, or flesh diet. The long, helpless infancy and the dependence through the years of childhood have forged, it would seem, *ab origine* a maternal bond that expresses itself in persistence of family and in formal recognition of kinship, system of kinship being perhaps the first basis of social organization. When humans lost the oestrous cycle is unknown; its weakening and loss is probably a feature of domestication, and it may have occurred early in the history of man, eldest of the domesticated creatures.

Built into the biologic nature of man therefore appear to be qualities tending to maximize geographic expansiveness, vigorous reproduction, and a bent to social development. His extreme food range favored numerical increase; I question, for instance, any assumptions of sporadic or very sparse populations of Paleolithic man in any lands he had occupied. The dominant and continuous role of woman in caring for the family suggests further inferences. Maternal duties prescribed as sedentary a life as possible. Her collecting of food and other primary materials was on the short tether of her dependent offspring. Hers also was the care of what had been collected in excess of immediate need, the problem of storage, hers the direction toward homemaking and furnishing. To the "nature" of woman we may perhaps ascribe an original social grouping, a cluster of kindred households, in which some stayed home to watch over bairns and baggage while others ranged afield. Baby-sitting may be one of the most ancient of human institutions.

Implicit in this interpretation of the nature of man and primordial society, as based on his trend to sedentary life and clustering, are territoriality, the provision of stores against season of lack, and probably a tendency to monogamy. These traits are familiar enough among numerous animals, and there is no reason for denying them to primitive man. Shifts of population imposed by seasons do not mean wandering, homeless habits; nomadism is an advanced and specialized mode of life. Folk who stuffed or starved, who took no heed of the morrow, could not have possessed the Earth or laid the foundations of human culture. To the ancestral folk we may rather ascribe practical-minded economy of effort.

Their success in survival and in dispersal into greatly differing habitats tells of ability to derive and communicate sensible judgements from changing circumstances.

The culture of man is herewith considered as in the main a continuum from the beginning; such is its treatment by archeology. The record of artifacts is much greater, more continuous and begins earlier than do his recovered skeletal remains. Thereby hangs the still-argued question of human evolution, about which divergent views are unreconciled. If culture was transmitted and advanced in time and space as the archeologic record indicates, there would appear to be a linked history of a mankind that includes all the specific and generic hominid classifications of physical anthropology. Man, *sensu latiore,* therefore may conceivably be one large species complex, from archaic to modern forms, always capable of interbreeding and intercommunication. Variation occurred by long geographic isolation, blending usually when different stocks met. The former is accepted; the latter seems assured to some and is rejected by others, the Mount Carmel series of skulls being thus notoriously in dispute.

Neanderthal man, poor fellow, has had a rough time of it. He invented the Mousterian culture, a major advance which appears to have been derived from two anterior culture lines. The Abbe Breuil has credited him with ceremonial cults that show a developed religious belief and spiritual ceremonial (Breuil and Lantier, 1951, chap. 18). Boyd, in his serologic classification of mankind (1950), the only system available on a genetic basis, has surmised that Neanderthal is ancestral to a Paleo-European race. There is no basis for holding Neanderthal man as mentally inferior or as unable to cope with the late Pleistocene changes of European climate. Yet there remains aversion to admitting him to our ancestry. The sad confusion of physical anthropology is partly the result of its meager knowledge of hereditary factors, but also to *Homo's* readiness to crossbreed, a trait of his domestication and a break with the conservatism of the instinctive.

We are groping in the obscurity of a dim past; it may be better to consider cultural growth throughout human time as proceeding by invention, borrowing, and blending of learning, rather than by evolution of human brain, until we know more of biological evolution in man. The little that we have of skeletal remains is subject to unreconciled evaluations; the record of his work is less equivocal. The question is not, could Peking man have left the artifacts attributed to him, as has been the subject of debate, but did he—that is, do the bones belong with the tools?

When primordial man began to spread over the Earth, he knew little, but what he had learned was by tested and transmitted experience; he cannot have been fear-ridden but rather, at least in his successful kinds, was venturesome, ready to try out his abilities in new surroundings. More and more he imposed himself on his animal competitors and impressed his mark on the lands he inhabited. Wherever he settled, he came to stay unless the climate changed too adversely or the spreading sea drove him back.

Climatic Changes and Their Effects on Man

The age of man is also the Ice Age. Man may have witnessed its beginning; we perhaps are still living in an interglacial phase. His growth of learning and his expansion over the Earth have taken place during a geologic period of extreme instability of climates and also of extreme simultaneous climatic contrast. His span has been cast within a period of high environmental tensions. Spreading icecaps caused the ocean to shrink back from the shallow continental margins, their waning to spread the seas over coastal plains. With lowered sea levels, rivers trenched their valley floors below coastal lowlands; as sea level rose, streams flooded and aggraded their valleys. Glacial and Recent time have been governed by some sort of climatic pendulum, varying in amplitude of swing, but affecting land and sea in all latitudes, and life in most areas. The effects have been most felt in the Northern Hemisphere, with its large continental masses, wide plains, high mountain ranges, and broad plateaus. Millions of square miles of land were alternately buried under ice and exposed; here, also the shallow seas upon the continental shelf spread and shrank most broadly.

This time of recurrent changes of atmosphere, land, and sea gave advantage to plastic, mobile, and prolific organisms, to plants and animals that could colonize newly available bodies of land, that had progeny some of which withstood the stresses of climatic change. The time was favorable for biologic evolution, for mutants suited to a changed environment, for hybrids formed by mingling, as on ecologic frontiers. To this period has been assigned the origin of many annual plant species dependent on heavy seed production for success (Ames, 1939). Adaptive variations in human stocks, aided by sufficiently isolating episodes of Earth history, have also been inferred.[1]

The duration of the Ice Age and of its stages has not been determined. The old guess of a million years over all it still convenient. The four glacial and three interglacial stages may have general validity; there are doubts that they were strictly in phase in all continents. In

---

1. As most recently by Coon, 1953.

North America the relations of the several continental icecaps to the phases of Rocky Mountain glaciation, and of the latter to the Pacific mountains, are only inferred, as is the tie-in of pluvial stages in our Southwest. That great lakes and permanent streams existed in many of the present dry lands of the world is certain; that these pluvial phases of intermediate latitudes correspond to glacial ones in high latitudes and altitudes is in considerable part certain, but only in a few cases has a pluvial state been securely tied to a contemporaneous glacial stage. The promising long-range correlation of Pleistocene events by eustatic marine terraces and their dependent alluvial terraces is as yet only well started. Except for northwestern Europe, the calendar of the later geologic past is still very uncertain. The student of further human time, anxious for an absolute chronology, is at the moment relying widely on the ingenious astronomical calendar of Milakkovitch and Zeuner as an acceptable span for the Ice Age as a whole and for its divisions. It is not acceptable, however, to meteorology and climatology.[2] Slowly and bit by bit only are we likely to see the pieces fall into their proper order; nothing is gained by assurance as to what is insecure.

The newer meteorology is interesting itself in the dynamics of climatic change (Shapley, 1953; Mannerfelt et al., 1949). Changes in the general circulation pattern have been inferred as conveying, in times of glacial advance, more and more frequent masses of moist, relatively warm air into high latitudes and thereby also increasing the amount of cloud cover. The importance now attached to condensation nuclei has directed attention again to the possible significance of volcanic dust. Synoptic climatological data are being examined for partial models in contemporary conditions as conducive to glaciation and deglaciation (Leighly, 1949, pp. 133-34). To the student of

the human past, reserve is again indicated in making large climatic reconstructions. Such cautions, I should suggest, with reserve also to my competence to offer them, with regard to the following:

It is misleading to generalize glacial stages as cold and interglacial ones as warm. The developing phases of glaciation probably required relatively warm moist air, and decline may have been by the dominance of cold dry air over the ice margins. The times of climatic change may thus not coincide with the change from glacial advance to deglaciation. We may hazard the inference that developing glaciation is associated with low contrast of regional climates; regression of ice and beginning of an interglacial phase probably are connected (although not in each case) with accentuated contrast or "continentality" of climates. One interglacial did not repeat necessarily the features of another; nor must one glacial phase duplicate another. We need only note the difference in centers of continental glaciation, of direction of growth of ice lobes, of terminal moraine-building, of structure of till and of fluvioglacial components to see the individuality of climates of glacial stages. In North America, in contrast to Europe, there is very little indication of a periglacial cold zone of tundra and of permafrost in front of the continental icecaps. Questionable also is the loess thesis of dust as whipped up from bare ground and deposited in beds by wind, these surfaces somehow becoming vegetated by a cold steppe plant cover.

The events of the last deglaciation and of the "postglacial" are intelligible as yet only in part. A priori it is reasonable to consider that the contemporary pattern of climates had

2. Shapley, 1953; Willet, 1950; Simpson, G. C., 1934, 1940.

become more or less established before the last ice retreat began. Lesser later local climatic oscillations have been found but have been improperly extended and exaggerated, however, in archeological literature. In the pollen studies of bogs of northwestern Europe, the term "climatic optimum" was introduced innocently to note a poleward and mountainward extension of moderate proportions for certain plants not occurring at the same time over the entire area. Possibly this expansion of range means that there were sunnier summers and fall seasons, permitting the setting and maturing of seed for such plants somewhat beyond their prior and present range; that is, under more "continental" and less "maritime" weather conditions. This modest and expectable variation of a local climate in the high latitudes and at the changing sea borders of North Atlantic Europe has been construed by some students of prehistory into a sort of climatic golden age, existent at more or less the same time in distant parts of the world, without regard to dynamics or patterns of climates. We might well be spared such naively nominal climatic constructions as have been running riot through interpretations of prehistory and even of historic time.

The appearance or disappearance, increase or decrease, of particular plants and animals may not spell out obligatory climatic change, as has been so freely inferred. Plants differ greatly in rate of dispersal, in pioneering ability, in having routes available for their spread, and in other ways that may enter into an unstable ecologic association, as on the oft-shifted stage of Pleistocene and Recent physiography. The intervention of man and animals has also occurred to disturb the balance. The appearance and fading of pines in an area, characteristic in many bog pollen columns, may tell nothing of climatic change: pines are notorious early colonizers, establishing themselves freely in mineral soils and open situations and yielding to other trees as shading and organic cover of ground increase. Deer thrive on browse; they increase wherever palatable twigs become abundant, in brush lands and with young tree growth; ecologic factors of disturbance other than climate may determine the food available to them and the numbers found in archeologic remains.

The penetration of man to the New World is involved in the question of past and present climates. The origin and growth of the dominant doctrine of a first peopling of the Western Hemisphere in postglacial time is beyond our present objective, but it was not based on valid knowledge of climatic history. The postglacial and present climatic pattern is one of extremes rarely reached or exceeded in the past of the Earth. Passage by land within this time across Siberia, Alaska, and Canada demanded specialized advanced skills in survival under great and long cold comparable to those known to Eskimo and Athabascan, an excessive postulate for many of the primitive people of the New World. Relatively mild climates did prevail in high latitudes at times during the Pleistocene. At such times in both directions between Old and New World, massive migrations took place of animals incapable of living on tundras, animals that are attractive game for man. If man was then living in eastern Asia, nothing hindered him from migrating along with such nonboreal mammals. The question is of fundamental interest, because it asks whether man in the New World, within a very few thousands years, achieved independently a culture growth comparable and curiously parallel to that of the Old, which required a much greater span. There is thus also the inference that our more primitive aborigines passed the high latitudes during more genial climes rather than that they lost subsequently numerous useful skills.

Fire

Speech, tools, and fire are the tripod of culture and have been so, we think, from the beginning. About the hearth, the home and workshop are centered. Space heating under shelter, as a rock overhang, made possible living in inclement climates; cooking made palatable many plant products; industrial innovators experimented with heat treatment of wood, bone, and minerals. About the fireplace, social life took form, and the exchange of ideas was fostered. The availability of fuel has been one of the main factors determining the location of clustered habitation.

Even to Paleolithic man, occupant of the Earth for all but the last 1 or 2 percent of human time, must be conceded gradual deformation of vegetation by fire. His fuel needs were supplied by dead wood, drifted or fallen, and also by the stripping of bark and bast that caused trees to die and become available as fuel supply. The setting or escape of fire about camp sites cleared away small and young growth, stimulated annual plants, aided in collecting, and became elaborated in time into the fire drive, a formally organized procedure among the cultures of the Upper Paleolithic *grande chasse* and of their New World counterpart.

Inferentially, modern primitive peoples illustrate the ancient practices in all parts of the world. Burning, as a practice facilitating collecting and hunting, by insensible stages became a device to improve the yield of desired animals and plants. Deliberate management of their range by burning to increase food supply is apparent among hunting and collecting peoples, in widely separated areas, but has had little study. Mature woody growth provides less food for man and ground animals than do fire-disturbed sites, with protein-rich young growth and stimulated seed production, accessible at ground levels. Game yields are usually greatest where the vegetation is kept in an immediate state of ecologic succession. With agricultural and pastoral peoples, burning in preparation for planting and for the increase of pasture has been nearly universal until lately.

The gradually cumulative modifications of vegetation may become large as to selection of kind and as to aspect of the plant cover. Pyrophytes include woody monocotyledons, such as palms, which do not depend on a vulnerable cambium tissue, trees insulated by thick corky bark, trees and shrubs able to reproduce by sprouting, and plants with thick, hard-shelled seeds aided in germination by heat. Loss of organic matter on and in the soil may shift advantage to forms that germinate well in mineral soils, as the numerous conifers. Precocity is advantageous. The assemblages consequent upon fires are usually characterized by a reduced number of species, even by the dominance of few and single species. Minor elements in a natural flora, originally mainly confined to accidentally disturbed and exposed situations, such as windfalls and eroding slopes, have opened to them by recurrent burning the chance to spread and multiply. In most cases the shift is from mesophytic to less exacting, more xeric, forms, to those that do not require ample soil moisture and can tolerate at all times full exposure to sun. In the long run the scales are tipped against the great, slowly maturing plants—the trees (a park land of mature trees may be the last stand of what was a complete woodland). Our eastern woodlands, at the time of white settlement, seem largely to have been in process of change to park lands. Early accounts stress the open stands of trees, as indicated by the comment that one could drive a coach from seaboard to the Mississippi River over almost any favoring terrain. The "forest primeval" is exceptional. In the end the success in a land occupied by man of whatever cultural level goes to the annuals and short-lived perennials, able to seed heavily or to reproduce by rhizome and tuber. This grossly drawn sketch may introduce the matter of processes resulting in what is called ecologically a secondary fire association, or subclimax, if it has historical persistence.

The climatic origin of grasslands rests on a poorly founded hypothesis. In the first place, the individual great grasslands extend over long climatic gradients from wet to dry and grade on their driest margins into brush and scrub. Woody growth occurs in them where there are breaks in the general surface, as in the Cross Timbers of our Southwest. Woody plants establish themselves freely in grasslands if fire protection is given: the prairies and steppes are suited to the growth of the trees and shrubs native to adjacent lands but may lack them. An individual grassland may extend across varied parent-materials. Their most common quality is that they are upland plains, having periods of dry weather long enough to dry out the surface of the ground, which accumulate a sufficient amount of burnable matter to feed and spread a fire. Their position and limits are determined by relief; nor do they extend into arid lands or those having a continuously wet ground surface. Fires may sweep indefinitely across a surface of low relief but are checked shortly at barriers of broken terrain, the checking being more abrupt if the barrier is sunk below the general surface. The inference is that origin and preservation of grasslands are due, in the main, to burning and that they are in fact great and, in some cases, ancient cultural features.

In other instances simplified woodlands, such as the pine woods of our Southeast, *palmares* in tropical savannas, are pyrophytic deformations; there are numerous vegetational alternatives other than the formation of grassland by recurrent burning. Wherever primitive man has had the opportunity to turn fire loose on a land, he seems to have done so, from time immemorial; it is only civilized societies that have undertaken to stop fires.

In areas controlled by customary burning, a near-ecologic equilibrium may have been attained, a biotic recombination maintained by similarly repeated human intervention. This is not destructive exploitation. The surface of the ground remains protected by growing cover, the absorption of rain and snow is undiminished, and the loss of moisture from ground to atmosphere possibly is reduced. Microclimatic differences between woodland and grassland are established as effect if not as cause, and some are implicit in the Shelter Belt Project.

Our modern civilization demands fire control for the protection of its property. American forestry was begun as a remedy for the devastation by careless lumbering at a time when dreadful holocausts almost automatically followed logging, as in the Great Lakes states. Foresters have made a first principle of fire suppression. Complete protection, however, accumulates tinder year by year; the longer the accumulation, the greater is the fire hazard and the more severe the fire when it results. Stockmen are vociferous about the loss of grazing areas to brush under such protection of the public lands. Here and there, carefully controlled light burning is beginning to find acceptance in range and forest management. It is being applied to long-leaf pine reproduction in Southeastern states and to some extent for grazing in western range management. In effect, the question is now being raised whether well-regulated fires may not have an ecologic role beneficent to modern man, as they did in older days.

Peasant and Pastoral Ways

The next revolutionary intervention of man in the natural order came as he selected certain plants and animals to be taken under his care, to be reproduced, and to be bred into domesticated forms increasingly dependent on him for survival. Their adaptation to serve human wants runs counter, as a rule, to the processes of natural selection. New lines and processes of organic evolution were entrained, widening the gap between wild and domestic forms. The natural land became deformed, as to biota, surface, and soil, into unstable cultural landscapes.

Conventionally, agricultural origins are placed at the beginning of Neolithic time, but it is obvious that the earliest archeologic record of the Neolithic presents a picture of an accomplished domestication of plants and animals, of peasant and pastoral life resembling basic conditions that may still be met in some parts of the Near East.

Three premises as to the origin of agriculture seem to me to be necessary: (1) That this new mode of life was sedentary and that it arose out of an earlier sedentary society. Under most conditions, and especially among primitive agriculturists, the planted land must be watched over continuously against plant predators. (2) That planting and domestication did not start from hunger but from surplus and leisure. Famine-haunted folk lack the opportunity and incentive for the slow and continuing selection of domesticated forms. Village communities in comfortable circumstances are indicated for such progressive steps. (3) Primitive agriculture is located in woodlands. Even the pioneer American farmer hardly invaded the grasslands until the second quarter of the past century. His fields were clearings won by deadening, usually by girdling, the trees. The larger the trees, the easier the task; brush required grubbing and cutting; sod stopped his advance until he had plows capable of ripping through the matted grass roots. The forest litter he cleaned up by

occasional burning; the dead trunks hardly interfered with his planting. The American pioneer learned and followed Indian practices. It is curious that scholars, because they carried into their thinking the tidy fields of the European plowman and the felling of trees by ax, have so often thought that forests repelled agriculture and that open lands invited it.

The oldest form of tillage is by digging, often but usually improperly called "hoe culture." This was the only mode known in the New World, in Negro Africa, and in the Pacific islands. It gave rise, at an advanced level, to the gardens and horticulture of Monsoon Asia and perhaps of the Mediterranean. Its modern tools are spade, fork, and hoe, all derived from ancient forms. In tropical America this form of tillage is known as the *conuco,* in Mexico as the *milpa,* in the latter case a planting of seeds of maize, squash, beans, and perhaps other annuals. The conuco is stocked mainly by root and stem cuttings, a perennial garden plot. Recently, the revival of the Old Norse term "swithe," or "swidden," has been proposed (Izikowitz, 1951, p. 7 n.; Conklin, 1954).

Such a plot begins by deadening tree growth, followed toward the end of a dry period by burning, the ashes serving as quick fertilizer. The cleared space then is well stocked with a diverse assemblage of useful plants, grown as tiers of vegetation if moisture and fertility are adequate. In the maize-beans-squash complex the squash vines spread over the ground, the cornstalks grow tall, and the beans climb up the cornstalk. Thus the ground is well protected by plant cover, with good interception of the falling rain. In each conuco a high diversity of plants may be cared for, ranging from low herbs to shrubs, such as cotton and manioc, to trees entangled with cultivated climbers. The seeming disorder is actually a very full use of light and moisture, an admirable ecologic substitution by man perhaps equivalent to the natural cover also in the protection given to the surface of the ground. In the tropical conuco an irregular patch is dug into at convenient spots and at almost any time to set out or collect different plants, the planted surface at no time being wholly dug over. Digging roots and replanting may be going on at the same time. Our notions of a harvest season when the whole crop is taken off the field are inapplicable. In the conucos something may be gathered on almost any day through the year. The same plant may yield pot and salad greens, pollen-rich flowers, immature fruit, and ripened fruit; garden and field are one, and numerous domestic uses may be served by each plant. Such multiple population of the tilled space makes possible the highest yields per unit of surface, to which may be added the comments that this system has developed plants of highest productivity, such as bananas, yams, and manioc, and that food production is by no means the only utility of many such plants.

The planting systems really do not deserve the invidious terms given them, such as "slash and burn" or "shifting agriculture." The abandonment of the planting after a time to the resprouting and reseeding wild woody growth is a form of rotation by which the soil is replenished by nutriments carried up from deep-rooted trees and shrubs, to be spread over the ground as litter. Such use of the land is freed from the limitations imposed on the plowed field by terrain. That it may give good yields on steep and broken slopes is not an argument against the method, which gives much better protection against soil erosion than does any plowing. It is also in these cultures that we find that systems of terracing of slopes have been established.

Some of the faults charged against the system derive from the late impact from our own culture, such as providing axes and machetes by which sprouts and bush may be kept whacked out instead of letting the land rest under regrowth, the replacement of subsistence crops by money crops, the world-wide spurt in population, and the demand for manufactured goods which is designated as rising standard of living. Nor do I claim that under this primitive planting man could go on forever growing his necessities without depleting the soil; but rather that, in its basic procedure and crop assemblages, this system has been most conservative of fertility at high levels of yield; that, being protective and intensive, we might consider it as being fully suited to the physical and cultural conditions of the areas where it exists. Our Western know-how is directed to land use over a short run of years and is not the wisdom of the primitive peasant rooted to his ancestral lands.

Our attitudes toward farming stem from the other ancient trunk whence spring the sowers, reapers, and mowers; the plowmen, dairymen, shepherds, the herdsmen. This is the complex already well represented in the earliest Neolothic sites of the Near East. The interest of this culture is directed especially toward seed production of annuals, cereal grasses in particular. The seedbed is carefully prepared beforehand to minimize weed growth and provide a light cover of well-worked soil in which the small seeds germinate. An evenly worked and smooth surface contrasts with the hit-or-miss piling of earth mounds, "hills" in the American farm vernacular, characteristic of conuco and milpa. Instead of a diversity of plants, the prepared ground receives the seed of one kind. (Western India is a significant exception.) The crop is not further cultivated and stands to maturity, when it is reaped at one time. After the harvest the field may lie fallow until the next season. The tillage implement is the plow, in second place, the harrow, both used to get the field ready for sowing. Seeding traditionally is by broadcasting, harvesting by cutting blades.

Herd animals, meat cattle, sheep, goats, horses, asses, camels, are either original or very early in this system. The keeping of grazing and browsing animals is basic. All of them are milked or have been so in the past. In my estimation milking is an original practice and quality of their domestication and continued to be in many cases their first economic utility; meat and hides, the product of surplus animals only.

The over-all picture is in great contrast to that of the planting cultures; regular, elongated fields minimize turning the animals that pull the plow; fields are cultivated in the off season, in part to keep them free of volunteer growth; fields are fallowed but not abandoned, the harvest season is crowded into the end of the annual growth period; thereafter, stock is pastured on stubble and fallow; land unsuited or not needed for the plow is used as range on which the stock grazes and browses under watch of herdboys or herdsmen.

This complex spread from its Near Eastern cradle mainly in three directions, changing its character under changed environments and by increase of population.

1. Spreading into the steppes of Eurasia, the culture lost its tillage and became completely pastoral, with true nomadism. This is controversial, but the evidence seems to me to show that all domestication of the herd animals (except for reindeer) was effected by sedentary agriculturists living between India and the Mediterranean and also that the great, single continuous area in which milking was practiced includes all the nomadic peoples, mainly as a fringe about the milking seed-farmers. It has also been pointed out that nomadic cultures depend on agricultural peoples for some of their needs and thus lacking a self-contained economy, can hardly have originated independently.

2. The drift of the Celtic, Germanic, and Slavic peoples westward (out of southwestern and western Asia?) through the northern European plain appears to have brought them to their historic seats predominantly as cattle- and horse-raisers. Their movement was into lands of cooler and shorter summers and of higher humidity, in which wheat and barley did poorly. An acceptable thesis is that, in southwestern Asia, rye and probably oats were weed grasses growing in fields of barley and wheat. They were harvested together and not separated by winnowing. In the westward movement of seed farmers across Europe, the weed grains did better and the noble grain less well. The cooler and wetter the summers, the less wheat and barley did the sower reap and the more of rye and oat seeds, which gradually became domesticated by succeeding where the originally planted kinds failed.

Northwestern and Central Europe appear to be the home of our principal hay and pasture grasses and clovers. As the stock-raising colonists deadened and burned over tracts of woodland, native grasses and clovers spontaneously took possession of the openings. These were held and enlarged by repetition of burning and cutting. Meadow and pasture, from the agricultural beginnings, were more important here than plowland. Even the latter, by pasturing the rye fields and the feeding of oat straw and grain, were part of animal husbandry. Here, as nowhere else did the common farmer concern himself with producing feed for his stock. He was first a husbandman; he cut hay to store for winter feed and cured it at considerable trouble; he stabled his animals over the inclement season, or stall-fed them through the year; the dung-hill provided dressing for field and meadow. House, barn, and stable were fused into one structure. The prosperity of farmstead and village was measured by its livestock rather than by arable land.

The resultant pattern of land use, which carries through from the earliest times, as recovered by archeology in Denmark and northern Germany, was highly conservative of soil fertility. The animal husbandry maintained so effective a ground cover that northern Europe has known very little soil erosion. Animal manure and compost provided adequate return of fertility to the soil. Man pretty well established a closed ecologic cycle. It was probably here that man first undertook to till the heavy soils. Clayey soils, rich in plant food but deficient in drainage, are widespread in the lowlands, partly due to climatic conditions, partly a legacy of the Ice Age. The modern plow with share, moldboard, and colter had either its origin or a major development here for turning real furrows to secure better aeration and drainage. Beneficial in northwestern and Central Europe, it was later to become an instrument of serious loss elsewhere.

3. The spread of sowing and herding cultures westward along both sides of the Mediterranean required no major climatic readjustment. Wheat and barley continued to be the staple grains; sheep and goats were of greater economic importance than cattle and horses. Qualities of the environment that characterized the Near East were accentuated to the west; valleys lie imbedded in mountainous terrain, the uplands are underlain by and developed out of limestone, and, to the south of the Mediterranean, aridity becomes prevalent. The hazard of drought lay ever upon the Near Eastern homeland and on the colonial regions to the west and south. No break between farmer and herdsman is discernible at any time; as the village Arab of today is related to the Bedouin, the environmental specialization may have been present from the beginning; flocks on the mountains and dry lands, fields where moisture sufficed and soil was adequate.

That the lands about the Mediterranean have become worn and frayed by the usage to which they have been subjected has long been recognized, though not much is known as to when and how. The eastern and southern Mediterranean uplands especially are largely of limestone, attractive as to soil fertility but, by their nature, without deep original mantle of soil or showing the usual gradation of subsoil into bedrock and thus are very vulnerable to erosion. The less suited the land was or became to plow cultivation, the greater the shift to pastoral economy. Thus a downslope migration of tillage characterized, in time, the retreating limits of the fields, and more and more land became range for goats, sheep, and asses. Repeatedly, prolonged droughts must have speeded the downslope shift, hillside fields suffering most, and with failing vegetation cover becoming more subject to washing when rains came.

Thus we come again to the question of climatic change as against attrition of surface and increased xerophytism of vegetation by human disturbance and, in particular, to what is called the "desertification" of North Africa and the expansion of the Sahara. A case for directional change in the pattern of atmospheric circulation has been inferred from archeology and faunal changes. I am doubtful that it is a good case within the time of agricultural and pastoral occupation. Another view is that the progressive reduction of plant cover by man has affected soil and ground-surface climate unfavorably. Largely, and possibly wholly, the deterioration of the borders of the dry lands may have been caused by adverse, cumulative effects of man's activities. From archeologic work we need much more information as to whether human occupation has been failing in such areas over a long time, or whether it has happened at defined intervals, and also whether, if such intervals are noted, they may have a cultural rather than an environmental (climatic) basis.

No protective herbaceous flora became established around the shores of the Mediterranean on pastures and meadows as was the case in the north. Flocks and herds grazed during the short season of soft, new grass but most of the year browsed on woody growth. The more palatable feed was eaten first and increasingly eliminated; goats and asses got along on range that had dropped below the support levels required by more exacting livestock. As is presently true in the western United States, each prolonged drought must have left the range depleted, its carrying capacity reduced, and recovery of cover less likely. Natural balance between plants and animals is rarely re-established under such exploitation, since man will try to save his herd rather than their range. A large and long deterioration of the range may therefore fully account for the poor and xerophytic flora and fauna without postulating progressive climatic desiccation, for the kinds of life that survive under overuse of the land are the most undemanding inhabitants.

Comparative studies of North Africa and of the American Southwest and northern Mexico are needed to throw light on the supposed "desiccation" of the Old World. We know the dates of introduction of cattle and sheep to the American ranges and can determine rate and kind of change of vegetation and surface. The present desolate shifting-sand area that lies between the Hopi villages and the Colorado River was such good pasture land late in the eighteenth century that Father Escalante, returning from his canyon exploration, rested his travel-worn animals there to regain flesh. The effects of Navaho sheep-herding in little more than a century and mainly in the last sixty years are well documented. Lower Cali-

fornia and Sonora are climatic homologues of the western Sahara. Against the desolation of the latter, the lands about the Gulf of California are a riot of bloom in spring and green through summer. Their diversity, in kind and form, of plant and of animal life is high, and the numbers are large. When Leo Waibel came from his African studies to Sonora and Arizona, he remarked: "But your deserts are not plant deserts." Nor do we have hammadas or ergs, though geologic and meteorologic conditions may be similar. The principal difference may be that we have had no millennial, or even centuries-long, overstocking of our arid, semiarid, and subhumid lands. The scant life and even the rock and sand surfaces of the Old World deserts may record long attrition by man in climatic tension zones.

Impact of Civilization in Antiquity and the Middle Ages

Have the elder civilizations fallen because their lands deteriorated? Ellsworth Huntington read adverse climatic change into each such failure; at the other extreme, political loss of competence has been asserted as sufficient. Intimate knowledge of historical sources, archeologic sites, biogeography and ecology, and the processes of geomorphology must be fused in patient field studies, so that we may read the changes in habitability through human time for the lands in which civilization first took form.

The rise of civilizations has been accomplished and sustained by the development of powerful and elaborately organized states with a drive to territorial expansion, by commerce in bulk and to distant parts, by monetary economy, and by the growth of cities. Capital cities, port cities by sea and river, and garrison towns drew to themselves population and products from near and far. The ways of the country became subordinated to the demands of the cities, the *citizen* distinct from the *miserabilis plebs.* The containment of community by locally available resources gave way to the introduction of goods, especially foodstuffs, regulated by purchasing, distributing, or taxing power.

Thereby removal of resource from place of origin to place of demand tended to set up growing disturbance of whatever ecologic equilibrium had been maintained by the older rural communities sustained directly within their metes. The economic history of antiquity shows repeated shifts in the areas of supply of raw materials that are not explained by political events but raise unanswered questions as to decline of fertility, destruction of plant cover, and incidence of soil erosion. What, for instance, happened to Arabia Felix, Numidia, Mauretania, to the interior Lusitania that has become the frayed Spanish Extremadura of today? When and at whose hands did the forests disappear that furnished ship and house timbers, wood for burning lime, the charcoal for smelting ores, and urban fuel needs? Are political disasters sufficient to account for the failure of the civilization that depended on irrigation and drainage engineering? How much of the wide deterioration of Mediterranean and Near Eastern land came during or after the time of strong political and commercial organization? For ancient and medieval history our knowledge as to what happened to the land remains too largely blank, except for the central and northern European periphery. The written documents, the testimony of the archeologic sites, have not often been interpreted by observation of the physical condition of the locality as it is and comparison with what it was.

The aspect of the Mediterranean landscapes was greatly changed by classical civilization through the introduction of plants out of the East. Victor Hehn first described Italy as wearing a dress of an alien vegetation, and though he carried the theme of plant introduction out of the East too far, his study (1886) of the Mediterranean lands through antiquity is not only memorable but retains much validity. The westward dispersal of vine, olive, fig, the stone fruits, bread wheat, rice, and many ornamentals and some shade trees was due in part or in whole to the spread of Greco-Roman civilization, to which the Arabs added sugar cane, date palm, cotton, some of the citrus fruits, and other items.

European Overseas Colonization

When European nations ventured forth across the Atlantic, it was to trade or raid, the distinction often determined by the opportunity. In Africa and Asia the European posts and factories pretty well continued in this tradition through the eighteenth century. In the New World the same initial activities soon turned into permanent settlement of Old World forms and stocks. Columbus, searching only for a trade route, started the first overseas empire. Spain stumbled into colonization, and the other nations acquired stakes they hoped might equal the Spanish territorial claim. The Casa de Contratacion, or House of Trade, at Seville, the main Atlantic port, became the Spanish colonial office. The conquistadores came not to settle but to make their fortunes and return home, and much the same was true for the earlier adventurers from other nations. Soldiers and adventurers rather than peasants and artisans made up the first arrivals, and few brought their women. Only in New England did settlement begin with a representative assortment of people, and only here were the new communities transplanted from the homeland without great alteration.

The first colony, Santo Domingo, set in large measure the pattern of colonization. It began with trade, including ornaments of gold. The quest for gold brought forced labor and the dying-off of the natives, and this, in turn, slave-hunting and importation of black slaves. Decline of natives brought food shortages and wide abandonment of conucos. Cattle and hogs were pastured on the lately tilled surfaces; and Spaniards, lacking labor to do gold-placering, became stock ranchers. Some turned to cutting dyewoods. Of the numerous European plants introduced to supply accustomed wants, a few, sugar cane, cassia, and ginger, proved moderately profitable for export, and some of the hesitant beginnings became the first tropical plantations. One hope of fortune failing, another was tried; the stumbling into empire was under way by men who had scarcely any vision of founding a new homeland.

What then happened to the lands of the New World in the three colonial centuries? In the first place, the aboriginal populations in contact with Europeans nearly everywhere declined greatly or were extinguished. Especially in the tropical lowlands, with the most notable exception of Yucatan, the native faded away, and in many cases the land was quickly repossessed by forest growth. The once heavily populous lands of eastern Panama and northwestern Colombia, much of the lowland country of Mexico, both on the Pacific and Gulf sides, became emptied in a fery few years, was retaken by jungle and forest, and in considerable part remains such to the present. The highlands of Mexico, of Central America, and of the Andean lands declined in population greatly through the sixteenth and perhaps well through the seventeenth century, with slow, gradual recovery in the eighteenth. The total population, white and other, of the areas under European control was, I think, less at the end of the eighteenth century than at the time of discovery. Only in British and French West

Indian islands were dense rural populations built up.

It is hardly an exaggeration to say that the early Europeans supported themselves on Indian fields. An attractive place to live for a European would ordinarily have been such for an Indian. In the Spanish colonies, unlike the English and French, the earlier grants were not of land titles but of Indian communities to serve colonist and crown. In crops and their tillage the colonists of all nations largely used the Indian ways, with the diversion of part of the field crop to animal feed. Only in the Northeast, most of all in our Middle Colonies, were native and European crops fused into a conservative plow-and-animal husbandry, with field rotation, manuring, and marl dressing. The Middle Colonies of the eighteenth century appear to have compared favorably with the best farming practices of western Europe.

Sugar cane, first and foremost of the tropical plantations, as a closely planted giant grass, gave satisfactory protection to the surface of the land. The removal of cane from the land did reduce fertility unless the waste was properly returned to the canefields. The most conservative practices known are from the British islands, where cane waste was fed to cattle kept in pens, and manuring was customary and heavy. Bagasse was of little value as fuel in this period because of the light crushing rollers used for extracting cane juice; thus the colonial sugar mills were heavy wood users, especially for boiling sugar. The exhaustion of wood supply became a serious problem in the island of Haiti within the sixteenth century.

Other plantation crops—tobacco, indigo, cotton, and coffee—held more serious erosion hazards, partly because they were planted in rows and given clean cultivation, partly because they made use of steeper slopes and thinner soils. The worst offender was tobacco, grown on land that was kept bared to the rains and nourished by the wood ashes of burned clearings. Its cultivation met with greatest success in our Upper South, resulted in rapidly shifting clearings because of soil depletion, and caused the first serious soil erosion in our country. Virginia, Maryland, and North Carolina show to the present the damages of tobacco culture of colonial and early post-colonial times. Southern Ohio and eastern Missouri repeated the story before the middle of the nineteenth century.

As had happened in Haiti, sharp decline of native populations brought elsewhere abandonment of cleared and tilled land and thereby opportunity to the stockman. The plants that pioneer in former fields which are left untilled for reasons other than because of decline of fertility include forms, especially annuals, of high palatability, grasses, amaranths, chenopods, and legumes. Such is the main explanation for the quick appearance of stock ranches, of *ganado mayor* and *menor,* in the former Indian agricultural lands all over Spanish America. Cattle, horses, and hogs thrived in tropical lowland as well as in highland areas. Sheep-raising flourished most in early years on the highlands of New Spain and Peru, where Indian population had shrunk. Spanish stock, trespassing upon Indian plantings, both in lowland and in highland, afflicted the natives and depressed their chances of recovery (Simpson, L., 1952). In the wide savannas stockmen took over the native habits of burning.

The Spaniards passed in a few years from the trading and looting of metals to successful prospecting, at which they became so adept that it is still said that the good mines of today are the *antiguas* of colonial working. When mines were abandoned, it was less often due to the working-out of the ore bodies than to inability to cope with water in shafts and to the exhaustion of the necessary fuel and timber. A good illustration has been worked out for Parral in Mexico (West, 1949). Zacatecas, today in the midst of a high sparse grassland, was in colonial times a woodland of oak and pine and, at lower levels, of mesquite. About Andean mines the scant wood was soon exhausted, necessitating recourse to cutting mats of *tola* heath and even the clumps of coarse *ichu* (stipa) grass. Quite commonly the old mining *reales* of North and South America are surrounded by a broad zone of reduced and impoverished vegetation. The effects were increased by the concentration of pack and work animals in the mines, with resultant overpasturing. Similar attrition took place about towns and cities, through timber-cutting, charcoal- and lime-burning, and overpasturing. The first viceroy of New Spain warned his successor in 1546 of the depletion of wood about the city of Mexico.

I have used mainly examples from Spanish America for the colonial times, partly because I am most familiar with this record. However, attrition was more sensible here because of mines and urban concentrations and because, for cultural and climatic reasons, the vegetation cover was less.

Last Frontiers of Settlement

The surges of migration of the nineteenth century are family history for many of us. Never before did, and never again may, the white man expand his settlements as in that brief span that began in the later eighteenth century and ended with the first World War. The prelude was in the eighteenth century, not only as a result of the industrial revolution as begun in England, but also in a less heralded agricultural revolution over Western and Central Europe. The spread of potato-growing, the development of beets and turnips as field crops, rotation of fields with clover, innovations in tillage, improved livestock breeds—all joined to raise agricultural production to new levels in western Europe. The new agriculture was brought to our Middle Colonies by a massive immigration of capable European farmers and here further transformed by adding maize to the small grains—clover rotation. Thus was built on both sides of the North Atlantic a balanced animal husbandry of increased yield of human and animal foods. Urban and rural growth alike went into vigorous upswing around the turn of the eighteenth century. The youth of the countryside poured into the rising industrial cities but also emigrated, especially from Central Europe into Pennsylvania, into Hungarian and Moldavian lands repossessed from the Turks and into South Russia gained from the Tartars. The last *Völkerwanderung* was under way and soon edging onto the grasslands.

The year 1800 brought a new cotton to the world market, previously an obscure annual variant known to us as Mexican Upland cotton, still uncertainly understood as to how it got into our South. Cleaned by the new gin, its profitable production rocketed. The rapidly advancing frontier of cotton-planting was moved westward from Georgia to Texas within the first half of the century. This movement was a more southerly and even greater parallel to the earlier westward drive of the tobacco frontier. Both swept away the woodlands and the Indians, including the farming tribes. The new cotton, like tobacco, a clean cultivated row crop and a cash crop, bared the fields to surface wash, especially in winter. The southern upland soils gradually lost their organic horizons, color, and protection: gullies began to be noted even before the Civil War. Guano and Chilean nitrate and soon southern rock phosphate were applied increasingly to the wasting soils. Eugene Hilgard told the history of cotton in our South tersely and well in the United States Census of 1880. As I write, across from my window stands the building bearing his name and the inscription: TO RESCUE FOR HUMAN SOCIETY THE NATIVE VALUES OF RURAL LIFE. It was in wasting cotton fields that Hilgard learned soil science and thought about a rural society that had become hitched wholly to world commerce. Meantime the mill towns of England, the Continent, and New England grew lustily; with them, machine industries, transport facilities, and the overseas shipment of food.

The next great American frontier may be conveniently and reasonably dated by the opening of the Erie Canal in 1825, provisioning the cities with grain and meat on both sides of the North Atlantic, first by canal and river, soon followed by the railroad. The earlier frontiers had been pushed from the Atlantic Seaboard to and beyond the Mississippi by the cultivation of tropical plants in extratropical lands, were dominantly monocultural, preferred woodlands, and relied mainly on hand labor. For them the term "plantation culture" was not inapt. The last thrust, from the Mohawk Valley to the Mississippi, was West European as to agricultural system, rural values, settlers, and largely as to crops.

By the time of the Civil War, the first great phase of the northern westward movement had crossed the Missouri River into Kansas and Nebraska. New England spilled over by way of the Great Lakes, especially along the northern fringe of prairies against the North Woods. New York and Baltimore were gateways for the masses of Continental emigrants hurrying to seek new homes beyond the Alleghenies. The migrant streams mingled as they overspread the Mississippi Valley, land of promise unequaled in the history of our kindred. These settlers were fit to the task: they were good husbandmen and artisans. They came to put down their roots, and the gracious country towns, farmsteads, and rural churches still bear witness to the homemaking way of life they brought and kept. At last they had land of their own, and it was good. They took care of their land, and it did well by them; surplus rather than substance of the soil provided the foodstuffs that moved to eastern markets. Steel plows that cut through the sod, east-west railroads, and cheap lumber from the white-pine forests of the Great Lakes unlocked the fertility of the prairies; the first great plowing-up of the grasslands was under way.

Many prairie counties reached their maximum population in less than a generation, many of them before the beginning of the Civil War. The surplus, another youthful generation, moved on farther west or sought fortune in the growing cities. Thus, toward the end of the century the Trans-Missouri grassy plains had been plowed up to and into the lands of drought hazard. Here the Corn Belt husbandry broke down, especially because of the great drought of the early nineties, and the Wheat Belt took form, a monocultural and unbalanced derivative. I well remember parties of landlookers going out from my native Missouri county, first to central Kansas and Nebraska, then to the Red River Valley, and finally even to the Panhandle of Texas and the prairies of Manitoba. The local newspapers "back home" still carry news from these daughter-colonies, and still those who long ago moved west are returned "home" at the last to lie in native soil.

The development of the Middle West did exact its price of natural resources. The white-pine stands of the Great Lakes were destroyed to build the farms and towns of the Corn Belt; the logged-over lands suffered dreadful burning. As husbandry gave way westward to wheat-growing, the land was looked on less as homestead and more as speculation, to be cropped heavily and continuously for grain, without benefit of rotation and manuring, and to be sold at an advantageous price, perhaps to reinvest in new and undepleted land.

The history of the extratropical grasslands elsewhere in the world is much like our own and differs little in period and pace. Southern Russia, the Pampas, Australia, and South Africa repeat largely the history of the American West. The industrial revolution was made possible by the plowing-up of the great nontropical grasslands of the world. So also was the intensification of agriculture in western Europe, benefiting from the importation of cheap overseas feedstuffs, grains, their byproducts of milling (note the term "shipstuff"), oil-seed meals. Food and feed were cheap in and about the centers of industry, partly because the fertility of the new lands of the world was exported to them without reckoning the maintenance of resource.

At the turn of the century serious concern developed about the adequacy of resources for industrial civilization. The conservation movement was born. It originated in the United States, where the depletion of lately virgin lands gave warning that we were drawing recklessly on a diminishing natural capital. It is to be remembered that this awareness came, not to men living in the midst of the industrial and commercial centers of the older countrysides, but to foresters who witnessed devastation about the Great Lakes, to geologists who had worked in the iron and copper ranges of the Great Lakes and prospected the West in pioneer days, to naturalists who lived through the winning of the West.

The Ever Dynamic Economy

As a native of the nineteenth century, I have been an amazed and bewildered witness of the change of tempo that started with the first World War, was given an additional whirl on the second, and still continues to accelerate. The worry of the earlier part of the century was that we might not use our natural resources thriftily; it has given way to easy confidence in the capacities of technologic advance without limit. The natural scientists were and still may be conservation-minded; the physical scientists and engineers today are often of the lineage of Daedalus, inventing ever more daring reorganizations of matter and, in consequence, whether they desire it or not, of social institutions. Social science eyes the attainments of physical science enviously and hopes for similar competence and authority in reordering the world. Progress is the common watchword of our age, its motor-innovating techniques, its objective the ever expanding "dynamic economy," with ever increasing input of energy. Capacity to produce and capacity to consume are the twin spirals of the new age which is to have no end, if war can be eliminated. The measure of progress is "standard of living," a term that the English language has contributed to the vernaculars of the world. An American industrialist says, roundly, that our principal problem now is to accelerate obsolescence, which remark was anticipated at the end of the past century by Eduard Hahn (1900) when he thought that industrialization depended on the production of junk.

Need we ask ourselves whether there still is the problem of limited resources, of an ecologic balance that we disturb or disregard at the peril of the future? Was Wordsworth of the early industrial age farsighted when he said that "getting and spending we lay waste our powers"? Are our newly found powers to transform the world, so successful in the short run of the last years, proper and wise beyond the tenure of those now living? To what end are we committing the world to increasing momentum of change?

The steeply increasing production of late years is due only in part to better recovery, more efficient use of energy, and substitution of abundant for scarce materials. Mainly we have been learning how to deplete more rapidly the resources known to be accessible to us. Must we not admit that very much of what we call production is extraction?

Even the so-called "renewable resources" are not being renewed. Despite better utilization and substitution, timber growth is falling farther behind use and loss, inferior stands and kinds are being exploited, and woodland deterioration is spreading. Much of the world is in a state of famine, without known means of remedy or substitution.

Commercial agriculture requires ample working capital and depends in high degree on mechanization and fertilization. A late estimate assigns a fourth of the net income of our farms to the purchase of durable farm equipment. The more farming becomes industry and business, the less remains of the older husbandry in which man lived in balance with his land. We speak with satisfaction of releasing rural population from farm to urban living and count the savings of man-hours in units of farm product and of acres. In some areas the farmer is becoming a town dweller, moving his equipment to the land for brief periods of planting, cultivating, and harvest. Farm garden, orchard, stable, barn, barnyards and woodlots are disappearing in many parts, the farm families as dependent as their city cousins on grocer, butcher, baker, milkman, and fuel services. Where the farm is in fact capital in land and improvements, requiring bookkeeping on current assets and liabilities, the agriculturist becomes an operator of an outdoor factory of specialized products and is concerned with maximizing the profits of the current year and the next. Increasing need of working capital requires increased monetary returns; this is perhaps all we mean by "intensive" or "scientific" farming, which is in greater and greater degree extractive.

The current agricultural surpluses are not proof that food production has ceased to be a problem or will cease to be the major problem of the world. Our output has been secured at unconsidered costs and risks by the objective of immediate profit, which has replaced the older attitudes of living with the land. The change got under way especially as motors replaced draft animals. Land formerly used for oats and other feed crops became available to grow more corn, soybeans, cotton, and other crops largely sold and shipped. The traditional corn-oats-clover rotation, protection of the surface and maintaining nitrogen balance, began to break down. Soybeans, moderately planted in the twenties and then largely for hay, developed into a major seed crop, aided by heavy governmental benefit payments as soil-building, which they are not. Soil-depleting and soil-exposing crops were given strong impetus in the shift to mechanized farming; less of the better land is used for pasture and hay; less animal and green manure is returned to fields. The fixation of nitrogen by clover has come to be considered too slow; it "pays better" to put the land into corn, beans, and cotton and to apply nitrogen from bag or tank. Dressing the soil with commercial nitrogen makes it possible to plant more closely, thus doubling the number of corn and other plants to the acre at nearly the same tillage cost. Stimulation of plant growth by nitrogen brings increased need of additional phosphorus and potash. In the last ten years the Corn Belt has more or less caught up with the Cotton Belt in the purchase of commercial fertilizer. The more valuable the land, the greater the investment in farm machinery, the more profitable the application of more and more commercial fertilizers.

The so-called row crops, which are the principal cash crops, demand cultivation during much of their period of growth. They give therefore indifferent protection to the surface while growing and almost none after they are harvested. They are ill suited to being followed by a winter cover crop. The organic color is fading from much of our best-grade farm lands. Rains and melting snow float away more and more of the top soil. There is little concern as long as we can plow more deeply and buy more fertilizer. Governmental restriction of acreage for individual crops has been an inducement to apply more fertilizer to the permitted acreage and to plant the rest in uncontrolled but usually also cash crops. Our commercial agriculture, except what remains in animal husbandry such as dairying, is kept expanding by increasing overdraft on the fertility of our soils. Its limits are set by the economically available sources of purchased nitrogen, phosphorus, potassium, and sulfur.

Since Columbus, the spread of European culture has been continuous and cumulative, borne by immediate self-interest, as in mercantilist economy, but sustained also by a sense of civilizing mission redefined from time to time. In the spirit of the present, this mission is to "develop the underdeveloped" parts of the world, material good and spiritual good now having become one. It is our current faith that the ways of the West are the ways that are best for the rest of the world. Our own ever growing needs for raw materials have driven the search for metals and petroleum to the ends of the Earth in order to move them into the stream of world commerce. Some beneficial measure of industry and transport facility thereby accrues to distant places of origin. We also wish to be benefactors by increasing food supply where food is inadequate and by diverting people from rural to industrial life, because such is our way, to which we should like to bring others.

The road we are laying out for the world is paved with good intentions, but do we know where it leads? On the material side we are hastening the depletion of resources. Our programs of agricultural aid pay little attention to native ways and products. Instead of going out to learn what their experiences and preferences are, we go forth to introduce our ways and consider backward what is not according to our pattern. Spade and hoe and mixed plantings are an affront to our faith in progress. We promote mechanization. At the least, we hold, others should be taught to use steel plows that turn neat furrows, though we have no idea how long the soil will stay on well-plowed slopes planted to annuals. We want more fields of maize, rice, beans of kinds familiar to us, products amenable to statistical determination and available for commercial distribution. To increase production, we prescribe dressing with commercial fertilizers. In unnoticed contrast to our own experience these are to be applied in large measure to lands of low productivity and perhaps of low effectiveness of fertilizers. Industrialization is recommended to take care of the surplus populations. We present and recommend to the world a blueprint of what works well with us at the moment, heedless that we may be destroying wise and durable native systems of living with the land. The modern industrial mood (I hesitate to add intellectual mood) is insensitive to other ways and values.

For the present, living beyond one's means has become civic virtue, increase of "output" the goal of society. The prophets of a new world by material progress may be stopped by economic limits of physical matter. They may fail because people grow tired of getting and spending as measure and mode of living. They may be checked because men come to fear the requisite growing power of government over the individual and the community. The high moments of history have come not when man was most concerned with the comforts and displays of the flesh but when his spirit was moved to grow in grace. What we need more perhaps is an ethic and aesthetic under which man, practicing the qualities of prudence and moderation, may indeed pass on to posterity a good Earth.

REFERENCES

Ames, Oakes. 1939. *Economic Annuals and Human Cultures.* Cambridge, Mass.: Botanical Museum of Harvard University. 153 pp.

Boyd, W. C. 1950. *Genetics and the Races of Man.* Boston: Little, Brown and Co. 453 pp.

Breuil, Henri, and Lantier, Raymon. 1951. *Les Hommes de la pierre ancienne.* Paris: Payot. 334 pp.

Conklin, Harold C. 1954. "An Ethnoecological Approach to Shifting Agriculture," *Transactions of the New York Academy of Sciences,* Series II, XVII, No. 2, 133-42.

Coon, Carleton. 1953. "Climate and Race," pp. 13-34 in Shapley, Harlow (ed.), *Climatic Change.* Cambridge, Mass.: Harvard University Press. 318 pp.

Hahn, Eduard. 1900. *Wirtschaft der Welt am Ausgang des neunzebuten Jahrhunderts.* Heidelberg: C. Winter, 320 pp.

Hehn, Victor. 1888. *Wanderings of Plants and Animals from Their First Home.* Ed. James S. Stallybrass. London: Swan Sonnenschein and Co. 523 pp.

Izikowitz, Karl Gustav. 1951. *Lamet: Hill Peasants in French Indochina.* ("Etnologiska Studier," No. 17.) Göteborg: Etnografiska Museet. 375 pp.

Leighly, John B. 1949. "On Continentality and Glaciation," pp. 133-46 in Mannerfelt, Carl M:son, *et al.* (eds.), *Glaciers and Climate (Symposium Dedicated to Hans W:son Ahlmann as a Tribute from the Swedish Society for Anthropology and Geography). (Geografiska Annaler,* Vol. XXXI, Häfte 1-4) 383 pp.

Mannerfelt, Carl M:son, *et al.* (eds.). 1949. *Glaciers and Climate (Symposium Dedicated to Hans W:son Ahlmann as a Tribute from the Swedish Society for Anthropology and Geography). (Geografiska Annaler,* Vol. XXXI, Häfte 1-4) 383 pp.

Marsh, George P. 1864. *Man and Nature.* New York: Charles Scribner and Co.; London: Sampson, Low and Son. 577 pp.

——————— 1874. *The Earth as Modified by Human Action.* New York: Scribner, Armstrong and Co. 656 pp. (2d ed., 1885. New York: C. Scribner's Sons. 629 pp.)

Shapley, Harlow (ed.). 1953. *Climate change.* Cambridge, Mass.: Harvard University Press. 318 pp.

Simpson, G. C. 1934. "World Climate during the Quaternary Period," *Quarterly Journal of the Royal Meteorological Society,* LX, No. 257 (October), 425-78.

——————— 1940. "Possible Causes of Change in Climate and Their Limitations," *Proceedings of the Linnean Society of London,* CLII, Part II (April), 190-219.

Simpson, Lesley B. 1952. *Exploitation of Land in Central Mexico in the Sixteenth Century.* ("Ibero-Americana," No. 36). Berkeley: University of California Press. 92 pp.

West, Robert C. 1949. *The Mining Community in Northern New Spain: The Parral Mining District.* ("Ibero-Americana," No. 30.) Berkeley: University of California Press. 169 pp.

Willett, H. C. 1950. "The General Circulation at the Last (Wurm) Glacial Maximum," *Geografiska Annaler,* XXXII. Häfte 3-4, 179-87.

Man Adapting: His Limitations and Potentialities

Rene Jules Dubos

In a cave not far from here archeologists have discovered a group of sandals 11,000 years old. These sandals, which were made by Stone Age people, could nevertheless fit twentieth-century human beings. They symbolize the unbroken continuity of human life.

On the other hand, the phrase "Man as the Measure" implies that human beings prize their individuality above everything else. Admittedly, the large agglomerations of the modern world can be likened to anthills or beehives; each person in them has a limited function and returns to rest at a particular place in the evening—as if he were but an anonymous and interchangeable unit in an immense colony. In fact, however, no two human beings are entirely alike.

We find it easy to believe that we have much in common because we come from lands and cultures that have been in contact for many thousand years. But we really never forget that we differ in geographical origin, in physical and mental attributes, in social and religious allegiances. Almost unconsciously we thus give a dual meaning to the abstract word "man." We use it to symbolize a paradox inherent in the human species—the biological unity of mankind and the experiential diversity of human life.

Irrespective of origin, all men are fundamentally similar in structure, requirements, and mental attributes; yet individual human beings and human civilizations differ profoundly from place to place and from time to time. Social planning—whether for the next fifty years or fifty thousand years—must take into account these two complementary aspects of man's natural biological unity and experiential diversity.

The Unchanging Attributes of *Homo sapiens*

The prehistoric and historic events of the human adventure in the Americas provide convincing evidence of the unity of mankind.

As far as can be judged, various populations of ancient man began to move into the American continent during the late paleolithic period—some 20,000 years ago. After their initial penetration, they rapidly spread over the whole continent, but they remained almost completely isolated from the rest of mankind until the seventeenth century. During that period of isolation, they progressively developed several great cultures, exquisitely adapted to particular aspects of American nature. The Incas in the Peruvian Andes; the Mayas in the tropical forests of Central America; the Pueblos in the semi-desertic Southwest; the Plains and Forest Indians of North America; the Eskimos in the frozen North, all developed cultures that were completely independent of the earlier and contemporary ones in Africa, Asia and Europe. Yet the achievements of the American aborigines —their temples, sculptures, pottery, basketwork, and even more remarkably their love songs and their legends of creation—are very meaningful to all of us who come from cultures originating from outside the Americas. Obviously, the most fundamental and universal characteristics of the human mind were fully developed by the time ancient man first penetrated the American continent.

During the past five centuries, many waves of immigration brought other human races into intimate contact with the various tribes of American aborigines. Interbreeding resulted in many varied and highly successful racial mixtures. The genetic and physiological compatibility between races that had been separated for so many thousand years confirms the cultural evidence that all human beings originally derive from the same evolutionary stock.

In several parts of the world, there still exist today small tribes who have had no significant contact with modern civilization and whose ways of life have hardly changed since the Stone Age. Yet experience has repeatedly shown that the members of these tribes, born and raised under extremely primitive conditions, adapt rapidly to modern life. An infant born from a culturally backward people, but adopted very early in life by a more advanced cultural group, rapidly takes on the behavioural characteristics of his foster society and commonly rejects the culture of his natural parents.

Cro-Magnon man, who lived as a hunter 30,000 years ago, long before the emergence of agriculture and of village life, was probably indistinguishable from modern man mentally as well as biologically. He stood upright like us, had the same body shape, and his cranial size was at least as large as ours; the implements he made still fit our hands; the urges that shaped his familial activities and his tribal organization are still operative in us; his paintings, sculptures, and other artifacts deeply move us, symbolically and esthetically.

These anthropological findings are of great importance for the social planner because they prove that modern man retains many biological and mental characteristics of his remote ancestors. His physiological needs and drives, his responses to environmental stimuli, his potentialities and limitations, are still determined by the 20,000 pairs of genes that governed human life when ancient man was a paleolithic hunter or a neolithic farmer. Two examples will suffice to illustrate how profoundly the physiological and behavioral responses of modern man are still conditioned by the genetic endowment that defines the species *Homo sapiens.*

Ancient man naturally lived in intimate contact with nature and his activities were therefore governed by natural rhythms, such as the daily change from light to darkness and the recurrence of the seasons. As a result, his bodily and mental functions exhibited diurnal, lunar, and seasonal cycles linked to those of the cosmic order. Remarkably enough, these biological cycles persist in modern man even where the physical environment has been rendered almost uniform and constant by technological control. Rapid changes of latitude during jet air travel cause physiological disturbances because the body cannot adapt rapidly enough to the new day-night rhythms. These disturbances of jet travel are not subjective; they originate from the fact that the secretion of hormones proceeds according to a biological time clock which remains geared—for a variable period of time—on the day-night rhythm of the place of usual residence. Similarly, modern life may create physiological disturbances when it carries the day into the night with electric illumination and maintains the same temperature and food supply throughout the year. The urban environment continuously changes but man's body machine continues to function in accordance with the cosmic events under which evolution took place.

The so-called fight and flight response is another type of innate and subconscious activity which is governed by very ancient emotional and physiological forces still at work in modern man. When prehistoric man encountered a human stranger or a wild beast, certain hormonal processes placed his body in readiness for combat or for escape. Today, the same physiological processes are still set in motion under circumstances that modern man symbolizes as a threat, for example, in the course of personal conflicts at the office or at a cocktail party. In most cases, such responses are no longer useful and may even be deleterious; but they persist nevertheless, as do numerous other mechanisms and structures originating from the evolutionary past.

Admittedly, the fact that modern man is constantly moving into new environments seems to indicate that he has enlarged the range of his biological adaptability and is escaping from the bondage of his past. But this is only an appearance. Wherever he goes, and whatever he does, man is successful only to the extent that he functions under environmental conditions not drastically different from the ones under which he evolved. He can climb Mount Everest and fly at high altitudes only because he has learned to protect himself against cold and to carry an adequate oxygen supply. He moves in outer space and at the bottom of the sea only if he remains within enclosures that duplicate certain physicochemical characteristics of the terrestrial environment. Even the Eskimos, who appear so well adapted to the Arctic winter, in reality cannot long resist intense cold. Sheltered in their igloos or clothed in their parkas, they live almost a tropical life!

Thus man does not really "master" the environment. What he does is to create sheltered environments within which he controls local conditions. Such control has been achieved for temperature, food, water, and oxygen tension, and a few other obvious physicochemical requirements. But human life is affected by many other factors that are poorly understood, and often not controllable at all. For example, the biological cycles mentioned earlier became inextricably woven in the human fabric during evolutionary times, and they still link modern man to cosmic events. The dissociation of modern life from these natural cycles is likely to exert some deleterious effects on the human organism. In fact, man is likely to suffer from many of the new environmental forces he has set in motion because he has not encountered them in his evolutionary past. He may develop some tolerance against environmental pollution, severe crowding, constant exposure to intense sensory stimuli, and the regimentation of life in a completely mechanized world. But one can anticipate that this tolerance will have deleterious consequences for the human race in the long run.

In brief, the genetic endowment of *Homo sapiens* has changed only in minor details since the Stone Age, and there is no chance that it can be significantly or usefully modified in the foreseeable future. This genetic stability defines the potentialities and requirements of man's nature. It also determines the physiological limits beyond which human life cannot be safely altered by social and technological innovations. The frontiers of technology are in final analysis less significant for the life of man than his own frontiers, which are determined by the genetic constitution he acquired during his evolutionary past.

The Potentialities of Man

While the fundamental traits of *Homo sapiens* are permanent and universal, man's ways of life and the structures of his societies differ from one area to another and change endlessly with time. The biological evolution of man's body and mind almost came to a standstill some 50,000 years ago, but his sociocultural evolution now transforms human life at an accelerated rate.

Civilizations have their primary origin in the various responses to environmental stimuli. This versatility of response, in turn, is a consequence of the wide range of differences that exist among human beings. Except for identical twins, no two individual persons are genetically alike. Furthermore, the physical and psychic character of each person is profoundly influenced by environmental forces. From nutrition to education, from the topography of the land to urban design, countless influences play a role in determining the expression of the genetic endowment and thus in shaping the body and the mind of man.

Sociocultural forces are ultimately derived of course from man's biological nature, but through a feedback process they continuously alter his body, his mind, and his social patterns. This feedback accounts for the paradox that experiential man and his ways of life are continuously changing even though his genetic make-up is so remarkably stable.

Contrary to popular belief, genes do not determine the traits of a person; they merely govern his responses to the physical and social environment. Furthermore, not all the genes of a person are active at all times. Through complex mechanisms that are only now being recognized, environmental stimuli determine which parts of the genetic equipment are repressed and which parts are activated. Thus each individual person is as much the product of the environment as of his genetic endowment. Human beings perceive the world, and respond to it, not through the whole spectrum of their potentialities, but only through the areas of this spectrum that have been made functional by environmental stimulation. The life experiences determine what parts of the genetic endowment are converted into functional attributes.

The conditioning of the physical and mental personality by the environment has of course long been recognized. Winston Churchill was aware of its importance for human life when he urged that the House of Commons, damaged by German bombs during the war, be rebuilt as exactly as possible in its original form instead of being replaced by a modern and more efficient building. He feared that changing the physical appearance and organization of the House might alter the character of parliamentary debates and therefore of English democracy. In his words, "We shape our buildings and then they shape us."

The environment and ways of life determine in fact not only the conditions under which men function, but also the kind of persons their descendants will become. In fact, environmental factors have their most profound and most lasting effects when they impinge on the young organism during the formative phases of its development. Suffice it to mention as an example the acceleration in growth and in sexual development that is occurring at present in the Western world as well as in the Oriental countries that have become Westernized. Japanese teenagers are now very much taller than their parents not as a result of genetic changes, but probably because they are better protected against malnutrition and childhood infections.

The experiences of early life are of particular importance because the human body and especially the brain are incompletely differentiated at the time of birth. The infant develops his physical and mental individuality as he responds to the stimuli that impinge on him during growth. To a very large extent, the physical appearance and mental characteristics of the adult are thus the products of the responses made to the total environment during the formative years. In other words, anatomic structures and physical performance, as well as behavioral patterns, are molded by the surroundings and the conditions of life during childhood; furthermore, the effects of such early influences commonly persist throughout the whole life span. For example, a child brought up in Florence is constantly exposed to the sights, sounds, and smells characteristic of this beautiful city; his development is conditioned by the stimuli derived from palaces, churches, and parks. He may not be aware of the responses aroused in him by these repeated experiences. But they become part of his biological make-up and render him lastingly different from what he would have become had he developed in London, Paris, or New York.

From all points of view, the child is truly the father of the man. Most aspects of human life are governed by a kind of biological

Freudianism, because socially and individually, the responses of human beings to the conditions of the present are always conditioned by the biological remembrance of things past.

Environment for Man

Each person has a wide range of innate potentialities that remain untapped. Whether physical or mental, these potentialities can become expressed only to the extent that circumstances are favorable to their existential manifestation. Society thus plays a large role in the unfolding and development of man's nature.

One can take it for granted that the latent potentialities of human beings have a better chance to become actualized when the social environment is sufficiently diversified to provide a variety of stimulating experiences, especially for the young. As more persons find the opportunity to express their biological endowment under diversified conditions, society becomes richer in experiences and civilizations continue to unfold. In contrast, if the surroundings and ways of life are highly stereotyped, the only components of man's nature that flourish are those adapted to the narrow range of prevailing conditions.

The lesson to be derived from the story of biological evolution is that man has been so successful because he is the least specialized creature on earth; he is indeed the most adaptable. He can hunt or farm, be a meat eater or a vegetarian, live in the mountains or by the seashore, be a loner or engage in teamwork, function in a free democracy or in a totalitarian state.

History shows, however, that societies which were once efficient because they were highly specialized rapidly collapsed when conditions changed. A high specialized society is rarely adaptable. Adaptability is essential for social as well as for biological success. Therein lies the danger of the standardization and regimentation so prevalent in modern life. We must shun uniformity of environment as much as absolute conformity in behavior.

At the present time, unfortunately, the creeping monotony of our technological culture goes hand in hand with the monotony of our behavior, taste, patterns of education and of mass communication. And yet it is certain that we can exploit the richness of man's nature only if we make a deliberate effort to create as many diversified environments as possible. This may result in some loss of efficiency, but the more important goal is to provide the many kinds of soil that will permit the germination of the seeds now dormant in man's nature. Diversity of social environment constitutes in fact a crucial aspect of functionalism, whether in the planning of cities, the design of dwellings, or the management of life. So far as possible, the duplication of uniformity must yield to the organization of diversity.

Irrespective of genetic endowment, a child who grows in a city slum will differ as an adult from one who has spent most of his life within the sheltering cocoon of a modern apartment house, or from one who has participated in the chores of a family farm. Unfortunately, awareness of the fact that surroundings exert a profound effect on human life is based largely on untutored observations and has not yet been converted into scientific knowledge.

Environmental factors obviously condition all aspects of human life, but nobody really knows which factors are influential or how they work. The problem, however, is not hopeless. Experiments have revealed that in animals, also, early influences condition growth, longevity, behavior, resistance to stress, and learning ability. The effects exerted on human life by early influences can therefore be studied through the use of experimental models, much as is being done for other types of biological problems. The knowledge thus acquired will certainly help in the rational management of society.

The population avalanche and the universal trend toward urbanization will, needless to say, affect all aspects of future life. By the end of this century, most human beings will be born, will live, and will reproduce within the confines of megalopolis. Until now, cities have constantly grown and renewed themselves through the influx of people originating from rural areas or from primitive countries. Very soon, however, and for the first time in history, this transfusion of new blood will come to an end; the human race will reproduce itself almost exclusively out of persons city-born and city-bred. To a very large extent, the future will therefore depend upon our ability to create urban environments having the proper biological qualities.

One of the most important unknowns in this regard is the effect of population pressure. There is no immediate danger that the United States will experience famine or even a decrease in the standard of living as a result of population pressure. We suffer from overpopulation, nevertheless, because human life is affected by determinants that transcend technology and economics.

Unwittingly we tend to regard ourselves and our fellow men as things, rather than as human beings. We do not recognize any danger in crowding as long as we can produce enough food for physical growth, and enough goods for economic growth. Yet overpopulation can destroy the quality of human life through many mechanisms such as traffic jams, water shortages, and environmental pollution; spreading urban and suburban blight; deterioration in professional and social services; destruction of beaches, parks, and other recreational facilities; restrictions on personal freedom owing to the increased need for central controls; the narrowing of horizons as classes and ethnic groups become more segregated, with the attendant deepening of racial tensions.

Paradoxically the dangers of overpopulation will be increased by the extreme adaptability of the human race. Human beings can become adapted to almost anything—polluted air, treeless avenues, starless skies, aggressive behavior, the rat-race of overcompetitive societies, even life in concentration camps. But in one way or another, we have to pay later for the adjustment we make to undesirable conditions. The cost includes for example increase in chronic diseases and decadence of human values. Congested environments, even though polluted, ugly, and heartless, are compatible with economic growth and with political power. But they damage the physical and spiritual aspects of human life.

The adaptability of human beings to environments that are compatible with organic life, but destructive of human values, creates difficult problems for community planning. We have to determine simultaneously on the one hand the kind and intensity of stimulation required for individual and social development; on the other hand, the levels of stimulation and adaptation that are potentially dangerous for the future.

In brief, it is obvious that the technological factors, such as supplies of food, power, or natural resources, that are required for the operation of the body machine and of the industrial establishment are not the only factors to be considered in determining optimum population size. Just as important for maintaining *human* life is an environment in which it is possible to satisfy the longing for quiet, privacy, independence, initiative, and open space. These are not frills or luxuries but constitute real biological necessities. They will be in short supply long before there is a critical shortage of the sources of energy and materials that keep the human machine going and industry expanding.

A Science of Environmental Biomedicine

The contributions of science to human welfare and especially to the future of mankind will be severely restricted until a systematic effort is made to study the biological effects of the environmental factors that affect life in the modern world; for example, the varied aspects of biological and chemical pollution; the constant and unavoidable exposure to the stimuli of urban and industrial civilization; the emotional trauma and often the solitude of life in congested cities; the monotony and tensions of regimented life; the boredom arising from automated work and indeed from compulsory leisure.

Unfortunately, such a science of environmental biomedicine does not yet exist. Its development might be initiated through the study of a few specific problems. The following can serve as illustrations of such environmental problems having a direct relevance to human life and amenable to scientific analysis.

a. The lasting effects of early influences, i.e., of the effects exerted on the organism during the formative stages of its development.

b. The delayed and indirect effects of environmental pollutants.

c. The distant consequences of exposure to subcritical levels of potentially injurious substances and stimuli.

d. The effects of crowding on hormonal activities, on resistance to stresses, and on behavioral patterns.

e. The range of adaptive potentialities.

f. The effects of housing conditions on the development of sense organs and of various physiological processes.

For any one of these problems, it is possible to imagine experimental models duplicating one aspect or another of actual human situations. It must be emphasized, however, that unusual research facilities will be required for the development and use of such experimental models. The following items immediately come to mind.

a. Experimental animals of known genetic structure and of controlled experiential past.

b. Quarters for maintaining animals under a wide range of conditions throughout their whole life span and indeed for several generations.

c. Large enclosures for maintaining animal populations of various sizes and densities, exposed to different types of stimuli.

d. Equipment (using telemetry) for observing and measuring responses without disturbing the experimental system.

e. Equipment for recording, retrieving, and analyzing the complex data to be derived from the study of large populations and multifactorial systems.

The mere listing of these facilities points to the need for new types of institution with a special organization of highly integrated personnel. What is required is nothing less than a bold imaginative departure to create a new science of environmental biomedicine.

Conclusions

A scientific philosophy of mankind can be derived from the knowledge that man's nature encompasses two aspects that are radially different, yet complementary. On the one hand, *Homo sapiens* has existed as a species with a well-defined genetic endowment for some 50,-000 years. On the other hand, man has continued to unfold his latent potentialities ever since that time. This ancient lineage accounts for the biological limitations of mankind and also for its immense phenotypic diversity.

The biological limitations inherent in the human species create a collective responsibility for directing technological development in such a manner as to make it compatible with human survival and welfare. The wide range of phenotypic potentialities gives to each person the chance to select or create his surroundings and way of life and thus to develop along the lines he chooses.

The existentialist faith that "man makes himself" implies of course the willingness to decide and the courage to be. But it also demands that action be guided by a deep scientific knowledge of man's nature. For this reason, we must create a new science of environmental biomedicine that will help in predicting and hopefully in controlling man's responses to the environmental conditions prevailing in the modern world.

New kinds of scientific institution must be developed to study the effects that surroundings and ways of life exert on physical and mental development, especially during the formative years of life.

The characteristics of individual persons, and of societies, are largely determined by feedback reactions between man's nature and environmental forces. Since man has much freedom in selecting and creating his environment, as well as his ways of life, he can determine by such decisions what he and his descendants will become. In this light, man can truly "make himself" consciously and willfully. He has the privilege of responsible choice for his destiny—probably the noblest, and a unique, attribute of the human condition.

UPI

Quantity and Quality of Man

On the basis of the best available estimates there are slightly under 3.6 billion individuals in the human population at this time. Whether this is a good thing or not depends on its relation to available environmental resources. What the "resources" in question are depends in turn on the environmental niche and the unique needs of our species. The capacity of the environment to support a population of a particular species usually fluctuates to a greater or lesser degree, and the "quality" of the population is one indication of how well that population is following or "tracking" the environmental fluctuation. In other words, we might expect successful tracking, or lack of it, to be reflected in the life span, and the physical, physiological, and psychological condition of individuals in the population.

It is hard to identify, intuitively, any unique optimum relationship between population size and environmental carrying capacity. Maximum realization of individual potential (a common measure of "good" in human existence) is likely to be found in populations which are below the level where the resources of the environment are fully utilized. As a result we might regard it as a good thing if our population lagged a little behind during a period when the environmental carrying capacity was expanding. Such a tracking error would have unpleasant consequences, however, if the environmental carrying capacity began to decline.

The process by which a population tracks its environment has, in some instances, been found to be controlled directly by the supply of essential resources, particularly food. Local or regional food shortages may cause emigration, or reduction of life expectancy and reproductive rates through malnutrition or starvation. Less acute food shortages are associated with stunted growth, poor health, and susceptibility to disease. Shortages of other resources, or accumulation of toxic metabolites may have effects paralleling those of food shortages.

Among vertebrates, a refinement in the method of environmental tracking, involving social interactions, is common. The social interactions function in relation to perceived relationships with other individuals in respect to certain resources. Space is the most common of these. This is another way of saying that tracking involves perception of crowding, and some consequent behaviors which tend to oppose or reduce crowding.

There is a general phenomenon here. Vertebrates of many sorts respond to crowding through interactions which stimulate unrest, and force emigration, among younger individuals who have not yet become socially established. As population density rises to the environmental carrying capacity more and more young are forced out to colonize marginal, high-risk habitats. There they are eliminated with few, if any, descendents.

One consequence of events of the last 50 years is that human young have become imprisoned. Economic, social, and political restrictions, and the disappearance of uncolonized frontier areas exclude the possibility of emigration in any geographic sense.

Human culture and technology complicate and modify the expression of innate human traits without eliminating them. With no opportunity for emigration in the geographic sense, desperate human beings may attempt emigration by development of "countercultures." The resultant innovation may be productive, but difficult, and it is well to inquire into the consequences of total frustration of such emigration-inducing behavioral mechanisms as might exist in man.

Many experiments have been conducted in which populations of rodents have had to endure crowding without any opportunity for emigration. The consequences are well documented, though often misinterpreted. The misinterpretation involves a confusion of pathological malignancy with normal function. Emigration-inducing behaviors become perverted in the absence of an opportunity for emigration, and many interpretations have been analagous to an evaluation of the normal function of an intestine solely on the basis of its behavior when obstructed. The pathological syndrome of confined experimental rodent populations is, however, consistent, and pertinent when we are concerned with the quality of existence for man today. In such populations the pressure on the young becomes intense, with results that are deleterious to the population as a whole. Many young may be killed. Others may survive only in maimed physical, physiological, or psychological conditions. There is a rising disruption of the social structure. There is increased incidence of violent combat, and abnormal sexual and maternal behavior. Eventually there are signs of extreme physiological stress.

The long-continuing increase in the numbers of human beings on this finite earth has undergone extreme acceleration in recent decades. An increasing proportion of men are existing in increasingly dense aggregations where they undergo increasingly intense and increasingly frequent social contacts. An increasing proportion of mankind exists in marginal habitats (where the resource base is relatively low). The objective of this chapter is to inquire into the present condition of the human population. In developed nations men now experience crowding plus abundance of resources. In underdeveloped nations men experience resource scarcity, but they experience crowding also. The reader is asked to read the selections which follow with a mind alert for evidence of both resource shortage and the generalized vertebrate overcrowding syndrome. What is the present overall quality of human life, and what trends are implied by the present day quality of individual men and their lives? Does a significant proportion of human beings now exist at densities irreconcilable with either the innate qualities of vertebrate man, or the most desirable of human aspirations? Do the technology and economics of the twentieth century combine to obscure the situation? And are technology and the whole matrix of our culture conspiring to act like a plug in a volcano, accentuating the lag of the population in responding to a surpassed and possibly declining environmental carrying capacity, thus building the intensity of the eruption to come?

Degrees of Concentration: A Note On World Population Distribution

David Grigg

Many writers have commented upon the extremely uneven distribution of population over the earth's surface. Vidal de la Blache, for example, estimated that in 1913 two-thirds of the world's population was to be found upon only one-seventh of its area.[1] In 1937 C. B. Fawcett[2] identified four Major Human Regions, defined as large continuous areas where the population density exceeded the world mean of 40 per square mile; these regions contained "about three-fourths of the world's population on little more than one-eighth of its total land area" (Table I).

TABLE I

**Major human regions, c. 1930-35
(after Fawcett, 1937)**

Region	Area[a]			Population	
	Million square miles	Percentage total land area	Percentage total land area	Millions	Percentage total world population
Far East	1.7	2.9	3.4	500	25.0
India and Ceylon	1.0	1.7	2.0	350	17.5
Europe	2.8	4.9	5.6	520	26.5
Eastern North America	1.9	3.3	3.8	100	5.0
	7.4	12.8	14.8	1470	74.0

[a]Fawcett did not indicate what he took as total land area. Column 2 has been calculated on the assumption that it was 57 million square miles, column 3 on the basis of 50 million square miles.

TABLE II
Degrees of population concentration, 1953 (after Woytinsky)

Region	Area		Population		
	Thousand square miles	Percentage of world land area	Millions	Percentage of world total	Density per square mile
Densely Populated					
Eastern China	1,000	1.75	400	16.6	400
Korea and Japan	200	0.35	100	4.1	500
Indochina, Philippines, Indonesia	250	0.45	120	5.0	480
India and Ceylon	1,000	1.75	410	17.1	410
Western and Central Europe	1,200	2.05	370	15.4	310
North Eastern United States	500	0.9	90	3.75	180
Central regions of U.S.A., coastal region of North and South America, Nigeria, Egypt	100	0.1	160	6.6	160
Total Densely Populated	4,250	7.35	1650	68.6	388
Moderately Populated Areas					
Central U.S.A., South America,[a] U.S.S.R.,[b] Central China, North India	17,500	30.5	710	29.6	41
Sparsely Populated Areas[c]	15,650	27.5	40	1.6	3
Uninhabitable Areas[d]	20,000	34.9	—	—	—
World	57,400	100	2400	99.8	42

[a]Excluding densely populated areas and tropical forests.

[b]Excluding densely populated area in west and sparsely populated tundra, taiga, and Arctic regions.

[c]Semi-arid, taiga, tropical forests.

[d]Arctic and tundra regions, desert.

SOURCE: W.S. and E.S. Woytinsky, *World Population and Production, Trends and Outlook*, 1953, p. 41.

Fawcett's "Europe" included western Russia, Turkey, the Caucasus, the Levant and North Africa, whilst the Far East included not only Japan, Korea and eastern China, but also Southeast Asia.

In 1953, W. S. and E. S. Woytinsky[3] made a more ambitious attempt to estimate the degree of concentration (Table II).

The Woytinskys, like Fawcett, distinguished four major population concentrations (Table III).

The four major regions do not include Southeast Asia; together, however, they comprise 57 percent of the world's population upon 6.8 percent of its area. If the other densely populated areas are added in then 68.6

percent of the population lived upon 7.15 percent of the area. On the other hand, "sparsely populated" and "uninhabited areas" together formed 62.4 percent of the earth's surface but had less than 1 percent of the population.

Table IV and Fig. I show the attempt of the present author to measure the degree of population concentration. It is based upon the map of population density in the first volume of *The Times Atlas of the World.* As a first step, the area of the earth's surface occupied by different density classes was measured from the map, and the results are presented by regions in Table IV. In 1960, world population was 3000 million and the mean density, if Antarctica is included, 52 per square mile. Fig. I,

TABLE III
Major population concentration, 1950 (after Woytinsky)

Region	Area		Population		
	Million square miles	Percentage total land area	Millions	Percentage world total	Density per square miles
East Asia	1.2	2.1	500	20.8	416
India and Ceylon	1.0	1.7	410	17.1	410
Europe	1.2	2.1	370	15.4	310
Northeastern U.S.A.	0.5	0.9	90	3.75	180
	3.9	6.8	1370	57.05	351

TABLE IV
Percentage of continental land surfaces occupied by different population density classes

Region	Population per square mile						
	0-4.9	5-49	50 and over	50-99	100 and over	100-249	250 and over
Africa	49.5	34.9	15.6	8.4	7.2	5.5	1.7
Europe	21.8	15.0	63.2	25.2	38.0	20.5	17.5
U.S.S.R.	62.6	21.7	15.7	11.9	3.8	3.1	0.7
Asia	50.4	14.4	35.2	16.2	19.0	3.7	15.3
North America (including Greenland)	74.1	17.6	8.3	6.8	1.5	0.6	0.9
Latin America	62.2	28.2	9.6	4.2	5.4	5.2	0.2
Oceania	90.9	7.8	1.24	1.0	0.24	0.08	0.16
World (excluding Antarctica and certain uninhabited Arctic islands)	59.5	23.1	17.4	9.0	8.4	4.2	4.2
World (including Antarctica)	63.2	21.1	15.9	8.2	7.5	3.75	3.75

TABLE V
Major population concentrations, (over 50 per square mile) c. 1960

Region	Area		Population		
	Million square miles	Percentage of total land area	Millions	Percentage of world population	Density per square mile
East Asia	1.56	2.7	760	25.3	486
South Asia	0.99	1.7	500	16.6	505
Europe	2.35	4.1	500	16.6	212
Eastern North America	0.62	1.09	130	4.3	209
	5.52	9.59	1890	62.8	342
Other areas with population exceeding 50 per square mile	3.5	6.2	400-500	13.3-16.6	114-142
All areas exceeding 50 per square mile	9.02	15.79	2290-2390	76.1-79.4	254-264

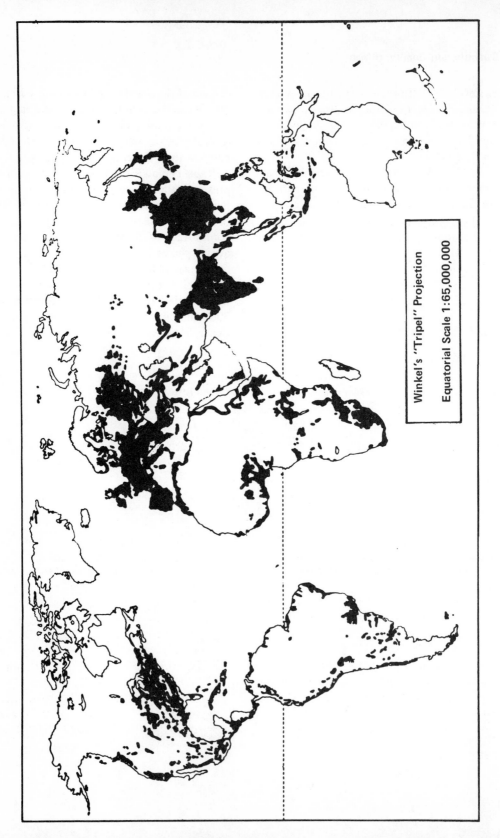

Winkel's "Tripel" Projection

Equatorial Scale 1:65,000,000

Figure 1.

based upon the *Times* map, shows areas which exceeded 50 per square mile. Four major continuous areas can be distinguished:

1. East Asia: eastern China, Japan, Korea, Formosa;

2. South Asia: India, Pakistan and Ceylon;

3. Europe and western Russia, but not the Caucasus, Turkey, North Africa or the Levant;

4. Eastern North America: eastern U.S.A. and the adjacent parts of Canada.

The area of these four concentrations was calculated directly from the map. The population of these areas was estimated from United Nations Demographic Handbooks and articles on population distribution. The population data refer to about 1960 (Table V).

It will be seen that the four major concentrations as defined in Fig. I contain 63 percent of the world's population upon one-tenth of its area. If, however, the population in all the other areas marked in Fig. I are added, then between three-quarters and four-fifths of the world's population occupy 16 percent of the earth's land surface.

REFERENCES

1. P. Vidal de la Blache, *Principles of Human Geography,* Constable, London, 1965, p. 28.
2. C. B. Fawcett, "The changing distribution of population," *Scottish Geographical Magazine,* vol. 53, No. 6, 1937, pp. 361-73.
3. W. S. and E. S. Woytinsky, *World Population and Production Trends and Outlook,* New York, The Twentieth Century Fund, 1953, p. 41.
The author wishes to express his gratitude to Miss Margaret Wilkes for her assistance.

Population Density and the Style of Social Life
Nathan Keyfitz

To see how population density can occur at a rudimentary level of culture and what its effects may be, think of hunting groups with given apparatus, say spears or bows and arrows. Following Clifford Geertz (1963) and Julian Steward (1955) we observe that if the animals which they hunt move in herds, as do caribou, then large groups of hunters can pursue them. When they find a herd and attack successfully, there will be food for all. This herding of the animal prey is reflected in the gathering together of men, and communities can be large. If, on the other hand, the prey consists of small animals spread through a forest and caught one at a time, then men will have to spread out correspondingly; the large community cannot come into existence; human life will be lived out in isolated families. This latter condition, says Steward, applies to the Bushman and Eskimo, who have little in common but the dispersion of their game. Larger groups appeared among the Athabaskans and Algonkians of Canada and probably the prehorse plains' bison hunters.

This primary fact of dispersion or concentration will determine other circumstances. Isolated families cannot evolve the division of labor that is possible in bigger communities. In larger groups, specialization is likely to arise as some individuals become more adept at making spears, others at sighting herds, others at the tasks of surrounding the prey. Specialization will mean a variety of occupations, and insofar as men are made by their work, a variety of men. Some of the occupations will have more prestige than others, perhaps because they require rarer skills. Even at this primitive level, where money may not exist in any form, the notion of a market for talent and a corresponding prestige hierarchy has a possible bearing.

Prestige is a source of power. Ambitious individuals can use occupational prestige to gain further power, especially if they have organizational ability. This suggests an interaction of economic and political phases, never entirely separated in real life. My only point here is that both the economy and the polity are more elaborated in large communities than in small, and hence more to be looked for in tribes living off herds of big game than in those living on dispersed small animals. The sociability of the animal, so to speak, permits a higher degree of sociability in the men; density causes density. To go one step more, and exaggerate somewhat, the animals have created the economic and political structures in the human group.

Human history and ecology did not stop at the hunting culture. The great change in society was the invention of agriculture, which even in the form of neolithic shifting cultivation permits much higher density than does hunting. Is the discovery of planting and tilling the cause of greater human density? Or

was the causation the other way, the density coming first and forcing men to utilize their environment more intensively? Fortunately we do not need to stop here to investigate this question of meta-history. The important fact is that agriculturalists can produce a surplus, which hunters can rarely manage. Robert M. Adams (1965) has described the agricultural base of the early cities. The farmer can grow enough for himself and his family and have, let us say, 5% left over. Once this technological achievement occurs, then 5% of the population can live in cities. It becomes worth-while for a ruler to dominate the farmer, to collect the food as booty or taxes, and use it to support an army and a court. The patriarch becomes a prince.

Some of the troubles, as well as the glories, of civilization are implicit in the first cities, however small they may have been by modern standards. The total number of people which could congregate was limited, because strong political organization was needed to dominate a countryside, and an organization that tried to spread too far would be diluted and lose its control; the ancient empires often did outreach themselves in this way and fell apart. Physically, the area of control could not be too extended, since the transport of grain by ox-cart, the means used in the land empires of Asia, has natural limits set by the fact that the ox has to be fueled from the cargo. Among pre-machine cities, Rome did attain a population of nearly a million, but this was by virtue of extraordinarily competent, and harsh, organization of the lands around the Mediterranean, and by the use of sea transport for the movement of North African and Balkan grain.

Long before ecology or sociology became formal studies, a North African writer and politician called Ibn Khaldhum described with the utmost clarity how the population that could be concentrated in the capital city of an empire increased when the skill of the ruler and the discipline of his army and tax collectors enabled him to dominate a larger area of countryside, and how the population of the same city diminished when the rule was weaker so that the outlying provinces could successfully revolt from the exactions of the capital.

I have referred, then, to several levels of density—the dispersed hunters, the larger hunting group, the agriculturalists, the pre-industrial city of landlords and princes organizing the countryside and living off the proceeds.

The city which constituted the capital of a despotic empire or of one of its provinces is not the only historical type. In Europe cities grew up specifically released from feudal ties, exempted from the domination of princes or landholders, their sustenance obtained by trade, religious, or entertainment functions, their independence assured by a sworn brotherhood of armed merchants. Not having to oppress a peasantry in order to secure their food, they could be loose in their internal arrangements; a medieval proverb says that "City air makes men free." The typical modern western city lives by a great extension of these same nonpolitical functions, and especially by manufacturing with mechanical power. Far from having to squeeze its food from the countryside, the city has become an autonomous economic force. Today, the countryside wants city goods more than the city wants food. The concentrated population of cities, which in the preindustrial empires was parasitic, has now become incredibly productive. Exploitation, if that is the right word, goes the other way from that of Ibn Khaldhun's account; rural legislators tax the cities to maintain support prices for grains, butter and other foods. Today's pattern, at least in the United States, England, and other western

countries, is that men are more productive in dense settlements than in sparse ones.

The increase of cities, especially the increase of very large cities, is to be seen on all continents. Not only in the rich countries as foci of industry and trade, but in the poorest, to which industry has hardly come, the cities are expanding. In fact, during the 1950's the urban populations of developed countries increased by 25%, while those of under-developed countries increased by 55%. The increase of poor, dense populations was twice as rapid as that of rich ones, Bourgeois-Pichat (1966) tells us.

How could that 55% increase occur, if what I have said about the preindustrial city being dependent on the limited surplus of a countryside is true? The surplus food of the Asian peasantry did not increase by 55% in the 1950's; it hardly increased at all. How can Djakarta be five times as large as it was before World War II, and three times as populous as ancient Rome at the highest point of its imperial power? Djakarta has not much more industry than Rome had. Its weak civil or military domination of an island territory, in some degree democratic, cannot compare in extractive power with the iron rule of Rome. The answer, of course, is that it draws food from foreign territories, including the United States; some of it paid for with the export of raw materials; some of it borrowed; some as gifts.

Unable or unwilling to exploit its own peasantry, the large contemporary nonindustrial city more and more bases itself ecologically on the fields of the American west, together with the ships and harbors which link those fields with its massive populations. Population in the Asian countryside itself is growing beyond food supplies; far from having a surplus to ship to the city, the peasant is himself hungry.

Once the local countryside can no longer produce enough food for its own inhabitants so that these must be supported by foreign food, they tend naturally to gather into such seaport cities as Djakarta, Calcutta, and Rio de Janeiro, as close as possible to the spot where the boats will discharge their cargoes of American, or occasionally Burmese or Cambodian, grain. If people are to be fed from abroad it is cheaper to have them at the seaports than dispersed through the countryside. At the present time the United States is shipping about 800,000 tons of grain per month to India alone. At the Asian standard of about one pound per person per day, this is enough for 40,000,000 people to live on; that number happens to be about equal to all the citizens of India living in the seacoast cities. If population continues to increase in the countryside and food does not, one can expect further flight to those cities.

One could say much about what density and size will do to the condition of dependence of those cities. We know that their inhabitants tend to perform services rather than make goods. The services have the function of distributing the claim to the food shipments, the dominant ideal being to give employment rather than to get work done. Some studies have indicated that the new migrants to the cities retain links with the countryside. Others show that the simple and traditional patterns of association in the countryside are transferred to the city, which thereby seems like a number of contiguous villages, lacking only their fields and their crops. These dense cities of rural culture are a new phenomenon in the world.

For some quite different concomitants of density, shown in their most accentuated form, we must go to those world cities of the 19th century which were ecological consequences of the railroad and steamship—New York, London, Paris, Berlin. In the 20th-century West a process of dispersal has occurred; cities produced by the automobile are less dense than those produced by the railroad and street car. We are getting strip cities, of which the best known is Megalopolis, the name Jean Gottman (1961) gave to Boston-New York-Washington.

The industrial city of the 19th century as well as the strip city of the late 20th century intensifies competition on many levels. We must not only find a livelihood, we must find a life, each of us for himself, in the crowded city. This search for a tolerable physical and moral existence preoccupies every city dweller, and it has drastic consequences for urban society as a whole. Just as Darwin saw the animal or plant adapting to a niche in which it is partly sheltered from competition, so the sociologist Durkheim (1960) sees the city man restlessly searching for, and adapting to, a niche constituted by a specialized occupation and specialized personality. During the strike of airplane mechanics I was part of an undifferentiated mass seeking tickets at an airline counter. If one had to face daily the direct competition of millions of people, the struggle would so weigh on each of us that existence would be impossible for our spirits as well as our bodies. One's niche may be teacher, stock broker, or truck driver; it requires skills that others lack, or involves work that others do not want to do. It gives each a place with a certain minimum of predictable security. We are under constant inducement to better our position, and we seek to do so by further specialization.

Now the electronics engineer in Chicago, say, has to concern himself, at most, with the competition of other electronics engineers. But he does not even have to cope with them, at least in the short run. There are a hundred specialties within the field of electronics, and within each of these recognized specialties an individual practitioner, through his own tastes and capacities, can make himself unique. People in a particular plant come to depend on him. If the city is, on the one side, a jungle of potentially infinite and destroying competition, on the other, it shows a nearly infinite capacity of its members to differentiate themselves, to become useful to one another, to become needed.

The differentiated citizen can afford to be tolerant of those he meets, even to like them. This could not be true if competition were more direct, with individuals as personally ambitious as we know them to be in western countries. The struggle for upward mobility, characterizing all developed societies, can only through the process of specialization avoid the harshness of personal character that the blast of full competition would create.

The differentiation is only possible within an economic space that is honeycombed with organizations that are themselves competing, at a supra-individual level, and have their own lives, usually longer than those of individual men. The plants and firms live among a host of other organizations which serve varied interests—trade unions, professional societies, sporting clubs.

Corresponding to the infinite shades and gradations of personalities and types of work, spread through a complex social space, an unprecedented sensitivity to symbol systems comes into existence in the city. The contemporary mathematician, or biologist, or sociologist, along with such other products of city culture as the banker, the store manager, or the traffic analyst, each has his own characteristic set of symbols and has to cope with unprecedented variety, subtlety, and sheer mass of material. Typically 8 or 10 years of intense training, for most of us only possible between the ages of about 15 and 25, are the necessary means to develop the sensitivity, the awareness of issues, the minimum basic storehouse of

facts in a given field. You only put up with me, if you do, because you suppose that I have an extensive and powerful storehouse of facts in my own field, which is the mathematics of population. This reciprocal imputation of subtle, mysterious, and extensive knowledge is what permits mutual respect in the more specialized residents of the city. We do not know just what it is that the other man knows, but we assume he knows something and is capable of doing his own job with reasonable competence, whether it is embryology or pants cutting.

Such a basis of respect is characteristically metropolitan. In a society of smaller volume such as a village, each gets to know all about the few score or the few hundred people with whom he will have contact in the course of a lifetime. He knows them as whole people, is concerned with literally everything about them.

Each of us as city people has contact in a day with as many individuals as the villager meets in the course of a lifetime; this includes store clerks, bus conductors, taxi drivers, students, colleagues, theatre ushers, not to mention those we pass as we walk or drive along the street. It would destroy us if we had to react to every one of them as people. We want to know about each of them only enough to cover his particular relationship with us. We care only that the bus conductor is an authorized employee of the company and will take our fare and drop us at the corner of Madison and 42nd Street. Whether he is happily married with four children or a debauched bachelor, whether he is Presbyterian or Catholic, we never inquire. His uniform tells us everything about him that we need to know. It is mere personal whim on our part if we even look at his face.

The well adapted citizen of the high income metropolis has learned to protect himself against its potentially infinitely varied stimulation. In some measure he becomes blase; whatever happens he has seen something more exciting. He becomes absent-minded and dulls his recollection of gross stimuli and even his perception of them in order to accentuate his capacity to react to the subtler issues and symbols of his own business, professional, technical, or scholarly life. He goes out of his way to cultivate ignorance of fields outside his own. Whereas constant full exposure to what the city offers and demands would weary and frustrate him, by protecting himself sufficiently against stimuli he need show only a slight antipathy or even be perfectly good-natured. This is the nature of urban contacts, suggestively portrayed by Georg Simmel (1964). Note that this characterization applies only to those members of the city who have adapted to city life over two or three generations, and as a result are suitably educated, and are productive enough to command the facilities of the city. They do not apply to the recent migrant to Chicago from the rural south, or to Calcutta from inland Bengal.

The ultimate refuge against the pressures of the metropolis is flight. A quiet place in the country becomes the ideal of all and, in one form or another, the seasonal recuperation of most. But with the acceleration of the population growth, and especially with the improvement of transport through the private automobile, that quiet place in the country, the most precious of resources, is bound to become scarcer. We not only are a larger population, and able to get out of the city more easily, but a larger fraction of us has the means to travel. We are 197 million people in the 3.6 million square miles of the United States. Deducting the areas of the cities, superhighways, lakes, deserts (both natural and those made by man, for instance by mining operations), we could still each have 4 acres of countryside—which

means just enough to get out of the sight of one another. But not very long ago the United States was growing at 1.5% per year, a pace which would double our numbers each 45 years. It would halve the 4 acres in 45 years, quarter it in 90 years. I mention these figures only to show that within the lifetime of children now born, at an increase of 1.5% per year, dispersion would become impossible; no amount of redeployment would enable us to get out of sight of one another.

Simultaneous with the increase in numbers, the advance of technology makes each of us more mobile, and requires more space—especially highways—to be set aside merely for facilitating our movement. Aside from this, if effective density is counted in units or potential contact rather than in people per square mile, we increase our effective density merely in improving transportation. Man is unlike other creatures in drastically remaking his means of locomotion. United States' automobile and truck registrations were 86 million in 1964, and will pass the 100 million mark by the end of the 1960's. Nearly 10 million new vehicles are being put on the road each year, offset by only two-thirds that number of scrappings; this fact alone would tend to crowd us even if we were the same number of people. Year by year we are both more in numbers and are moving faster, always within a fixed people-container, the terrestrial area of the United States. The result is rising pressure and temperature, apparently under the operation of laws analogous to those governing the behaviour of gases.

I do not know how we shall respond. Will we build up higher capacity for discretion and reserve? Will we develop the sort of etiquette of noninterference with our neighbors—silence, dignity, and good humor—that helps make life tolerable on a long submarine voyage? I knew a charming family in Paris during the housing shortage who had two and a half rooms, counting the bathroom, for five people. They managed the situation well, and despite proximity each was able to have his private thoughts without interference. The life required a degree of self-discipline that not all of us could furnish. Events on the city streets this hot summer do not suggest that our civilization is moving toward those standards of reserve, discretion, and respect for the rights of others that would make greater density acceptable.

In fact, the response of society to higher density is usually the very opposite of reserve and respect for the privacy of individuals; it is rather interference and planning. The frontier had no traffic lights or parking regulations. People did not have to regulate their activities by the activities of others. If the clocks and watches of frontier families were randomly in error, little inconvenience would have resulted; but if those of a modern city were wrong, say by 2 hours, all its activities would be brought to a halt. Everything we do interlocks with what others do. The frontier needed no zoning bylaws or building standards. The necessity for all these forms of planning has come with density. Richard Meier (1962) describes an arrangement of the city of Madras in South India, one of the growing harbor cities I have spoken of, such that 100 million people could live in it, but life would have to be planned in the most excruciating detail. People would not even be allowed to own bicycles, simply because parking individually owned bicycles would tie up the streets. Movement would be restricted, and all that was necessary would be provided by mass transportation.

The density continuum from the frontier to Meier's imaginary Madras is also a planning continuum; density and planning seem to be positively correlated.

Less clear is the degree in which freedom in the West has declined with planning, and hence with density. George Orwell's *1984* is inconceivable without high population density, supplemented by closed circuit television and other devices to eliminate privacy. It exhibits in extreme form an historical process by which the State has been extending its power at the expense of the Church, the Family, and the Local Community, a process extending over 150 years.

Though the trend to State power has accompanied the increase of density, I do not know whether, up to now, density, even combined with the march of technology, and combined with State control, has diminished the individual's effective freedom on balance. After all, the individual benefits from the State and from the increasing wealth that has gone along with density and with planning, including clocks and traffic lights. In the United States the gain to freedom through rising average wealth may offset the attack on freedom through State measures. In the USSR increasing wealth seems to be bringing about a loosening of totalitarian structures.

A convincing account of the relation between sparseness and abundance on the one side and national character on the other is presented by David Potter (1954). The wealth of the economy, first on the frontier and then in the cities, encouraged mobility and individual success as an ideal. The real wealth of the country—and this arose from spaciousness in the largely agricultural epoch when national character was being formed—was great enough that a mobile, optimistic, ambitious, and generous attitude would have scope for success and would often succeed. One man who succeeded with these traits encouraged the growth of similar qualities in others. In the United States, sparsely settled during most of its history, when men were scarce and resources plentiful, labor was sought after by employers rather than the other way round. This gave the common man confidence in himself; self-confidence made him open-handed and he taught his children open-handedness.

Plenty also encouraged freedom—if there are enough goods for everyone, let each one take the part of the patrimony he wants or can earn. If there are enough seats on the train each passenger can choose the one he wants without supervision; a shortage of seats compels a system of reservations, which is to say planning. Abundance brings men to potential economic equality—at least in the sense that each can hope to find the resources that will make his labor fruitful. By virtue of the same fact, it inclines them to political equality. Density and poverty make for the opposite of democracy—as Wittfogel (1957) argues in his study of hydraulic civilization.

If density tends to shackle us, and wealth to free us, then the question of what will happen in the United States in the 21st century that is only 34 years away is still an open question. I do not know enough of the conditions of the present race between exhaustion of raw materials on the one hand and technology on the other to make a firm statement, but suppose technology wins and we become much denser but also much richer. For us the wealth worked counter to the density. But what about those dense societies which have little prospect of attaining the wealth of the United States and whose density, combined with their traditional agricultural techniques, already places them near the point of starvation?

I started this discussion with the pre-industrial environment, and went on to those contemporary preindustrial cities clustered around the coasts of the underdeveloped world in which ports, built for the export of now obsolete colonial tropical products, are operating at capacity to unload cargoes of food. Here are larger cities than Ibn Khaldhun contemplated. If those city dwellers oppress any peasantry it is that of the United States. But it is not that of the United States either, since American productivity has increased faster in agriculture than in industry these last years. The labor cost of that food to the American farmer is at an all-time low, and in any case, the city-dweller is paying for PL480 shipments, for which the Prairie farmer gets spendable cash. I said that if poor peasant populations are to be fed by imported grain rather than by grain from their own countryside, then it is convenient for them to bring their mouths to the harbors where the grain is unloaded, rather than having the grain carried to them dispersed over the countryside, and that this is the basis for much contemporary urban growth.

How different the case would be if the United States had given fertilizer factories rather than wheat. Then more grain could sprout throughout the South Asian mainland, and there would be no reason for the peasants to accumulate in the city in numbers beyond the industrial jobs available.

I have spoken of the problem of privacy in the dense industrial city. The problem is accentuated in the cities of Asia, where millions of similar beings exist in close contact, without the shelter provided to each through specialization and the division of labor. In the West we each come to have a special claim on the production of others through our own specialized production, especially through the voluminous contribution which each of us can make in our highly capitalized society. We have some degree of uniqueness of personality through all those devices by which we differentiate ourselves, some of them arising out of our work, but others quite separate: sports and hobbies, for instance, which the affluent society lavishly supports and equips. Education, also a result of our wealth, inducts us into elaborate, differentiated symbol systems and into that etiquette of restraint and reserve which mitigates the closeness of city living.

The citizen of the poor crowded city is in all these respects disadvantaged. He has neither specialized work, nor capital to make his work productive, nor hobbies to reinforce a personal identity, nor education to make him sensitive to complex symbol systems. Here is the hurt of density without its benefits.

The only defense for the poverty-stricken urbanite in the tropics is through the retention of some of his rural habits; he may manage a kind of village existence, with village ecological relations complete in all respects except that the village fields are absent. The village within the city of Calcutta or Madras may well have its own temple and traditional service occupations: sweepers, watchmen, priests, headmen. Some physical production by village methods goes on, and there are tanners, potters, and makers of bullock-carts—but very few farmers. Each compressed village in the city may have some residue of the administrative structure of the village its inhabitants have left, and village factionalism need not be forgotten on the move to the city. The village in the city may have a rate of growth more rapid than the old village, because its death rate may be low, its birth rate remains almost up to rural levels, and the city is swelled by new entrants from the countryside. It is hungry, as the rural village often was, but now its hunger is a matter of high political importance, affecting both national and international politics.

Our picture now is of two dense agglomerations of mankind facing one another, both urban, one rich and one poor. America takes its cities into the countryside—it becomes an urban society through and through. Asia remains rural, even to parts of the dense agglomerations at the seaports—it brings its rural culture into the cities. In a sense, the rich city has called the poor one into existence, first by DDT and medicine which lower the death rate of India from perhaps 35 per thousand to 20 per thousand, and then by the provision of food.

In a quite different sense, Europe had earlier contributed to Asian population. The industrial revolution of Europe and America demanded raw materials, and this demand was translated into a demand for people who would produce its goods. The needs of Europe for sugar, spices, and rubber brought into existence large populations, for example in Indonesia; Java grew from under 5 million in 1815 to 40 million by World War II. And when our technical advance, especially Western synthetics, enabled us to make the things that formerly could only be produced by tropical sunshine and tropical labor, those populations were left high and dry. They are functionless in relation to the Western industrial machine which brought them into existence, but they keep growing nonetheless.

Western government and electorates sense the tragic state of affairs, and, at least vaguely, feel responsible for this aftermath of colonialism; therefore we provide food and other kinds of aid. But such are the dilemmas of doing good in this difficult world that each shipment of food draws more people to the seaports, and we arrive at nothing more constructive than a larger population than before dependent on shipments of food. If the need for food is a temporary emergency, philanthropy is highly recommended, but the condition of tropical agriculture and population seems to be chronic rather than acute.

And yet the aid cannot be stopped. As the populations of these port cities grow so does the problem of producing and shipping food to them. But so also do the economic, political, and moral problems of cutting off the aid.

When famine was the work of God, whose acts man lacked the technical capacity to offset, such issues did not arise.

The only escape is through the economic development of the countries of Asia, Africa, and Latin America. With increase of income arises the sort of communication system through which people can receive and act on signals in regard to the size of their families. Americans since 1957 have come to understand that their families had been too large—signals reach them to this effect through the price system and through the difficulties of placing children in college and in a job. For underdeveloped people such a signalling system of prices and costs is not in existence, and messages that create a desirable feedback and permit automatic control do not carry.

On the other hand, the growth of population for many reasons itself inhibits the development process which could solve the population problem. That is why some students of the matter think that the control of population should be tackled directly. Each point by which the birth rate falls makes the process of saving and investment and hence development that much easier.

I have spoken of two sorts of dense agglomerations of people looking at one another across the oceans. The one is wealthy, modern, productive, highly differentiated by occupation, handling complex symbol systems, dominating the environment. The other is poor, traditional, nonproductive except for services that are not badly wanted, less differentiated by occupation, illiterate or barely literate, highly dependent on the environment. Our rich American cities contain a minority which is taking refuge from rural poverty. Some Asian cities consist of a majority of such refuges.

I have refrained from reference to the individual human tragedy of the multiplying homeless sidewalk-dwellers of Calcutta and other cities of the tropics. No one can say that their plight is irremediable; evidently a number of countries are achieving development today, including Hong Kong, Taiwan, Mexico, Turkey. On the other hand, I see no grounds for the facile optimism that declares that development is inevitable for all. To turn the despairing kind of density into the affluent kind requires that three issues be squarely met: food supplies must be assured, industry established, population controlled. Only in some of the underdeveloped countries are these seen as key issues and seriously tackled.

Is the hardship of life in the crowded and poor city itself a stimulus to effective action? Do density and poverty make for greater sensitivity to the real problems, and greater judiciousness in their treatment? Not necessarily; especially not for the miserable newcomers, the first ill-adapted migrant generation to the city. In the slums of first settlement, whether in 19th century London and Paris or 20th century Chicago and Calcutta, city mobs can be readily aroused by their troubles to action and to violence, but they do not necessarily see the root of their frustrations and the way to overcome them. Penetrating analysis does not guide mob action. The crowd, mobilized by some incident, acts with a violence out of all proportion to the event that excited its anger. It streams through a city street, stops to throw bottles at the police who reply with tear gas, overturns automobiles, is finally dispersed by a National Guard armed with bayonets. Far from being a disappearing relic of the past, it is with us both in the temperate zone and in the tropics, in wealthy countries and in poor ones. Food riots occur in Bombay and civil rights riots in Chicago, New York and Cleveland. This ultimate manifestation of population density, which colors the social history of all continents, is a challenge to learn more about the causes of tension and frustration in city life.

REFERENCES

Adams, R. M. 1965. *Land Behind Baghdad.* University of Chicago Press, Chicago, Ill.

Bourgeois-Pichat, Jean. 1966. *Population Growth and Development.* International Conciliation, No. 556, January. Carnegie Endowment for International Peace.

Durkheim, Emile. 1960. *De la division du travail social.* 7th ed. Chap. III, La solidarite due a la division du travail ou organique, Presses Universitaires de France, Paris. pp. 79-102.

Geertz, Clifford. 1963. *Agricultural Involution: The Process of Ecological Change in Indonesia.* University of California Press, Berkeley, Calif.

Gottman, Jean. 1961. *Megalopolis: The Urbanized Northeastern Seaboard of the United States.* Twentieth Century Fund, New York.

Hayley, Amos H. 1950. *Human Ecology: A Theory of Community Structure.* Ronald Press, New York.

Meier, R. L. 1962. Relations of technology to the design of very large cities. In *India's Urban Future* (Roy Turner Ed.). University of California Press, Berkeley, Calif. pp. 229-323.

Potter, David M. 1954. *People of Plenty: Economic Abundance and the American Character.* University of Chicago Press, Chicago, Ill.

Riesman, David 1950. *The Lonely Crowd: A Study of the Changing American Character.* University of Chicago Press, Chicago, Ill.

Simmel, Georg. 1964. The Metropolis and Mental Life. In *The Sociology of Georg Simmel* (edited and translated by Kurt H. Wolff). Free Press of Glencoe, Collier-Macmillan, London. pp. 409-424.

Steward, J. 1955. *Theory of Culture Change.* University of Illinois Press, Urbana, Ill.

Wirth, Louis. 1964. Urbanism as a Way of Life. In *On Cities and Social Life.* University of Chicago Press, Chicago, Ill. pp. 60-83.

Wittfogel, Karl A. 1957. *Oriental Despotism: A Comparative Study of Total Power.* Yale University Press, New Haven, Conn.

Wolff, Kurt H. (Ed.) 1964. *The Sociology of Georg Simmel.* Collier-Macmillan, London.

Overpopulation and Mental Health
George M. Carstairs

Before I left Edinburgh for Toronto I looked into my office to see if there was anything important in the mail. There wasn't. However, the visit was not entirely wasted because I had the opportunity of picking up the student newspaper. In its centre-page spread I saw a current report on the student scenes where the action was believed to be at its hottest. There were reports from Berkeley, Paris, India, Germany and Toronto. It was in the plane, reading the students' newspaper, that I learned about the interesting district called Yorkville which I gather is the Haight-Ashbury of the Toronto scene.

I have been impressed by the fact that the centres of most vigorous student protest have not been the small private universities. They have been those institutions like Berkeley, and Columbia, and the University of Rome with an enrollment of 70,000, and the Sorbonne with an enrollment of 80,000, and West Berlin. It has always been in extremely large universities that the intense protests have begun. But the protests have only partly been due to the fact that the university student body has increased much more rapidly than has the staff or the teaching facilities. Mere numbers do not trigger protest, but numbers may well have something to do with it.

My main purpose here is not to discuss over-population in the universities but to consider whether rapid increases in world population will have an effect upon mental health. One cannot separate mental health from

physical health because if a person is malnourished and sick it is much more difficult to keep
a level mind, a level outlook, on the circumstances of life. We are anxious first of all to
know whether we will even be able to feed the
increasing population in some parts of the
world. Having kept them alive, will we be able
to keep them physically healthy? Once that
anxiety has been allayed our next consideration concerns the quality of life they are going
to lead in new crowded populations.

The first thing we note is that human beings
have shown a remarkable capacity for adaptation. Think of the extremes in natural human
societies, from north where the Eskimo lives in
a tiny community sometimes many days
march from the next tiny community, to
southern India or Indonesia where landscapes
simply teem with humanity in any direction
you look. Yet members of both these kinds of
community feel at home, feel that this is how
one is meant to live, and would be excruciatingly uncomfortable if put in the opposite
situation. They have adapted then to a remarkable degree, but at a certain cost—the cost of
settling for varying levels of nutrition, of
physical hardship, of acceptance of disease.
These have been accepted traditionally over
many generations because in primitive societies there has been no awareness that there is
any alternative. Indeed there was no other alternative for them until quite recent times.
There is a falling death rate, particularly of
infants born in such communities, which is
seen as a good alternative to high mortality
rates.

In former centuries the vast majority of
mankind could not indulge in the luxury of
aspiring to a high standard of living. Simply to
survive into late adulthood, at the same level
of subsistence as one's forefathers, was good
fortune enough. From the time of the earliest
pre-historic civilizations to the present day, in
almost every human society, only the privileged elite have been in a position to cultivate
their sensibilities, and to expand the boundaries of human experience and understanding.
In London, as recently as the beginning of the
present century, the very chances of survival
through early infancy were more than twice as
high for the children of the rich as for the
children of the poor. Today, throughout the
world, such survival has become more generally attainable for rich and poor alike; and
now, for the first time in the history of mankind, education, self-awareness, and the aspiration for a meaningful and satisfying life-
experience are being shared to an increasing
extent by whole populations.

Inevitably, once the killing diseases and the
threat of starvation have been averted, people
become increasingly aware of, and discon-
tented with, minor forms of discomfort or unhappiness. One of the striking changes in morbidity, in both highly developed and in
developing countries during recent decades,
has been the apparent increase in neurosis and
psychosomatic disorders. These functional illnesses—which some people would prefer to
regard as manifestations of "problems of living" rather than of disease—have long been
recognized among the privileged classes. Already in 1689 Thomas Sydenham declared
that half of his non-febrile patients (that is, a
sixth of his total practice) were hysterical, and
in 1733 George Cheyne, in his book entitled
"The English Disease," stated that a third of
his patients were neurotic. But both Sydenham
and Cheyne were fashionable physicians, most
of whose clientele was drawn from the wealthy
minority of the English society of their day.
Sydenham himself observed that hysteria was
commoner among women of the leisured
classes than among those who had to toil. It is
only in the present day that the working
classes have been in a position to enjoy the

luxury of being neurotic, but recent surveys, both in the East[1] and in the West,[2] have shown that already the rates for almost every form of mental illness are highest among the socio-economically under-privileged sections of contemporary societies.

It must be emphasized that the very marked increase in the "visibility" of mental disorders in most countries of the world is partly due to the better control of infections and other serious physical illnesses. Neurosis is a by-product of a raised level of expectation of the quality of life-experience; it can, therefore, act as "divine discontent," a spur towards the further enhancement of the standard of living—provided, of course, that steps *can* be taken to remedy the adverse environmental factors to which the symptoms of neurosis have drawn our attention.

Here we are confronted by a vital question: what will be the consequences, for mental health, of a continuing massive increase in human population?

As yet, the science of human behaviour is not sufficiently developed to be able to answer this question with precision, or even with confidence. Nevertheless it is possible to learn from studies of animals, both in their natural environment and under experimental conditions, and to note certain regularly occurring consequences of severe overcrowding; with due caution, one can infer some similar repercussions of overcrowding in man. There are also a number of direct observations, in human populations, of the inter-relationships between overcrowding and certain indices of mental health, from which we can predict with greater confidence the likely consequences of overcrowding on a still larger scale.

Studies of Animal Behaviour

At first sight, it might seem that much could be learned from observations on species such as lemmings, or voles, which are subject to periodic fluctuations of population size. There is still a good deal of controversy among naturalists as to whether these fluctuations are essentially determined by rather gross environmental factors of food supply or infection, or whether social interactions also play an important role. In recent years the work of ethologists has taught us a great deal about the interaction of innate, biological propensities and *learning experiences,* in many species. At a relatively crude level, this can be shown in the modification of the animals' adrenal size and activity. The adrenals play an essential role in the animal's response to stress, whether by fighting or by taking flight. There is a conspicuous difference between the size of the adrenals in wild rats and in rats which have been bred for generations in captivity, the latter having much smaller adrenal glands. When wild rats are caged, and allowed to breed, a diminution in adrenal size becomes apparent in a few generations.

In colonies in which there is a great deal of fighting the mean size of the rats' adrenals increases by up to 30 percent and this is true both of the aggressors and the victims. Observations in nature have shown marked diminution in adrenal size when rat populations are depleted; similar findings have been reported on an overcrowded herd of deer.

Adrenal activity is stimulated by social interaction especially by the challenge of attack and the need for counter-attack in self-defence. It is an interesting finding that the quality of the stress response takes on a different character for the animal which is victorious in the contest. Such an animal can go from strength to strength, able to fight one battle after another and in the intervals of fighting its sexual potency is also at a high level. In contrast, an animal which undergoes a series of defeats becomes debilitated, even although suffering no obvious physical injury, and is sexually less active. The biologist, S. A. Barnett[3] has shown that prolonged exposure to even moderate hostility leads to weakness and death; he has epitomised this reaction as follows: "evidently the bodily response to humiliation resembles, in some ways, that to danger to life or limb." Usually the loser in such contests is able to survive by escaping from the scene of battle and thereafter refraining from challenging its victor; but there are situations both in the wild and in the captive state where animals are unable to escape, and are repeatedly confronted by the threat of a contest in which they are doomed to defeat. Observing such caged rats, Barnett reported that quite often when a rat had been engaged in combat and got the worst of it, that rat would drop dead. Sometimes this happened within hours of when it had been introduced into this threatening environment. Sometimes it happened some days later after prolonged exposure to a succession of threats. Barnett performed post-mortems on the dead rats and found that there was no gross injury, no loss of blood or wounding to account for the sudden death. He did histological studies of their adrenal glands and established to his satisfaction that it was in the first instance due to a massive over-secretion of the medulla of the adrenal gland. In the second instance, it was due to an exhaustion of the cortical cells of the adrenal gland. It seems that exposure to humiliation and defeat can be as physiologically wounding in its way as actual physical trauma.

An analogy may be found in observations on the toxicity of amphetamine drugs, whose action is similar to that of adrenaline, the secretion of the medulla of the adrenal gland. A relatively small dose of amphetamine will prove fatal to a rat which is confined in a cage with many other rats, whereas a rat which is kept in isolation can survive doses of amphetamine up to four times greater. It is presumed that the effect of the drug is greatly enhanced, in the former situation, by the numerous stressful interactions with the other rats, each of which stimulates the output of more adrenaline until complete exhaustion supervenes.

These, of course, represent extremes of over-stimulation. Many species of animals and birds have evolved self-protective behaviour patterns to ensure that such extremes will not occur. Typical of these behaviour patterns is the "peck order" or status hierarchy, by virtue of which a group of animals who meet each other regularly first fight each other, and then mutually agree on a rank-order of ascendancy after which the animal of inferior status invariably concedes in the face of a challenge from those above him in rank. More detailed studies[4,5] have shown that status hierarchies can be either *absolute,* where every member of a group of animals invariably remains in the same position in relation to each of his fellows, or *relative,* in which, under different circumstances of time or place, the individual's respective degrees of ascendancy over each other may change. Absolute status hierarchy is most likely to be found where all the animals in a group share the same living-space, and it becomes most clearly defined when that space is a restricted one. Under such circumstances, Barnett has shown that adrenal size becomes inversely correlated with height in the social hierarchy.

Relative dominance is seen most clearly in animals which have individual territories. When on their home ground, they are often able to vanquish an intruder and compel him to retreat, whereas if they are challenged by the same individual on *his* home territory they in turn will admit defeat. It seems that not only birds, but most mammals (including man) exhibit this kind of territorial behaviour. Not only football teams, but all of us, tend to perform best on our home ground—mental as well as physical—and to resist anyone who ventures to challenge us there. Naturalists have recognized in territorial behaviour, and in the varying degrees of dominance associated with the centre and the periphery of the territory, a self-regulating mechanism which ensures an optimal degree of dispersion of the species. It has also been noted that if several rats are introduced at the same time into a strange environment they coexist amicably; but if strangers are subsequently added they fight the stranger.

When animals such as domestic cats, which customarily enjoy quite a wide range of movement, are crowded together in a limited space there tends to emerge one particularly tyrannical "despot" who holds all the others in fear, and also one or more whom Leyhausen[6] terms "pariahs," at the bottom of the status hierarchy. These unfortunate creatures, he observes, are "driven to frenzy and all kinds of neurotic behaviour by continuous and pitiless attack by all the others." Although these "pariahs" bear the severest brunt, the whole community of cats held in such close confinement is seen to suffer. These cats "seldom relax, they never look at ease, and there is continuous hissing, growling and even fighting. Play stops altogether, and locomotion and exercise are reduced to a minimum." (Ibid.)

This clearly represents a pathological social situation, in which overcrowding and confinement conspire to accentuate disturbing confrontations between individuals. In Hamburg some years ago the rats became a plague to such a point that a special campaign had to be mounted in the sewers of the city. Biologists took advantage of this experiment of nature to study the rats before and after the extermination campaign. The campaign actually reduced their numbers to something like one-tenth of what they had been at the height of the plague. It was noticed that weight for weight there was a very significant reduction in the size of the adrenal glands when the numbers of the rat population were diminished by the extermination programme. A precisely similar finding was reported by another naturalist who studied a herd of deer which increased in numbers till it was overpopulating its territory and then had to be artificially reduced. Again there was an increase in the size of the adrenal with the increase in population, and a diminution when the numbers in the territory were reduced.

A comparative psychologist, John Calhoun,[7] studying the behaviour of colonies of rats under different degrees of over-population observed similar changes in their customary inter-relationships. Where overcrowding was most marked, the enforced social interactions were seen to interfere with the satisfaction of quite basic biological needs such as feeding, nest-building and the care of the young. Normally mother rats whose nest is disturbed will carry their young one by one to a place of safety, but in overcrowded pens this behaviour pattern was lost, and the rat's maternal care became so faulty that in one experiment 80 percent and in another 96 percent of all the young died before reaching maturity. Among the males, some became ascendant over their fellows but others showed a number of disturbances of behaviour, of which two patterns were particularly striking: some males appeared to opt out of sexual and social interaction altogether, skulking alone on the periphery of the group, while others became morbidly hypersexual, mounting female rats, whether receptive or not, whenever they could do so without being attacked by one of the ascendant males. The latter type of action is very unusual among wild mammals. These hyperactive rats contravened many of the norms of behaviour of their group, even becoming cannibal towards the young of their own kind. Christian, observing mice, showed that with overcrowding the reproduction rate is lowered; there are stillbirths and failure of lactation, and hence infant deaths. There is also a delayed effect; the next generation shows faulty maternal behaviour.

These observations on rats and cats must be regarded as having only marginal relevance to human experience and human behaviour. We come a little nearer the human when we read of studies carried out by biologists on primates. For many years, in fact from 1932 till almost 1965, an old text of Zuckerman's[8] was the chief authority on primate behaviour. (It is alleged that a London bookseller advertised it as dealing with the sex life of bishops.) It was the report of very careful observations on primate behaviour carried out in a succession of extended visits to zoos in South Africa where he observed baboons and chimpanzees in captivity. He pictured their behaviour as being dominated by competition, particularly competition for sexual dominance within the little group, and fighting. It is only in quite recent years that Schaller, Washburn, De Vore,[9] Jane Goodall and other ethologists have reported on how these species behave in their natural wild state. We now realize that accurate though Zuckerman's observations were, they reflected a totally distorted picture of the natural behaviour of these apes. It was a picture of the behaviour of apes confined in a space much more crowded than their natural habitat. Even gorillas seem to engage in a very few combats and relatively few threatening behaviourisms in their natural state. On the other hand observations on packs of monkeys in India which inhabit the fringe of jungles near human villages indicate an increase in the frequency of combative encounters as the packs increase in size.

Observations on Humans

What about human behaviour in confined conditions? We do have some direct reports from survivors of concentration camps among whom are the outstanding psychologist Bettelheim[10] of Chicago, and the psychiatrist Eitinger[11] of Norway. Their candid reports of what it is like to be in a concentration camp with an extreme improbability of coming out alive make rather bitter reading because they show how even the most upright and courageous people tended to show a deterioration in their usual humanitarian values. It was exceptionally rare for anyone to be able to maintain his standards under these extreme conditions.

Others have described conditions of life in the slightly less rigorous but still very crowded and confined conditions of prisoner-of-war camps. Two biologists who have shown a special interest in this field have been Paul Leyhausen[6] and Konrad Lorenz, both of whom endured several years of internment in prisoner-of-war camps. Lorenz, who writes with such marvellous eloquence, has described his own experiences in a Canadian prisoner-of-war camp. He tells very candidly how, after a longish confinement, he found the slightest mannerisms of his companions in the officers' huts to be unendurably irritating. However, he had the presence of mind to recognize the irrationality of his bad temper and when he found himself seething he would leave the company of his companions, walk to the barbed wire and look out across the acres of snow and empty land until his temperature subsided and he was fit to rejoin his fellows. The experiences of Leyhausen and Lorenz have been corroborated by other medical and psychiatric witnesses.[10,12,13]

These too, like the observations on caged cats and rats, were instances of extreme conditions, and yet one has to realize that there are many impoverished groups in the world whose conditions of life today are scarcely better. In theory, of course, they can escape from their surroundings, but in practice the "culture of poverty" can induce a sense of despair of ever being able to escape.[14] One is tempted to draw an analogy between the rat which is subjected to a series of physical defeats, or the "pariahs" in an overcrowded colony of cats, and the members of problem families in our city slums who display a seeming inability to make a successful social adaptation. It appears that social institutions and transmitted value systems can create a sense of confinement no less demoralizing than the bars of a cage. People can be confined just as effectively by growing up in an impoverished social milieu as they can by iron bars. There is a sense of incapability to escape when one has been denied the opportunity to develop one's intellectual faculties, to develop the imagination, to develop skills or capacities which would enable one to break away from these belittling surroundings.

Many years ago, Faris and Dunham[15] drew attention to the ecological concentration of certain forms of mental illness in those parts of a large city where both overcrowding and social disorganization—or *anomie* as Durkheim[16] had earlier described it—were most marked. Subsequent research has challenged Dunham's specific contention that schizophrenia is generated by the conditions of life in a socially disorganized community, but many other studies have confirmed his demonstration that alcoholism, illegitimacy, divorce, delinquency, and numerous other forms of social pathology are most prevalent in such areas.

This remains, however, an interesting contrast, in the social correlates of two particular manifestations of social pathology, namely *suicide* and *attempted suicide,* at least as they are observed in cities of the Western world. Suicide rates are highest in areas where many people live in a state of *social isolation,* bereft of the support family, or of any other primary group. On the other hand studies of attempted suicide have shown that the most important social correlate is *overcrowding.* Typically, the person who makes a non-fatal suicidal gesture has been harassed beyond endurance by recurrent friction within the domestic group, in cramped and overcrowded premises. Here too, as in the instance of rats' dose-resistance to amphetamine, one can see the mutual reinforcement of multiple factors. A majority of those who attempt suicide are relatively young men and women, who often have had a bad start in life with unstable or absent parent-figures. These patients tend to experience great difficulty in their turn, in forming stable interpersonal relationships: they are often at the same time demanding and inconsiderate towards others, and yet themselves emotionally immature and dependent. Their deficiencies prompt them to seek out partners from whom they hope to derive support, but all too often the partner whom they select is handicapped in much the same way; so far from meeting each other's dependency needs, these unfortunates only succeed in making each other's state even worse than before. Often, too, they turn to drink or drugs to allay their need for dependence and this in turn further impoverishes their ability to form rewarding personal relationships.[17] During recent years many countries have been obliged to take stock of increasing rates of alcoholism, delinquency and attempted suicide, indicating that an increasing number of citizens in our large cities feel alienated from the goals and the rewards to which their fellow citizens aspire, and alienated so profoundly that they despair of ever being able to get back into the mainstream of humanity.

Alienation and despair are the product of extreme situations, such as those, as I have noted, that were realized in the grotesque, doomed societies of the Nazi concentration camps. Many, if not most, of the inmates of such camps found themselves surrendering their customary standards of behaviour and their values, becoming completely disoriented by the inhuman conditions under which they were forced to live.[11]

There have been crises, in the course of human history, when quite large sectors of mankind experienced this sense of alienation from participation in the life of their fellow-countrymen. Sometimes after prolonged deprivation their discontents have exploded in outbreaks of revolution, as a result of which a new social order has been created; but at other times leaderless masses of the dispossessed have shown themselves only too ready to become the dupes of mentally unstable yet charismatic demagogues, who promised them a magical deliverance from their miseries. As Norman Cohn[18] shows, they become the leaders of milennial cults, promising the dawn of a golden age. An interesting thing about these milennial cults, which have recurred over and over again in European history, is that they begin with a preaching of brotherhood, generosity, sharing, saying that the riches of the earth are going to be shared among us all. It is a naive Utopian optimism. They waited for a magical answer to all their problems, and of course the magical answer was denied, the dream remained unfulfilled.

Indeed, quite soon the authorities, the Establishment of the day, began to take repressive measures against the dupes of these demagogic leaders, and then came a clash. The repressive forces were seen as hostile, threatening, demoniacal, and the movement which began with brotherhood, peace, and charity, inevitably ended in bloodshed, and sometimes in extremely violent bloodshed. One sees this story repeated over and over again.

Such an incident, though in a new guise, occurred after the Second World War. "Cargo cults" emerged in the islands of Polynesia and Indonesia. Cargo cults are rather like the milennial movements of the Middle Ages in Europe. There was a preaching of a magical solution to the problems of the poor and underprivileged. The name derived from the belief that a cargo airplane was going to land and pour out a cornucopia of all the things that they had witnessed with the sudden intrusion of Western troops. Jeeps, radios, washing machines were going to be delivered in abundance. These cults began with kindliness, naive hopes and innocent expectations, and ended with frustration, anger and bloodshed.

A similar phenomenon has occurred repeatedly in modern times, when the pace of political change has out-stripped a society's capacity to meet the newly aroused expectations of its members. When, because of increasing overpopulation, the standards of living actually decline at the very time when people's aspirations have been raised, the stage is set for further outbreaks of collective irrationality and violence.[19] This is the predicament of many developing areas today: India, Indonesia, South America, West Africa.

Now, going back to the beginning of my talk, I remind you that many of the movements of student protest in recent years have also begun generously, with concepts of brotherhood and concern for their fellow men. And this is still true of large elements in them. It is true of the work that many American students have done to further the cause of integration. It is true of the impulse that led them to protest against the war in Vietnam. For quite a time it was maintained in this style. They talked of flower power and of love-ins, feeling that this generous spirit would prevail and would be recognized and accepted. There is just a hint now, perhaps especially since the unhappy events of the Chicago Convention, that there has been a turn in this milennial movement. I notice, consulting the student newspaper from Edinburgh, that in certain quarters the slogan of "Flower-Power" is being replaced by a new ugly slogan, "Kill the pig." You can see the sequence. The repressive forces, the men in blue with truncheons, have appeared and are now being portrayed by some as the forces of evil, an exaggerated, demoniacal, bad object whom it is permitted to hate. And I think we have to be on our guard against this.

It is not numbers only, it is in addition the crash of high expectations frustrated. This gives rise to anger and, if we don't look out, to violence and bloodshed. In that sequence, I suggest, is the real threat of uncontrolled overpopulation to mental health. It is imperative that we recognize the gravity of this threat, because mankind today possesses weapons of such destructive power that the world cannot afford to risk outbreaks of mass violence; and yet the lesson of history points to just such a disaster, unless population control can be achieved before vast human communities degenerate into the semblance of concentration camp inmates, if not to that of Zuckerman's pathologically belligerent apes.

REFERENCES

1. — Lin, T. Y., (1953). A study of the incidence of mental disorder in Chinese and other cultures. *Psychiatry, 16,* 313.

2. — Srole, L., Langner, T. S., Michael, S. T., Opler, M. K. and Rennie, T. A. C. (1962). *Mental Health in the Metropolis.* New York: McGraw-Hill.

3. — Barnett, S. A. (1964). The biology of aggression. *Lancet,* ii. 803.

4. — Wynne-Edwards, V. C. (1962). *Animal Dispersion in Relation to Social Behaviour.* Edinburgh and London: Oliver and Boyd.

5. — Leyhausen, P. (1965a). The communal organization of solitary mammals. *Sym. Roy. Soc. Lond., 14,* 249.

6. — Leyhausen, P. (1965b). The sane community—a density problem? *Discovery,* September 1965.

7. — Calhoun, J. B. (1963). Population density and social pathology. (In: Duhl, L. J. *The Urban Condition.* New York: Basic Books).

8. — Zuckerman, P. G. (1932). *The Social Life of Monkeys and Apes.* London: Kegan Paul.

9. — De Vore, I. (1965). *Primate Behaviour.* New York: Treubner King.

10. — Bettelheim, B. (1963). Individual and mass behaviour in extreme situations. *J. Abnormal and Soc. Psychol., 58,* 417.

11. — Eitinger, L. (1961). *Concentration Camp Survivors in Norway and Israel.* Allen and Unwin.

12. — Cochrane, A. L. (1946). Notes on the psychology of prisoners of war. *Brit. Med. J., i,* 282.

13. — Gibbens, T. C. N. (1947). *The Psychology and Psychopathology of the Prisoners of War.* M.D. thesis, University of London.

14. — Lewis, Oscar. (1959). *Five Families: Mexican Case Studies in the Culture of Poverty.* New York: Basic Books.

15. — Faris, R. E. L. and Dunham, H. W. (1939). *Mental Disorders in Urban Areas.* Chicago University Press.

16. — Durkheim, E. (1897). *Le Suicide.* Paris.

17. — Kessel, W. I. N. and McCulloch, J. W. (1966). Repeated Acts of Self-Poisoning and Self-Injury. *Proc. Roy. Soc. Med., 59,* 89.

18. — Cohn, N. (1957). *The Pursuit of the Millenium.* London: Secher and Warburg.

19. — Worsely, P. (1957). *The Trumpet Shall Sound.* London: MacGibbon and Kee.

Casualties of Our Time
Amasa B. Ford

"It is changes that are chiefly responsible for diseases, especially the greatest changes, the violent alterations both in the seasons and in other things."
— Hippocrates

Violent alterations in the human environment have occurred at an increasing rate since the beginning of the Industrial Revolution. From the late 18th century, dislocation from the land, turbulent crowding in growing cities, and the economic deprivations of factory life affected the health of the people. Old diseases like tuberculosis flared up, and new sources of death and disability developed, such as the industrial injuries which were incurred by inexperienced hands attempting to master new machinery. Great changes, visible in a man's lifetime, gave motives for new laws and institutions. Social hygiene, with tardy assistance from therapeutic medicine, brought effective measures to bear on the new health problems, while hospitals, asylums, and other institutions were established to take the place of the now obsolete welfare systems of farm and village. Economic and technological changes thus produced specific new kinds of casualties, along with new resources for coping with them.

In the present century, the rate of change has accelerated. Many old problems have been mastered, but new ones have arisen. "Poor laws" and workhouses have gone the way of phthisis and chlorosis. Now we must ask whether general hospitals can cope with increasing drug addiction among alienated youth or how health departments can protect the public against cigarettes and overeating. But before we can restructure our health services we must assess what we know about the particular health needs of today. Because established social institutions have great inertia,

change is slow and tends to lag behind need. New problems, therefore, call for special attention, since they foreshadow future needs.

The purpose of this report is to identify sources of death and major disability which are new or are of new importance in developed countries in the two decades since World War II. Using examples from Great Britain and the United States, we make some estimates of how people are being affected.

Certain casualties result from immediate causes, such as the toxic effects of a new drug or the increased use of motor vehicles. Many, possibly more, take the form of major disability resulting from conditions that have complex origins. Examples are the extended survival of old people with chronic disease and the social alienation of young people. A rough classification by cause will serve as an outline, since an understanding of how disease originates is the most reasonable basis for control and prevention. Effects on health may be produced by changes in population, by technological developments, by new factors in medicine, or by shifting social and cultural patterns. These categories, however, should not obscure the fact that specific casualties may result from multiple causes.

Signs of Change

Prosperity and life expectancy have reached unprecedented levels in developed countries during the past 20 years, but there are indications that we may be approaching the limit of effectiveness of current methods of disease control and prevention.

In the early 1950's infant mortality rates ceased to improve at the rate which had prevailed for many years. In the decade 1946-56, rates had decreased 46 percent in the United States and 45 percent in England and Wales. In the subsequent 10 years, they decreased only 16 and 22 percent, respectively. The estimated average length of life, which in effect is inversely related to infant mortality, has increased to 70.1 years in the United States and 68.4 years in England and Wales, but the rates of increase since 1956 have been less than a fifth of what they were in the previous 10 years. The progressive reduction of deaths in the first year of life, which has been one of the finest fruits of social and medical development for over 100 years, has been arrested.

Similar changes are taking place in other mortality rates. Until the mid-1950's the trend of the overall death rate in the United States was downward. Since then it has leveled off, while in England and Wales, as well, the general mortality rate has been almost stationary for all age groups since about 1956. Some continued improvement in death rates from major infectious diseases has been counterbalanced by substantial increases in overall mortality from such conditions as ischemic (arteriosclerotic) heart disease, malignancies, motor vehicle accidents, pneumonia, and cirrhosis of the liver.

In addition to these broad changes in mortality trends, some kinds of morbidity have been moving counter to the economic tide in these two decades of affluence. Although employment and consumers' expenditures have continued to rise, so has "sickness-absence" from work. While a greater proportion of young people attend colleges and universities every year, crime rates, drug addiction, and illegitimate births are increasing more rapidly among teenagers and young adults than in any other age group.

These trends may be temporary eddies in a stream of general progress or the first signs of rocks ahead. In either case, they warn that medicine and public health are facing new

tasks for which old methods are not adequate. Let us examine some specific problems.

Population Changes

The numbers of persons aged 65 and over in the United States increased from 10.4 million in 1946 to 18.5 million in 1966. This increase of 78 percent was twice that of the general population during the same period (Table 19 in reference 1). The rate of increase for those aged 85 and older was even greater: 111 percent in the same period. These trends have altered the age structure of the population. Since disability increases rapidly with age, an increase in numbers of old persons means a disproportionate increase in those who need care (Table 1).

TABLE 1
Measures of disability among older persons in the United States in the 1960s 49

	Age			
	65-74		75+	
Status of persons	Rate (per 1000)	Number (1000's)	Rate (per 1000)	Number (1000's)
Persons in institutions				
In nursing and personal care homes	8	90	57	355
In geriatric and chronic disease hospitals and wards	2	19	6	37
In long-stay mental hospitals	9	99	11	67
Disability among persons not in institutions				
Blind or unable to read newsprint	24	263	96	565
Deaf or unable to hear conversation	162	1808	317	1904
With one or more chronic condition and:				
Unable to work or keep house	135*	2370*		
Confined to house	47*	821*		
Need help in getting around	65*	1139*		

*Postraumatic amnesia for more than 24 hours.

Thus, in the United States there are 700,000 older persons in institutions, and this number has been increasing by an average of 15,000 a year for the past 20 years. There are, moreover, from two to four times as many disabled older persons living at home, and their numbers have likewise been growing more rapidly than other groups in the population.

Population migration within a country, especially when it occurs rapidly, is likely to be followed by social maladjustment with adverse effects on health. An example is the massive postwar movement of Negroes from the rural south and whites from the impoverished areas of the western Appalachians into the northern cities of the United States. Over the past 20 years, between 0.5 and 2 million persons have migrated from farms every year, while the farm population has declined from 18.0 to 6.4 percent (Table 24 in reference 1). The nonwhite population of 19 of the 21 largest cities in the United States increased by over a third between 1950 and 1960. The resulting strain on health and welfare services is evidenced by the fact that during this decade 13 of these large cities experienced rising infant mortality rates among nonwhite residents.[2] In five of these cities rates increased among whites as well. By 1966 the situation had improved in some cities, but in four, the infant mortality rates among nonwhites continued to rise, while in one the rate among whites was again higher than in 1960. This drastic population movement has affected more than infant mortality. It has been a groundswell under the urban violence which has periodically swept through American cities in the past 10 years. Migrants into the cities have transferred from their rural environment educational, social, and nutritional deficiencies which have accentuated contrasts in health between the poor in the center cities and well-to-do suburbanites. Recently enacted programs have so far not been able to reduce the casualties of this latest wave of migration into the cities of North America.[3]

Technological Change

Rapid technological change has become so familiar in developed countries over the past 20 years as to blunt our perception of what is happening. Greater ease of living tempts us to overlook costs in human health. Whereas the effects of some changes, such as environmental contamination, are complex and may take years to assess, the impact of the automobile is unmistakable. We can begin to count these casualties now.

The annual crude death rate from motor vehicle accidents in England and Wales increased from 95 per million in 1946-47 to 110 per million in 1956-57 and, even more rapidly, to 152 per million in 1966-67 (Table 8 in reference 4). About 40 percent of these deaths were attributed to skull fracture or head injury. Many of those who died in this way were old people, but there was also a disproportionate mortality among young men (Fig. 1). The death of a young man entails the social loss of his productive years. Still more costly is the increasing number of young people who sur-

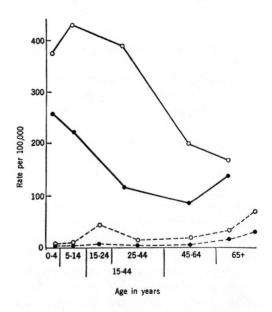

Figure 1.

vive head injuries to live for years with residual disability.

A rough estimate of the morbidity resulting from head injuries can be based on records of hospital discharges in England and Wales. During the 8 years for which data are available, the numbers of discharges with a diagnosis of head injury have increased at an annual rate of almost 5 percent. Of these injuries, 46 percent resulted from traffic accidents (Table 18 in reference 5). The estimates in Table 2 are probably high, since the relationship between "discharges" and numbers of patients is unknown. The actual figures, however, would not be likely to be less than two-thirds of those presented (50 percent readmission rate in a year). Thus, at least 1000 disabled people are being added to the population every year from this source, and the rates continue to increase.

TABLE 2
Head injury (International Classification of Diseases codes N800, N801, N803, N804, and N850-N856), discharges from hospitals in England and Wales, 1966 (Table 19 in references 5; 50). Items other than total are estimates.

Item	Number	Percent of total
Total such discharges	110,000	100
Major injury*	7,700	7.0
Died in hospital of injury	3,630	3.3
Survived with functional recovery	2,530	2.3
Survived with handicap; able to do limited work	880	0.8
Survived with handicap; unable to work	660	0.6

Ecological Casualties

The casualties of motor vehicle accidents can be attributed to an immediate cause. But for several other forms of death and disability which may result indirectly from advancing technology, a cause and effect relationship is more difficult to establish. Public attention has

recently been directed to the possible effects of environmental pollution on human health.[6] Scientific evidence is beginning to clarify the dangers.[7] Some of the most evident environmental pollutants which have been accumulating at an increased rate during the past 20 years are pesticide residues in soil and food chains, fallout of isotopes with long half-lives, and an increased burden of organic and mineral wastes in surface water.

Air pollution is a form of contamination which has been studied particularly intensively in the past 10 years, and its relationship to emphysema and chronic bronchitis is becoming more evident. These diseases are major causes of sickness-absence, chronic disability, and mortality in industrial countries. In the United States and the Netherlands death rates from these conditions have more than doubled since 1950.[8,9]

Links between chronic bronchitis and enphysema on the one hand and air pollution on the other are becoming firmer. Sudden increases in respiratory illness and death were dramatically evident in the London smog of 1952, when an excess mortality of 4000 persons was recorded in a 5-week period.[10] Subsequent work has shown that changes in the atmospheric concentration of oxidants, carbon monoxide, sulfur dioxide, and oxides of nitrogen are significantly related to hospital admission rates and length of stay for respiratory and circulatory conditions.[11] Symptoms of chronic pulmonary disease have been found more frequently in areas of greater air pollution.[12] Anatomic emphysema increases in prevalence above the age of 40 and has been shown to be more common in a sample from a heavily industrialized urban community with high air pollution than in one from a prairie-agricultural city with much lower pollution.[13] As the epidemiologic picture of chronic pulmonary disease developed, the "cause," like that of other chronic conditions, appears to be the interaction of several environmental and host factors, including age, climate, and smoking as well as air pollution.[14]

Efforts to make even a rough estimate of casualties due to air pollution are beset with difficulties. Etiology is complex, nomenclature has changed, and bronchitis and emphysema are both diagnoses which are frequently linked with other diseases. Still, the fact remains that 18,763 more deaths were attributed to these two causes in the United States in 1966 than 10 years earlier—an increase of almost two and a half times, and one observed at all ages above 35.[9] If only one half of these increased deaths were attributable to air pollution, this would still amount to nearly 1000 additional deaths a year. Chronic bronchitis and emphysema are typical chronic diseases, in which years of increasing disability usually precede death. The price of a polluted atmosphere in terms of illness and reduced functional capacity may be greater than its cost in years lost through premature death.

In addition to air pollution, new methods of food preparation and handling, massive distribution of detergents, pesticides, and antibiotics, and increasing military and industrial use of radioactive materials all pose possible threats to human health by environmental contamination. Whether strontium-90 in the bones or chlorinated hydrocarbon residues in other tissues will eventually produce casualties cannot be predicted now, but these processes call for close surveilance. The fact that mortality from heart disease and from major malignancies of the digestive, respiratory, and urinary tracts all show positive correlations with degree of urbanization in the United States suggests that environmental pollution may contribute to other conditions besides chronic pulmonary disease.[15]

Curiously, the casualties of technological change are exceptionally difficult to enumer-ate, although they arise from the practical application of science. Possibly 1000 additional deaths from chronic pulmonary disease a year —and what else? The very speed of change makes evaluation difficult. Intensive study of the pathogenesis of those conditions which show increasing prevalence is needed, as well as improved methods of monitoring contamination of the environment.

Casualties of Medical Progress

Doctors have always been viewed by the rest of mankind with respect, tinged with suspicion, since they appear to have special access to the secrets of life and death. As medicine has become more scientific it has become more effective, but also, in some ways, more dangerous. New drugs which control or cure previously fatal disease can also injure and kill. New diagnostic methods permit the detection of hidden disease, but accidents can occur in their use. The very fact that once-doomed lives can be saved means more surviving invalids.

Since doctors are trained to be unbiased observers, they should be watching closely and reacting promptly to unintended harm resulting from new methods which they adopt. That a healthy self-correction does in fact operate in medicine has been demonstrated repeatedly in the past 20 years. An example is the story of retrolental fibroplasia. Increased blindness among premature infants, first noted in the 1930's, was found to be associated with prematurity in 1942[16] and with oxygen therapy in 1949.[17] Oxygen was soon pinpointed as the injurious agent, and a decade of rigid oxygen restriction (1955-64) followed. As a result, the number of new cases has dropped sharply, although hundreds of persons remain blind, with only a 3.4 percent chance of regaining sight.[18] More recent studies, however, suggest that the neurological sequels of prematurity increased during that decade and that they may have resulted from insufficient oxygen. The use of oxygen for premature babies will evidently have to be still more precisely regulated before we can gain its benefits without incurring new casualties.[19]

The pace of introducing and distributing new drugs has become so rapid that control of harmful effects inevitably lags. In the decade 1955-64 production of vitamins more than doubled in the United States, while that of penicillin increased more than five times (Table 105 in reference 1). Ten years ago, Welch estimated that there were 17 to 20 million individuals in the United States (about 10 percent of the population) who may react to contact with antibiotics.[20] Hospital surveys show that 10 to 18 percent of hospitalized patients who receive drugs develop reactions to them.[21] The mechanisms by which therapeutic drugs may cause death or disability include disturbance of body defense mechanisms, cell injury, imbalance of essential materials, genetic disturbances, chemical carcinogenesis, and change in microbial ecology.[22]

An informal network of clinical observations, laboratory testing, and legal requirements provides much information about the adverse effects of drugs.[22,23] Epidemiologic information, however, from which the incidence and risk of casualty can be estimated is extremely limited. The effects of two specific drugs, thalidomide and chloramphenicol, have been studied in terms of population and can serve as examples, but these only suggest the magnitude of the problem.

Absent and distorted limbs and other severe congenital anomalies can be produced by the sedative thalidomide if it is administered in the first trimester of pregnancy. This dramatic example of drug toxicity has received publicity, the drug has been withdrawn, and only a limited cohort of malformed children

remain as reminders of the risks inherent in the pace of change in modern drug usage. A systematic search of the records of 23 hospitals in Hamburg revealed 139 cases of thalidomide-type of malformation in the years 1958-63. The drug was introduced in 1957 in Germany and elsewhere,[24] and the first case of malformation occurred that year. The rate of thalidomide-type malformation rose rapidly from 0.2 per 1000 births in 1958 to 3.1 per 1000 births in the first 6 months of 1962. Thalidomide was withdrawn in November 1961. Eight months later, in the last 6 months of 1962, the birth rate of malformed infants had fallen to 0.5 per 1000, and only one such case was recorded in these hospitals in 1963.[25] A survey of Canadian infants in 1963 disclosed 117 cases, and 869 were known to investigators by 1965.[25,26]

The toxicity of chloramphenicol has been even more thoroughly documented over the past 20 years. In spite of all that is known about it, this drug continues to be widely used, for both sound and trivial purposes. Chloramphenicol was introduced in 1948; the first warnings of its association with aplastic anemia were published in 1952; and the peak number of cases was reported in 1959. The mechanism by which this (and other drugs) produce this irreversible and fatal suppression of the bone marrow is not known, but the epidemiologic association is very strong. For example, of 771 cases of aplastic anemia reported to a registry of drug reactions in the United States (1956-66), 43.8 percent had received chloramphenicol, and 45.5 percent of these had received no other drug.[27] A detailed epidemiologic study has been made in the state of California, where all fatal cases of aplastic anemia in an 18-month period (60 cases) were analyzed for exposure to the drug and considered in relation to drug sales and distribution. The risk of fatal aplastic anemia within a year of exposure to chloramphenicol is estimated to lie between 1 : 40,800 and 1: 24,400, which is 13 times the risk for persons not exposed to this drug.[28]

Thus, there is a measurable toll of chronic illness and death to be set off against the presumably larger sum of lives saved and suffering relieved by the use of one antibiotic. For other drug-induced diseases we have no comparable estimates, but the very number of such conditions which have been identified calls for greater caution in the use of drugs than many practitioners now exercise. Among the diseases resulting from the use of modern drugs are such well-recognized clinical problems as peptic ulcer and reactivation of tuberculosis after treatment with steroid drugs, persistent Parkinson's disease after treatment with chlorpromazine, and damage to the eighth (auditory) nerve from streptomycin.

Diagnostic Risks

The adverse effects of therapeutic drugs may be a necessary price, but it is never one to be paid complacently. Casualties resulting from diagnostic procedures are still less acceptable. No method of diagnosis, including medical consultation and simple admission to a hospital, is without risk. Techniques which involve the introduction of instruments and other substances into the body are used increasingly. These include needle puncture of arteries, veins, the spinal canal, and most external and internal organs; catheterization of blood vessels, the heart, and the urinary tract; x-ray contrast visualization of organs and vessels by means of injected substances; biopsy of every tissue in the body; and local and general anesthesia given for diagnostic purposes. The major hazards of these procedures are mechanical trauma to tissues or organs, anoxia of brain or heart, embolism, spread of tumor cells, infection, drug reactions, and disabling anxiety.

Radiographic visualization of the kidneys and urinary tract by intravenously injected substances is a common and well-established diagnostic measure. This procedure has a safety record which does credit to medicine and the pharmaceutical industry. A summary in 1954 of 3.8 million such procedures reported a mortality rate of 0.008 percent.[29] No permanent disabilities are described in recent series, although transient side reactions, such as nausea and vomiting, are reported in 5 to 25 percent.[30]

The use of arteriography (injection of contrast material into arteries) has expanded greatly in recent years. With this technique, the dangers of anoxia and embolism are added to those of drug reaction and infection. The incidence of severe permanent disability resulting from these procedures is reported to range from 0 to 12 percent.[31,32] In the centers with the most experience it is probably about 0.5 percent.[33] The total number of such procedures cannot be estimated, but active hospitals perform 400 to 500 a year. Techniques are constantly changing, and safety improves with experience.[32]

Prior to cardiac surgery, the heart and related vessels are usually studied by introducing a catheter to measure pressures, to take blood samples, and to inject contrast materials. Experience in 16 major cardiac-catheterization laboratories in the United States over a 2-year period was summarized in 1968.[34] The average incidence of major complications was 3.59 percent and the mortality rate 0.44 percent, based on 55 deaths. The incidence of major residual disability (estimated from the text) was 0.16 percent. If we assume conservatively that the average number of procedures for all 513 cardiac laboratories in the United States (during 1961) is half what is reported here, there would be approximately 100,000 such procedures, with 160 new cases of permanent disability and 440 deaths from this source every year. This is a high price to pay for information, but it is important to remember that the greatest risks occur in infants with severe, sometimes life-threatening congenital heart disease. For such patients, accurate diagnosis and skillful surgery may offer the only chance.

Although the advance of science in medicine is sometimes said to have simplified the doctor's task and to have reduced the need for experience and judgment, the facts just presented indicate the opposite. Many drugs and many diagnostic procedures can be used with little fear of adverse reactions, but new dilemmas are constantly being presented to the physician. Is it better medical practice to give cautious advice to the family of a child with a heart murmur or to insist on thorough investigation, perhaps including cardiac catheterization with its attendant risks? What, apart from specific diseases like typhoid fever, are the indications for the use of chloramphenicol? How serious must the threat to life be before a drug which carries a known risk of fatal adverse reaction is used? How much must a physician know about a new drug like thalidomide before he starts to prescribe it for his patients? Surely the situation calls for more, and perhaps different, technical training, continued throughout professional life. Beyond this, the doctor needs as much as ever—perhaps more than ever—broad awareness of human values and sound personal judgment.

Extended Survival

Old ethical issues have been sharpened by more effective scientific methods, but the recently increased power to extend human life raises an essentially new question which has profound social, ethical, and religious implications.

New kinds of therapy which have been developed in the past 20 years permit life to continue in spite of potentially lethal disease. Following the discovery of insulin, for example, other essential hormones have been identified and made available so that life can be sustained after total removal or destruction of the adrenal glands, the pituitary, or the thyroid. Orthopedics, neurosurgery, and cardiac surgery are now able to accomplish the complete or partial repair of certain congenital abnormalities which, untreated, are incompatible with life. New prostheses, mechanical and transplanted, are coming into use, although still in very small numbers of cases. Above all, a wide range of effective new antibiotics makes possible the control of infection which in the past terminated the lives of the disabled, the elderly, and those with chronic disease.

The question of euthanasia has in the past been largely an academic one, since there was little support among either the general public or the medical profession for assigning to doctors or anyone else the responsibility for actively terminating life. Now, however, the possibility arises with increasing frequency of permitting a patient to die by withholding treatment which could prolong life, sometimes a painful and distorted life.

An example of the dilemmas posed to physicians and society is the treatment of spina bifida cystica. This is a congenital anomaly in which the lower end of the spinal canal fails to close and a meningomyelocele, containing nerves and spinal fluid, is present. Complications include progressive hydrocephalus, sometimes producing mental retardation, and various degrees of neurological deficit in the legs and lower trunk. Largely because of infection, mortality in the first year of life was 88 percent before modern treatment was developed, and the survivors were those who had the least deformity.[35] From 1955 through 1962, deaths from spina bifida and hydrocephalus in England and Wales occurred at a constant rate of about 1200 a year. After the introduction of early surgery (in the first 24 hours of life), the number of deaths fell rapidly to 815 in 1966. The introduction of this lifesaving procedure is too recent to assess ultimate effects. Table 3, however, suggests that 358 children with mental and physical handicaps are being added to the population every year, most of whom would have died as infants prior to 1963. As these disabled children advance through the school years, more places will be required for them in special schools, and many will need permanent care at public expense.

A second, less apparent by-product of improved treatment is the increased genetic burden produced when infants or children with potentially lethal heritable disorders are enabled to survive and reproduce. The clearest example is that of congenital pyloric stenosis, which can be effectively cured by a simple operation, allowing a normal life for persons who might otherwise have died in infancy. The resulting increase in numbers of such cases does not pose serious burdens of treatment or care.[36] However, there are other more serious and common diseases, such as diabetes, in which heredity plays a definite part, and in which modern therapy makes possible a higher reproductive rate. More research is needed before we can estimate the numbers of casualties to be expected from this source.[37]

Social and Cultural Change

The past 20 years have witnessed the maturing of an essentially new social order, aptly termed the affluent society. Economic productivity and automation have brought us to a point from which the total eradication of poverty can be glimpsed as a practical goal.[38] Even now, large segments of the population of developed countries enjoy a superabundance of food and physical conveniences more lavish than those which were available to the tiny minorities of rulers and nobility in the past. Citizens of the United States have an average of over 3000 calories of food energy available daily, which is twice the minimum required to sustain life.[39] The average energy expenditure of the American factory worker is probably less than twice the rate of resting metabolism and far below the level of effort required to gain a living by hunting or farming.[40] Of a sample of British civil servants, 10 percent report that they do not even climb stairs in the course of a normal day.[41]

Obesity has become a major health problem, highly associated with increased morbidity and mortality.[42] It seems logically related to increased calorie supply and reduced energy expenditure,[43] but information is lacking about whether the prevalence of obesity is increasing.[44] Social and economic change have clearly produced some casualties in this way, but the numbers may be static.

Arteriosclerotic heart disease, on the other hand, has become increasingly prominent as the leading cause of death in advanced countries, with prevalence and incidence still rising. Among the several factors which define the high-risk population for this disease are obesity, high consumption of saturated fats, and physical inactivity.[45] In the 10 years 1957-1966, the death rate in the United States for arteriosclerotic heart disease increased 10 percent, while the death rate from all causes decreased 0.7 percent. During this period, the numbers of persons whose deaths were attributed to this condition increased by an average of 17,425 each year in the United States and 3,750 in England and Wales. The causes of these casualties are not yet plain, but the epidemic is occurring specifically in the "advanced" countries and must be connected with social and possibly with psychological changes related to industrial and economic growth.

Young Casualties

Young people are at the vortex of modern social change. Whether they initiate change or react to it, they are paying an increasingly heavy toll. Popular opinions about previously unacceptable kinds of sexual behavior are rapidly changing,[46] and the availability of effective methods of contraception has contributed to rapid changes in sexual relationships among young people. The ultimate psychological and social benefit of more honest attitudes toward basic human functions could well be great, but there are penalties along the way. Whether the birth of an illegitimate child can be classed as a casualty might be questioned, but the fact is that the rate of such births is rapidly rising (Tables 4 and 5). The data shown are incomplete, but they do demonstrate trends. To be sure, in England and Wales, two-thirds of the births to girls aged 15 to 19 are subsequently legitimated by marriage, but it is also true that two-thirds of the births to girls of this age are conceived out of wedlock. In spite of penicillin and its successors, which offer potentially effective means for eliminating bacterial venereal disease, the battle against syphilis shows doubtful gains, while a new wave of gonorrheal infections is evident in England and Wales (Table 6) and in the United States.[47] Young people are affected predominantly. In 1967, age-specific rates show that new cases of venereal disease increase rapidly among males from 0.03 per 100,000 below age 16 to 2.85 at ages 16 and 17, to 9.33 at 18 and 19, and to 12.58 at 20 to 24. Above age 25, the rate drops back to 5.31 per 100,000. A treated case of gonorrhea in a young man, of course, does not produce disability. But figures on reported cases of venereal disease are only the tip of the iceberg. One estimate puts the reservoir of persons needing treatment for syphilis in the United States at 1,200,000.[47] Gonorrhea is notoriously difficult to detect in women; the causative organism can develop resistance to antibiotics; and the incidence of nonspecific urethritis in males, sometimes complicated by arthritis and eye involvement, is also increasing.[48] Late complications of syphilis and gonorrhea do not show signs of increasing, but there is a lag of several years in their development.

Modern youth is dedicated to change, whether in the form of volunteer work in developing countries, reform of universities, or extreme shifts of fashion in personal appearance; compliance and conformity are intolerable. Not all can stand the pace. Some fall by the way and are injured or disabled. Young victims of the automobile have been referred to above. Addiction to drugs among teen-agers and young adults has been rising at startling rates in England and Wales. Reported rates of crime are also increasing, most rapidly among young people (Fig. 2). Comparable figures for the United States are difficult to produce because of variations in states' laws and because of the concentration of narcotics users in large coastal cities, but there is no doubt that juvenile crime rates and the use of addictive narcotics are increasing rapidly in this country also. From 1950 to 1960 the numbers of persons in correctional institutions in the United States increased 27 percent, compared with a population increase of 19 percent. A quarter of these—99,021 individuals in 1960—were under the age of 25, and the rate of increase in this group was higher than in any other, namely 36 percent (Table 229 in reference 1). As with increases in illegitimacy and venereal disease, it is difficult to enumerate exactly how

TABLE 3
Outcome of cases of spina bifida,
England and Wales, 1963-67 51

Outcome	Number	Percent of total
Born alive 1963-67	6000	*
Surviving in 1968 under 5 years old	2500	*
Surviving in 1968 age 5 to 10 years	1250	*
Surviving in 1968 age 10 to 15 years	200	*
Born alive in 1966	1200	100
Expected to survive to 5 years	480	40
Mentally and physically normal	132	11
Mentally normal, physical handicaps	216	18
Educationally subnormal	96	8
Profoundly retarded	48	4

*Not applied.

TABLE 4
"Maternities conceived outside marriage,"
England and Wales 4

Year	Annual number per 1000 unmarried females of age		
	15—19	20—24	25—44
1938	11.8	32.6	18.6
1952-55	15.7	42.8	25.3
1956	19.0	48.6	28.9
1960	24.0	58.0	35.5
1964	30.3	68.0	42.5
1967	35.9	67.6	46.4

TABLE 5
"Estimated illegitimacy," United States 52

Year	Annual number per 1000 unmarried females of age		
	15—19	20—24	25—44
1940	7.4	9.5	7.1
1950	12.6	21.3	14.1
1960	15.7	40.3	21.8
1966	17.5	40.8	23.6

TABLE 6
New cases of venereal disease, England and Wales 48

Disease	Average annual number in 1000's for years		
	1953—57	1958—19	1963—67
Syphilis	5.3	4.2	3.8
Gonorrhea	19.9	33.1	37.9

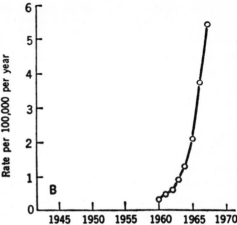

Figure 2.

Conclusions

The roll of casualties of our time is incomplete. Among those numbered in hundreds every year we have counted invalid survivors of spina bifida, patients accidentally injured during cardia catheterization, and those disabled by reactions to such drugs as chloramphenicol. Rising casualties numbering thousands annually result from the health environment surrounding certain infants born in our cities, from the vulnerability of young people to head injuries, drug addiction, and crime, many more young people become physically disabled and socially dependent each year because of early involvement in crime and drug addiction. The term "dropout," however, has melancholy overtones. Present methods of prevention and control are simply not working, penal systems and drug and venereal disease treatment centers are already overloaded and often out of date. Like the casualties of war, these are the most costly kind because of the years left to live and the likelihood of relapse.

and from chronic lung disease associated with air pollution. Increasing numbers, in the tens of thousands every year, suffer or die from arteriosclerotic heart disease or are disabled by the frailties of age. Other casualties may be on the way: additional victims of environmental pollution, more infants surviving with genetic defects, more casualties of affluence, made useless by automation or retired from boring work, more artificially supported survivors, and more casualties of new drugs. Though these numbers may in a sense be outweighed by a rising standard of living, better education, less work, and less discomfort, they are surely enough to cause concern.

These casualties raise some fundamental questions. Do we understand the real reasons why death rates now are scarcely falling? How well do we grasp the meaning of increased sickness-absence or the use of drugs?

In practical terms, are we prepared to cope with the immediate future? First, do we have adequate information? Are our systems of surveillance and detection flexible enough to keep up with the rate of change? Who is responsible for keeping track of what is happening and for finding out causes? Who acts on the knowledge we already have and will acquire? How can we prevent new casualties? Who needs to be educated and who will be the teachers?

Disabled and dying people need help immediately. Are we training the workers, developing the services, and setting aside the funds that are already needed to make a decent life for tens of thousands of disabled old people and thousands of invalid children and young adults? Finally, what value do these needs have in relation to our other social purposes? Some casualties have occurred while we had our eyes fixed on goals other than human welfare. Concentrated attention directed to these unfortunate fellow-beings may help to get some important priorities sorted out.

REFERENCES AND NOTES

1. U.S. Bureau of the Census, *Statistical Abstract of the United States: 1967* (U.S. Government Printing Office, Washington, D.C., ed. 88, 1966).
2. S. Shapiro, E. R. Schlesinger, R. E. Nesbitt Jr., *Infant, Perinatal, Maternal and Childhood Mortality in the United States* (Harvard Univ. Press, Cambridge, Mass., 1968), p. 83.
3. J. R. Hochstim, A. Athanasopoulos, J. H. Larkins, *Amer. J. Public Health* 58, 1815 (1968); *National Advisory Commission on Civil Disorders, Report* (U.S. Government Printing Office, Washington, D.C., 1968), pp. 136-139.
4. General Register Office, The Registrar General's *Statistical Review of England and Wales, 1966* (H. M. Stationery Office, London, 1967).
5. Ministry of Health and General Register Office, *Report on Hospital Inpatient Enquiry, 1965* (H. M. Stationery Office, London, 1968).
6. R. Carson, *Silent Spring* (Hamilton, London, 1963); P. H. Abelson, *Amer. J. Public Health* 58, 2043 (1968).
7. J. R. Goldsmith, *Arch. Environ. Health* 18, 710 (1969).
8. K. Biersteker, *ibid.,* p. 531; L. Breslow, *ibid.,* 8, 24 (1963); I. M. Moriyama, *Public Health Rep. 78* (1963), p. 743.
9. A. J. Klebba, *Mortality from Diseases Associated with Smoking,* PHS Publ. No. 1000, Ser. 20, No. 4 (U.S. Government Printing Office, Washington, D.C., 1966).
10. W. P. D. Logan, *Lancet* 1953-I, 336 (1953).
11. T. D. Sterling, S. V. Pollack, J. Weinkam, *Arch. Environ. Health* 18, 485 (1969).
12. B. G. Ferris and J. L. Whittenberger, *N. Engl. J. Med.* 275, 1413 (1966).
13. S. Ishikawa, D. M. Bowen, V. Fisher, J. P. Wyatt, *Arch. Environ. Health* 18, 660 (1969).
14. E. J. Cassell, M. D. Lebowitz, I. M. Mountain, H. T. Lee, D. J. Thompson, D. W. Wolter, J. R. McCarroll, *ibid.,* p. 523.
15. N. E. Manos, *Comparative Mortality Among Metropolitan Areas of the United States, 1949-51,* PHS Publ. No 562 (U.S. Government Printing Office, Washington, D.C., 1957): E. A. Duffey and R. E. Carroll, *United States Metropolitan Mortality 1959-61,* PHS Publ. No. 999-AP39 (U.S. Government Printing Office, Washington, D.C., 1967).
16. T. L. Terry, *Amer. J. Ophthalmol.* 25, 203 (1942).
17. V. E. Kinsey and L. Zacharias, *J. Amer. Med. Ass.* 139, 572 (1949).
18. E. Rogot, I. D. Goldberg, H. Goldstein, *J. Chronic Dis.* 19, 179 (1966).
19. W. A. Silverman, *Amer. J. Public Health* 58, 2009 (1968).
20. H. Welch, C. N. Lewis, H. I. Weinstein, B. B. Beckman, *Antibiot. Med. Clin. Ther.* 4, 800 (1957): H. Welch, *J. Amer. Med. Ass.* 170, 2093 (1959).
21. Anonymous, *Brit. Med. J.* 1, 527 (1969).
22. W. Modell, *Annu. Rev. Pharmacol.* 5, 285 (1965).
23. R. H. Moser, *Diseases of Medical Progress* (Thomas, Springfield, Ill., ed. 2, 1964).
24. F. H. Happold, *Medicine at Risk* (Queen Anne Press, London, 1967), pp. 9-10.
25. W. Lenz, *Ann. N.Y. Acad. Sci.* 123, 228 (1965).
26. J. F. Webb, *Can. Med. Ass. J.* 89, 987 (1963).
27. M. M. Wintrobe, *J. Roy. Coll. Physicians London* 3, 99 (1969).
28. R. O. Wallerstein, P. K. Condit, C. K. Kasper, J. W. Brown, F. R. Morrison, *J. Amer. Med. Ass.* 208, 2045 (1969); K. M. Smick, P. K. Condit, R. L. Proctor, V. Sutcher, *J. Chronic Dis.* 17, 899 (1964).

29. E. P. Prendergrass, P. J. Hodes, R. L. Tondreau, C. C. Powell, E. D. Burdick *Acta Radiol. Suppl.* 116, 84 (1954).

30. M. P. Small, and J. F. Glenn, *J. Urol.* 99, 223, 1968; S. H. Macht, R. H. Williams, P. S. Lawrence, *Amer. J. Roentgenol.* 98, 79 (1966).

31. S. Kuramoto, M. Watanabe, K. Yoshimura, Y. Takamiya, Y. Ohnaka, T. Lee, K. Iwai, *Kurume Med. J.* 14, 95 (1967): E. K. Lang, *Acta Radiol. Diag.* 5, 296 (1966); T. Södermark and P. Mindus, *Acta Med. Scand.* 183, 177 (1968); J. D. Mortensen, *Circulation* 35, 1118 (1967); G. Haut and K. Amplatz, *Surgery* 63, 594 (1968).

32. J. McGuire and T. Chou, *Amer. Heart J.* 73, 293 (1967).

33. L. V. Perrett and W. J. Cladicott, *Austral. Radiol.* 11, 107 (1967); W. K. Haas, W. S. Fields, R. R. North, I. I. Kricheff, N. E. Chase, R. B. Bauer, *J. Amer. Med. Ass.* 203, 961 (1968).

34. E. Braunwald and J. H. C. Swan, *Circulation* 37, suppl. 4 (1968).

35. Anonymous, *Lancet* 1969-II, 34 (1969).

36. C. O. Carter, *Human Heredity* (Penguin Books, Harmondsworth, Middlesex, England, 1963).

37. K. Mather, *Human Diversity* (Oliver and Boyd, London, 1964).

38. Anonymous, *The Economist,* 10 May 1969, p. 19.

39. A. Keys, K. Brozek, A. Henschel, O. Michelsen, H. L. Taylor, *Biology of Human Starvation* (Univ. of Minnesota Press, Minneapolis, 1950).

40. A. B. Ford and H. K. Hellerstein, *Circulation* 20, 537 (1959).

41. J. N. Morris, unpublished data.

42. H. H. Marks, *Bull. N.Y. Acad. Med.* 36, 296 (1960).

43. J. Mayer, *Physiol. Rev.* 33, 472 (1953).

44. National Center for Chronic Disease Control, *Obesity and Health* (U.S. Department of Health, Education, and Welfare, Washington, D.C., 1966).

45. J. N. Morris, *Uses of Epidemiology* (Livingstone, Edinburgh, 1967), pp. 172-182.

46. Anonymous, *Time,* 6 June 1969, pp. 18-19.

47. W. J. Brown, in *Preventive Medicine,* D. W. Clark and B. MacMahon, Eds. (Churchill, London, 1967), pp. 525-538.

48. Chief Medical Officer, *On the State of the Public Health, 1947-1967* (H. M. Stationery Office, London, 1948-1968).

49. G. S. Wunderlich, *Characteristics of Residents in Institutions for the Aged and Chronically Ill,* PHS Publ. No. 1000, Ser. 12, No. 2 (U.S. Government Printing Office, Washington, D.C., 1965); C. A. Taube, *Characteristics of Patients in Mental Hospitals,* PHS Publ. No. 1000, Ser. 12, No. 3 (U.S. Government Printing Office, Washington, D.C., 1965); A. L. Jackson, *Prevalence of Selected Impairments,* PHS Publ. No. 1000, Ser. 10, No. 48 (U.S. Government Printing Office, Washington, D.C., 1968); C. S. Wilder, *Limitation of Activity and Mobility Due to Chronic Conditions,* PHS Publ. No. 1000, Ser. 10, No. 45 (U.S. Government Printing Office, Washington, D.C., 1968).

50. W. Lewin, *Brit. Med. J.* 1, 465 (1968); P. S. London, *Ann. Roy. Coll. Surg. Engl.* 41, 460 (1967).

51. J. Lorber, *Medical Officer* 19, 213 (1968).

52. U.S. Department of Health, Education and Welfare, Public Health Service, *Vital Statistics of the United States, 1940-1966* (U.S. Government Printing Office, Washington, D.C.).

53. Home Office, *Annual Criminal Statistics* (H.M. Stationery Office, London, 1945-67); reports by H.M. Government of the U.K. to the United Nations.

54. I thank Professor J. N. Morris, Professor of Public Health at the London School of Hygiene and Tropical Medicine, for guidance and assistance. The report was written in his department during my sabbatical leave, with financial support from the United States Public Health Service and the Milbank Memorial Fund.

The Consumer
Marge Piercy

My eyes catch and stick
as I wade in bellysoft heat.
Tree of miniature chocolates filled with li-
queurs,
trees of earrings tinkling in the mink wind,
of Bach oratorios spinning light at 33 1/3,
tree of Thailand silks murmuring changes.
Pluck, eat and grow heavy.
Choose and buy and transcend.
Your taste defines you.
From each hair a wine bottle dangles.
A toaster is strung through my nose.
An elevator is installed in my spine.
The mouth of the empire
eats onward through the apple of all.
Armies of brown men
are roasting into coffee beans,
are melted into chocolate,
are pounded into copper.
Their blood is refined into oil,
black river oozing rainbows
of affluence.
Their bodies shrink
into grains of rice.
I have lost my knees.
I am the soft mouth of the caterpillar.
Men and landscapes are my food
and I grow fat and blind.

UPI

Metabolites

Spokesmen for the New Left have characterized concern with pollution as an establishment plot to divert concern from problems of discrimination, war and poverty. The Daughters of the American Revolution believe concern with pollution is a communist plot to subvert free enterprise. Irrespective of the political ends to which it may be put, pollution represents the accumulation of metabolites in the human culture tube.

The analogy is informative, rather than contemptuous. Earth is the culture vessel for man, and both local and worldwide pollution problems represent excess of waste products, by-products, spillovers and pheromones of the growing human population. If the planetary ecosystem represents a most complex medium, the metabolites of technological man represent a fantastically varied mix. The only place where the analogy really breaks down is on the question of whether the culture was especially intended for *Homo sapiens* or, for that matter, whether the whole thing involves any intent whatsoever.

In simple cultures population growth ceases, and population decline and extinction frequently result from the absence of recycling mechanisms and consequent accumulation of toxic metabolites. In more complex cases it may be a matter of the balancing mechanisms breaking down or proving inadequate. Earth (the biosphere) is the most complex culture known, and an anthropocentric view is to some extent justified because the course of events is greatly influenced by human attitudes.

Human attitudes commonly involve some delusions which contribute greatly to the pollution problem. One such attitude is the faith that the biosphere (or at least its capacity to dilute) is infinitely large. On this assumption rests the belief that human population, and per capita consumption (with equivalent production of biological and technological metabolites) can grow indefinitely. Another contributory delusion is belief in the existence of technological solutions to all problems, with the consequence that no trend in the environment can be irreversible

if we want to reverse it badly enough. A third delusion, to which our economic and political institutions remain true despite defection of a few of the sciences, is that the earth really was *intended* for man. This justifies the policy of giving economic and political concerns precedence over concern with environmental deterioration.

Unfortunately, stubborn delusions are able to persist side by side with contrary realities. Significant pollution is no longer restricted to areas in and around, or even downstream from human feed lots (cities). Pollution-wise, if not politics-wise, "one world" status has been reached for a number of pollutants including radionuclides, dust, CO_2, persistent pesticides, lead, ice and cloud nuclei, and polychlorinated biphenyls. Not that their distribution is uniform. Even the processes of physical distribution are complex and outside our present knowledge on both local and worldwide scales, and many are subject to biological concentration and transmission. Pollutants commonly crop up in unexpected concentrations at unexpected places.

Our concern with the problem involves some contradictions more weighty than those expressed by the New Left and the DAR. In much of the world, the machinery for translating human needs into action involves profit as a motivating agent. Thus the manufacturers of our competition pheromones (pesticides) piously claim that their products derive justification from human hunger pangs, when the proximate motive is actually economic gain. For the manufacturers, the real value of the humanitarian stance is that it is useful in public relations. That is, the humanitarian argument aids in obstructing efforts, even those by governments, to restrict the flow of pesticides into the ecosystem.

The profit motive may contribute to our strange system of prescribing pesticide use (recommendations are largely made by salesmen who work on a commission basis). Should we really discriminate against medical doctors by denying them commissions on their prescriptions?

Again, we indulge in a surprising degree of "idealism" with respect to release of substances of all sorts into the biosphere. Any chemical thus released is considered innocent of ill effects until proven guilty. Along the same lines, the courts of law often refuse to consider claims for pollution damage unless *economic loss* can be demonstrated. Noise, for example, is regarded as a mere irritant which should be tolerated in the interests of the economy. Noise may interfere with the learning process in schools, may cause emotional or physical disease, may deaden our hearing and our joy in living, yet the average farm tractor continues to produce one hundred and seven decibels, we fail to enforce anti-noise ordinances, and we rush ahead with development of the supersonic transport, even on the weakest of economic grounds.

The pesticide case is a telling example of the way in which economic forces clash with environmental sanity. Warnings on pesticide pollution were sounded more than 20 years ago, yet between 1948 and 1967 worldwide use of pesticides, including the most persistent and toxic, increased sevenfold. In 1970 we are

UPI

allowing manufacturers of chlorinated hydrocarbons to block restrictive regulations, despite the already visible evidence of worldwide disaster, and the knowledge that the effective peak of biospheric contamination from material already
released will not be reached for another 15 to 75 years even if further manufacture
is stopped immediately.

If our economic system is one pillar of support for pollution, another is
ignorance and the tendency to underestimate the complexity of ecosystems. We
do know that some environments and some species are more sensitive than others.
We know that the responses of whole ecosystems to particular pollutants, or
combinations of pollutants, are not predictable on the basis of laboratory studies
of a few species. We are, in short, able to make no firm predictions in specific
cases. This ignorance has yet to produce much humility or caution. One reason
for this is compartmentalization in the education of those who influence our
destinies. We tend to educate scientists as specialists, and politicians as scientific
ignoramuses. Ecologically the two are about equivalent. We really have no idea
what the eventual outcome of the DDT episode will be, partly because neither
scientists, nor politicians, can accomplish the necessary synthesis of information
and action.

Allow me a second example of the evils of narrow expertise. A recent analysis
of the problem of sulfur dioxide pollution by generating plants concluded that
palliative measures might cost between 500 million and two billion dollars yearly
through the year 2000, and ended on the cheery note that by that time the
problem would be solved the fossil-fueled generators would be replaced by nuclear
power plants. This is a happy solution from the point of view of a middle-aged
fossil-fuels expert. It is less happy when one is younger and reads that the
commitment to a crash program of conversion to nuclear plants has been made,
even though the problems of waste disposal, environmental contamination during
normal operation, and prevention of major nuclear accidents, have not been
satisfactorily solved and may prove incapable of satisfactory solution!

There is one absolutely essential point with regard to the pollution question.
Pollution is part of the environmental cost of maintaining the present human
population and supporting its growth in numbers and consumption of resources.
There may be technological palliatives for most or all pollution problems, but we
are learning that each such technological "solution" carries its own added cost
in some alternate environmental damage, as well as in dollars. Are we doomed
to blunder ever deeper into the technological morass when the real question is
simply whether we choose to stabilize the human population now instead of later?

Pollution is a subject on which the interest and concern of the public is easily
aroused. The evidence to date suggests, however, that the public is easily pacified.
Even in the absence of particularly good placebos, public concern is not long
sustained where the issues are complex and the problems are not personally acute.
We seem to learn little from past errors, and economics still rules over ecology
in the halls of government. Such pollution legislation as is passed is often delayed,

unenforced, or evaded. The international nature of pollution has yet to produce significant international cooperation. These are not good omens for a happy solution of the much more intransigent, and inclusive, problem of too many individuals in the culture tube.

The Smell of Cologne
Samuel Taylor Coleridge (1772-1834)

In Koln, a town of monks and bones,
And pavements fanged with murderous stones,
And rags and hags, and hideous wenches,
I counted two and seventy stenches,
All well defined, and several stinks!
Ye nymphs that reign o'er sewers and sinks,
The river Rhine, it is well known,
Doth wash your city of Cologne;
But tell me, Nymphs, what power divine
Shall henceforth wash the river Rhine?

Radioactivity and Fallout: The Model Pollution
George M. Woodwell

Ecologists are so thoroughly accustomed to playing Cassandra, predicting doom, that we hardly give credit for even major postponements of the day of doom. Doom has so many components these days that it is difficult to sort out which one is more important at any moment: the crisis of the dollar, the war, the cities, the pollution, the population, or the fact that the students are going to pot. What is encouraging and central to a symposium entitled "Challenge for Survival" is the fact that there is now a broad and growing consensus that accepts the simple truth that the size of the human population is a key and that the ultimate amelioration of these crises depends on establishing some sort of equilibrium between population and resources. Establishing

any such equilibrium is difficult, almost hopelessly so, not only because of the difficulties of controlling the numbers of people but also because a burgeoning, and in some ways, malignant technology both increases certain types of resources and simultaneously destroys other essential ones. Thus an expanding technology offers ever cheaper power and transportation but also threatens to degrade the environment in diverse ways, one of the most important being with wastes that are biologically active.

My objective is to summarize the broader scientific aspects of the problems associated with release of certain types of toxic wastes into the environment. Atomic energy provides a model, unfinished, rough-hewn in many of its parts, polished in others, but overall a brilliant example of the marshalling of scientific and political talent on an international scale to mitigate a world-wide pollution problem. That problem has not been solved, but there is hope that it can be and there are many lessons from it. I propose to examine the radiation-pollution problem and to show how it can be used to contribute to a solution of certain other analogous pollution problems, especially those we have now with pesticides, which appear to be the world's most dangerous pollutants.

First, the attitudes that allowed world-wide contamination of the earth with radioactivity and which now allow other even more serious pollutions are important and still dominant, although weakening. The most important assumption that led to the problems with radioactivity is the assumption of dilution. Toxic materials released into the environment are widely assumed to be diluted to innocuousness. If there are local effects from the toxicity, they are transient; an abundant and vigorous nature repairing any damage within at most a few years—or the effects are accepted as small cost for technological progress. The assumption is practical, and as long as the environment is very large in comparison with the quantity of the release, "dilution" appears to occur. At least the material "disappears."

So thoroughly ingrained is this philosophy that its corollary, the right to pollute, has become a second major philosophical and, for somewhat different reasons, legal assumption: we tend to require detailed scientific proof of direct, personal damage to man as a prerequisite for even considering restriction of any right to pollute.

On the basis of these two assumptions—dilution and the assumption of the right to pollute until proof of damage—we, man all over the world, are now well embarked on a program of releasing unmeasured quantities of different kinds of biologically active substances into the general environment. It is one of the spectacular contradictions of our time that in the age of science we should be entering blindly on a thousand, unplanned, uncontrolled, unmonitored, unguided, largely unrestrained, and totally unscientific experiments with the whole world as the subject and survival at hazard.

The assumption of dilution, so easy to make, so cheap, so comforting, so much a part of human nature, is a trap. Biologically active materials released into the biosphere travel in patterns that are surprisingly well known. A major contribution of atomic energy has been definition of these patterns, using as tracers the radioactivity in fallout from bomb tests. Intensive studies began in earnest only after the series of bomb tests in the Pacific in the spring of 1954 known as "Castle." This was the series, some of you will recall, that included the test that dropped fallout on Rongelap Atoll, exposing its inhabitants, about 65, to an estimated 175 r (500 r is widely accepted as the mean lethal exposure for man). A Japanese fishing vessel, the *Lucky Dragon,* and its crew were also caught in the fallout; and for several months tuna caught in the Pacific and landed in Japan were sufficiently radioactive that authorities would not allow them to be sold. In the eyes of the world this series of misfortunes was "proof of damage" and frightening proof of a new-found capacity to degrade the environment in places far removed from those affected directly by the blast and the ionizing radiation accompanying the bomb. Some of the radioactivity from these tests is, of course, still circulating through the biosphere. It was this series of frightening events, publicized widely at the time and pursued doggedly over the years by many knowledgeable and deeply concerned scientists, that triggered sufficient public interest in the problems of world-wide pollution with radioactivity to mount a really significant international research program tracing these patterns (UNSCEAR, 1958,

1962, 1964) and that ultimately resulted in the treaty among the more civilized nations banning nuclear tests in the atmosphere.

From the research on man-made radioactivity we have learned these things, all relevant and even basic to the greater problems of pollution that we are just now beginning to recognize as a major "challenge to survival:"

First, particulate matter, introduced into the lower atmosphere, enters air currents that move around the world in periods of 15 to 25 days in the middle latitudes, sometimes less.

Second, the half-time of residence (time for one-half of the material to be removed) of particulate matter carried in these currents ranges between a few days and a month, the same general range as the time to travel around the world. Thus, it is no surprise these days to measure the radioactive cloud produced by tests in central Asia on several successive trips around the world.

Third, such material tends to be removed from air and deposited on the ground by precipitation, more in the early precipitation in any storm than later.

Fourth, the patterns apply to any particulate matter entering the air currents of the troposphere. Some fraction of the pollen, for instance, that is released by plants close to the ground surface enters these patterns and is transported in air and deposited in precipitation (Gatz and Dingle, 1965).

Hardly less important, we know that certain radionuclides are accumulated from the environment into the tissues of plants and animals, where they may be concentrated in high degree, concentration factors of a hundred to a thousand-fold and higher being common. But the patterns of movement of radionuclides through ecological systems are not capable of simple generalization: each radionuclide travels its own peculiar path. Thus ^{137}Cs behaves in ways that are similar to K and tends to be accumulated in muscle with greatest accumulations in older organisms and in carnivores. The accumulation of ^{137}Cs in Eskimos through the lichen-caribou food chain is well known. However, ^{90}Sr is accumulated in bones, and herbivores get most of it; ^{137}I, in thyroids; ^{55}Fe, in blood and elsewhere. Discovery of these patterns has required intensive research on each substance, both to trace its movement through the patterns of air and water circulation and to discover its pathways through living systems (Åberg and Hungate, 1967; Polikarpov, 1966).

But this is only part of the story. It documents the fact that dilution into an infinite environment is not a safe assumption, but what of the effects? Again the story is not a simple one, each substance presenting its own peculiar set of hazards. With ionizing radiation the problem seems reasonably straightforward although it has never seemed very straightforward to those charged with developing standards of safety. There is general agreement that the principal hazard is a direct hazard to man through damage to the genetic material, "mutations," which are, for the practical purposes of this discussion, all deleterious. Man is most vulnerable because radiation causes an increase in the frequency of deleterious mutations, in the jargon, an increase in the "genetic load," by adding to the numbers of genetically determined unfortunates. If man is protected from this hazard, levels of man-made radiation in nature will almost certainly be so low as to have no significant effects on other organisms, because in these species, unlike man, genetic unfortunates are removed by selection. Increasing mutation rates will probably not increase rates of evolution as so often assumed. Thus the hazard of ionizing radiation, released from whatever source, is first a direct hazard to man; this is not so for many other toxic substances as we shall see in a moment.

But again, there is no simple answer to the question of how much radiation is "safe." It would be very convenient if there were some threshold below which ionizing radiation has no effect. A considerable weight of evidence suggests that there is no threshold for production of mutations; even very low exposures increase mutation frequencies slightly. On the other hand, there is clear benefit from radiation exposures in medicine, and it is not possible to enjoy the benefits of nuclear power without some small increase in radiation exposure for workers in the plant. We must arrive at a compromise in exposing people directly, limiting direct exposures around nuclear plants systematically and vigorously to those well below levels that increase genetic hazards appreciably. But the problems raised by the cycling of radioactive wastes are different. It is more difficult to anticipate exposure, more difficult to make even a reasonable guess as to how much of any nuclide will appear in human tissues. And due to the movement of air and water around the world, the problem for long-lived materials suddenly becomes not merely a local one, confined to the Hudson River Estuary, the Columbia River, the Irish Sea, or the Bay of Biscay, all water bodies now receiving appreciable radioactive wastes, but a world-wide one. Each increment of waste enters a world-wide pool of that material. Once we recognize that many pollution problems are world problems, not local ones, then we can approach them systematically estimating the totals that we are willing to have in the biosphere at one time. Radioactive materials decay at some constant rate, often defined as "half-life," the time for half the activity to be lost. Thus there is a constant rate of removal of any radioactive substance from the environment due simply to physical decay. If we know that a quantity (A) of a substance is released into the environment regularly and that at the end of time (t) the fraction $\frac{A_t}{A_o}$ = R remains, then the amount

(S) present after n units of time will be:

$$S = A \frac{(1 - R^n)}{1 - r}$$

R is related to half-life ($t_{1/2}$) by:

$$R = e^{-0.693 \frac{t}{t_{1/2}}}$$

where t is the period for which the retention rate is derived and t and $t_{1/2}$ are in the same units. The equilibrium concentration will be given by assuming that $n \rightarrow \infty$, when

$$S = \frac{A}{1-R}$$

Thus the equilibrium concentration of something with a half-life in the range of 1 year will be about twice the annual release. If the half-life is 10 years, the equilibrium will be about 15 times the annual release. A material with a half-life of 30 years, such as ^{137}Cs, will achieve an equilibrium in the environment that is about 50 times the annual release. If the substance is sequestered in forms that make it unavailable for circulation through living systems, then the equilibrium calculated on the basis of physical half-life will simply state the maximum that could be circulating. The actual amount will depend on the efficiency of the mechanism that sequesters the radioactivity. For example, ^{137}Cs circulates freely in biological systems, but tends to be fixed in certain micaceous clays and effectively removed from further circulation. The rate of fixation in clays, however, is not easily estimated.

What is clear is that there is need to relate the input of toxic wastes to their world-wide equilibriums and these equilibriums to the changes such concentrations will induce in the biosphere. With ionizing radiation man is more vulnerable than his environment. The most serious radiation hazards will ultimately arise from gases such as tritium and ^{85}Kr

which are long-lived, produced in appreciable quantities, and are difficult to contain. The problems may be aggravated as atomic energy becomes more widely used by industry because the interests of long-term safety and industrial profits are not always coincident. And there will be other proposals such as the one to build a new Isthmian canal with nuclear explosives; and there will be continuing military pressure to test bombs in the atmosphere again. There is much more evidence, however, to suggest that we know how to control these problems than there is that we recognize and can control other analogous ones.

There are probably large quantities of many different kinds of metabolites of civilization circulating in the biosphere right now, substances that are released in large quantities, are biologically active, have long half-lives, and therefore present problems that are analogous to those presented by ionizing radiation. The persistent pesticides, however, seem to be by far the greatest problem, but not strangely enough, by poisoning man. With pesticides there is a hazard to man, but we have taken pains to insulate man's food chain by elaborate regulations. By and large these regulations are effective, although not always *(New York Times,* Sunday, 7 April 1968). What has happened is that we have at least temporarily protected man, but we have allowed virtually unlimited uses of long-lived pesticides in any place that will not contaminate human food produced in agriculture. The result has been the accumulation in the biosphere of concentrations of persistent pesticides, especially DDT, that are quite clearly degrading ecological systems all over the world. The extent of the changes are far from clear. What is clear is that certain carnivorous and scavenging birds the world over are suffering rapid reductions in reproductive capacity that seem clearly related to pesticide burdens. This ap-

plies even to oceanic birds such as the Bermuda petrel that do not ever come into contact with man or with sprayed areas (Wurster and Wingate, 1968). But there is ample reason to believe that many groups in addition to birds are effected, including oceanic fisheries and perhaps even phytoplankton, the basis of all oceanic food chains (Wurster, 1968). The trend, if allowed to continue, will follow the pattern set by eutrophication of Lake Erie and numerous other smaller lakes, now being followed rapidly by Lake Michigan. While we may be able to afford to lose lakes in this way, we cannot afford to lose the oceans.

It is hardly surprising that we have an acute problem with pesticides when one recognizes that probably very close to 100% of the world production of DDT is distributed in places where it can move freely through the various cycles of the biosphere. There is good evidence to support the assumption that the chemical degradation of pesticides follows the same pattern as radioactive decay: the amount decaying is proportional to the concentration (Hamaker, 1966). This means that we can use the same arithmetic to estimate the total quantity that will be circulating in the biosphere at equilibrium if we know the half-life and the rate of release.

The half-life of pesticides in nature is not easily estimated. Studies of agricultural soils indicate a half-time for disappearance of residues of DDT in the range of 2-4 years. But undoubtedly this reflects several types of losses including erosion, vaporization, co-distillation with water, and leaching into the water table as well as chemical degradation. When organic soils are present, residues tend to remain for many years (Woodwell and Martin, 1964; and others). An estimate of half-life of 10 years for DDT residues in biological systems seems minimum; their persistence may be considerably longer.

The annual production of pesticides in the world appears not to be tabulated. The U.S. Tariff Commission reports U.S. production of DDT. Between 1957 and 1967 production ranged between 99 and 179 million pounds, with the highest in 1963. In 1967 production was about 103 million pounds. The mean for the 11-year period was about 147 million pounds. If U.S. production is 75% of world production, then we might assume with some justification that the world equilibrium would be based on an annual release of 200 million lb. into the biosphere. The total amount of DDT residues circulating in the biosphere would then be about 3 billion lb., but this equilibrium would be approached only after 75 years. If we had used DDT at this high rate since 1946, we would now have about 1.5 billion lb. in the biosphere or about one-half the total we would have if the residues came to equilibrium under these conditions. Thus we can expect far greater changes in the world's biota than we have seen so far if we continue using these long-lived pesticides.

My colleagues will be quick to point out that DDT production in the United States has dropped in the past year, and there is some evidence of a downward trend in a period when total use of pesticides is increasing abruptly. With more than a billion pounds of DDT now cycling in the biosphere, and with abundant evidence of effects, there can hardly be a downward trend in use that is steep enough to avoid irreparable changes in the earth's biota, changes that can only be deleterious to our long-term interests in survival.

These observations simply point once more to the fact that the world is now small and we must tidy it up if we intend to continue using it for very long. The history of ionizing radiation is an example of a pollution whose hazards we have appraised and whose sources we have controlled. Pesticides are an example of the opposite: failure to appraise the hazards; and now so much political and industrial power supporting the status quo that we may not be able to control it before we have lost an important fraction of the earth's biota, driving the whole earth a significant step down the path that Lake Erie has followed.

To return to my original point, survival demands control of population, but it also demands some much more stringent and unpopular limitations on technology. It is true that there is some ultimate limit on the amount of power that can be produced by atomic energy without exposing man to unacceptable levels of radioactivity. It is also true that there is a limit on the amount of persistent pesticides we can tolerate circulating through the biosphere. Testing these limits as we are doing at the moment is a strange and dangerous game, suggesting that we have not yet learned that the *Challenge* is for *Survival.*

REFERENCES

Aberg, B., and F. P. Hungate (eds.). *Radioecological Concentration Processes* (Proc. Intern. Symp., Stockholm, 25-29 April 1966), Pergamon Press, London, 1040 p.

Gatz, D. F., and A. N. Dingle. 1965. Air cleansing by convective storms. In: *Radioactive Fallout from Nuclear Weapons Tests,* A. W. Klement, Jr. (ed.), Div. Tech. Information, Oak Ridge, Tenn., p. 566-581.

Hamaker, J. W. 1966. Mathematical prediction of cumulative levels of pesticides in soil. In: *Organic Pesticides in the Environment,* Advances in Chemistry Series, No. 60, R. F. Gould (ed.), American Chemical Society, Washington, D.C., p. 122-145.

New York Times. Sunday, 7 April 1968. Montana Dairies Shut in Poisoning. Polikarpov, G. C. 1966. *Radioecology of Aquatic Organisms* (translated from the Russian by Scripta Technica, Ltd.), English translation, V. Schultz and A. W. Klement, Jr. (eds.), Reinhold, New York, 314 p.

Report of the United Nations Scientific Committee on the Effects of Atomic Radiation. 1958. General Assembly, Official Records: 13th Session, Suppl. No. 17.

Report of the United Nations Scientific Committee on the Effects of Atomic Radiation. 1962. General Assembly, Official Records: 17th Session, Suppl. No. 16.

Report of the United Nations Scientific Committee on the Effects of Atomic Radiation. 1964. General Assembly, Official Records: 19th Session, Suppl. No. 14.

Woodwell, G. M. and F. T. Martin. 1964. Persistence of DDT in soils of heavily sprayed forest stands. *Science,* 145: 481-483.

Wurster, C. F. 1968. DDT reduces photosynthesis by marine phytoplankton. *Science,* 159: 1474-1475.

Wurster, C. F. and D. B. Wingate. 1968. DDT residues and declining reproduction in the Bermuda petrel. *Science,* 159: 979-981.

Pesticide Pollution
Clarence Cottam

Like a spaceship the planet Earth has a finite air supply, vast and subject to being refreshed and recycled. Even man's immense use of oxygen over the past highly industrialized century has made only a dent in the available total; the recycling system has been able to produce almost as much elemental oxygen as man has used.

As any good spaceship must be, Earth has also been self-sufficient in many ways. Working solely with light and heat from the sun, its system has provided a renewable food and shelter supply for a pyramid of living organisms, at the top of which man has elected to place himself. But the ship has been sabotaged.

In 1874 a German chemistry student named Othmar Zeidler unwittingly dropped a rock into the planet's intricate living machinery. The rock rattled around harmlessly for 65 years. Then it jammed and began to chew up the gears, and today our spaceship's life support systems are showing signs of breaking down. Under heavy stress in any case from other human activities, living organisms are disappearing in droves, leaving great gaps in the pyramid. Some of these gaps serve, as far as we know, "only" to impoverish the human mind and spirit; others threaten the entire biological structure.

Perhaps the least published effect of Zeid-ler's discovery—now known the world over by the acronym DDT—will be the most damaging in the long run. Scattered scientific reports seem to indicate that the air-freshening system of the Earth, consisting mainly of green plant plankton in the oceans, is breaking down. Although once oxygen production from the oceans was almost able to keep up with oxygen consumption by man, now there are strong signs that production is declining even more under the influence of DDT and a half dozen much more poisonous relatives.

By far the majority of synthetic organic chemicals was synthesized more in an effort to learn the mechanics of organic reactions than to develop a new substance. So it was with Zeidler's synthetic. Zeidler learned how to make DDT, described its basic properties, then put it on a dusty back shelf along with other useless curiosities.

In 1939 the Swiss chemist Paul Mueller was taking part in a concerted effort to find an insecticide that could halt the ravages of the imported Colorado potato beetle. Zeidler's creation was one of thousands of chemicals tested for insect-killing properties. Mueller found it potent. Either on contact or after ingestion, minute quantities scrambled an insect's nerve endings and caused death. The small amounts that could be expected to reach a human through a treated crop proved unable to strike the human dead on the spot; and as far as the researchers were concerned, these two considerations meant that their hunt for an ideal pesticide was successful.

Shortly after Mueller's discovery the U.S. armed forces began to use the chemical to control insect disease vectors, and the popularity of DDT was assured. By 1948, when Mueller was awarded the Nobel Prize for his work, one of mankind's greatest assaults on his environment was well under way.

At first satisfaction with DDT knew no bounds. During World War II it generated wild enthusiasm. It controlled malaria-carrying mosquitoes, typhus-carrying lice, and the arthropod vectors of some two dozen other diseases. It has been extravagantly estimated that the use of DDT in this war averted about 10 million deaths and 200 million cases of pest-borne illness. To many people after the war, DDT seemed the final answer to control of agricultural pests. It was cheap, effective in microscopic amounts, and, most important of all, it was persistent.

No one knows precisely how long DDT takes to break down in nature, but it has been estimated that it has a half-life of at least 15 years. This means that half a given application will have broken down in 15 years, half the remainder in another 15 years, and so on. One application indeed goes a long way—just what a tragically long way should have been realized, but no one wanted to be a Jeremiah. With the mix of childish enthusiasm and arrogance typical of the technological revolution in America, DDT by the thousand tons was spewed over the surface of our planet.

Few bothered to ask what effect a chemical with such demonstrated biological potency could have on living organisms other than insects, or considered that not all insects harm man—that some, in fact, are beneficial. If such

simple questions were so poorly considered, it could hardly be expected that DDT's effect on the "balance of nature" would be seriously weighed. The fledgling science of ecology was then only beginning to publish the idea that no member of a community of life, or ecosystem, is independent of the others; that no matter how useless or annoying the smallest creature might seem, it has a function in the planet's system; that before eliminating any organism, therefore, it might be wise to consider what that function is. Meanwhile, DDT in dust, spray, and fog continued to blanket the land. If in a year or so an application washed into the soil out of range of target insects or ran off into drainage ditches, more DDT was applied. If mosquitoes bothered suburban taxpayers, out came a fogging truck to spread DDT. By the time the first small sign of trouble appeared in 1949, DDT had been applied to every crop from tomatoes to timber and was treated around people's homes almost as carelessly as oatmeal. Then someone noticed DDT in milk.

Now, inasmuch as milk involves children, it constitutes a sensitive part of the American political anatomy. So people paid some attention to this small fact. How did DDT, an insecticide, get into cows? In a demonstration of several of its more obvious faults, traces of DDT dust and spray intended for fields under crops had drifted to neighboring pastures. Dust from croplands blown into pastures carried more DDT. There its persistence allowed it to accumulate on alfalfa, and cows ate the alfalfa. DDT is practically insoluble in water, but it is soluble in various hydrocarbons and in animal fat. Traces of DDT, not detectable on the alfalfa, are concentrated by the cow in fatty tissues and, of course, in the butterfat of its milk.

After a few such early instances, it was decided that perhaps DDT should be used more carefully—"in moderation on crops, only in amounts sufficient to get the job done" as one

UPI

agricultural bulletin advised in 1950. No one suggested not using it at all.

In the two decades since, while bits of evidence have shown DDT to be a global pollutant, production of DDT in the United States has grown to an industry with a sales peak in 1962 of nearly $30 million, declining to about $20 million in 1969. Chemists, following the lead of DDT's molecular structure, went even farther by developing a whole class of DDT-like compounds called chlorinated hydrocarbons, many of which are much more poisonous than DDT. All these poisons pose substantially the same problems as DDT even if they are not yet used on the same scale.

The evidence was assembled for the public by the late Rachel Carson in her bestseller *Silent Spring,* published in 1962. This book, which in the light of more recent evidence and discussion seems a model of reason and restraint, sparked hysterical denunciations of the author, her scientific credentials, conclusions, and even writing style. Her basic message was: Man's assault on the living environment with chemicals, principally DDT but including a large supporting cast, had succeeded in poisoning the environment for higher creatures such as birds, fish, and mammals (presumably including man). This assault had succeeded in tearing askew the delicate ecosystems of the earth that ultimately keep man alive. But—a cruel irony—it was on the verge of failing utterly to control insect pests.

Complex higher animals have generation times ranging up to man's 20-odd years and generally produce few offspring per female. So the trial and error of evolution can proceed only slowly. Insects, on the other hand, may have generation times of only a few days or weeks, each female producing huge numbers of offspring. Consequently, if 99 percent of a population of flies is killed by DDT, the 1 percent surviving because of a slight inborn resistance can produce a new crop of flies very quickly. The new crop, all stemming from resistant flies, may have up to 2 percent able to survive a DDT application. And so it goes—each generation a bit more resistant. By now scientists have been able to produce houseflies in the laboratory that can stand wetting down with a concentrated DDT solution!

Since the attacks on Miss Carson 7 years ago, action on a wide front has vindicated her. Michigan has banned the use of DDT except in a few minor circumstances. Wisconsin has placed stringent limitations on its use. Many states have restricted it and probably will ban it. Abroad, Australia has banned its use on pastures because Australian meat was not meeting residue requirements of the U.S. Food and Drug Administration. Sweden, which had given Mueller the Nobel Prize, has banned it.

It is too late to avoid serious consequences, however, and no one really knows what the final biological outcome will be. Enough DDT already has been released to play all sorts of havoc, yet there is even more lying in the soil to eventually seep into water supplies; therefore, even if the use of DDT were to halt tomorrow, concentrations of the poison in the water of the world would continue to rise before falling.

The various properties of DDT (and the chlorinated hydrocarbons in general tend to follow the pattern) may be summarized as follows:

*DDT is among the most pharmacologically active substances known. In many cases it behaves like hormones, those trigger chemicals secreted by organisms in tiny amounts to control their bodily functions.

*It is soluble in animal fat but not in water. Most of an organism's fluids are aqueous; only its fatty tissues can dissolve and retain DDT, so DDT tends to be deposited in these tissues.

*It is chemically stable. Chemical and biological forces that may encounter DDT in the environment break it down only with difficulty, and it breaks down spontaneously only very slowly.

*It is not easily metabolized in the systems of animals (an extension effect of its stability and its insolubility in most body fluids). Thus, once deposited in fat, it tends to stay there (half-life 15 years!). With continued ingestion of polluted food by an animal, DDT accumulates to ever higher levels. This leads to the phenomenon of biological magnification, discussed later.

*It is nonspecific. This potent substance hits friend as well as foe, often wiping out a predatory insect that has been able to control a pest better than DDT ever could. Consequently the target pest's population rises instead of drops. Even DDT's side effects share the shotgun approach, striking not at species or genera but whole orders of creatures.

The biological potency of the chlorinated hydrocarbons means that, whereas a particular dose may not poison an organism outright, it is capable of altering living systems in several harmful ways. Early signs indicate that chlorinated hydrocarbons, including DDT, can cause cancer and mutations. Autopsies of people dead of blood and liver diseases have shown higher than normal DDT residues. However, as the pesticide industry never fails to point out, no human deaths have been proven to be due to the direct effects of DDT under prescribed controls. But "prove it!" seems the most childish of all possible arguments in the face of a possible link.

The chemical's potency, its affinity for fat, its ability to accumulate, and its persistence all combine to suggest that there is no safety in tolerance levels set by the U.S. Food and Drug Administration unless these levels are zero. There is no such thing as a sensible degree of poisoning.

One fallacy in FDA tolerance levels is due to the fact that the effects of the chlorinated hydrocarbons may be additive. Tolerances for each are set, but a farmer often uses more than one chemical on a particular crop. Thus a vegetable may carry half the legal limit of four different chlorinated hydrocarbons—say, DDT, dieldrin, chlordane, and lindane for a hypothetical example. Animal tests have demonstrated that the effect is roughly like having twice the legal limit of any one chemical on the product. When a diet containing theoretically harmless levels of 2 parts per million each of DDT, lindane, toxaphene, chlordane, and methoxychlor—a total of 10 parts per million of chlorinated hydrocarbons—was fed to laboratory rats, liver damage resulted.

The phenomenon of biological magnification underlies another fallacy of "acceptable" tolerance levels. Americans who eat normal amounts of the normal American high-fat diet produced from farms using normal practices of insect control ingest "normal" levels of DDT in their food. These levels are not permitted to exceed the FDA's tolerances, so the food is "safe." Yet DDT accumulates in fatty tissues, and mother's milk, for example, is made from fatty tissue. A nursing baby receives a concentrate of the DDT its mother ate and thus ingests even higher levels. The FDA would consider mother's milk unsafe for human consumption were it a marketable item, so high is its content of DDT.

A clear illustration of biological magnification comes from the Green Bay area of DDT-laced Lake Michigan. There, the bottom muds were found to contain 0.014 parts per million of DDT. But tiny crustacea living in and on the mud absorbed and concentrated DDT in their bodies to a level of 0.41 ppm. Fish eating the polluted crustacea in large volumes accumulated 3 to 6 ppm of DDT in their bodies. Herring gulls eating large amounts of these fish accumulated the DDT from their diet until it reached *99 parts per million.* This amount is enough to kill them during times of food scarcity when they are forced to draw on their own poisonous fat reserves.

Biological magnification has economic effects, too, although other effects are much more serious. In the 1930's the people around the upper Great Lakes ate fish—beautiful big lake trout, whitefish, and other species; and commercial fishing was their livelihood in some places. Then sea lampreys and alewives appeared, let in around formerly impassible Niagara Falls by the Welland Canal, an ecological blunder in its own right. The parasitic lampreys wiped out the whitefish and trout industry. The alewives' population exploded to fill the void, nothing checking their increase but the physical size of the lake. Killed finally in millions by overcrowding, stinking windrows of alewives rimmed the lake as a memorial to man's meddling.

Years went by, a way to deal with the lamprey was found, and the lakes were safe again for commercially valuable fish and sport fish. Lake trout, coho salmon, and chinook salmon were stocked. As expected, the stocked fish found the environment empty of predators and competitors and full of prey in the form of alewives. All three species did well, but the coho waxed incredibly fat. The Lake Michigan fishery was reborn. It was estimated that the influx of sport fishermen alone would net the state $100 million a year. Released as fingerlings in the spring of 1966, the coho had grown to 15 or 20 pounds by the fall of 1967. Anglers went on something like a piscatorial gold rush. State officials dreamed rosy dreams. The 1968 fall season was anticipated with the eagerness usually reserved for Christmas by lakeshore dwellers, who were eating well again.

But DDT got there first. The FDA, finding DDT levels as high as 20 parts per million in coho harvested from the lake by commercial fishermen, seized the fish and forbade the sale of Lake Michigan coho. Few anglers wanted to go after fish the FDA said were poisonous. If the FDA's interim tolerance level of 5 ppm becomes permanent, it will eliminate the sale of about 80 percent of the commercial coho catch and severely limit sport fishing. Moreover, fingerlings apparently are killed by 20 ppm of DDT in their bodies, so natural spawning of all three stocked fish probably will be prevented until DDT levels in the lake fall considerably. With DDT's 15-year half-life, it could take 35 to 40 years before the residues in the fish get down to 5 ppm. The coho, chinook, and lake trout are at the top of a food chain that passes through the alewives from the microorganisms in the mud of the lake. DDT in the mud is magnified as it passes up the chain.

The principal arguments against DDT and its relatives must rest on their drastic alteration of the living environment. These alterations are subtle but profound and in the long run may threaten man's survival. p 134

One of the most dangerous areas of concern is the effect of DDT on oxygen production. Marine biologists subjected five common species of oceanic plant plankton to roughly the DDT levels found in the surface waters of the open ocean. Addition of the DDT was followed by a severe reduction in the rates of photosynthesis and cell division in all species. In one case the reduction in photosynthesis was more than 50 percent. Photosynthesis is the process by which green plants convert carbon dioxide, water, and the energy of sunlight into food for their own growth. Oxygen is a byproduct of this process. Between 50 and 70 percent of the Earth's oxygen is produced at sea by phytoplankton (plant plankton). Even without considering DDT, oxygen production is lagging behind oxygen consumption because of man's vast demands. If the effect found in the laboratory is widespread—and it is reasonable to believe it is—the results could be more disastrous than a nuclear war. Decreased conversion of carbon dioxide to oxygen, accom-

panied by the buildup of carbon dioxide from air pollutants, could result in planetary overheating due to the greenhouse, or heat-trapping, effect of this gas in the atmosphere. As a result, polar caps would melt, and sea levels would rise perhaps 200 feet. It can be hoped only that there is enough flexibility in the oxygen cycle to prevent such drastic consequences until the DDT in the oceans decomposes.

Even supposing that this long-term effect does not occur, a relatively short-term consequence probably cannot be avoided. Zooplankton (animal plankton) also is highly susceptible to DDT. Reduced amounts of zooplankton along with reduced photosynthesis and cell division in phytoplankton means a decrease in the amount of plankton available for organisms that eat it and a consequent drop in their populations, a drop in the population of organisms that feed on *them,* and so on. Plankton lies at the bottom of almost every marine food chain. It is unfortunate that this basis of marine productivity is being eroded just when optimists are proposing to feed ocean products to the increased billions of human population predicted in the next few decades.

A second area of concern, DDT's effect on hormones, has been receiving some publicity lately, due mainly to the fact that the bald eagle is the Union's symbol. Raptorial birds, of which the eagle is one, have been the first victims of the hormonal effects of DDT pollution because of their position at the top of food chains. In concentrations of a very few parts per million—in mammals and birds at least and perhaps in other animals—DDT stimulates the liver to produce enzymes that destroy many other enzymes and hormones. Among the destroyed substances is estradiol, which in birds regulates the withdrawal of calcium from bone for the manufacture of eggshell in the oviducts. Consequently these birds lay thin-shelled eggs—in some cases eggs with no shell at all. Such thin-shelled eggs cannot be incubated properly, so a potential eagle, hawk, or fish-eating bird is lost. Because of their position in their food chains, raptorial birds in the United States have been reduced to desperate straits by chlorinated hydrocarbon pollution. The once-numerous peregrine falcon is now extinct as a breeding bird in the East and is surviving only precariously elsewhere, partly because of breeding success depressed by DDT and its relatives and partly because of direct effects on these birds. The Everglades kite, pushed to the brink by man anyway, may be doomed to rapid extinction by pesticides. The Southern bald eagle is endangered, with perhaps only 600 birds surviving. The American osprey or fishhawk, once numerous, because of its feeding habits is especially liable to ingest DDT, and its population is in sharp decline and already at the danger point. The osprey feeds on fish; and fish, of course, are prime biological magnifiers.

Too many farmers dislike hawks and as a group are not inclined to mourn their demise. What once again goes unrealized, however, is the importance of the falconiforms in maintaining ecological balances. Birds of this order that do not eat mostly carrion, one valuable service, likely include rodents as a major item of their diet as another valuable service. One team of writers has estimated that, were the rodent control performed by these birds removed, mice would cover the United States from coast to coast 2-1/2 inches deep. This example is a statistical trick, but it indicates the important service that, if, not done by hawks and other predators, would have to be done by some agency of man's devising.

Hawks and eagles are merely the first birds to be affected by DDT's action on the liver. Brown pelicans are now in a dangerous decline, and other birds with less DDT in their diet must follow suit to some degree as they accumulate the chemical. Then there are the mammals to worry about. (Man, of course, is a mammal. What effect might DDT accumulation have on human biochemistry?)

The third area that greatly worries scientists is the possible mutagenicity of DDT. Whether DDT can cause mutations is not definitely known. It is known that many chemicals, including some that are widespread as pollutants, can cause changes in the gene structures of certain organisms, so biologists keep a sharp eye out for this phenomenon. (What effect might chlorinated hydrocarbons have on human genes?)

Obviously there are many unanswered questions about persistent pesticides, as well as demonstrable harmful effects.

What can concerned Americans do to protect themselves against DDT and other chlorinated hydrocarbons? It is impossible to escape the chemicals now. DDT has been found everywhere, even in the fat of Antarctic penguins. The best that can be done at this point is to totally halt the use everywhere of DDT and its relatives; dieldrin, aldrin, heptachlor, lindane, chlordane, and endrin are the principal offenders among the chlorinated hydrocarbons. The latter chemicals can be hundreds of times as poisonous as DDT, depending on the animal being poisoned. They have avoided being a worse problem only because they are more expensive so have not yet been used as widely. President Nixon's Environmental Quality Council has been considering the effect of the use of persistent chlorinated hydrocarbons on the environment in this country. If the Council decides that these chemicals are harmful, it has the duty to recommend to the President that he use the power of his office to halt their production and use. (It may be noted that 125 million pounds of DDT were produced in the United States in 1968. True, the bulk of it went abroad, but it will work its way back to us through the winds and waves, or through shock-waves in the planetary eco-system.)

Meanwhile there are many control methods for insects other than chlorinated hydrocar-bons, and the homeowner would be wise to seek them out. Japanese beetles, for instance, usually can be more effectively controlled with milky spore disease, a natural check on their population, than with chemicals. Often one application of MSD does a permanent job.

· If a homeowner wisely decides to dispose of his stock of chlorinated hydrocarbons, he should use foresight or he will do worse dam-age than he could by using them on his garden. They should not be flushed down the toilet; this is the fastest route to the aquatic environ-ment where they can begin their trip up the food chain. They should not be burned, be-cause aerosol cans would explode, and besides, higher temperatures than a simple bonfire are needed to break down the substances; they would either remain in the ash or go up in lethal smoke. Neither should they be thrown into the garbage can unless one knows for cer-tain they will go into an incinerator with tem-peratures of 1300°F or above. The best recom-mendation other than incineration over 1300°F is to bury pesticide containers in the backyard encased in the heaviest polyethylene bag obtainable and well knotted—or better yet, encased in concrete or asphalt—under *at least* 2 feet of soil.

Having urged his state and federal govern-ments to protect him from a poisoned environ-ment and having buried any chlorinated hy-drocarbons he may own, there is something more the citizen can do. He can stay alert to technological propaganda. Those who would develop and apply technology without suffi-cient foresight and at any cost are fond of casting their pet project as an irresistible force that we must all go along with, else we will be considered reactionary fossils or butterfly chasers.

Technology has but one justification: to serve man's needs for food, shelter, and cloth-ing so that he can be free to develop his unique assets—mind and spirit. Technology whose end result is an impoverished setting for the human mind—let alone technology that kills people—is worthless, a total failure. Man may survive the DDT blunder; he may survive the automobile; he may survive the invention of the hydrogen bomb. But there must be a limit to narrow escapes, and that limit will be reached soon, barring a change in the present attitude of awe and enthusiasm for every new product of technology.

Some Effects of Air Pollution on Our Environment
Vincent J. Schaefer

The rapid increase in air pollution is a fact that even the casual observer can see—and often smell. In some areas it is so bad that eye-watering conditions from smog are not un-common. This condition has happened and is actually getting worse despite considerable effort on the part of air pollution control offi-cials, industry, and the enforcement of a num-ber of local laws controlling trash burners, brush burning, and other practices.

Many observers are puzzled about this buildup of pollution especially when they notice that visible plumes from chimneys and smoke stacks are rarely seen except from electric power plants, steel and pulp mills, cement plants, and some chemical plants.

One of the major sources of air pollution consists of invisible plumes of particulates so small, as they emerge from the combustion chamber, chemical reaction, or gaseous vapor source, that they are optically invisible. Such particles have cross sections less than 0.1μ. One source of such particles is the automobile. When in good operating condition, the effluent from the auto exhaust pipe is quite invisible. However, if one measures the number of particles emitted by an idling automobile, it is the order of one hundred billion (1×10^{11}) particles per second. Another potent source of invisible particles may actually result from an air pollution law which is directed at the control of visible smoke plumes. While this law was designed to force industrial plants to install electrical precipitators, scrubbers, and other smoke control devices, it is possible in some instances to pass the effluent from an industrial process through a hot flame (an afterburner) to vaporize it and thus as with the automobile, the pollution plume becomes invisible. The concentration of tiny particles is so high, however, that agglomeration often occurs and the knowledgeable observer will detect the plume downwind of the offending source. Under such conditions the use of the afterburner is particularly bad since in addition to making the particles much smaller than they would normally be, they then have a longer residence time in the atmosphere because of their smaller size. Also, the afterburner will generate nitrogen oxide, a poisonous gas which also serves as one of the catalysts for particle growth involving unburned gasoline vapor.

Although some persons believe that unless pollution is curbed in the near future we will run out of breathable air, I believe that other problems will confront us before that happens.

The human body is a highly resistant mechanism to airborne particles. If this were not the case, I do not see how smokers could live! In the process of smoking, the individual insults his lungs with a concentration of at least ten million smoke particles per cubic centimeter. This is a concentration that is 10 to 100 times greater than is encountered in a very badly polluted urban area like Los Angeles or New York City. While there is increasing evidence that the smoker is, in fact, shortening his life by the act of smoking, there are many contradictory facts about smoking which require more understanding about this complex question.

In considering this problem I have called the cigarette a "synergistic reactor." By this term I mean the following: when a cigarette is smoked, there is a very hot zone at the site of the burning tobacco. When the smoker inhales, this burning zone increases in temperature as more air ventilates and intensifies the burning of the tobacco. If the cigarette is smoked in air which contains pollution particles, many of these particles (with a concentration of 10,000 to 100,000 per cubic centimeter or more) are drawn through the burning zone and vaporized. Thus, besides receiving the products of combustion of the tobacco and paper of the cigarette, an additional load comprised of a wide variety of chemical substances is also taken into the lungs. Through vaporization these chemical substances are now in a highly reactable condition with the lungs virtually serving as a test tube; the concentration of gaseous vapors being so high that many reactions can take place and consequently a host of new chemicals may form. These new precipitates being small are readily absorbed, dissolved, or precipitated on the moist surface of the lungs.

When one considers the nearly infinite variety of substances which float in the air of the urban environment, is it any wonder that confused information is an inherent part of the health records of smokers in an urban region?

One of the disturbing aspects of the increase in air pollution over the past decade is that it has apparently increased by nearly an order of magnitude in the area upwind of our cities. This tenfold increase in particulates in areas which previously were characterized as clean "country" air has been measured in northern Arizona near the San Francisco Peaks, in northwestern Wyoming at the Old Faithful area of Yellowstone National Park, and in the Adirondack Mountains of northeastern New York.

When the country air becomes contaminated, then it can no longer dilute the pollution sources to the degree which once was possible.

During the past 3 years we have measured the concentrations of particulates in many parts of the contiguous United States (Schaefer, 1969). In eight transcontinental flights which encompassed most of the major cities of the country and the majority of the clean and polluted regions of the country in between, we have been able to gain a very broad view of the degree to which polluted air covers the country. These findings show the extent to which pollution sources spread their pall over large areas of the continent and clearly shows the fallacy of the "air shed" idea. Unlike water which is primarily controlled by gravity and thus can be related to a particular drainage system often called a water shed, the air and its load of particulates is not controlled by geographical barriers in most instances as it moves rapidly from one region to another, controlled primarily by pressure systems and the weather accompanying them. It is only during periods of clear, quiet weather that local inversions intensify the pollution loads and thus cause local concern as the concentration of particles builds downward from the lid of the temperature inversion and the general public becomes aware of the pollution haze and is sometimes frightened by it.

It is the measurable increase in the continental and global levels of particulates which concerns me at the present time, since I believe certain components of the polluted air may affect us in a more subtle way which may become a more serious problem than the foul smell and eye-watering components we now associate with polluted air.

For a quarter of a century I have been observing, studying, and measuring the characteristics of atmospheric clouds. In the mid-1940's, supercooled clouds were frequently observed in the northeastern United States. During the period 1946-52, we conducted more than 150 flight studies of supercooled clouds in eastern New York, mainly over the Adirondack Mountains. During the past 5 years, supercooled clouds have become relatively rare in this same region while low-level ice crystal clouds (false cirrus) are of common occurrence.

This disappearance of supercooled clouds has been accompanied by the occurrence of a strange kind of precipitation which I have called "misty" rain and "dusty" snow. The mist consists of water droplets of about 0.050-cm diameter, so small that they tend to drift down rather than fall.

The "dusty" snow has a slightly larger cross section, but the droplet from a melted crystal has about the same size as the misty rain. Thus, it is likely that some of the mist originated as snow but melted as it fell into warm air. When this type of snow is falling, only a thin dust-like layer of snow accumulates on the ground.

I believe the origin of both of these forms of precipitation is produced by air pollution. A superabundance of both cloud nuclei and nuclei for ice crystal formation are commonly observed in polluted air. The water vapor in the air collects on such particles, but since there are so many of them (often 10 to 20 times more cloud nuclei and hundreds of times more ice nuclei), the particles are so numerous and so small that they inhibit the precipitation process by stabilizing the clouds. Such stable clouds prevent sunshine from reaching the earth and thus may effectively change some of the dynamics of weather systems. Whether such effects will eventually cause changes in climate can only be determined by much more intensive research.

The type of air pollution which produces a large increase in the concentration of cloud nuclei could be almost any smoke from the burning of organic materials such as garbage, wood, paper, the effluent from pulp mills, a majority of chemical plants, and electric power plants emitting sulfur residues. On the other hand, only a few sources of ice crystals have been identified. A very definite increase in such nuclei may be found in the smoke from steel plants (Chagnon, 1969). By far the greatest and most ubiquitous source is the automobile. The most effective source is the auto whose exhaust is quite invisible. I have recently found that an auto equipped with the so-called antipollution devices is as effective a source of potential ice nuclei as a car without such a control mechanism. If anything, it appears to produce even more nuclei! The material responsible for the production of ice crystals is the submicroscopic residues produced by the burning of gasoline. This mechanism and the results have been previously described in some detail (Schaefer, 1966; Hogan, 1967; Schaefer, 1968a and b) and will not be repeated here.

If our studies continue to show the increase in occurrence of over-seeded clouds, the persistence of clouds for longer periods of time due to their stabilization in areas of polluted air, there is a strong possibility that such conditions will lead to serious environmental and ecological problems resulting from this inadvertent modification of the weather and precipitation.

I hope I am wrong about these mechanisms and their consequences. However, the more data I accumulate and the more observations I make, the more the evidence increases that some major effects in the atmosphere are occurring over hundreds of thousands of square miles. Since the weather systems of our planet are interconnected on a global scale, these effects may lead to an ever-increasing impact on the climatic patterns of the world. While such effects are not necessarily irreversible, it would require major changes in the present trend of our scientific and technological developments to reverse the present situation.

Only an educated and aroused public is likely to demand a change in this deterioration of our environment. There are some hopeful signs that the public is aware of some of these abuses and dangers. Many more efforts like the "Conversation in the Disciplines" in which we have participated are needed. I hope this will happen; otherwise we will encounter a host of serious problems within the next generation.

REFERENCES

Chagnon, S. A. 1969. La Porte weather anomaly, fact or fiction? *Bull. Amer. Meteorol. Soc.,* 49:4.

Hogan, A. 1967. Ice nuclei from direct action of iodine vapor with vapors from leaded gasoline. *Science,* 158:800.

Schaefer, V. J. 1966. Ice nuclei from automobile exhaust and iodine vapor. *Science,* 154:1555.

————. 1968a. Ice nuclei from auto exhaust and organic vapors. *J. Appl. Meteorol.,* 7:113.

————. 1968b. The effect of a trace of iodine on ice nucleation measurements. *J. Rech. Atmos.,* 3:181.

————. 1969. The inadvertent modification of the atmosphere by air pollution. *Bull. Amer. Meteorol. Soc.,* 50:199.

Winds, Pollution, and the Wilderness
Lester Machta

Wilderness stands as a splendid sanctuary from human encroachment on the ground, but no barriers can exclude wind-borne pollution. Man can clean up the atmosphere if he is willing to pay the price, but we foresee no way of modifying the winds to spare a favored region.

The concentrations of air pollutants are almost always smaller far from a city, as in a wilderness, than nearby. But this absolute level of contamination is only one measure of comparison of urban to wilderness harm. Society accepts urban pollution and the use of pesticides because these grew slowly and seem inextricably bound to a higher standard of living. Some of the modern technology, such as pesticides, have been used in the wilderness. But, in contrast to the city dweller, man-made air-borne pollutants from afar yield very few benefits to balance their cost to those enjoying a wilderness.

It is often not appreciated that gaseous and fine particulate matter injected into the atmosphere remains there until scavenging processes or chemical reactions remove it. Proof of the pervasiveness of air-borne pollution as a global problem was dramatically illustrated by radioactive fallout over every part of the earth, land and sea, pole to pole, from nuclear tests exploded at very few locations. Most commonly, pollutants are washed out of the air in rain and snow, but this contaminates soil, lakes, and rivers, usually remote from the source of the pollution. The reason some imagine that the atmosphere can be used as a boundless sewer is its tremendous volume, which swallows the pollutants and usually dilutes them to acceptable levels. The atmosphere is indeed enormous, containing about five million trillion tons of air. But as we have discovered in the past few decades, even this vast envelope of air is insufficient to cope with man's ability to produce ever larger amounts of pollutants.

Our early atmosphere contained virtually no oxygen, but much more carbon dioxide than now exists. Green life converted carbon dioxide to its "waste," oxygen, producing the present 20 per cent oxygen. But man, particularly during the past 80 years of industrialization, has now reversed the process. His burning of fossil fuels—coal and oil—stored from the past has now raised the present carbon dioxide content of the air over 10 per cent since the end of the last century. Each year the clean atmosphere increases its carbon dioxide content by about one-quarter of a per cent, according to records at Mauna Loa, Hawaii, and at Antarctica. This growth rate is only half the estimated release rate from burning of fuels; the remaining half of the output is absorbed mainly by the oceans. More carbon dioxide in the air is not necessarily detrimental; trees and other plants can grow faster in the presence of higher concentrations.

There is special concern about the growth of carbon dioxide in the air because of its vital importance in establishing the heat balance of the atmosphere. As the carbon dioxide increases so should the temperature in the lower atmosphere, by blanketing outgoing radiation from the earth—the "greenhouse" effect. Fortunately, recent calculations suggest that the warming from increased carbon dioxide should be very small. During the first half of the twentieth century, there has been, in fact, a gradual small warming, but this trend has been replaced over the past twenty years by a slight cooling. It is speculated that this latter effect may also be attributed to pollution in the form of increased dustiness which, by reflecting and scattering sunlight back to space, cools the lower atmosphere. The greater dustiness may have counteracted the warming due to increased carbon dioxide. Some observations in non-urban environments show increases with time in atmospheric turbidity or dustiness; similar measurements in urban regions reveal higher turbidity than nearby rural areas. One must be cautious in necessarily blaming climatic trends on human activities; there have also been climatic fluctuations before 1900 when human intervention was less likely.

The more important aspect of air pollution to a wilderness comes from sources immediately upwind, rather than those undergoing global dilution. Certain topographic and weather features might provide some protection for the wilderness, but, in general, there is little to impede the air-borne plume from a factory chimney or a DDT aerial spray blowing to an otherwise inaccessible region. The wind speeds in the lower atmosphere typically blow 5 to 25 miles per hour. Thus wilderness areas many tens of miles from a city or farm can be affected by wind-borne pollution within a few hours. During this travel time, the pollutants are diluted and often depleted. Dilution is accompanied by both horizontal and vertical mixing. Horizontal mixing reduces concentrations near a plume center, but a wider area contains low concentrations. Vertical mixing of contaminants from low altitude sources carries pollution aloft and is always beneficial in reducing ground level concentrations. The wind direction is variable. Thus, pollution from a single source may reach a wilderness only infrequently. But as industrialization grows, pesticide usage increases, and new roads ring the wilderness, the frequency of successively higher levels of pollution can be expected, since winds from many directions will bring pollution to the wilderness.

Topographic barriers such as mountains and special weather patterns frequently impede atmospheric mixing. It is clear that the air within a valley cannot mix horizontally beyond the valley walls. This is evident to aircraft passengers observing smoke from chimneys meander within the natural terrain channels, like river beds. Some of the more infamous air pollution episodes have occurred within or downwind of industrialized valleys: the Donora, Pennsylvania, and the Trail, British Columbia, cases, for example.

But of greater significance, because it changes, is the lid that nature can impose to impede vertical mixing. Normally, the temperature decreases with altitude; the greater the decrease the more vigorous the up-and-down mixing. But under certain weather situations one finds a temperature increase with height (called an inversion), sometimes at the ground but more often aloft, lasting for many days. This inversion inhibits upward mixing through it. When the mixing lid lies below about 5,000 feet, the pollution from many of the current massive sources cannot be diluted adequately to maintain acceptable air quality levels at the ground. This latter circumstance is common to the California coastal area and is one of the prime reasons why Los Angeles smog is so notorious.

There is a world-wide "lid" to up-down mixing called the tropopause, located at about 35,000 feet over the United States. About 75 per cent of the mass of the atmosphere lies below 35,000 feet. Pollutants released to the air take a few days to a few weeks to mix up to the tropopause, and in that time the horizontal winds keep the air in roughly the same band of latitudes as it started in. This is unfortunate because both the maximum industrialization and population lie in the temperate latitudes, say between 25° and 60°N. Air pollutants may stay in the same latitudinal band blowing around the world from west to east, mixing with three-quarters of the depth of air for up to a few months. For these meteorological reasons, not all of the atmospheric volume is available to dilute wastes released during the preceding few months.

The combination of channelling within valleys and imperfect vertical mixing produces particularly undesirable air pollution conditions. It has been found that some of the oxidants, components of smog created photochemically in the Los Angeles basin and contained below the mixing lid, channel through the Rim Forest of the San Bernardino Pass. Here one finds ponderosa pine with burned tips identical to those produced from controlled exposure to observed oxidant levels. The Rim Forest lies 4,400 feet above and four miles north of the city of San Bernardino, and over fifty miles from downtown Los Angeles. As another example, scientists at La Jolla, California, note a marked increase in carbon dioxide as a murky pall approaches from the north, capped below the mixing lid and sliding along the coastal mountain range. The source of the pollution again is the Los Angeles Basin, eighty miles to the north.

While the weather and the temperature inversions along the California coast are somewhat unique, other meteorological patterns can create persistent inversions over almost any part of the United States. The slow moving, light wind, high pressure weather system has been recognized as an air pollution culprit over both the eastern and western United States. In the east it is most common in the southern Appalachian Mountains, over the Great Smoky Mountains, in fact, and in the west over the Great Basin. The Smoky Mountains were named in colonial times because of the blue haze produced by atmospheric reactions involving turpenes from the trees. Nowadays, man's wastes add to those of nature. The prevalence of reduced visibility due to either source is a consequence of weather conditions in the southern Appalachian Mountains. The eastern high pressure systems or anticyclones are most frequent in October. Before the advent of pollution, man would eagerly look forward to their onset in October as beautiful Indian Summers. Today, the sluggish anticyclones can be recognized by the air traveler by a layer of dirty air with a sharp top covering many thousands of square miles for several days. Since there are usually few clouds associated with an anticyclone, the abundant sunshine may produce photochemical smog of the Los Angeles variety to add to other pollutants. The light winds drift about aimlessly rather than follow the prevailing flow. A wilderness may be covered by air with attendant pollution from many diverse sources tens to hundreds of miles away.

The downwind decrease of pollutant concentration results from chemical reactions as well as from dilution. These atmospheric reactions which transform a given pollutant to another substance are complex and, by and large, imperfectly understood. Lest one imagine that such reactions are necessarily beneficial, one ought quickly to note at least two which are not. The photochemical oxidants of the Los Angeles smog start out as oxides of nitrogen, hydrocarbons, and other compounds, none of which are necessarily as toxic as the ozone, and other oxidants which they and sunlight produce. Sulfur dioxide oxidizes and its products react with water to form sulfuric acid. In many ways, the acid is more deleterious than the original gaseous sulfur dioxide and water.

The prime means by which the atmosphere cleanses itself is through precipitation scavenging—the removal of impurities by rain and snow. The deposition of pollutants in rainwater can cause damage far from the source.

For example, in southern Sweden during the past ten years, the rainwater and the lakes and rivers which it feeds have exhibited a dramatic increase of acidity. The once prosperous salmon industry is now almost extinct and there is concern over ancient stone monuments. The cause is thought to be sulfuric acid in rainwater from sulfur dioxide coming from the heavily industrialized northwest Europe, probably over 100 miles away. Certain contaminants, like sulfates and artificial radioactivity, are observed in rainwater all over the world. They and others have been changed from an air to a soil or water contaminant and can enter humans and animals via food rather than inhalation. A second and largely unexplored mode of removal of dust from the atmosphere is filtration of air passing through forests. Conifers with their sticky needles are especially effective filters.

The details of the mechanism of washout of contaminants have been, and continue to be, a lively subject of research. Particles are removed with precipitation by impaction and agglomeration which are controlled by the relative sizes and by electrical and chemical properties of the raindrops or snowflakes and the foreign particles. It may be surprising that some gases are also removed by falling water droplets. After pollutants have mixed vertically through many thousands of feet, it is found that, on the average, the lifetime of a typical pollutant particle is about two to four weeks. Within this time, if you will recall, the pollution from the industrialized temperate latitudes can remain in the same geographical band. Thus, the wastes from man's activities both contaminate the air and fall back to earth in the same general band of latitude in which they started.

The transport of pollution to a living wilderness is not purely theoretical; living things will respond, most often unfavorably, to such intrusion. The varied response of flora and fauna to pollution could occupy many pages in this journal; for our purposes, a few examples may be cited.

Michael Treshow in the August 1968 volume of Phytopathology states:

"The historical record, as well as our own studies, have demonstrated how air pollution can entirely destroy vast acreages of sensitive vegetation. One classic illustration of devastation wrought by air pollutants is in the Copper Basin area of Tennessee, where about 7,000 acres of once rich deciduous forest were completely denuded and another 17,000 acres replaced with grassland species following the destruction of the native forest species. Gully erosion stripped away the soil, and even the climate was altered. The mean monthly maximum air temperature was up to 3.5F higher, and rainfall nearly 10 per cent less, in the bare area. Timber and watersheds were similarly decimated in areas surrounding the town of Trail, B. C.; in Ontario, Canada; and in Anaconda, Mont., where sulfur oxides again were present in concentrations lethal to dominant conifers. In another incident, fluorides were reported to have been responsible for needle burning and death of ponderosa pine over a 50-square-mile area near Spokane, Wash. Pine was eliminated as a dominant species in the immediate vicinity of the operations."

Sometimes the identification of pollution as the subtle damaging agent requires detective work. Dochinger reports that a white pine disease called chlorotic dwarf in the Northeast, Central, and Lake States has puzzled pathologists for the past 60 years. Certain trees in a stand appear to be congenital runts; the disease has been attributed to fungi, unfavorable soil, and a virus. Only recently has he demonstrated the damage to be due to air pollution.

Examples of the damage to fauna are equally numerous. In this case, it is not only the effluents from urbanization and industrialization but the pesticides and herbicides which are equally if not more at fault. Many animals and birds are capable of concentrating certain pollutants found in low concentrations in the environment. The widespread grazing habits of reindeer have introduced far more cesium-137, a product of nuclear testing, into their muscle tissue than was expected. Dead and dying robins had between three and twenty times more DDT in the residues of their brains as in the leaves that formed the source of food for the earthworms that, in turn, were eaten by the robins.

A Government report, "Restoring the Quality of Our Environment," states: "Such herbicides, though providing beneficial effects in agricultural regions, may produce serious adverse effects in natural communities, in state and national parks, and other areas dedicated to recreational purposes."

The tale of destruction from air pollution can continue almost endlessly. But it must be recognized that in the foreseeable future the solution lies in pollution control, not weather control. One cannot erect fences around the wilderness to restrain the winds. The cure for air pollution in the wilderness is the same as elsewhere: limit the emissions at the sources.

Environmental Noise Pollution: A New Threat to Sanity
Donald F. Anthrop

The sources of noise today seem almost limitless. From the kitchen in the modern home comes a cacophony that would require ear defenders in industry to prevent hearing loss. In a series of measurements made in one kitchen, a dishwasher raised the noise level in the center of the kitchen from 56 to 85 decibels, while the garbage disposal raised it to more than 90 decibels. A food blender produces about 93 decibels. Power lawn mowers and leaf rakers, outside air conditioners, and power tools such as saws contribute to the noise in the home. But for most Americans, construction and transportation sources, particularly trucks, motorcycles, sports cars, private airplanes and helicopters as well as commercial jets and military aircraft, are the most serious offenders.

Construction Noise

Particularly in large cities, construction noise is a very substantial and seemingly continuous nuisance. This noise can be substantially reduced with existing technology and without great cost. In December 1967, Citizens for a Quieter City in New York demonstrated a muffled air compressor developed in Great Britain and used there for the past five years which reduced the noise level from 86 to 79 decibels at a distance of 25 feet. The compressor is enclosed in a plastic housing lined with foam plastic. This organization also demonstrated a muffled jack hammer which produced significantly fewer decibels. Tests at the British Building Research Station have shown that jack hammer noise can be muffled considerably without any great impairment of performance. Many European cities are already using muffled jack hammers and air compressors equipped with sound attenuating devices. Some of the techniques that can be employed were illustrated by the Diesel Construction Company in the construction of a 52-story office building in lower Manhattan. Foundation blasting was muffled with special steel wire mesh blankets, demolition was done during late hours and weekends when few people were in the area, and steel beams were welded rather than riveted together.

Motor Vehicle Noise

Transportation constitutes the principal source of noise in most American cities. There are now 81 million privately-owned passenger cars in the United States compared with 25.5 million at the end of World War II. Each year seven million of these wear out or are junked, but in the past few years an average of 10 million new ones have been produced or imported each year. Thus, the number of automobiles is increasing at the rate of nearly four per cent a year. As if 81 million automobiles didn't create enough congestion and noise, there are also 2.4 million motorcycles and 16.5 million trucks.

Motor vehicle noise has been primarily an urban problem. In a recent study of noise in Boston schools, a mean reading of 78 decibels* was recorded in a school playground in downtown Boston. In Wellesley, a suburb of Boston, the noise level in the school playground was only 58. Thus, children in the city school were exposed to a noise intensity 100 times greater than the suburban Wellesley children. But the rapid increase in the number of motor vehicles, the production of larger and noisier trucks, the construction of the interstate highway system, and the exodus of people from city to suburb has increasingly brought noise pollution to suburban areas and the countryside.

One of the most comprehensive noise surveys ever made was the London survey in 1961. Noise measurements were made at 540 locations in central London, and 1,400 residents at those locations were interviewed. At 84 per cent of the points traffic noise predominated. About one-third of the people specifically mentioned motor vehicle noise as a major irritant. Furthermore, traffic noise appeared to be as important an annoyance as all other noises together.

A number of surveys have established beyond doubt that the noise problem near high-speed highways arises principally from trucks, motorcycles, and sports cars. In 1964 the California Highway Patrol conducted a series of tests along California highways in which noise levels of 25,351 passenger cars, 4,656 gasoline trucks, and 5,838 diesel trucks were measured. Noise levels of the passenger cars, measured 50 feet from the road, varied between 65 and 86 decibels with the average falling at about 76. On the other hand, noise levels for diesel trucks ranged from 68 to 99 decibels with the average at about 87.

Anti-Noise Laws

The results of these various surveys demonstrate quite clearly that in order to achieve quieter living conditions, cities must reduce motor vehicle noise. Yet governments at all levels have thus far failed to achieve any meaningful reductions. In 1965 the State of New York enacted a law limiting the noise a motor vehicle can produce at a distance of 50 feet to 88 decibels while traveling 35 miles per hour. In 1967 California enacted legislation which sets a limit of 92 decibels for motorcycles and trucks of three tons gross or more, traveling at speeds above 35 miles per hour. All other motor vehicles are limited to 86 decibels. That these limits are much too high is suggested by the fact that in 1961 California hired an accoustical consulting firm to make a survey of motor vehicle noise and to recommend limits consistent with existing technology and currently available noise measuring techniques. The firm recommended maximum limits of 87 decibels for trucks and motorcycles and 77 for other motor vehicles.

Even these lower limits were deemed to be easily attainable with existing technology.

*Decibel: a unit for measuring the volume of sound, equal to the logarithm of the ratio of the intensity of the sound to the intensity of an arbitrarily chosen standard sound.

Furthermore, no valid argument has been advanced to justify higher noise limits for motorcycles than for passenger cars. There is no reason why a 50-horsepower motorcycle should be allowed to make as much noise as four 300-horsepower Cadillacs. Yet the new California noise law permits precisely this situation. Worse yet, the law is not being enforced, particularly with respect to motorcycles, which have become a real threat to sanity in city and back-country alike.

A substantial fraction of the motorcycles being operated in California today have altered mufflers or no mufflers at all, and many bear no registration plates. All such motorcycles are being operated in violation of the state motor vehicle code irrespective of any noise laws. The unwillingness of some local governments to use the tools already at their disposal to achieve quieter communities is a hindrance to the enactment of more effective noise control legislation.

The future of our cities depends in no small measure on how successful we are in reducing traffic noise and congestion. Three approaches are open to us: (1) reduce the noise at the source; (2) eliminate the source through the use of quiet, underground mass transit systems; (3) reduce the noise near freeways by depressing the roadway or constructing a sound barrier along the right-of-way.

While highway design features can greatly reduce the noise nuisance in communities near freeways, motor vehicle noise in our existing cities can only be reduced by quieting or eliminating the source. The origins of noise in motor vehicles are primarily direct radiation from the exhaust, inlet, engine, transmission and tires, and complex vibrations of the outer surfaces of the vehicle. The exhaust is the predominant source of noise in an unsilenced internal combustion engine. It has been demonstrated that complete silencing of the exhaust of a 10-ton diesel truck by means of a series of large mufflers reduces the noise 10 to 15 decibels in the low frequency range. Silencing of the engine inlet produces a smaller noise reduction but over a wider frequency range.

Control of noise produced by the engine structure is somewhat more difficult. One approach has been to build an acoustically-lined enclosure around the engine. In any case, the noise level of nearly all motor vehicles could be reduced by 10 to 15 decibels in the near future at small cost.

Finally, substitution of electrically or steam-powered vehicles for the internal combustion engine would not only result in a major reduction of urban air pollution, but would enormously reduce traffic noise. The brightest hope for the future clearly lies in such vehicles coupled with underground systems for the movement of goods.

Aircraft Noise

Since there are now nearly 1,200 jet airliners, about an equal number of piston aircraft, and more than 100,000 private airplanes in service in the United States, the aircraft noise problem has become very widespread. Today millions of Americans are affected by this aural assault: Congressmen Benjamin Rosenthal and Herbert Tenzer whose Long Island communities lie under the flight paths for La Guardia and Kennedy have warned that the mood of their constituents has become one of desperation, not just unhappiness.

The courts have held that insofar as the operation of aircraft is concerned, the federal government has preempted the field. A 1963 ordinance of Hempstead, Long Island which regulated the altitude and flight path of aircraft while over the city was ruled invalid in a 1967 court suit. Ordinances such as the recent one passed by the city of Santa Barbara banning supersonic flights over the city also would probably be declared invalid in a court test.*

Noise levels in some communities near our major airports have become so intolerable that many residents cannot continue to live in those communities. Lawsuits totalling $200 million are pending in the courts. A few people have been awarded damages where it was shown that property values had declined or where some directly measurable economic penalty had been incurred. But generally, the private citizen has been able to get little compensation for the abuse he suffered. Recently the airport operators, who consider the reduction of aircraft noise to be primarily the responsibility of the manufacturers, charged the airlines and manufacturers with smokescreen tactics on the noise abatement problem and withdrew from the industry-wide National Aircraft Noise Abatement Council.

One does not have to be directly under the flight path of a large jet on take-off in order to receive an ear-splitting roar. When a 707-320B jet is four miles from the point of brake release at the end of the runway it has attained an altitude of about 800 feet and the noise level on the ground one-half mile on either side of the flight path is approximately 85 decibels.

Federal Efforts

Federal officials should not be surprised by the magnitude of the present problem. In 1952 President Truman received a report, "The Airport and Its Neighbors," from his Airport Commission. The Commission said greater consideration should have been given residents living in an area when airports were first built and that civil and military officials should make much greater efforts to reduce take-off noise over residential areas.

But federal officials are just now beginning to do something about the problem. In August 1968, President Johnson signed into law a measure requiring the Federal Aviation Administration to undertake control and abatement of aircraft noise. The FAA was not particularly eager to have this responsibility, for the law appears to make the FAA liable for damage suits arising from aircraft noise.

The FAA has initiated noise control procedures at some airports, but until quieter engines are built, there is not a great deal it can do with regard to jet transport noise. The noise control procedures that have been implemented are directed almost solely at reducing the noise level in communities lying directly under the flight path while the plane is at low altitude. While reductions have been achieved in such communities, the result has often been to spread the noise around to other communities. This is precisely what has occurred at the Washington, D. C. National Airport where the FAA requires departing aircraft to climb

---

*The intensity of a sound is a purely physical attribute and can be measured by acoustical apparatus without the presence of a human observer. If we listen to a sound whose intensity is gradually increased, the sensation we call loudness increases also, but loudness is by no means proportional to intensity. Rather the loudness sensation is more nearly proportional to the logarithm of intensity. Partly for this reason, a logarithmic scale is used for the expression of acoustic intensities, and intensities are measured relative to the intensity at the threshold of audible sound, namely 10^{-16} watts/cm^2. The unit is the decibel, and the intensity level, I.L., is then defined by

$$\text{I.L.} = \frac{10 \log I}{10^{-16}} = 10 \log I + 160$$

If the intensity of a sound is 10^{-16} watts/cm^2 (the faintest audible sound), its intensity level is zero decibels. A sound whose intensity is 10^{-15} watts/cm^2 is 10 times as intense as this threshold, and its intensity level is 10 decibels (abbreviated 10 dB); a sound whose intensity is 10^{-14} watts/cm^2 is 100 times as intense as the threshold, and its intensity level is 20 dB; etc. A sound whose intensity is 10^{-10} watts/cm^2 has an intensity level of 60 dB and is 100 times as intense as a sound of 110^{-12} watts/cm^2 which has an intensity level of 40 dB. However, the 60 dB sound is not 100 times as loud as the one of 40 dB. Since intensity level is, by definition, a logarithmic function, the sensation of loudness is approximately proportional to the intensity level in decibels. Consequently, the loudness of the 60 dB sound exceeds that of the 40 dB sound by the same amount that the loudness of the 40 dB sound exceeds that of a 20 dB sound.

as quickly as possible to 1,500 feet and then cut back the power and follow the Potomac River northward. Flights over the White House, the Capitol, the Washington Monument, and the U. S. Naval Observatory are prohibited. But since Washington National Airport is just across the Potomac River from the Lincoln Memorial, central Washington is still bombarded by the constant roar of jets, and communities such as Georgetown are now directly under the flight path. Why should residents of Georgetown be subjected to the noise while congressmen on Capitol Hill are protected from the din? If the congressional office buildings rather than residential communities were under the flight path, Congress would long ago have taken steps to end the nuisance.

The solution to the aircraft noise problem in the District of Columbia is to close Washington National Airport. Few people presently use Dulles because it is so far from the city, but it would be much more attractive if a rapid transit system connected the airport with downtown Washington. Furthermore, a substantial percentage of the traffic at Washington National Airport consists of Washington-New York and Washington-Boston commuter service. If high-speed rail service were available between these points, this traffic could be almost eliminated.

Getting at the Source

While flight procedures can bring relief to some communities, the only solution to the aircraft noise problem lies in quieting or eliminating the source. NASA is financing research and development to develop a new "quiet engine." Preliminary tests indicate the new quiet engine will reduce take-off noise by 15 decibels. In static tests with a Pratt and Whitney J-57 engine, Boeing claims to have obtained a noise reduction of nearly 40 decibels by use of acoustical linings in the engine. There are reports that the proposed European airbus will use advanced engines which will produce a 75 decibel noise level on take-off. For comparison, the Boeing 707-320B in normal operation (that is, in the absence of FAA noise control procedures) produces about 107 decibels on take-off. Clearly, then, the manufacturers can build quieter aircraft if they are forced to do so. When can we expect some relief? Manufacturers say the giant Boeing 747, scheduled for late 1969, is already in production and cannot be fitted with new engines even if they were available. The 747 is expected to produce a 100 decibel noise level on take-off. The airlines argue that to retro-fit existing turbojets with the new quiet engine would cost $6 million per plane and that they cannot afford it. Thus, if the present trend continues, we cannot expect any relief before the late 1970s. But by that time any noise reduction will be partly offset by the doubling of air traffic expected between now and 1975. The fact is that the present exasperating noise problem exists because the aircraft manufacturers and the airlines have operated on the basis of their own short-range economic interests and have failed to devote the efforts and resources needed to solve it.

If the already grave situation is not to become worse, some bold steps will have to be taken:

1. The federal government should provide a greatly increased funding level for quiet engine research so that take-off noise will be reduced by 40, not 20 decibels.

2. Whenever a substantially quieter engine is developed, the FAA should require existing aircraft to be retro-fitted with the new engine. If the airlines cannot afford the cost without increasing fares, then fares should be increased. The small percentage of the population that uses the airlines should be required to

assume part of the burden for providing a livable environment for the millions of people who suffer from the noise but derive no economic benefit from it.

3. Particularly in densely populated areas such as the Northeast Corridor, the Chicago-Pittsburgh region, and the San Diego-San Francisco corridor, high-speed rail transportation could substantially reduce air traffic.

4. Future airports should be planned according to the principles used at Dulles International Airport and the new one now being planned for Dallas where 18,000 acres are being purchased to prevent encroachment of residential dwellings.

5. New airports should be located 20 or 30 miles from the metropolitan area, as Dulles International Airport is, and serviced by high-speed surface transportation.

FAA Authority

When Congress passed the Aircraft Noise Abatement Act, the FAA was clearly given authority to regulate noise from private planes. Yet the FAA has so far done nothing about this growing menace and has indicated little interest in doing anything. The light planes of today are more powerful and far noisier than they were a decade ago. Worse, there are a lot more of them. One can reasonably ask why a single businessman in an executive plane should be allowed to create a noise nuisance that irritates literally thousands of people in the communities along his flight path? Furthermore, noise from private planes is becoming a frequent intruder into the solitude of national parks and wilderness areas.

The present noise levels produced by light aircraft are quite unnecessary. The FAA should prohibit private planes from flying below 8,000 feet over populated areas and should require that all private planes be equipped with mufflers and acoustical materials to reduce engine noise.

While a feeble first step has been taken to reduce the noise produced by civilian aircraft, the deafening roar of military planes continues unabated, for the FAA does not have jurisdiction over military planes or flight operations. The Department of Defense has made no effort to develop quieter jet aircraft, claiming that it cannot afford the weight penalty that quieter engines would impose. Instead of making a serious effort to reduce noise levels in communities near military installations, Defense has embarked upon a public relations campaign to convince the American public that they should not only tolerate but welcome this assault on their eardrums because the military establishment is defending them. This country's military brass seems quite willing to destroy our environment in the name of defending it.

A case in point is the Alameda Naval Air Station which lies adjacent to the city of Oakland, California in the very heart of a metropolitan area. Over 1.75 million people live within 12 miles of the runway. Berkeley and Oakland residents frequently find themselves rudely awakened early Sunday morning by jets streaking over the East Bay hills with afterburners blazing. If an aroused public demands the closure of some of these poorly situated installations, perhaps the Defense Department will be motivated to develop quieter aircraft.

Sonic Boom and the SST

The worst is yet to come when—and if—Boeing's supersonic transport (SST), built with federal financing, goes into service in the 1970s. Whenever a plane flies faster than the speed of sound (about 344 meters per second) it generates shock waves which trail out behind the plane on both sides of its path. When these shock waves intercept the earth, they produce the thunderclap we call "sonic boom." Typically the boom is felt along a belt that extends 40 miles on each side of the plane's flight path. The severity of the boom depends on the plane's size and altitude, but there is no known way to eliminate the boom itself. There exists a common misconception that this sonic boom is produced only once when the plane first exceeds the speed of sound. In fact, it is produced continuously along the plane's path while it is in supersonic flight.

The whole SST program places in serious question the commitment of the FAA, the Department of Transportation and Congress to noise reduction. Thus far, Congress has appropriated $653 million for SST. Worse yet, on July 11, 1968 the Senate defeated an amendment to the Aircraft Noise Abatement Act which would have prohibited the SST from flying at supersonic speeds across continental America. The proponents of SST in Congress argued that prohibition of overland flights was unnecessary because the FAA probably would not permit such flights anyway. But the very fact that Congress was unwilling to legislate against sonic boom indicates overland flights by the SST are anticipated. And since the FAA is the agency responsible for the direction and funding of the entire SST development program, asking it to regulate sonic boom is like putting the fox in the chicken coop. The attitude of the Department of Transportation on the sonic boom issue is illustrated in a statement made by Major General Jewell C. Maxwell, the chief of the SST program: "We believe that people in time will come to accept the sonic boom as they have the rather unpleasant side effects which have accompanied other advances in transportation."

This is a myth which so far has survived scientific evidence to the contrary. Aircraft noise studies have shown that people become more intolerant of jet aircraft as the number of fly-overs or the duration of each fly-over is increased.

In order to assess public acceptance of sonic boom, the FAA conducted tests in Oklahoma City in 1964. During a six-month period, 1,253 supersonic flights were made over the city. Oklahoma City was one of the most favorable locations the FAA could have chosen to get public acceptance of sonic boom since nearly one-third of the city's residents depend on the aviation industry for their living. Furthermore, no sonic booms were made at night— the really critical test. Yet 27 per cent of the people said they could never learn to live with the sonic boom and over 4,900 residents filed damage claims against the FAA. Most people found the booms more irritating at the end of the tests than at the beginning.

Operation of the SST over continental United States would not only shatter the solitude of nearly every park and wilderness area in the country, but could do extensive damage to some of these places as well. Between August 11 and December 22, 1966 some 83 sonic booms, several of which caused extensive damage, were recorded in Canyon de Chelly National Monument, Arizona. One of these booms loosened an estimated 80 tons of rock which fell on ancient Indian cliff dwellings and caused irreparable damage. Damage has also been reported in Bryce Canyon National Park, Utah.

Canada has already banned the operation of supersonic aircraft over its provinces. Both Switzerland and West Germany have indicated they will prohibit supersonic flights within their borders if their citizens complain.

Boondoggle Program

The whole SST program is an economic boondoggle, the prime beneficiary of which is the aircraft manufacturing industry. The FAA has committed $1.3 billion or about 83 per cent of the estimated development cost and Congress has already appropriated half of this amount. But low cost estimates and delays in the program now indicate the cost to the federal government will be at least $3.5 billion before the first plane is sold. The FAA talks glowingly of estimated sales between $20 and $48 billion, but not long ago the Institute of Defense Analysis issued a report which indicated that if supersonic travel were restricted to overwater flights, there would be a market for only 279 planes and the whole project would become an economic disaster.

Even if the SST is initially operated at supersonic speeds only on overwater flights, mounting economic pressures to expand the market for the plane will almost certainly result in overland routes across the United States. Former Transportation Secretary Alan Boyd has said: "I think it will be entirely possible to operate a route over the Plains area and possibly across the Canadian border without discomfort or inconvenience to people on the ground."

The operation of such a route would reduce the flying time between Chicago and San Francisco only about 30 minutes. If supersonic flight on overland routes is not restricted, 150 SSTs may be in domestic operation by 1990. Must 50 million people be subjected to perhaps 30 booms a day so that a few can reduce their travel time by 30 minutes?

While the abatement of much of the noise that presently plagues our society is in part a technical problem, both the impetus and the money for solving it must come from the political arena, and the sonic boom problem is entirely political. A quieter society will only be achieved when a concerned public demands a new system of priorities from the politicians.

The Myth of the Peaceful Atom
Richard Curtis and Elizabeth Hogan

"What is past is past, and the damage we may already have done to future generations cannot be rescinded, but we cannot shirk the compelling responsibility to determine if the course we are following is one we should be following."

So said Senator Thruston B. Morton of Kentucky on February 29, 1968, upon introducing into Congress a resolution calling for comprehensive review of federal participation in the atomic energy power program. Admitting he had been remiss in informing himself on this "grave danger," Morton said he had now looked more deeply into nuclear power safety and was "dismayed at some of the things I have found—warnings and facts from highly qualified people who firmly believe that we have moved too fast and without proper safeguards into an atomic power age."

Senator Morton's resolution on nuclear power was by no means the only one before Congress in 1968. Indeed, more than two dozen legislators urged investigation and re-evaluation of this program. This fact may come as a surprise to much of the public, for the belief is widespread that the nuclear reactors being built to generate electricity for our cities are safe, reliable, and pollution-free. But a rapidly growing number of physicists, biologists, engineers, public health officials, and even staff members of the Atomic Energy Commission itself—the government bureau responsible for regulation of this force—have been expressing serious misgivings about the planned proliferation of nuclear power plants. In fact, some have indicated that nuclear power, which Supreme Court Justices William

O. Douglas and Hugo L. Black described as "the most deadly, the most dangerous process that man has ever conceived," represents the gravest pollution threat yet to our environment.

As of June, 1968, 15 commercial nuclear power plants were operating or operable within the United States, producing about one per cent of our current electrical output. The government, however, has been promoting a plan by which 25 per cent of our electric power will be generated by the atom by 1980, and half by the year 2000. To meet this goal, 87 more plants are under construction or on the drawing boards. Although atomic power and reactor technology are still imperfect sciences, saturated with hazards and unknowns, these reactors are going up in close proximity to heavy population concentrations. Most of them will be of a size never previously attempted by scientists and engineers. They are, in effect, gigantic nuclear experiments.

As most readers will recall, atomic reactors are designed to use the tremendous heat generated by splitting atoms. They are fueled with a concentrated form of uranium stored in thin-walled tubes bound together to form subassemblies. These are placed in the reactor's core, separated by control rods that absorb neutrons and thus help regulate chain reactions of splitting atoms. When the rods are withdrawn, the chain reactions intensify, producing enormous quantities of heat. Coolant circulated through the fuel elements in the reactor core carries the heat away to heat-exchange systems, where water is brought to a boil. The resultant steam is employed to turn electricity-generating turbines.

Stated in this condensed fashion, the process sounds innocuous enough. Unfortunately, however, heat is not the only form of energy produced by atomic fission. Another is radioactivity. During the course of operation,

the fuel assemblies and other components in the reactor's core become intensely radioactive. Some of the fission by-products have been described as a million to a billion times more toxic than any known industrial chemical. Some 200 radioactive isotopes are produced as by-products of reactor operation, and the amount of just one of them, strontium-90, accumulated in a reactor of even modest (100-200 megawatt) size, after it has been operative for six months, is equal to what would be produced by the explosion of a bomb 190 times more powerful than the one dropped on Hiroshima.

Huge concentrations of radioactive material are also to be found in nuclear fuel-reprocessing plants. Because the intense radioactivity in a reactor core eventually interferes with the fuel's efficiency, the spent fuel assemblies must be removed from time to time and replaced by new, uncontaminated ones. The old ones are transported to reprocessing plants where the contaminants are separated from the salvageable fuel as well as from plutonium, a valuable by-product. Since no satisfactory means have been found for neutralizing or for safely releasing into the environment the radioactive liquid containing the contaminants, it must be stored until it is no longer dangerous. Thus, reprocessing plants and storage areas are immense repositories of "hot" and "dirty" material. Furthermore, routes between nuclear power plants and the reprocessing facility carry traffic bearing high quantities of such material.

Even from this glimpse it will be apparent that public and environmental safety depend on the flawless containment of radioactivity every step of the way. For, owing to the incredible potency of fission products, even the slightest leakage is harmful and a massive release would be catastrophic. The fundamental question, then, is how heavily can we rely on human wisdom, care, and engineering to hold this peril under absolute control?

Abundant evidence points to the conclusion that we cannot rely on it at all.

The hazards of peaceful atomic power fall into two broad categories: the threat of violent, massive releases of radioactivity or that of slow, but deadly, seepage of harmful products into the environment.

Nuclear physicists assure us that reactors cannot explode like atomic bombs because the complex apparatus for detonating an atomic warhead is absent. This fact, however, is of little consolation when it is realized that only a conventional explosion, which ruptures the reactor mechanism and its containment structure, could produce havoc on a scale eclipsing any industrial accident on record or any single act of war, including the atomic destruction of Hiroshima or Nagasaki.

There are numerous ways in which such an explosion can take place in a reactor. For example, liquid sodium, which is used in some reactors as a coolant, is a devilishly tricky element that under certain circumstances burns violently on contact with air. Accidental exposure of sodium could initiate a chain of reactions: rupturing fuel assemblies, damaging components and shielding, and destroying primary and secondary emergency safeguards. If coolant is lost, as it could be in some types of reactors, fuel could melt and recongeal, forming "puddles" that could explode upon reaching a critical size. If these explosions are forceful enough, and safeguards fail, some of the fission products could be released outside the

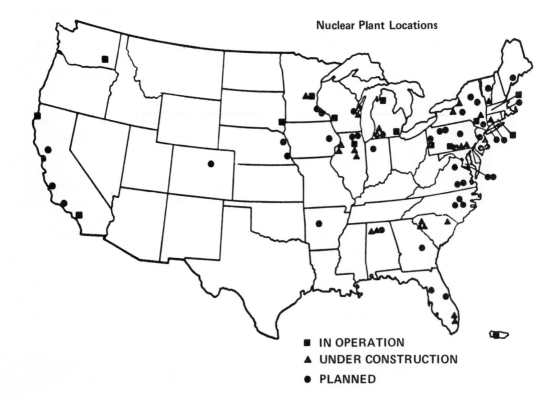

Nuclear Plant Locations

■ IN OPERATION
▲ UNDER CONSTRUCTION
● PLANNED

plant and into the environment in the form of a gas or a cloud of fine radioactive particles. Under not uncommon atmospheric conditions such as an "inversion," in which a layer of warm air keeps a cooler layer from rising, a blanket of radioactivity could spread insidiously over the countryside. Another possibility is that fission products could be carried out of the reactor and into a city's watershed, for all reactors are being built on lakes, rivers, or other bodies of water for cooling purposes.

What would be the toll of such a calamity?

In 1957 the Atomic Energy Commission issued a study (designated Wash. —740), largely prepared by the Brookhaven National Laboratory, that attempted to assess the probabilities of such "incidents" and the potential consequences. Some of its findings were stupefying: From the explosion of a 100-200 megawatt reactor, as many as 3,400 people could be killed, 43,000 injured, and as much as 7 billion dollars of property damage done. People could be killed at distances up to 15 miles and injured up to 25. Land contamination could extend for far greater distances: agricultural quarantines might prevail over an area of 150,000 square miles, more than the combined areas of Pennsylvania, New York, and New Jersey.

The awful significance of these figures is difficult to comprehend. By way of comparison, we might look at one of the worst industrial accidents of modern times: the Texas City disaster of 1947 when a ship loaded with ammonium nitrate fertilizer exploded, virtually leveling the city, killing 561 people, and causing an estimated $67 million worth of damage. Appalling as this catastrophe was, however, it does not begin to approach the potential havoc that would be wreaked by a nuclear explosion occurring in one of the plants now being constructed close to several American cities.

The scientists and engineers who produced the Brookhaven Report optimistically ventured to give high odds against such an occurrence, asserting that the structures, systems, and safeguards of atomic plants were so engineered as to render it practically incredible. At the same time, though, the report was replete with such statements as:

"The cumulative effect of radiation on physical and chemical properties of materials, after long periods of time, is largely unknown."

"Much remains to be learned about the characteristics and behavior of nuclear systems."

"It is important to recognize that the magnitudes of many of the crucial factors in this study are not quantitatively established, either by theoretical and experimental data or adequate experience."

Even if the report had been founded on more substantial understanding of natural and technical processes, many of the grounds on which the Brookhaven team based its conclusions are shaky at best.

For one thing, all of us are familiar with technological disasters that have occurred against fantastically high odds: the sinking of the "unsinkable" *Titanic,* or the November 9, 1965, "blackout" of the northeastern United States, for example. The latter happening illustrates how an "incredible" event can occur in the electric utility field, most experts agreeing that the chain of circumstances that brought it about was so improbable that the odds against it defy calculation.

Congressional testimony given in 1967 by Dr. David Okrent, a former chairman of the AEC's Advisory Committee on Reactor Safeguards, demonstrated that fate is not always a respecter of enormously adverse odds. "We do have on record cases where, for example, an applicant, appearing before an atomic safety and licensing board, stated that a mathematical impossibility had occurred; namely, one tornado took out five separate power lines to a reactor. If one calculated strictly on the basis of probability and multiplied the probability for one line five times, you get a very small number indeed," said Dr. Okrent, "but it happened."

A disturbing number of reactor accidents have occurred—with sheer luck playing an important part in averting catastrophe—that seem to have been the product of incredible coincidences. On October 10, 1957, for instance, the Number One Pile (reactor) at the Windscale Works in England malfunctioned, spewing fission products over so much territory that authorities had to seize all milk and growing foodstuffs in a 400-square-mile area around the plant. A British report on the incident stated that *all* of the reactor's containment features had failed. And, closer to home, a meltdown of fuel in the Fermi reactor in Lagoona Beach, Michigan, in October, 1966, came within an ace of turning into a nuclear "runaway." An explosive release of radioactive materials was averted, but the failures of Fermi's safeguards made the event, in the words of Sheldon Novick in *Scientist and Citizen,* "a bit worse than the 'maximum credible accident.' "

The atomic industry has attempted to design components and safeguards so that failure of one vital system in a plant will not affect another, resulting in a "house of cards" collapse. However, two highly regarded authorities, Theos J. Thompson and J. G. Beckerley, in a book on reactor safety advise us not to place too much faith in claims of independent safeguards: "A structure as complex as a reactor and involving as many phenomena is likely to have relatively few completely independent components." Many manufacturers and utility operators have resisted the idea of producing "redundant safeguards" on the grounds of excessive cost.

Investigations of reactor breakdowns usually disclose a number of small, seemingly unrelated failures, which snowballed into one big one. A design flaw or a human error, a component failure here, an instrumentation failure there—all may coincide to contribute to the total event. Thompson and Beckerley, examining several atomic plant accidents, pinpointed 13 different contributing causes in three of the accidents that had occurred up to the time of their 1964 study.

Among the many factors contributing to reactor accidents, the human element is the most difficult to quantify. And perhaps for that reason, it has been largely overlooked in the AEC's assessments of reactor safety. Yet, a private researcher of nuclear accidents, Dr. Donald Oken, M.D., Associate Director of the Psychosomatic and Psychiatric Institute of Michael Reese Hospital in Chicago, reported: "A review of reports of past criticality and reactor incidents and discussions held with some of the health personnel in charge reveal a number of striking peculiarities in the behaviour of many of those involved—in which they almost literally asked for trouble."

AEC annuals are full of reports of human negligence: 3,844 pounds of uranium hexafluoride lost owing to an error in opening a cylinder; a $220,000 fire in a reactor because of accidental tripping of valves by electricians during previous maintenance work; numerous vehicular accidents involving transport of nuclear materials. None of these accidents led to disaster, but who will warrant that, with the projected proliferation of power plants and satellite industries in the coming decade, a moment's misjudgment will not trigger a nightmare? Perhaps worse, the likelihood of sabotage has scarcely been weighed, despite a number of incidents and threats.

It should be apparent that if men are to build safe, successful reactors, the whole level of industrial workmanship, engineering, inspection, and quality control must be raised well above prevailing levels. The more sophisticated the technology, the more precise the correspondence between the subtlest gradations of care or negligence and that technology's success or failure. When meters, grams, and seconds are no longer good enough, and specifications call for millimeters, milligrams, and milliseconds, the demands made on men, material, and machinery are accordingly intensified. Minute lapses that might be tolerable in a conventional industrial procedure will wreck the more exacting one. And when the technology is not only exacting but hazardous in the extreme, then a trivial oversight, a minor defect, a moment's inattention may spell doom.

While there is little doubt that American technology is the most refined on earth, there is ample reason to believe that it has more than met its match in the seemingly insurmountable problems posed by the peaceful atom. Societies of professional engineers, and others concerned with establishing technical and safety criteria for the nuclear industry, have described between 2,800 and 5,000 technical standards that are necessary for a typical reactor power plant in such areas as materials, testing, design, electrical gear, instrumentation, plant equipment, and processes. Yet, due to the rapidity with which the nuclear industry has developed, as of March, 1967, only about 100 of these had been passed on and approved for use.

It is not surprising, then, to learn that serious technical difficulties are turning up in reactor after reactor. At the Big Rock Point Nuclear Plant, a relatively small reactor near Charlevoix, Michigan, control rods were found sticking in position, studs failing or cracked, screws jostled out of place and into key mechanisms, a valve malfunctioning for more than a dozen reasons, foreign material lodging in critical moving parts, and welds cracked on every one of sixteen screws holding two components in place. A reactor at Humboldt Bay in California manifested cracks in the tubes containing fuel: in order to keep costs down, stainless steel had been used instead of a more reliable alloy. The Oyster Creek plant in New Jersey showed cracks in 123 of 137 fuel tubes, and welding defects at every point where tubes and control-rod housings were joined around the reactor's vessel. Reactors in Wisconsin, Minnesota, Connecticut, Puerto Rico, New York, and elsewhere have experienced innumerable operating difficulties, and some, such as the $55 million Hallam plant in Nebraska, have been forced to shut down for good, owing to plant malfunction.

Chilling parallels can be drawn between failures in nuclear utility technology and in the nuclear submarine program. In October, 1962, Vice Admiral Hyman G. Rickover, Director of AEC's Division of Naval Reactors, took the atomic industry to task in a speech in New York City:

"It is not well enough understood that conventional components of advanced systems must necessarily meet higher standards. Yet it should be obvious that failures that would be trivial if they occurred in a conventional application will have serious consequences in a nuclear plant because here radioactivity is involved. . . ."

Rickover went on to cite defective welds, forging materials substituted without authorization, violations of official specifications, poor inspection techniques, small and seemingly "unimportant" parts left out of components, faulty brazing of wires, and more. "I assure you," he declared, "I am not exaggerating the situation; in fact, I have understated it. For every case I have given, I could cite a dozen more."

The following April, the U.S. atomic submarine *Thresher,* while undergoing a deep test dive some 200 miles off the Cape Cod coast, went down with 112 naval personnel and 17 civilians and never came up again. Subsequent investigation revealed that the sub suffered from many of the same ailments described in Rickover's speech. "It is extremely unfortunate," said Senator John O. Pastore, chairman of the joint congressional committee that held hearings on the disaster, "that this tragedy had to occur to bring a number of unsatisfactory conditions into the open." We must now ask if the same will one day be said about a power plant near one of our large cities.

If a major reactor catastrophe did occur there is good reason to believe that the consequences would be far worse than even the dismaying toll suggested by the 1957 Brookhaven Report, for a number of developments since then have made the threat considerably more formidable.

The Brookhaven Report's accident statistics, for instance, pertained to a reactor of between 100 and 200 megawatts. But while the 15 reactors currently operating in the United States average about 186 megawatts, the 87 plants going up or planned for the next decade are many times that size. Thirty-one under construction average about 726 megawatts; 42 in the planning stage average 832; 14 more, planned but without reactors ordered, will average 904. Some, such as those slated for Illinois, California, Alabama, and New York anticipate capacities of more than 1,000 megawatts. Con Edison has just announced it intends to build four units of 1,000 megawatts each on Long Island Sound near New Rochelle in teeming Westchester County—four nuclear reactors, each with a capacity five to ten times that of the reactor described in the Brookhaven Report.

These facilities will accordingly contain more uranium fuel, and because it is costly to replace spent fuel assemblies (this delicate and dangerous process can take six weeks or longer), the new reactors are designed to operate without fuel replacement far beyond the six months posited in the Brookhaven Report. As a result, the buildup of toxic fission products in tomorrow's reactors will be far greater than at present, and an accident occurring close to the end of the "fuel cycle" in such a plant could release fantastic amounts of radioactive material.

Most serious of all, perhaps, is that tomorrow's reactors are now slated for location in close proximity to population concentrations. While the Brookhaven Report had its hypothetical reactor situated about 30 miles from a major city, many of tomorrow's atomic plants will be much closer. Although the AEC has drafted "guidelines" for siting reactors, the Commission has failed to make utilities adhere to them. In 1967, Clifford K. Beck, AEC's Deputy Director of Regulation, admitted to the Joint Committee on Atomic Energy that nuclear plants in Connecticut, California, New York, and other locations "have been approved with lower distances than our general guides would have indicated when they were approved."

Also, we must remember that while a reactor may not be near the legal boundaries of a metropolis, it may lie close to a population center. Thus, while Con Edison's Indian Point plant is 24 miles from New York City (two more plants are now being built there), it is within 10 miles of an estimated population of 155,510. It need only be recalled that the Brookhaven Report foresaw people being killed by a major radioactive release at distances up to 15 miles to realize the significance of these figures.

In a recent study of nuclear plant siting made by W. K. Davis and J. E. Robb of San Francisco's Bechtel Corporation, the locations of 42 nuclear power plants (some proposed, some now operable) were examined with respect to population centers inhabited by 25,000 residents or more. Their findings are unnerving: only *two* plants in operation or planned are more than 30 miles from a population center. Of the rest, 14 are between 20 and 27 miles away, 15 between 10 and 16 miles, and 11 between 1 and 9 miles.

Is it necessary to build atomic plants so big and so close? The answer has to do with economics. The larger a facility is, the lower the unit cost of construction and operation and the cheaper the electricity. The longer the fuel cycle, the fewer the expensive shutdowns while spent fuel assemblies are replaced. The closer the plant is to the consumer, the lower the cost of rights-of-way, power lines, and other transmission equipment.

On a few occasions an aroused public has successfully opposed the situation of plants near population centers. When the Pacific Gas and Electric Company persisted in trying to build a reactor squarely over earthquake faults in an area of known seismic activity—the site was Bodega Head, north of San Francisco—a courageous conservation group forced the company to back down. It has been suggested, though, that the group might not have won had not the Alaskan earthquake of 1964, occurring while the fight was going on, underscored the recklessness of the utility's scheme.

Announcement by Con Edison at the end of 1962 of its proposal to build a large nuclear plant in Ravenswood, Queens, close to the center of New York City brought a storm of frightened and angry protest. Although the utility's chairman noted, "We are confident that a nuclear plant can be built in Long Island City, or in Times Square for that matter, without hazard to our own employees working in the plant or to the community," David E. Lilienthal, the former head of the AEC, had a contrary opinion, declaring he "would not

dream of living in Queens if a huge nuclear plant were located there." Outraged citizens and a number of noted scientists prevailed.

For the most part, however, the battle has been a losing one. Con Edison, for example, after its defeat in the Ravenswood fight, has just announced an interest in building a reactor on Welfare Island, literally a stone's throw from midtown Manhattan. Also, New York's Governor Nelson Rockefeller has gone on record advocating an $8 billion electric power expansion program based extensively on nuclear energy. The state legislature approved of the program, and in 1968, voted to bolster the plan with state subsidies.

Some of the deepest concern about the size and location of atomic plants has been expressed by members of the AEC themselves. "The actual experience with reactors in general is still quite limited," said Harold Price, AEC's Director of Regulation, in 1967 congressional hearings, "and with large reactors of the type now being considered, it is nonexistent. Therefore, because there would be a large number of people close by and because of lack of experience, it is . . . a matter of judgment and prudent at present to locate reactors where the protection of distance will be present."

Price's statement is mild compared to that made in the same hearings by Nunzio J. Palladino, Chairman of the AEC's Advisory Committee on Reactor Safeguards for 1967, and Dr. David Okrent, former Chairman for 1966: "the ACRS believes that placing large nuclear reactors close to population centers will require considerable further improvements in safety, and that *none of the large power reactors now under construction is considered suitable for location in metropolitan areas* [our italics]."

The threat of a nuclear plant catastrophe constitutes only half of the double jeopardy in which atomic power has placed us. For even if no such calamity occurs, the gradual exhaustion of what one scientist terms our environmental "radiation budget," due to unavoidable releases of radioactivity during normal operation of nuclear facilities, poses an equal and possibly more insidious threat to all living things on earth.

Most of the fission products created in a reactor are trapped. Contaminated solids, liquids, and gases are isolated, allowed to decay for a short period of time, then concentrated and shipped in drums to storage areas. These are called "high-level wastes." But technology for retaining all radioactive contaminants is either unperfected or costly, and much material of low-level radioactivity is routinely released into the air or water at the reactor site. These releases are undertaken in such a way, we are told, as to insure dispersion or dilution sufficient to prevent any predictable human exposure above harmful levels. Thus, when atomic power advocates are asked about the dangers of contaminating the environment, they imply that the relatively small amounts of radioactive materials released under "planned" conditions are harmless.

This view is a myth.

In the first place, many waste radionuclides take an extraordinarily long time to decay. The half-life (the time it takes for half of an element's atoms to disintegrate through fission) of strontium-90, for instance, is more than 27 years. Thus, even though certain long-lived isotopes are widely dispersed in air or diluted in water, their radioactivity does not cease. It remains, and over a period of time accumulates. It is therefore not pertinent to talk about the safety of any single release of "hot" effluents into the environment. At issue, rather, is their duration and cumulative radioactivity.

Further, many radioactive elements taken into the body tend to build up in specific tissues and organs to which those isotopes are attracted, increasing by many times the exposure dosage in those local areas of the body. Iodine-131, for instance, seeks the thyroid gland; strontium-90 collects in the bones; cesium-137 accumulates in muscle. Many isotopes have long half-lives, some measurable in decades.

Two more factors controvert the view that carefully monitored releases of low-level radioactivity into the environment are not pernicious. First, there is apparently no radiation threshold below which harm is impossible. Any dose, however small, will take its toll of cell material, and that damage is irreversible. Second, it may take decades for organic damage, or generations for genetic damage, to manifest itself. In 1955, for example, two British doctors reported a case of skin cancer—ultimately fatal—that had taken forty-nine years to develop following fluoroscopic irradiation of a patient.

Still another problem has received inadequate attention. Man is by no means the only creature in whom radioactive isotopes concentrate. The dietary needs of all plant and animal life dictate intake of specific elements. These concentrate even in the lowest and most basic forms of life. They are then passed up food chains, from grass to cattle to milk to man, for example. As they progress up these chains, the concentrations often increase, sometimes by hundreds of thousands of times. And if these elements are radioactive. . . .

Take zinc-65, produced in a reactor when atomic particles interact with zinc in certain components. Scrutiny of the wildlife in a pond receiving runoff from the Savannah River Plant near Aiken, South Carolina, disclosed that while the water in that pond contained only infinitesimal traces of radioactive zinc-65, the algae that lived on the water had concentrated the isotope by nearly 6,000 times. The bones of bluegills, an omnivorous fish that feeds both on algae and on algae-eating fish, showed concentrations more than 8,200 times higher than the amount found in the water. Study of the Columbia River, on which the Hanford, Washington, reactor is located, revealed that while the radioactivity of the water was relatively insignificant: 1. the radioactivity of the river plankton was 2,000 times greater; 2. the radioactivity of the fish and ducks feeding on the plankton was 15,000 and 40,000 times greater, respectively; 3. the radioactivity of young swallows fed on insects caught by their parents near the river was 500,000 times greater; and 4. the radioactivity of the egg yolks of water birds was more than a million times greater.

Here then are clear illustrations of the ways in which almost undetectable traces of radioactivity in air, water, or soil may be progressively concentrated, so that by the time it ends up on man's plate or in his glass it is a tidy package of poison.

That nuclear facilities are producing dangerous buildups of radioisotopes in our environment can be amply documented. University of Nevada investigators, seeking a cause for concentrations of iodine-131 in cattle thyroids in wide areas of the western United States, concluded that "the principal known source of I-131 that could contribute to this level is exhaust gases from nuclear reactors and associated fuel-processing plants."

In his keynote address to the Health Physics Society Symposium at Atlanta, Georgia, early in 1968, AEC Commissioner Wilfred E. Johnson admitted that the release into the atmosphere of tritium and noble gases such as krypton-85 would present a potential problem in the future, and that, as yet, scientists had not devised a way of solving it. Krypton-85, although inert, has a 10-year half-life and tends to dissolve in fatty tissues, meaning fairly even distribution throughout the human body. Krypton-85 is particularly difficult to filter out of reactor discharges, and the accumulation of this element alone may exhaust as much as two-thirds of the "average" human's "radiation budget" for the coming century, based on the standards established by the National Committee on Radiation Protection and Measurement.

That "low-level" waste is a grossly deceptive term is obvious. In his book *Living with the Atom,* author Ritchie Calder in 1962 described an "audit" of environmental radiation that he and his colleagues, meeting at a symposium in Chicago, drew up to assess then current and future amounts of radioactivity released into atmosphere and water. Speculations covered the period 1955-65, and because atomic power plants were few and small during that time, the figures are more significant in relation to the future. Tallying "planned releases" of radiation from such sources as commercial and test reactors, nuclear ships, uranium mills, plutonium factories, and fuel-reprocessing plants, Calder's group came to a most disquieting conclusion: "By the time we had added up all the curies which might predictably be released, by all those peaceful uses, into the environment, it came to about 13 million curies per annum." A "curie" is a standard unit of radioactivity whose lethality can be appreciated from the fact that one trillionth of one curie of radioactive gas per cubic meter of air in a uranium mine is ten times higher than the official maximum permissible dose.

Calder's figures did not include fallout due to bomb testing and similar experiments, nor did they take into account possible reactor or nuclear transportation accidents. Above all, they did not include possible escape of stored high-level radioactive wastes, the implications of which were awesome to contemplate: "What kept nagging us was the question of waste disposal and of the remaining radioactivity which must not get loose. We were told that the dangerous waste, which is kept in storage, amounted to 10,000 million curies. If you wanted to play 'the numbers game' as an irresponsible exercise, you could divide this by the population of the world and find that it is over 3 curies for every individual."

Exactly what does Calder mean by "the question of waste disposal"?

It has been estimated that a ton of spent fuel in reprocessing will produce from forty to several hundred gallons of waste. This substance is a violently lethal mixture of short- and long-lived isotopes. It would take five cubic miles of water to dilute the waste from just *one* ton of fuel to a safe concentration. Or, if we permitted it to decay naturally until it reached the safe level—and the word "safe" is used advisedly—just one of the isotopes, strontium-90, would still be damaging to life 1,000 years from now, when it will have only one seventeen-billionth of its current potency.

There is no known way to reduce the toxicity of these isotopes; they must decay naturally, meaning *virtually perpetual containment.* Unfortunately, mankind has exhibited little skill in perpetual creations, and procedures for handling radioactive wastes leave everything to be desired. Formerly dumped in the ocean, the most common practice today is to store the concentrates in large steel tanks shielded by earth and concrete. This method has been employed for some twenty years, and about 80 million gallons of waste are now in storage in about 200 tanks. This "liquor" generates so much heat it boils by itself for years. Most of the inventory in these caldrons is waste from weapons production, but within thirty years, the accumulation from commercial nuclear power will soar if we embark upon the expansion program now being promoted by the AEC. Dr. Donald R. Chadwick, chief of the Division of Health of the U.S. Public Health Service, estimated in 1963 that the accumulated volume of waste material would come to two billion gallons by 1995.

It is not just the volume that fills one with sickening apprehension but the techniques of disposing of this material. David Lilienthal put his finger on the crux of the matter when he stated: "These huge quantities of radioactive wastes must somehow be removed from the reactors, must—without mishap—be put into containers that will never rupture; then these vast quantities of poisonous stuff must be moved either to a burial ground or to reprocessing and concentration plants, handled again, and disposed of, by burial or otherwise, with a risk of human error at every step." Nor can it be stressed strongly enough that we are not discussing a brief danger period of days, months, or years. We are talking of periods "longer," in the words of AEC Commissioner Wilfred E. Johnson, "than the history of most governments that the world has seen."

Yet already there are many instances of the failure of storage facilities. An article in an AEC publication has cited nine cases of tank failure out of 183 tanks located in Washington, South Carolina, and Idaho. And a passage in the AEC's authorizing legislation for 1968 called for funding of $2,500,000 for the replacement of failed and failing tanks in Richland, Washington. "There is no assurance," concluded the passage, "that the need for new waste storage tanks can be forestalled." If this is the case after twenty years of storage experience, it is beyond belief that this burden will be borne without some storage failures for centuries in the future. Remember too, that these waste-holding "tank farms" are vulnerable to natural catastrophes such as earthquakes, and to man-made ones such as sabotage.

Efforts are of course being made toward effective handling of the waste problems, but many technical barriers must still be overcome. It is unlikely they will all be overcome by the end of the century, when waste tanks will boil with 6 billion curies of strontium-90, 5.3 billion curies of cesium-137, 6.07 billion curies of prometheum-147, 10.1 billion curies of cerium-144, and millions of curies of other isotopes. The amount of strontium-90 alone is 30 times more than would be released by the nuclear war envisioned in a 1959 congressional hearing.

The burden that radioactive wastes place on future generations is cruel and may prove intolerable. Physicist Joel A. Snow stated it well when he wrote in *Scientist and Citizen:* "Over periods of hundreds of years it is impossible to ensure that society will remain responsive to the problems created by the legacy of nuclear waste which we have left behind."

"Legacy" is indeed a gracious way of describing the reality of this situation, for at the very least we are saddling our children and their descendants with perpetual custodianship of our atomic refuse, and at worst may be dooming them to the same agonizing afflictions and deaths suffered by those who survived Hiroshima. Radiation has been positively linked to cancer, leukemia, brain damage, infant mortality, cataracts, sterility, genetic defects and mutations, and general shortening of life.

The implications for the survival of mankind can be glimpsed by considering just one of these effects, the genetic. In a 1960 article, James F. Crow, Professor of Genetics at the University of Wisconsin School of Medicine and president of the Genetics Society of America, stated that for every roentgen of slow radiation—the kind we can expect to receive in increasing doses from peacetime nuclear activity—about five mutations will result per 100 million genes exposed, meaning that "after a number of generations of exposure to one roentgen per generation, about one in 8,000 . . . in each generation would have severe genetic defects attributable to the radiation."

The Atomic Energy Commission is aware of the many objections that have been raised to the atomic power program: why does it continue to encourage it? Unfortunately, the Commission must perform two conflicting roles. On the one hand, it is responsible for regulating the atomic power industry. But on the other, it has been charged by Congress to promote the use of nuclear energy by the utility industry. Because of its involvement in the highest priorities of national security, enormous power and legislative advantages have been vested in the AEC, enabling it to fulfill its role as promoter with almost unhampered success—while its effectiveness as regulator has gradually atrophied. The Commission consistently denies claims that atomic power is heading for troubled waters, optimistically reassuring critics that these plants are safe, clean neighbors.

The fact that there is no foundation for this optimism is emphasized by the insurance situation on atomic facilities. Despite the AEC's own assertion that as much as $7 billion in property damage could result from an atomic power plant catastrophe, the insurance industry, working through two pools, will put up no more than $74 million, or about one per cent, to indemnify equipment manufacturers and utility operators against damage suits from the public. The federal government will add up to $486 million more, but this still leaves more than $6 billion in property damages to be picked up by victims of a Brookhaven-sized accident. And no insurance company —not even Lloyds of London—will issue property insurance to individuals against radiation damage. If there is so little risk in atomic power plants, why is insurance so inadequate?

The knowledge that man must henceforth live in constant dread of a major nuclear plant accident is disturbing enough. But we must recognize that even if such calamities are averted, the slow saturation of our environment with radioactive wastes will nevertheless be raising the odds that you or your heirs will fall victim to one of a multitude of afflictions. There is no "threshold" exposure below which we can feel safe.

We have little time to reflect on our alternatives, for the moment must soon come when no reversal will be possible. Dr. L. P. Hatch of Brookhaven National Laboratory vividly made this point when he told the Joint Committee on Atomic Energy: "If we were to go on for 50 years in the atomic power industry, and find that we had reached an impasse, that we had been doing the wrong thing with the wastes and we would like to reconsider the disposal methods, it would be entirely too late, because the problem would exist and nothing could be done to change that fact for the next, say, 600 or a thousand years." To which might be added a sobering thought stated by Dr. David Price of the U.S. Public Health Service: "We all live under the haunting fear that something may corrupt the environment to the point where man joins the dinosaurs as an obsolete form of life. And what makes these thoughts all the more disturbing is the knowledge that our fate could perhaps be sealed twenty or more years before the development of symptoms."

What must be done to avert the perils of the peaceful atom? A number of plans have been put forward for stricter regulation of activities in the nuclear utility field, such as limiting the size of reactors or their proximity to population concentrations or building more safeguards. As sensible as these proposals appear on the surface, they fail to recognize a number of important realities: first, that such arrangements would probably be opposed by utility operators and the government due to their prohibitively high costs. Since our government seems to be committed to making atomic power plants competitive with conventionally fueled plants, and because businesses are in business for profit, it is hardly likely they would buy these answers. Second, the techni-

cal problems involved in containment of radioactivity have not been successfully overcome, and there is little likelihood they will be resolved in time to prevent immense and irrevocable harm to our environment. Third, the nature of business enterprise is unfortunately such that *perfect* policing of the atomic power industry is unachievable. As we have seen in the cases of other forms of pollution, the public spirit of men seeking profit from industrial processes does not always rise as high as the welfare of society requires. It is unwise to hope that stricter regulation would do the job.

What, then, is the answer? The only course may be to turn boldly away from atomic energy as a major source of electricity production, abandoning it as this nation has abandoned other costly but unsuccessful technological enterprises.

There is no doubt that, with this nation's demand for electricity doubling every decade, new power sources are urgently needed. Nor is there doubt that our conventional fuel reserves —coal, oil, and natural gas—are rapidly being consumed. Sufficient high-grade fossil fuel reserves exist, however, to carry us to the end of this century; and new techniques for recovering these fuels from secondary sources such as oil shale could extend the time even longer. Furthermore, advances in pollution abatement technology and revolutionary new techniques, now in development, for burning conventional fuels with high efficiency, could carry us well into the next century with the fossil fuels we have. This abundance, and potential abundance, gives us at least several decades to survey possible alternatives to atomic power, select the most promising, and develop them on an appropriate scale as alternatives to nuclear power. Solar energy, tidal power, heat from the earth's core, and even garbage and solid-waste incineration have to some degree been demonstrated as promising means of electricity generation. If we subsidized research and development of those fields as liberally as we have done atomic energy, some of them would undoubtedly prove to be what atomic energy once promised, without its deadly drawbacks.

Aside from the positive prospect of profitability in these new approaches, industry will have another powerful incentive for turning to them; namely, that atomic energy is proving to be quite the opposite of the cheap, everlasting resource envisioned at the outset of the atomic age. The prices of reactors and components and costs of construction and operation have soared in the last few years, greatly damaging nuclear power's position as a competitor with conventional fuels. If insurance premiums and other indirect subsidies are brought into line with realistic estimates of what it takes to make atomic energy both safe and economical, the atom might prove to be the most *expensive* form of energy yet devised—not the cheapest. In addition, because of our wasteful fuel policies, evidence indicates that sources of low-cost uranium will be exhausted before the turn of the century. Fuel-producing breeder reactors, in which the nuclear establishment has invested such high hopes for the creation of vast, new fuel supplies, have proven a distinct technological disappointment. Even if the problems plaguing this effort were overcome in the next ten or twenty years, it may still be too late to recoup the losses of nuclear fuel reserves brought about by prodigious misman-agement.

The proposal to abandon or severely curtail the use of atomic energy is clearly a difficult one to imagine. We have only to realize, however, that by pursuing our current civilian nuclear power program, we are jeopardizing every other industry in the country; in that light, this proposal becomes the only practical alternative. In short, the entire national community stands to benefit from the abandonment of a policy which seems to be leading us toward both environmental and economic disaster.

Man's incomplete understanding of many technological principles and natural forces is not necessarily to his discredit. Indeed, that he has erected empires despite his limited knowledge is to his glory. But that he pits this ignorance and uncertainty, and the fragile yet lethal technology he has woven out of them, against the uncertainties of nature, science, and human behavior—this may well be to his everlasting sorrow.

Suggested Additional Reading

Change, Hope and the Bomb. D. E. Lilienthal. Princeton University Press, Princeton, 1963.

Living With the Atom. R. Calder. University of Chicago Press, Chicago, 1962.

The Integrity of Science: A Report by the AAAS Committee on Science in the Promotion of Human Welfare. American Scientist, 53, June, 1965.

Thermal Pollution
LaMont C. Cole

Introduction

There are etymological objections to the expression "thermal pollution," but it is gaining adoption and I shall here accept it without comment as a descriptive term for unwanted heat energy accumulating in any phase of the environment.

Any isolated body drifting in space as the earth is must either increase continuously in temperature or it must dispose of all energy received from external sources or generated internally by one of two processes. The energy may either be stored in some potential form or it must be reradiated to space.

In the past the earth has stored some of the energy coming to it from outside. This was accomplished by living plants using solar energy to drive endothermic chemical reactions, thereby creating organic compounds some of which were ultimately deposited in sediments. This process had two very important effects: in protecting the organic compounds from oxidation it created a reservoir of oxygen in the atmosphere; and it also eliminated the necessity for reradiating to space the heat energy that would have been released by the oxidation of those compounds. Part of the stored organic matter gave rise to the fossil fuels, coal, oil, and natural gas, which we are so avidly burning today—and now the earth must at last radiate that heat energy back to space, or its temperature will increase.

Many of the most important problems currently facing man are ecological problems arising from the unrestrained growth of the human population and the resultant increasing strains being placed on the earth's life support system. Our seemingly insatiable demand for energy is one important source of strains on the earth's capacity to support life, and I propose here to examine it in very elementary terms.

The Earth's Sources of Energy

Well over 99.999% of the earth's annual energy income is from solar radiation and this must all be returned to space in the form of radiant heat. This has gone on forever, so to speak, and conditions on earth, since before the advent of man, have reflected the necessity of maintaining this balance of incoming and outgoing radiation. A ridiculously small proportion of the sunlight reaching us is used in photosynthesis by green plants. This energy powers all of the earth's microorganisms and animals and is converted to heat by their metabolism. In addition, man can obtain useful heat energy by burning organic matter in such forms as wood, straw, cattle dung, and garbage and other refuse. This is "free" energy in the sense that the earth would have reradiated it as heat even if man had not obtained useful work from it in the meantime. Physical labor performed by man and the work done by domestic animals are also means of utilizing this solar energy.

In addition, it is solar radiation that keeps the atmosphere and hydrosphere in motion. To the extent that man can utilize the energy of the winds, of falling water and of ocean currents, or can make direct use of sunlight, he can do so without imposing increased thermal stress on the total earth environment.

The earth has some other minor sources of energy that contribute to its natural radiation balance. The tides can be used to obtain useful energy and a plant in France is now producing electricity from this source. Also, tidal friction is very gradually slowing the rotation of the earth, thereby converting kinetic energy to heat which plays a small role in maintaining the earth's surface temperature before it is radiated to space. Heat is also emerging from the interior of the earth and current concepts attribute this heat to natural radioactivity. In some local areas this heat flux is concentrated, and in a few countries man utilizes it for such purposes as heating buildings and generating electricity. Italy generates more than 400,000 kw of electricity from geothermal heat and in the United States about 85,000 kw are derived from a geyser field north of San Francisco.

This then is the inventory of energy sources available to man without affecting the surface temperature of the earth or the quantity of heat energy that it must dispose of by radiation. It is important to note this because, at least in the "developed" nations, we actually seem to be regressing from the use of these natural energy sources. Certainly windmills, animal power, and manual labor are much less in evidence today than they were during my childhood, and useful work done by sailboats seems to be a thing of the past in our culture. When we burn fossil fuels or generate electricity by nuclear or thermonuclear reactions, we must inevitably impose an increased thermal stress on the earth environment and, to the extent that this heat is undesirable, it constitutes thermal pollution.

The Earth's Radiation Balance

I assume the mean temperature of the earth's surface to be 15 C or 288 K; this may be a degree too low or too high. In order to keep the discussion sufficiently simple-minded, I shall introduce three simplifications.

First, I shall treat the earth as though its entire surface was at a uniform temperature equal to its average value, thus ignoring temperature differences due to latitude, altitude, season, and time of day. The effect of this simplification is less drastic than one might expect. It is true if one measures radiation from several bodies at different temperatures and infers the mean temperature of the surfaces from the radiation, he will obtain values that are slightly too high. For example, if we have two otherwise identical areas, one at 0 C and the other at 50 C, the average of their combined radiation will equal that which would be given off by such an area at 28 C rather than at the true mean temperature of 25 C. For the earth as a whole, the error from this source is not likely to amount to even one degree, and the other two simplifications I shall make tend to cause small errors in the opposite direction.

Second, I assume that the earth radiates as

a blackbody or perfect radiator. This is probably nearly correct; and, in any case, the assumption is conservative for our purposes here. If the earth is actually a gray body rather than a black one, it will have to reach a somewhat higher temperature to dispose of the same amount of heat energy.

Finally, I assume that the earth radiates its energy to outer space which is at a temperature of 0 K. This is not quite correct because the portions of the sky occupied by the sun, moon, stars, and clouds of interstellar matter are at temperatures above absolute zero, and the earth's surface must therefore by very slightly warmer than it would otherwise be.

Accepting these simplifications, we can now turn to the Stefan-Boltzmann law of elementary physics which states that radiation is proportional to the fourth power of the absolute temperature, and easily calculate the amount of energy radiated from the earth to space.[1] At a temperature of 15 C (288 K) the total radiation from the earth turns out to be 2×10^{24} ergs/sec.

Or, looked at the other way around, if we know the quantity of heat that the earth's surface must get rid of by radiation, we can calculate what its surface temperature must be for it to do so. For example, various students of the subject have concluded that the mean temperature difference between glacial and ice-free periods is quite small, probably no more than 5 C (e.g., see Brooks, 1949). If we can assume that we are now about midway between these climatic types, then a rise of 3 C might be expected to melt the icecaps from Antarctica and Greenland thereby raising sea-level by some 100 m. This would drastically alter the world's coastlines as, for example, by

1. I take the surface area of the earth as 5.1×10^{18} cm^2 and the value of the Stefan constant as 5.67×10^{-5} erg cm^{-2} deg^{-4} sec^{-1}.

putting all of Florida under water, and drowning most of the world's major cities. To do this would require a 4.2% increase in the earth's heat budget (an increase of 8.44 x 10^{22} ergs/sec). By the same reasoning, a 4.1% decrease in the energy budget could be expected to bring on a new ice age. As we shall see presently, however, these things are not really quite so simple and predictable.

Man's Effect on the Heat Budget

The amount of energy now being produced by man, I take to be 5 x 10^{19} ergs/sec or 25/1000 of 1% of the total radiated by the earth. Strangely enough, I am quite happy with this perhaps brash attempt to estimate a difficult quantity. Putnam (1953) estimated the fuel burned by man 16 years ago at 3.3 x 10^{19} ergs/sec. If he was correct then, and if energy demand then had been growing at the rate it is now growing, man would now be using twice what I have estimated. Kardashev (1964) estimates the energy now produced by civilization at "over 4 x 10^{19} ergs per second" which is consistent with my estimate.

There is another way of getting at the figure. The rate of combustion of fossil fuels in the United States is accurately known, as is our rate of generating electricity. Our electrical production as of 1966 corresponds to one-fifth of our fossil fuel consumption. If we assume the same ratio for the entire earth, for which we do have credible figures on electrical generation, the world energy production turns out to be 4.4 x 10^{19} ergs/sec. With this many independent estimates converging, I am happy with the figure of 5 x 10^{19} ergs/sec.

This is such a tiny part of the earth's output of energy that it is evident that the heat released by man now has an absolutely insignificant direct effect on the average temperature of the earth's surface. Will this be true if

we go on increasing our demands for power? I have been hearing utility company officials assert that we must keep electrical generating capacity growing by 10% per year, but a more common projection is about 7%, which rate would double the capacity every 10 years. I am confident that nonelectrical uses of fuel for such purposes as heating, industry, and transportation are growing at least as rapidly, and I am told that the "developing" countries are going to continue to develop. Let us examine the consequences of an energy economy that continues to grow at 7% per year.

Waggoner (1966) considers it at least possible that a warming of 1 C would cause real changes in the boundaries between plant communities. For this to occur, the earth's energy budget would have to increase by about 3 x 10^{22} ergs/sec. How long would it take man to cause this at an increased energy production of 7% per year? The answer is 91 years.

As already mentioned, a rise of 3 C could, in the opinion of competent authorities, melt the ice caps and produce an earth geography such as has never been seen by man, and on which there would be much less dry land for man. This would take about 108 years to achieve by present trends. Let us rejoice that the United States Civil War did not embark on an energy-releasing spree as we are now doing.

The highest mean annual temperature of any spot on earth is believed to be 29.9 C (302.9 K) at Massawa on the Red Sea in Ethiopia. I think it is safe to assume that if the average temperature of the whole earth was raised to 30 C, it would become uninhabitable. This would take 130 years under our postulated conditions—the time from the start of the Victorian age in England to the present.

These calculations, rough as they are, make it clear that man is on a collision course with disaster if he tries to keep energy production growing by means that will impose an increased thermal stress on the earth. I have here ignored the fact that the fossil fuels, and probably uranium and thorium reserves also, would be exhausted before these drastic effects could be attained. The possibility exists, however, that a controlled fusion reaction will be achieved and bring these disastrous effects within the realm of possibility.

How can we avoid the consequences of the trends we are following? My answer would be to determine what level of human population the earth can support at a desirable standard of living without undergoing deterioration and then to move to achieve this steady state condition. I would like to see increased energy needs met by the direct utilization of solar energy. There are, however, visionary scientists, and policy makers who will listen to them, who will grasp at any straw to keep man's exploitation of the earth forever growing. In my mind I can hear them planning to air condition part of the earth for man and for whatever then will supply his food, while the rest of the earth is allowed to radiate the excess heat by attaining a very high temperature. They will consider putting generating plants and factories on the moon and other planets and reducing the heat stress on earth by reflecting solar radiation back to space. This latter possibility brings us to consideration of some secondary effects of our expanding energy budget.

Side Effects of Energy Use

The combustion of fuel releases not only heat but also frequently smoke and various chemicals, the most important of which are water and carbon dioxide, into the environment.

Smoke and other particulate matter in the atmosphere scatters and absorbs incoming solar radiation, thus reducing the amount of energy absorbed by the surface. It has little effect on the outgoing longwave radiation, so the net effect is a tendency to cool the earth's surface. Several large volcanic eruptions within historic times have caused a year or so of abnormally low temperatures all over the earth. Perhaps the most striking case was the mighty eruption of Mt. Tomboro in the Lesser Sunda Islands in 1815. The following year, 1816, was the famous "year without a summer," during which snow fell in Boston every month. Actual measurements following the 1912 eruption of Mt. Katmai in the Aleutians showed a reduction of about 20% in the solar radiation reaching the earth.

Water vapor and carbon dioxide have an effect opposite to that of smoke; they are transparent to sunlight but absorb the longwave radiation from the earth and, by reradiating some of it back to earth, tend to raise the surface temperature. Man's combustion of fossil fuels has caused a measurable increase in the CO_2 content of the atmosphere, and now increasing numbers of jet airplanes are releasing great quantities of both water vapor and CO_2 at high altitudes. This would tend to raise the earth's temperature. However, a phenomenon of increasing frequency is the coalescence of the contrails of jet airplanes into banks of cirrus clouds which will reflect some of the incoming solar radiation back to space. Obviously, we cannot now be certain of the ultimate effects of the materials we are releasing to the atmosphere, but they certainly have the potential for changing climates.

On a more local scale, we are using prodigious quantities of water for cooling industrial plants, especially electrical generating plants, and this use is expected to increase at least as rapidly as our energy use grows[2]—perhaps

2. Thermally more efficient electrical generating plants are considered possible: e.g., the "magnetohydrodynamic" generator (see Rosa and Hals, 1968).

more rapidly because nuclear plants waste more heat per kilowatt than plants burning fossil fuels. When heated water is discharged at a temperature above that of the air, we must expect an increase in the frequencies of mist and fog and, in winter, icing conditions.

Biological Effects of Thermal Pollution

The first and most obvious biological effect one thinks of is that bodies of water may become so hot that nothing can live in them. It is true that there are a few bacteria and blue-green algae that can grow in hot springs but even these are very unusual in water above 60 C. I know of only one case of a green alga living above 50 C—a species of *Protococcus* from Yellowstone Park. A few rotifers, nematodes, and protozoa have been found at above 50 C, and some, in a dried and encysted state, will survive much higher temperatures. In general, no higher organisms are to be expected actively living in water above about 35 C. I find it difficult to reconcile the data with the conclusion of Wurtz (1968, p. 139): "Water temperature would have to be increased to about 130 F (54.4 C) to destroy the microorganisms that are responsible for the self-purification capacity of a lake or stream." This statement is certainly incompatible with the recommendation of the National Technical Advisory Committee (1968) that: "All surface waters should be capable of supporting life forms of aesthetic value."

Another factor to be considered is the effect of temperature on the types of organisms present. Fish of any type are rare above 30 C. Diatoms, which are important members of aquatic food chains, decrease as temperature rises above 20 C, and they are gradually replaced by blue-green algae which are not important as food for animals, are often toxic, and are often the source of water blooms which kill the biota and make the water unfit for domestic use. Typically at about 30 C, diatoms and blue-greens are about equally represented and green algae exceed both. At 35 C, the diatoms are nearly gone, the greens are decreasing rapidly, and the blue-greens are assuming full dominance.

So far as animals are concerned, a body of water at a temperature of 30-35 C is essentially a biological desert. The green algae, which are well above their optimum temperature, can support a few types of cladoceran, amphiphod, and isopod crustaceans; bacteria and blue-green algae may abound; and mosquito larvae may do very well; rooted plants may grow in shallow regions; and a few crayfish, carp, goldfish, and catfish may endure. Largemouth bass can survive and grow at 32 C but they do not reproduce above about 24 C. A few other forms such as aquatic insects may inhabit the water as adults but may or may not be able to complete their life cycles there. Desirable game fishes such as Atlantic salmon, lake trout, northern pike, and walleyes require water below 10 C for reproduction.

Another effect of raising the temperature of water is to reduce the solubility of gases. The amount of oxygen dissolved in water in equilibrium with the atmosphere decreases by over 17% between 20 C and 30 C. At the same time, the need of organisms for oxygen increases. As a rule of thumb, metabolic rate approximately doubles for a 10 C increase in body temperature, although the effect is sometimes greater. Krogh (1914) found that the rate of development of frog eggs is about six times as great at 20 C as at 10 C. Similar effects have been noted in the rate of development of mosquitos.

Dissolved oxygen is often in critically short supply for aquatic organisms and increased temperature aggravates the situation. This may be partially compensated by more rapid diffusion of oxygen from the air and, during daylight hours, by photosynthesis. On the other hand, decay of organic matter and other oxidative processes such as the rusting of iron are more rapid at high temperatures. In polluted water the effect of biochemical oxygen demand (BOD) is more severe at high temperatures. The addition of heat to estuaries may be more critical than in bodies of freshwater because saltwater has a slightly lower specific heat and because oxygen is less soluble in saltwater.

In contrast to the situation with gases, salts become more soluble in water as the temperature increases. Chemical reactions become more rapid and increased evaporation may further increase the concentration of dissolved salts. At the same time the rate of exchange of substances between aquatic organisms and the medium increases. Toxins are likely to have greater effects, and parasites and diseases are more likely to break out and spread. In general, water at a temperature above the optimum places a strain on metabolic processes that may make adaptation to other environmental factors more difficult. For example, the Japanese oyster can tolerate a wider range of salinity in winter than in summer (Reid, 1961, p. 267).

In addition, if the water contains plant nutrients, objectionable growths of aquatic plants may be promoted by increased temperature—extreme cases leading to heavy mortality of fishes and other animals. It has been reported that polluted water from Lake Superior which does not support algal blooms there will do so if warmed to the temperature of Lake Erie.

Still other factors come into play when a body of water is thermally stratified. In a deep cold lake such as Cayuga where we are currently threatened with a huge nuclear generating plant, the lake water mixes, usually in May, so the entire lake is well oxygenated. As the surface warms, the lake stratifies with a level of light, warm water (the epilimnion)

floating with no appreciable mixing on a mass of dense cold water (the hypolimnion) in which the lake trout, their food organisms, and other things are living and consuming oxygen. This continues until the lake mixes again, usually in November, by which time the oxygen supply in the hypolimnion is seriously depleted. The power company plans to pump 750 million gallons per day from the hypolimnion at a temperature averaging perhaps 6 C and to discharge it at the surface at a temperature of about 21 C. They plan this despite a recommendation of the National Technical Advisory Committee (1968, p. 33) that: ". . . water for cooling should not be pumped from the hypolimnion to be discharged to the same body of water." The effect of this addition of heat on the average temperature of the lake will be trivially small, but the biological consequences can be out of all proportion to the amount of heating.

The heat will delay fall cooling of the epilimnion and hasten spring warming, so that the length of time the lake is stratified each year will be increased. The water from the hypolimnion is rich in available plant nutrients, and by warming it and discharging it in the lighted zone, the amount of plant growth will be increased. This means more organic matter sinking into the hypolimnion and using up oxygen when it decays. The threat to the welfare of the lake is very real.

Finally, we should comment on fluctuating temperatures. Many organisms can adapt to somewhat higher or lower temperatures if they have time. The water in reservoirs behind hydroelectric dams often becomes thermally stratified, and when it is released at the base of the dam, a stretch of cold stream is produced which can support a cold water fauna even in warm regions. But when water is only discharged during peak electrical generating hours, the stream becomes subject to severe temperature fluctuations that will exclude many sensitive organisms. Similarly, if fishes or other organisms acclimate to the warm discharge from a factory or power plant and congregate near it, they will be subjected to temperature shocks when the plant is shut down for maintenance or refueling.

Conclusion

Man cannot go on increasing his use of thermal energy without causing degradation of his environment, and if he is persistent enough, he will destroy himself. There are other energy sources that could be used, but no source can support an indefinitely growing population. As with so many other things, it is man's irresponsible proliferation in numbers that is the real heart of the problem. There is some population size that the earth could support indefinitely without undergoing deterioration, but people do not even want to consider what that number might be. I suspect that it is substantially below the present world population. One would think that any rational creature riding a space ship would take care not to damage or destroy the ship, but perhaps the word "rational" does not describe man.

REFERENCES

Brooks, C. E. P. 1949. *Climate Through the Ages.* Rev. ed. McGraw-Hill Book Co., New York.

Kardashev, N. S. 1964. Transmission of information by extraterrestrial civilizations. In: *Extraterrestrial Civilizations,* G. M. Tovmasyan (ed.). Trans. by National Aeronautics and Space Administration, Washington, D.C.

Krogh, A. 1914. On the influence of temperature on the rate of embryonic development. *Z. Allg. Physiol.,* 16: 163-177.

National Technical Advisory Committee. 1968. Water quality criteria. Report of the National Technical Advisory Committee to the Secretary of the Interior. Federal Water Pollution Control Administration, Washington, D.C.

Putnam, P. C. 1953. *Energy in the Future.* D. Van Nostrand Co., New York.

Reid, G. K. 1961. *Ecology of Inland Waters and Estuaries.* Reinhold Publishing Co., New York.

Rosa, R. J., and F. A. Hals. 1968. In defense of MHD. *Ind. Res.,* June 1968: 68-72.

Waggoner, P. E. 1966. Weather modification and the living environment. In: *Future Environments of North America,* F. F. Darling and J. F. Milton (eds.). Natural History Press, Garden City, N.Y.

Wurtz, C. B. 1968. Thermal pollution: The effect of the problem. In: *Environmental Problems,* B. R. Wilson (ed.). J. B. Lippincott Co., Philadelphia.

Effects of Pollution on the Structure and Physiology of Ecosystems
George M. Woodwell

The accumulation of various toxic substances in the biosphere is leading to complex changes in the structure and function of natural ecosystems. Although the changes are complex, they follow in aggregate patterns that are similar in many different ecosystems and are therefore broadly predictable. The patterns involve many changes but include especially simplification of the structure of both plant and animal communities, shifts in the ratio of gross production to total respiration, and loss of part or all of the inventory of nutrients. Despite the frequency with which various pollutants are causing such changes and the significance of the changes for all living systems,[1] only a few studies show details of the pattern of change clearly. These are studies of the effects of ionizing radiation, of persistent pesticides, and of eutrophication. The effects of radiation will be used here to show the pattern of changes in terrestrial plant communities and to show similarities with the effects of fire, oxides of sulfur, and herbicides. Effects of such pollutants as pesticides on the animal community are less conspicuous but quite parallel, which shows that the ecological effects of pollution correspond very closely to the general "strategy of ecosystem development" outlined by Odum[1] and that they can be anticipated in considerable detail.

The problems caused by pollution are of interest from two viewpoints. Practical people —toxicologists, engineers, health physicists, public health officials, intensive users of the environment—consider pollution primarily as a direct hazard to man. Others, no less concerned for human welfare but with less pressing public responsibilities, recognize that toxicity to humans is but one aspect of the pollution problem, the other being a threat to the maintenance of a biosphere suitable for life as we know it. The first viewpoint leads to emphasis on human food chains; the second leads to emphasis on human welfare insofar as it depends on the integrity of the diverse ecosystems of the earth, the living systems that appear to have built and now maintain the biosphere.

The food-chain problem is by far the simpler; it is amenable at least in part to the pragmatic, narrowly compartmentalized solutions that industrialized societies are good at. The best example of the toxicological approach is in control of mutagens, particularly the radionuclides. These present a specific, direct hazard to man. They are much more important to man than to other organisms. A slightly enhanced rate of mutation is a serious danger to man, who has developed through medical science elaborate ways of preserving a high fraction of the genetic defects in the population; it is trivial to the rest of the biota, in which genetic defects may be eliminated through selection. This is an important fact about pollution hazards—toxic substances that are principally mutagenic are usually of far greater direct hazard to man than to the rest of the earth's biota and must be considered first from the standpoint of their movement to man through food webs or other mechanisms and to a much lesser extent from that of their effects on the ecosystem through which they move. We have erred, as shown below, in assuming that all toxic substances should be treated this way.

Pollutants that affect other components of the earth's biota as well as man present a far greater problem. Their effects are chronic and may be cumulative in contrast to the effects of short-lived disturbances that are repaired by succession. We ask what effects such pollutants have on the structure of natural ecosystems and on biological diversity and what these changes mean to physiology, especially to mineral cycling and the long-term potential for sustaining life.

Although experience with pollution of various types is extensive and growing rapidly, only a limited number of detailed case history studies provide convincing control data that deal with the structure of ecosystems. One of the clearest and most detailed series of experiments in recent years has been focused on the ecological effects of radiation. These studies are especially useful because they allow cause and effect to be related quantitatively at the ecosystem level, which is difficult to do in nature. The question arises, however, whether the results from studies of ionizing radiation, a factor that is not usually considered to have played an important role in recent evolution, have any general application. The answer, somewhat surprisingly to many biologists, seems to be that they do. The ecological effects of radiation follow patterns that are known from other types of disturbances. The studies of radiation, because of their specificity, provide useful clues for examination of effects of other types of pollution for which evidence is much more fragmentary.

The effects of chronic irradiation of a late successional oak-pine forest have been studied at Brookhaven National Laboratory in New York. After 6 months' exposure to chronic irradiation from a ^{137}Cs source, five well-defined zones of modification of vegetation had been established. They have become more pronounced through 7 years of chronic irradiation. The zones were:

1) A central devastated zone, where exposures were > 200 R/day and no higher plants survived, although certain mosses and lichens survived up to exposures > 1000 R/day.

2) A sedge zone, where *Carex pensylvanica*[2] survived and ultimately formed a continuous cover (> 150 R/day).

3) A shrub zone in which two species of *Vaccinium* and one of *Gaylussacia* survived, with *Quercus ilicifolia* toward the outer limit of the circle where exposures were lowest (> 40 R/day).

4) An oak zone, the pine having been eliminated (> 16 R/day).

5) Oak-pine forest, where exposures were < 2 R/day, and there was no obvious change in the number of species, although small changes in rates of growth were measurable at exposures as low as 1 R/day.

The effect was a systematic dissection of the forest, strata being removed layer by layer. Trees were eliminated at low exposures, then the taller shrubs (*Gaylussacia baccata*), then the lower shrubs (*Vaccinium species*), then the herbs, and finally the lichens and mosses. Within these groups, it was evident that under irradiation an upright form of growth was a disadvantage. The trees did vary—the pines

(*Pinus rigida*) for instance were far more sensitive than the oaks without having a conspicuous tendency toward more upright growth, but all the trees were substantially more sensitive than the shrubs.[3] Within the shrub zone, tall forms were more sensitive; even within the lichen populations, foliose and fruticose lichens proved more sensitive than crustose lichens.[4]

The changes caused by chronic irradiation of herb communities in old fields show the same pattern—upright species are at a disadvantage. In one old field at Brookhaven, the frequency of low-growing plants increased along the gradient of increasing radiation intensity to 100 percent at > 1000 R/day.[5] Comparison of the sensitivity of the herb field with that of the forest, by whatever criterion, clearly shows the field to be more resistant than the forest. The exposure reducing diversity to 50 percent in the first year was ~1000 R/day for the field and 160 R/day for the forest, a greater than fivefold difference in sensitivity.[3]

The changes in these ecosystems under chronic irradiation are best summarized as changes in structure, although diversity, primary production, total respiration, and nutrient inventory are also involved. The changes are similar to the familiar ones along natural gradients of increasingly severe conditions, such as exposure on mountains, salt spray, and water availability. Along all these gradients the conspicuous change is a reduction of structure from forest toward communities dominated by certain shrubs, then, under more severe conditions, by certain herbs, and finally by low-growing plants, frequently mosses and lichens. Succession, insofar as it has played any role in the irradiated ecosystems, has simply reinforced this pattern, adding a very few hardy species and allowing expansion of the populations of more resistant indigenous species. The reasons for radiation's causing this pattern are still not clear,[3,6] but the pattern is a common one, not peculiar to ionizing radiation, despite the novelty of radiation exposures as high as these.

Its commonness is illustrated by the response to fire, one of the oldest and most important disruptions of nature. The oak-pine forests such as those on Long Island have, throughout their extensive range in eastern North America been subject in recent times to repeated burning. The changes in physiognomy of the vegetation follow the above pattern very closely—the forest is replaced by communities of shrubs, especially bear oak (*Quercus ilicifolia*), *Gaylussacia,* and *Vaccinium species.* This change is equivalent to that caused by chronic exposure to 40 R/day or more. Buell and Cantlon,[7] working on similar vegetation in New Jersey, showed that a further increase in the frequency of fires resulted in a differential reduction in taller shrubs first, and a substantial increase in the abundance of *Carex pensylvanica,* the same sedge now dominating the sedge zone of the irradiated forest. The parallel is detailed; radiation and repeated fires both reduce the structure of the forest in similar ways, favoring low-growing hardy species.

The similarity of response appears to extend to other vegetations as well. G. L. Miller, working with F. McCormick at the Savannah River Laboratory, has shown recently that the most radiation-resistant and fire-resistant species of 20-year-old fields are annuals and perennials characteristic of disturbed places.[8] An interesting sidelight of his study was the observation that the grass stage of long leaf pine (*Pinus palustris*), long considered a specific adaptation to the fires that maintain the southeastern savannahs, appears more resistant to radiation damage than the mature trees. At a total acute exposure of 2.1 kR (3 R/day), 85 percent of the grass-stage populations survived but only 55 percent of larger trees survived. Seasonal variation in sensitivity to radiation damage has been abundantly demonstrated,[9] and it would not be surprising to find that this variation is related to the ecology of the species. Again it appears that the response to radiation is not unique.

The species surviving high radiation exposure rates in the Brookhaven experiments are the ones commonly found in disturbed places, such as roadsides, gravel banks, and areas with nutrient deficient or unstable soil. In the forest they include *Comptonia peregrina* (the sweet fern), a decumbent spiny *Rubus,* and the lichens, especially *Cladonia cristatella.* In the old field one of the most conspicuously resistant species was *Digitaria sanguinalis* (crabgrass) among several other weedy species. Clearly these species are generalists in the sense that they survive a wide range of conditions, including exposure to high intensities of ionizing radiation—hardly a common experience in nature but apparently one that elicits a common response.

With this background one might predict that a similar pattern of devastation would result from such pollutants as oxides of sulfur released from smelting. The evidence is fragmentary, but Gorham and Gordon[10] found around the smelters in Sudbury, Ontario, a striking reduction in the number of species of higher plants along a gradient of 62 kilometers (39 miles). In different samples the number of species ranged from 19 to 31 at the more distant sites and dropped abruptly at 6.4 kilometers. At 1.6 kilometers, one of two randomly placed plots (20 by 2 meters) included only one species. They classified the damage in five categories, from "Not obvious" through "Moderate" to "Very severe." The tree canopy had been reduced or eliminated within 4.8 to 6.4 kilometers of the smelter, with only occasional sprouts of trees, seedlings, and successional herbs and shrubs remaining; this damage is equivalent to that produced by exposure to 40 R/day. The most resistant trees were, almost predictably to a botanist, red maple (*Quercus ruba*). Other species surviving in the zones of "Severe" and "Very severe" damage included *Sambucus pubens, Polygonum cilinode, Comptonia peregrina,* and *Epilobium angustifolium* (fire weed). The most sensitive plants appeared to be *Pinus strobus* and *Vaccinium myrtilloides.* The pine was reported no closer than 25.6 kilometers (16 miles), where it was chlorotic.

The example confirms the pattern of the change—first a reduction of diversity of the forest by elimination of sensitive species; then elimination of the tree canopy and survival of resistant shrubs and herbs widely recognized as "seral" or successional species or "generalists."

The effects of herbicides, despite their hope for specificity, fall into the same pattern, and it is no surprise that the extremely diverse forest canopies of Vietnam when sprayed repeatedly with herbicides are replaced over large areas by dense stands of species of bamboo.[11]

The mechanisms involved in producing this series of patterns in terrestrial ecosystems are not entirely clear. One mechanism that is almost certainly important is simply the ratio of gross production to respiration in different strata of the community. The size of trees has been shown to approach a limit set by the amount of surface area of stems and branches in proportion to the amount of leaf area.[12] The apparent reason is that, as a tree expands in size, the fraction of its total surface devoted to bark, which makes a major contribution to the respiration, expands more rapidly than does the photosynthetic area. Any chronic disturbance has a high probability of damaging the capacity for photosynthesis without reducing appreciably the total amount of respiration; therefore, large plants are more vulnerable than species requiring less total respiration. Thus chronic disturbances of widely different types favor plants that are small in stature, and any disturbance that tends to increase the amount of respiration in proportion to photosynthesis will aggravate this shift.

The shift in the structure of terrestrial plant communities toward shrubs, herbs, or mosses and lichens, involves changes in addition to those of structure and diversity. Simplification of the plant community involves also a reduction of the total standing crop of organic matter and a corresponding reduction in the total inventory of nutrient elements held within the system, a change that may have important long-term implications for the potential of the site to support life. The extent of such losses has been demonstrated by Bormann and his colleagues in the Hubbard Brook Forest in New Hampshire,[13] where all of the trees in a watershed were cut, the cut material was left to decay, and the losses of nutrients were monitored in the runoff. Total nitrogen losses in the first year were equivalent to twice the amount cycled in the system during a normal year. With the rise of nitrate ion in the runoff, concentrations of calcium, magnesium, sodium, and potassium ions rose severalfold, which caused eutrophication and even pollution of the streams fed by this watershed. The soil had little capacity to retain the nutrients that were locked in the biota once the higher plants had been killed. The total losses are not yet known, but early evidence indicates that they will be a high fraction of the nutrient inventory, which will cause a large reduction in the potential of the site for supporting living systems as complex as that destroyed—until nutrients accumulate again. Sources are limited; the principal source is erosion of primary minerals.

When the extent of the loss of nutrients that

accompanies a reduction in the structure of a plant community is recognized, it is not surprising to find depauperate vegetation in places subject to chronic disturbances. Extensive sections of central Long Island, for example, support a depauperate oak-pine forest in which the bear oak, *Quercus ilicifolia,* is the principal woody species. The cation content of an extremely dense stand of this common community, which has a biomass equivalent to that of the more diverse late successional forest that was burned much less recently and less intensively, would be about 60 percent that of the richer stand, despite the equivalence of standing crop. This means that the species, especially the bear oak, contain, and presumably require, lower concentrations of cations. This is an especially good example because the bear oak community is a long-lasting one in the fire succession and marks the transition from a high shrub community to forest. It has analogies elsewhere, such as the heath balds of the Great Smoky Mountains and certain bamboo thickets in Southeast Asia.

The potential of a site for supporting life depends heavily on the pool of nutrients available through breakdown of primary minerals and through recycling in the living portion of the ecosystem. Reduction of the structure of the system drains these pools in whole or in part; it puts leaks in the system. Any chronic pollution that affects the structure of ecosystems, especially the plant community, starts leaks and reduces the potential of the site for recovery. Reduction of the structure of forests in Southeast Asia by herbicides has dumped the nutrient pools of these large statured and extremely diverse forests. The nutrients are carried to the streams, which turn green with the algae that the nutrients support. Tschirley,[11] reporting his study of the effects of herbicides in Vietnam, recorded "surprise" and "pleasure" that fishing had improved in treated areas. If the herbicides are not toxic to fish, there should be little surprise at improved catches of certain kinds of fish in heavily enriched waters adjacent to herbicide-treated forests. The bamboo thickets that replace the forests also reflect the drastically lowered potential of these sites to support living systems. The time it takes to reestablish a forest with the original diversity depends on the availability of nutrients, and is probably very long in most lateritic soils.

In generalizing about pollution, I have concentrated on some of the grossest changes in the plant communities of terrestrial ecosystems. The emphasis on plants is appropriate because plants dominate terrestrial ecosystems. But not all pollutants affect plants directly; some have their principal effects on heterotrophs. What changes in the structure of animal communities are caused by such broadly toxic materials as most pesticides?

The general pattern of loss of structure is quite similar, although the structure of the animal communities is more difficult to chart. The transfer of energy appears to be one good criterion of structure. Various studies suggest that 10 to 20 percent of the energy entering the plant community is transferred directly to the animal community through herbivores.[14] Much of that energy, perhaps 50 percent or more, is used in respiration to support the herbivore population; some is transferred to the detritus food chain directly, and some, probably not more than 20 percent, is transferred to predators of the herbivores. In an evolutionarily and successionally mature community, this transfer of 10 to 20 percent per trophic level may occur two or three times to support carnivores, some highly specialized, such as certain eagles, hawks, and herons, others less specialized, such as gulls, ravens, rats, and people.

Changes in the plant community, such as its size, rate of energy fixation, and species, will affect the structure of the animal community as well. Introduction of a toxin specific for animals, such as a pesticide that is a generalized nerve toxin, will also topple the pyramid. Although the persistent pesticides are fat soluble and tend to accumulate in carnivores and reduce populations at the tops of food chains, they affect every trophic level, reducing reproductive capacity, almost certainly altering behavioral patterns, and disrupting the competitive relationships between species. Under these circumstances the highly specialized species, the obligate carnivores high in the trophic structure, are at a disadvantage because the food chain concentrates the toxin and, what is even more important, because the entire structure beneath them becomes unstable. Again the generalists or broad-niched species are favored, the gulls, rats, ravens, pigeons and, in a very narrow short-term sense, man. Thus the pesticides favor the herbivores, the very organisms they were invented to control.

Biological evolution has divided the resources of any site among a large variety of users—species—which, taken together, confer on that site the properties of a closely integrated system capable of conserving a diversity of life. The system has structure; its populations exist with certain definable, quantitative relationships to one another; it fixes energy and releases it at a measurable rate; and it contains an inventory of nutrients that is accumulated and recirculated, not lost. The system is far from static; it is subject, on a time scale very long compared with a human lifespan, to a continuing augmentive change through evolution; on a shorter time scale, it is subject to succession toward a more stable state after any disturbance. The successional patterns are themselves a product of the evolution of life, providing for systematic recovery from any acute disturbance. Without a detailed discussion of the theory of ecology, one can say that biological evolution following a

pattern approximating that outlined above, has built the earth's ecosystems, and that these systems have been the dominant influence on the earth throughout the span of human existence. The structure of these systems is now being changed all over the world. We know enough about the structure and function of these systems to predict the broad outline of the effects of pollution on both land and water. We know that as far as our interests in the next decades are concerned, pollution operates on the time scale of succession, not of evolution, and we cannot look to evolution to cure this set of problems. The loss of structure involves a shift away from complex arrangements of specialized species toward the generalists; away from forest, toward hardy shrubs and herbs; away from those phytoplankton of the open ocean that Wurster[15] proved so very sensitive to DDT, toward those algae of the sewage plants that are unaffected by almost everything including DDT and most fish; away from diversity in birds, plants, and fish toward monotony; away from tight nutrient circles toward very loose ones with terrestrial systems becoming overloaded; away from stability toward instability especially with regard to sizes of populations of small, rapidly reproducing organisms such as insects and rodents that compete with man; away from a world that runs itself through a self-augmentive, slowly moving evolution, to one that requires constant tinkering that is malignant in that each act of repair generates a need for further repairs to avert problems generated at compound interest.

This is the pattern, predictable in broad outline, aggravated by almost any pollutant. Once we recognize the pattern, we can begin to see the meaning of some of the changes occurring now in the earth's biota. We can see the demise of carnivorous birds and predict the demise of important fisheries. We can tell why, around industrial cities, hills that were once forested now are not; why each single species is important; and how the increase in the temperature of natural water bodies used to cool new reactors will, by augmenting respiration over photosynthesis, ultimately degrade the system and contribute to degradation of other interconnected ecosystems nearby. We can begin to speculate on where continued, exponential progress in this direction will lead; probably not to extinction—man will be around for a long time yet—but to a general degradation of the quality of life.

The solution? Fewer people, unpopular but increasing restrictions on technology (making it more and more expensive), and a concerted effort to tighten up human ecosystems to reduce their interactions with the rest of the earth on whose stability we all depend. This does not require foregoing nuclear energy; it requires that if we must dump heat, it should be dumped into civilization to enhance a respiration rate in a sewage plant or an agricultural ecosystem, not dumped outside civilization to affect that fraction of the earth's biota that sustains the earth as we know it. The question of what fraction that might be remains as one of the great issues, still scarcely considered by the scientific community.

References and Notes

1. E. P. Odum, *Science,* 164, 262 (1969).
2. Plant nomenclature follows that of M. L. Fernald in *Gray's Manual of Botany* (American Book, New York, ed. 8, 1950).
3. G. M. Woodwell, *Science,* 156, 461 (1967); G. M. Woodwell and A. L. Rebuck, *Ecol. Monogr.* 37, 53 (1967).
4. G. M. Woodwell and T. P. Gannutz, *Amer. J. Bot.* 54, 1210 (1967).
5. ———and J. K. Oostling, *Radiat. Bot.* 5, 205 (1965).
6. ———and R. H. Whittaker, *Quart. Rev. Biol.* 43, 42 (1968).
7. M. F. Buell and J. E. Cantlon, *Ecology* 34, 520 (1953).
8. G. L. Miller, thesis, Univ. of North Carolina (1968).
9. A. H. Sparrow, L. A. Schairer, R. C. Sparrow, W. F. Campbell, *Radiat. Bot.* 3, 169 (1963); F. G. Taylor, Jr., *ibid.* 6, 307 (1965).
10. E. Gorham and A. G. Gordon, *Can. J. Bot.* 38, 307 (1960); *Ibid.* p. 477; *ibid.* 41, 371 (1963).
11. F. H. Tschirley, *Science* 163, 779 (1969).
12. R. H. Whittaker and G. M. Woodwell, *Amer. J. Bot.* 54, 931 (1967).
13. F. H. Bormann, G. E. Likens, D. W. Fisher, R. S. Pierce, *Science* 159, 882 (1968).
14. These relationships have been summarized in detail by J. Phillipson [*Ecological Energetics* (St. Martin's Press, New York, 1966)]. See also L. B. Slobodkin, *Growth and Regulation of Animal Populations* (Holt, Rinehart and Winston, New York, 1961) and J. H. Ryther, *Science* 166, 72 (1969).
15. C. F. Wurster, *Science* 159, 1474 (1968).
16. G. M. Woodwell, *Ibid.* 138, 572 (1962).
17. Research carried out at Brookhaven National Laboratory under the auspices of the U. S. Atomic Energy Commission. Paper delivered at 11th International Botanical Congress, Seattle, Washington, on 26 August 1969 in the symposium "Ecological and Evolutionary Implications of Environmental Pollution."

Darker Woods
Elijah L. Jacobs

After a time the darkness slightly thinned;
Above a lane a strip of sky turned gray,
And trees near by took form, but farther away
The moonless night was dense. A breath of wind
Briefly stirred the branches. Under foot
The smooth, familiar ground was strange and rough;
The paths that ran, in daylight, straight enough
Went wandering now by pothole, stone, and root,
By dim and shapeless bushes. Scent and sound—
Some small thing scuttling through the underbrush,
A far hound's baying, the flutter of wings in the hush,
Decaying leaves in some dank bit of ground—
The darkness lapped them close. With casual ease
I owled along those dim and murky ways,
My night-sight coming slowly. I knew this maze.
I've darker woods to walk, God knows, than these.

UPI

We Have Only One Biosphere

We have examined the human population, evaluating the impact of mass society on men, and we have also considered the metabolites of mass technological man. Beyond these considerations, it is essential to take full recognition of the fact that the biosphere is a single ecosystem in which current conditions are more than the sum of human joy or misery, or the aggregate of pollutants. Our actions, large and small, cause adjustments throughout the ecosystem as a whole which may produce seemingly unrelated consequences, as well as magnifying or diminishing the immediate and more obvious results of human activity. Further, certain elements, and certain organisms, play essential or key roles. Human activities involving even minor interference at key points may lead to unexpected major changes.

The consequences of human activity certainly cannot all be negative in terms of the persistence and functioning of the biosphere, yet it is hard to discern, with confidence, those points where the long-term good of the biosphere (and thus of man) does in fact coincide with the short-term goals of men. This difficulty cannot serve, however, to justify either ignorant or arrogant attitudes. A "what's-good-for-man-is-good-for-the-biosphere" approach may get us somewhere, but that somewhere is unlikely to be where we would most like to go.

Part of our difficulty arises because we view effects of human activities in terms of individual life span and individual experience. To use an obvious example, concern for the land comes more readily to a farmer who will harvest a second crop in a few months than it does to a logger who knows that it will take up to two hundred years to generate a new timber crop on some inaccessible hillside. Similarly, it is difficult for us to grasp the vast scope and diversity of human activities. Worldwide changes in climate, worldwide dispersal of radionuclides, lead, DDT, polychlorinated biphenyls, and hundreds of other potent and potentially dangerous substances, a doubling or tripling of the rates of erosion of the continents, or melting and formation of ice sheets are so far removed from the scale of personal awareness that we consider them only with difficulty. Again, our benchmarks are taken from the world as we found it. Most human beings alive today were born after 1940. Yet by 1940 the natural communities of at least four fifths of the land area of the continental United States had either been totally obliterated (by construction or agriculture) or thoroughly decimated or disturbed (by grazing or logging). How many Americans alive today have seen a tall grass prairie? The more mesic grasslands of the northern hemisphere ceased to exist before most of us were born. How can we now estimate their value in their original state?

The conversion of complex natural communities to easily harvested monocultures of grain, or to suburbia, have easily calculable immediate benefits to man, but the long-term biosphere costs are not yet calculable. The problem of any intentional human husbandry of the biosphere as a whole is again complicated by the demand for specialization generated by our increasingly complex technology. Our awareness of possible consequences of our actions narrows with the focus of our interests, and our education. It is thus possible for climatologists to speak glibly of climate control in total unconcern for the possible impact on, and significance of, natural communities such as the coral reef complex, or the pelagic plankton and associated fisheries. An engineer may happily plan a dam across the Bering Strait in order to melt arctic ice and warm the climate so as to "grow oranges in Kamchatka," oblivious to students of pleistocene history who point out that an ice-free arctic ocean was probably the source of precipitation necessary for formation of the continental ice sheets. The possibility has been seriously raised, in fact, that thinning of the arctic ice shield is even now proceeding precipitously (possibly as a consequence of human activity) and that its disappearance might so change weather patterns as to simultaneously generate new continental glaciers and convert the wheat-growing former grasslands of the northern hemisphere into deserts. In the same vein, an economist may promote "development" of the Amazon basin in ignorance of the role, and vulnerability, of mycorrhizae, or the phenomenon of laterization which may turn "rich" jungle soil into rock when it is used for agricultural purposes.

The biosphere is surely sensitive to human activities, and man is inadvertently causing worldwide adjustments. Even those subregions which have proven least hospitable to man, the polar regions, tropical jungles, and deserts, are rapidly succumbing to the drive for additional resources. The natural communities of these difficult environments, and those indigenous men and cultures that have been part of some of them in the past, are proving to be the most fragile and sensitive to disruption.

Presumably our definition of "good" for the biosphere must encompass something better than a human feed lot and something other than thoughtless, and haphazard, progress towards a worldwide human analog of the present day chicken factory. The biosphere is a smoothly functioning system which maintains enormous richness and diversity. Without more than a fragmentary understanding of its total overall function we are happily shoveling sand into the gears in order to build nothing more substantial than sand castles.

The selections in this chapter are intended to demonstrate that we are influencing the planetary ecosystem as a whole, that even the most inaccessible or hostile environments are being severely modified, and that we are proceeding at random and in ignorance. There seems little reason to hope that the consequences of our activities will lead to a more stable, more diverse, or richer biosphere than that which now exists. There is every evidence to expect the contrary. The only "goal" toward which mankind as a whole appears to be moving is the generation and accommodation of a greater human population at a greater rate of consumption. That open-ended goal is equally open-ended in terms of deterioration and disruption of the only available planetary ecosystem.

Impacts of Man on the Biosphere
F. Fraser Darling

The manifold effects which the physical presence and activity of man have had on the face of the planet Earth through the relatively short period of man's history tend to be dynamic and interlocking, yet it is worthwhile attempting to classify them for a clearer understanding of temporal and spatial factors. The world as it existed before man used tools and fire was one of immense richness in natural resources, organic and inorganic. But to say that is to put the cart before the horse; natural resources were not resources at all until man was both present and able to make use of them. The ability to identify, reach and use natural resources has been a continuous process for man, and there is now a fair archaeological and historical understanding of the differing rates of exploitation in different parts of the world, the sudden changes of rate and style caused by changes in human condition and the enormously accelerated rate of change within the last century. It seems certain that man's ingenuity is running ahead of his wisdom. But the mistake must not be made of thinking men are wise if they express their notions with the constantly wagging finger and negatively-shaking head of Jeremiah. If civilization is a flower of evolution, it could not have grown had not man gained leisure to think and flexibility to act by actively managing to turn some of his environmental wealth to his own advantage beyond that of mere sustenance.

From the beginnings of civilization, man has altered environmental processes as he dug into the organic store of the planet's ecosystems. Even to light a fire of dead wood for keeping warm is to deflect a natural process of decay which would be humus-building, into the production of inorganic ash. For a long time man may not have been much else than the equivalent of an indigenous animal of limited change-producing activity, but even at the advent of the neolithic revolution, hunting and food-gathering man had changed parts of his world more or less unintentionally by the use of fire. This is a point to remember in considering the influence of man on the biosphere: as a species his impact goes far beyond the immediate contact. Fire would run and change vegetation complexes, fires would be used to help drive herds of game animals, a prodigious expenditure of organic matter for momentary expedience, and burning in season would be done to draw grazing animals to new grass, a path of inevitable impoverishment of habitat. Man was controlling the behavior of wild animals to some extent by these expedients which were slowly altering his habitat and theirs.

Men were so few and the world apparently so large that it would have been strange had man dwelt philosophically on his own numbers and the fate of natural riches. Even in the present century men speak proudly of their duty and their success in pushing back the wilderness. The student of the human condition, however, although he may hold to the philosophy of conservation, will accept the price of civilization which has been the loss of much natural wealth. When was the critical moment when man should have become conscious that a halt must be called to bald exploitation and exploitation must be matched with rehabilitation? It is possible that this moment has been reached now. The fear today is whether rehabilitation is possible or whether causes and consequences are setting up their own percussive oscillations to an extent beyond control.

It is worthwhile attempting some classification of man's impact on the biosphere, possibly in developmental and qualitative order which could be flexible and capable of being redistributed as an ecological diagram of interconnecting factors. It can be much expanded and given change of emphasis. Man's impact is not to be thought of as being wholly to the detriment of his ultimate welfare though it may be preponderantly so. Certain man-modified habitats may represent ecosystems of equal or greater production and wealth-building power than the natural conditions. Wealth-building in this connotation means the storing of organic capital as in a tropical forest or in a prairie soil or Chernozem.

An attempt has also been made to identify the new problems created by accelerated development and also to list the various measures taken until now by man to maintain the quality of the environment.

Food Getting

ANIMAL HUNTING WITH THE USE OF FIRE

Fire has been used for animal hunting, destroying forest incidentally and preventing much regeneration. But production of savannas as in parts of Africa with a rich spectrum of ungulate species may lead to a rich habitat for organic production. The savannas of Brazil and Guyana seem definitely to lose out as to their qualitative potential. Is this because of a limited array of ungulates as compared with Africa? Again, in North America, the Indians extended the prairie range of bison by burning forest and the resulting humid prairies maintained a very high rate of production, accumulating an immense store of rich soil.

Sometimes the terrain is also being burned intentionally to produce fresh young grass; this leads to impoverishment of the array of plants present originally, which upsets flexibility of the grazing complex in face of seasonal climatic variation. Examples are production of *Molinia* and *Scirpus* swards in the Scottish Highlands; and the gradual elimination of grass, giving way to gall acacia and lantana bush in parts of Africa and mesquite in the southwest of North America. Many important variants originate according to the frequency of burning and, in tropical countries particularly, the time of burning in relation to the dry and wet seasons.

SETTLED CULTIVATION

Settled cultivation, in opposition to shifting cultivation, leads to soil exhaustion if replenishment is not understood or impossible, or to possible aridity if active steps are not taken. The English "brecks" or broken lands are thought to have developed from neolithic cultivation. The dust bowl of the United States in the 1930s showed what continuous extractive cultivation without adequate replenishment could do when a cycle of drought years was encountered. Broadly, cultivation must manage to sustain the level of organic matter, or the soil either bakes or blows away. But many examples can be given where settled agriculture, under various ecological conditions, has been successfully carried out for centuries in the Far East, the Near East, the Mediterranean area and Central America. The ancient Negev demonstrated the efficacy of counterworks for the conservation of water; abandonment of these resulted in almost complete degradation of the habitat.

SHIFTING CULTIVATION

There tends to rise in the collective mind of conservation-minded people an antipathy toward shifting cultivation as being wholly deleterious to habitat. Examples could be given of this practice, however, that produce variety and valuable "edge effects" (that is, creation of ecotones) so long as the human population is low. The *chitemene* system in Central Africa is a good example: the gardens are not larger than half a hectare or so and, being surrounded by bush on fairly level terrain, soon return to bush after three to five years of use, and are not used for a further 40 years. Immediately after being relinquished these gardens are taken over by colonies of *Tatera* rats while the soil is loose and friable. The rats form a secondary protein crop for a year or two, when caught by small boys. Following this, the bush slowly recovers and replenishes the soil. Another good example of sophisticated and conservative shifting cultivation is given by Hanunoo agriculture in the Philippines.

Increasing human population causes the bush to be broken again too soon and deterioration of habitat follows. Shifting cultivation on steep slopes is almost wholly bad and, on some geological formations such as limestone, milpa cultivation may be quickly disastrous. Shifting cultivation is the bane of central and northern South America and land hunger with rapidly increasing human population creates a vicious circle from which it is difficult to break. Erosion is having long-term effects on far greater areas of country through interference with water relations.

IRRIGATION

Irrigation has a long history and is still a popular line of development, but in large-scale irrigation projects, improvement or maintenance of soil fertility over a long period raises many problems. As they are so often sited in arid areas where a high evaporation rate has kept salts in the soil, irrigation tends to redissolve the salts and deposit them again as a crystal crust. Re-use of irrigation water (as from the Colorado river) can mean good ground being made saline and useless at lower levels of operation. This also happens on Bear River, Utah, and on the great Sind scheme. Wise irrigation, or possibly fortunate irrigation, uses rain-water and, if this is adequate in supply, it can wash the more saline ground of the arid parts where the water is used. The Gesira scheme in the Sudan is an excellent example, where the country's economy has been much strengthened by nonsaline water from the Blue Nile (which runs at a higher level than the White Nile) which is taken off and spread over the triangle of the Gesira. The used water drains into the White Nile on the lower side of the triangle, the volume of which renders insignificant any salinization suffered by the Blue Nile water passing across the Gesira.

OVERGRAZING AND OVERBROWSING BY DOMESTICATED LIVESTOCK

Once domestication has occurred, sedentarization limits movements of stock and overgrazing is often the case. Different species of domestic stock adopted for different climatic conditions make for maintenance of habitat, but this can easily be upset by natural fluctuations of conditions. Technique of successful grazing has to be learned, and even when learned may not be generally followed. A modern research station can fall short, as for example during the reindeer era in Alaska when deterioration outpaced the research, and both the grazing and the reindeer population were dangerously affected. Modern examples of extractive grazing such as in parts of Africa, Australia, and New Zealand, and the shortgrass prairies of North America have rapidly caused deterioration of the soil.

NOMADISM OF PASTORAL ORIGIN

This deserves special mention because it can be so easily upset or broken, to the detriment of the habitat. Nomadism is at best a pursuit of the steppe regions of the world and produces least deterioration. It is utterly bound up with movement if habitat is to be conserved. Any undue lingering punishes the vegetational complex, reduces its numbers of species and therefore impoverishes it. The finest nomadic areas of the world were the Chernozem plains, like those of the Kuban where the Scythians roamed in the time of Herodotus. Their capacity for and dependence on movement exasperated Darius. Nomads depended on animal products for sustenance and warmth, and as protein consumers were an ecological elite, at the top of the food pyramid; if they increased in numbers, families, clans or tribes had to split off and move on. We know of these frightening nomadic emigrations in history. Less good steppe was used,

forest edges on mountain ranges were used, and there was an inevitable slow change, usually for the worse.

Conversion of Chernozem steppe from protein production, whether of livestock in the Ukraine and Kuban or bison in the American prairies, in order to increase production of wheat and maize, was a relinquishment by the human being of his ecologically aristocratic position. He now converted a complex community of grasses into a monoculture of one annual grass, wheat or maize, the seeds of which he could eat. This permitted a much larger number of mouths to be fed from the rich steppe by substituting human mouths for the grass-eating mouths of horses, cattle and sheep, but also meant a loss in reciprocation of the system as it had existed. Organic matter was stripped out as grain, for that commodity was not consumed on the steppe as had been the animal products, so there was profound loss of organic matter. The same thing happened in North America, when the Indian-bison continuum was replaced. The Chernozem soils of America and Europe have stood up well to the extractive agriculture imposed upon them and there is the likelihood that good agricultural methods will maintain them in the future, but a revolution of the plow of this kind always extends beyond those soils which are best able to bear it. Higher, poorer, and more arid steppe has been subjected to the dry farming of wheat with disastrous results.

Wherever sedentary populations have impinged on nomadic grazing there has been degradation of habitat for which the nomads are invariably blamed. Yet nomadism as a careful pastoral continuum is the least traumatic of human influences and as a form of husbandry utilizes areas which could not be utilized by man in any other way. It is essentially ecological in structure, relying on movement and seasonal use, a wide spectrum of grazing animals adapted to well-understood differences in habitat and producing much energy without loss of organic matter.

Nomadism is always unstable: apparently small political changes reduce it to habitat spoliation. The Masai in east Africa are a Nilotic tribe moved from a humid homeland into high, dry, steppe country. Yet their knowledge was good and accurate and their habitat did not deteriorate until the white man's veterinary skill broke the full incidence of mortality which formerly kept the herds to a proper level for the stock-carrying capacity of the ground. The Masai and their country are a good example of integration of nomadism with the indigenous animal population. Wild game existed alongside domestic herds, tolerated and respected. This harmony may well have originated in the Nilotic homeland, as in the Bahr-el-Ghazal, where the Dinka accept giraffe as part of the cattle herd. The Nilotics practice a restricted nomadism seasonally between the toich (grass flood plain) and the slightly elevated bush to which the herds return in the wet season. The Dinka's practical, empirical knowledge and acceptance of the ecosystem approach to living in his rich world is further exemplified by his tolerance of a small poisonous snake which inhabits his thatch and keeps insects within bounds. Even in this paradise of symbiotic cattle culture, the march of modernity and the hungry mouths elsewhere are demanding rice culture, the exclusion of game animals, and the distended but unsatisfied belly of the carbohydrate eater.

This section may be concluded with an example of nomadism having become allied with political power and made so powerful by law that a whole country was devastated. This is what happened with the transhumant society of the Meseta in Spain in the sixteenth century, to which Ferdinand and Isabella agreed for the profit of the Crown. Flocks of Merino sheep were driven through cultivated areas, and it even became illegal to fence against them. It took Spain nearly 200 years to beat the Meseta but the bare hills of Spain remain to this day.

Active Depletion of Renewable Natural Resources

DEFORESTATION

This has taken place from earliest times and is probably better documented than most other types of change of habitat. The earliest effect of restricted deforestation is one of enrichment of the habitat by providing change in a situation of little or no change. Parkland in forest is appreciated by many grazing animals. Primitive and later man, utterly surrounded by forest, arrives at a psychological state eventually when he must push back the forest and not be enfolded. The mediaeval horror sylvanum was very real, and should be remembered for a better understanding of history.

The Anglo-Saxon colonizing a fifth century Britain which had islands of plowland and pasture connected by roads in a sea of forest was not only knowledgeable in tools and woodwork, but ecologically. To create new arable ground in the long term he used the snouts of herds of swine, descendants of the European wild boar as a biological plow or grubber in the oak and beech forest. They garnered the pannage and by concentration of effort prevented regeneration. As actual felling took place, grass followed on the cultivation by the pigs and grass parklands or lawns were created, to be plowed as needed. In temperate, humid Britain, the changed biological system of forest to grassland with occasional trees made for little loss of organic matter and the output of free energy was probably not far behind that of the forest. Certainly it allowed more to be made available for human consumption.

The tendency to export natural resources leads to devastation, as has been documented many times. The oak forests of England did not all fall by the axes of the early Anglo-Saxons. The rise of England as a naval power involved heavy felling for smelting fuel. In the early seventeenth century the unrestricted felling of English timber was prohibited. This drove the iron smelters to the woods of western Scotland where the long sea lochs allowed ships laden with iron ore to penetrate deep into the forests of Scots pine and oak in the glens. Here devastation came quickly because of the high rainfall, acid soils and steep slopes. The Forestry Commission of the present day is still concerned with the repair of this damage, but at great cost because active humus has long disappeared from the once-timbered slopes.

Perhaps the onslaught on North American forests was the main reason for the rise of the principle of conservation. Fairfield Osborn said in 1948 of the United States:

"The story of our nation in the last century as regards the use of forests, grasslands, wildlife and water sources is the most violent and destructive of any written in the long history of civilization. The velocity of events is unparalleled."

This fact of weight and speed of destruction may find its absolution in the rise of the ideal and practice of conservation in the United States. Similarly, in Africa, an awareness of the need to conserve has often emerged as a result of destruction and denudation.

On a happier note, it may well be that the African Special Project carried out jointly by FAO and the International Union for Conservation of Nature and Natural Resources was the turning point in attitudes toward conservation in the rest of Africa; that and the Arusha Conference of 1961.

The felling of tropical forests has been a story of unrestricted extraction. The massive and impressive accumulation of organic matter presented by this oldest life form on the planet tends to mask its fragility. Like most forest forms, the tropical is primarily a photosynthetic factory of cellulose, with any protein as very secondary production. The tropical forest floor has been protected from the sun for thousands of years with the niches for decay and conversion so well filled that nutrition of the trees and the life of the canopy are bountiful. When the forest is felled the tender soil exposed to the sun oxidizes rapidly and disappears; as Omar Khayyám wrote:

"Like snow upon the desert's dusty face
Lighting a little hour or two—is gone."

The mahogany forests of Hispaniola have gone, those of Honduras are going and the results are only too well known.

Many countries, particularly within the tropics, have had their forest resources exploited without any provision for sustained yield management. It can be argued that because of local economics and political structures they had no alternative, and a good proportion is likely to resort to such expedience in the future. In order to survive they need ready cash to operate politically. Politics are peculiarly bound up with such expedience. It must be realized that political decisions are, and have been, a major factor affecting the natural environment.

DRAINAGE OF WETLANDS

In general, wetlands are highly productive of protein in the shape of mammals, birds, fish, and invertebrates, but not always in a form acceptable to the human being as food. Drainage has throughout history created navigable channels in various parts of the world and provided more agricultural land. The Netherlands and the fen country of England are good examples, and there are many such areas in the United States. Where the soil is alkaline peat it becomes highly fertile when dry enough to manage, but so contracts in volume that diking and pumping become necessary, for the land won is then below the level of the adjoining sea. The situation is vulnerable and the soil itself is so friable that it tends to blow. The question must always be whether the considerable capital works necessary can be maintained and whether the potential fertility of wetlands soil can pay for them. Once more the ecosystem is being changed from protein to largely carbohydrate grain production. Certainly more mouths can be fed on the "devaluated" product. The aesthetic loss of beautiful species, especially birds, as a result of drainage in wetlands is considerable. The United States has changed its former policy to some extent by allowing wetlands to refill and resume their former land form; for example, the Klamath marshes in northern California. It might be said here that allowing marshes to refill is only possible as a policy in a country affluent enough to be producing grain surpluses and where wildlife is beginning to have extrinsic values for recreation.

The Everglades national park in Florida provides a beautiful example of the complexity of wetland habitats and what can happen when water is deflected from these areas. This is essentially a wilderness and wildlife area on a flat oolite base slightly above sea level and tilting slightly southward. The Everglades depend on the seasonal passage southward of large quantities of water moving very slowly. During the wet season all wildlife disperses over the area and in the dry season it contracts to alligator holes. Alligator holes serve as survival reservoirs during the dry season. Drainage north of the park has deflected much of the water which should flow over the Everglades and the water table has thus been lowered. Further, a hole has been punched in the bottom of the system in the shape of a canal run-

ning up from the sea into the Everglades for the passage of pleasure boats. This both allows quicker loss of water and the occasional invasion of salt water. Finally, there has been a long period when alligators have been illegally hunted. Failing greater care of alligators, artificial alligator holes will have to be made to maintain many other species of plants and animals. The Everglades is a large national park, but is a vivid example of several thousand square miles being insufficient as an ecologically independent area.

Africa has many interior wetlands and one of the most promising lines of development in recent years has been the improved management and extended use of the fish resources of these wetlands. Zambia is an excellent example of the achievement of an improved standard of nutrition for a new industrial population from the freshwater fisheries of the numerous swamps. Yet the projected increase in human population has caused serious consideration as to whether such areas as the Kafue and Chambeshi flats should be drained to serve as wheat-growing lands. Not only would fishing be constricted as a result but the herds of game animals would also disappear, despite the fact that the red lechwe is an antelope moving in large herds, adapted to grazing inundated lands and potentially capable of management as a meat-producing animal.

OVERHUNTING OF DESIRABLE SPECIES

The disappearance or decrease in numbers of species has altered the ecosystem and the habitat. Knowledge of niche function is growing but there is much to learn about habitats in which a niche occupant has been removed. The elephant would provide a good comparative study because there are plenty of habitats from which it has recently been removed by overhunting and others in which it is more numerous than ever because of protection. In the latter situation production of grassland and removal of bush with the tsetse fly is being pushed back. The elephant is the agent of change but man's distortion of the elephant's way of life is the instigator of change. Not enough thinking has been done about the influence of reduction of game on the African savanna. Bison on the North American prairies were followed by the plow, so little is known of the change which actual reduction would have caused except that, on the fringes of prairie, as in Wisconsin, there is a coincident rise of even-age forest.

Intended extinction or eradication of species

The wolf is perhaps the best example because of its direct challenge to extended pastoralism. The supposed behavior of the wolf has become a psychological fiction which can still influence habitat. In the recent past, bounties were still given for wolves killed in the Arctic where human predation of the caribou was quite insufficient to crop the caribou population and to safeguard the range which, if overgrazed, might take a century to recover— longer than caribou could wait. There were excellent object lessons in the extinct reindeer populations on the west coast after 30 years of overgrazing of the ranges, but the psychological obsession about wolves had blinded people to the real cause. These overgrazed ranges can still be seen as different from the surrounding terrain when viewed from an airplane at 6,000 meters.

The necessity for frequent or continuous movement was described in relation to nomadic pastoralism. Wild animals have their natural behaviour patterns but not entirely of their own volition. There is the curious habit of "yarding" with several species of deer, treading around as a large herd in one small area in dead of winter. Food may be found further afield but they do not go to it, unless wolves or other predators stimulate them to move. How far lack of this function of the predator was responsible for the sudden decline in numbers of mule deer on the Kaibab plateau, Arizona, in 1916 will probably never be known. Mountain lions were still being killed in the Grand Canyon national park in 1950, so deep rooted is aversion to them. The influence on habitat of violent fluctuations due to lack of predator populations, which not only cut down numbers but keep the herds of their prey moving, is a matter for concern. Recent work on the moose/wolf ratio and equilibrium on Isle Royale in Lake Superior shows how a woodland habitat carrying these large deer can be maintained when man allows the wolf to do all the necessary culling.

Consequences of Mining and Industrial Processing

Toxic fumes and detritus

Smelting liberates toxic fumes which kill plants and sometimes animals. The sterilization of areas by copper smelting in Tennessee is well known. Aluminium smelting at the foot of the Great Glen in Scotland produced fluoride fumes which were blown northeastward up the glen by the prevailing winds. Grazing was affected and young cattle suffered. It took a court case and very dilatory action thereafter before scrubbers were installed to remove the hazard.

The lead mines and feldspar workings of Derbyshire, England, left areas of spoil with which the herbage was so impregnated that young stock could not be grazed on it. Lead poisoning affects the earth centuries after mining has ceased.

INTERFERENCE WITH NATURAL DRAINAGE

Coal and shale mining not only create conditions where subsidence occurs, but large quantities of detritus are dumped on the surface near the mine shaft. These vast dumps not only ruin the landscape aesthetically but also cause serious interference with surface drainage. At worst, these dumps are a menace to human communities as was experienced so tragically at Aberfan in Wales. The recent marriage of ecology and landscape architecture shows immense promise. The English counties of Durham and Northumberland have successfully initiated schemes for covering pit heaps with vegetation. The United Kingdom National Coal Board has done much opencast mining in the north of England and immense care is being taken to rehabilitate the landscape afterward, to the extent of 11 percent of the Board's income. Similar spectacular results can be found in several regions of the Federal Republic of Germany.

The strip mining and auger mining of the hills of Kentucky, U.S.A., are leaving eyesores of dereliction along the faces of the hills, and the watershed erosion which follows is causing serious repercussions below these hills on the good soils of the Ohio river country. It is all the more distressing to hear that the coal is being mined for sale at the lowest possible price to that model of rehabilitation policy, the Tennessee Valley Authority. One of the difficulties in much conservation endeavor is to stop different agencies within one nation working to each other's disadvantage. Once more there is a lack of adequate legislation: the process of law being largely by precedent, it can scarcely keep abreast of changes unimagined in the past.

DUMPING OF WASTES INTO RIVERS

This is an obvious and well-documented way of changing the biosphere. The trouble arises from the primitive notion that a river is a natural sewer. Paper-making is a modern industry—in terms of centuries—which has been responsible for fouling many rivers. The State of Maine, U.S.A., has a relatively low human population, but its rivers are polluted to an extent that precludes most of them receiving a run of salmon. Where human population is dense, sewage is the main problem. Knowledge is being gained all the time of how to deal with industrial wastes, but regulation lags. The recent increase in the use of pesticides has had a few tragic sequels in the poisoning of rivers and ditches. In comparison, the carelessness of shepherds when draining out baths of sheep dip is a small thing, but indicative of the thoughtlessness that spreads man's deleterious influence over habitat beyond the immediate vicinity. There has not been sufficient realization that much marine productivity on the continental shelf has an estuarine basis.

Consequences of Human Population Density

LOSSES AND GAINS IN PLANT AND ANIMAL LIFE

Obviously there must be losses when man becomes gregarious beyond the hunting and food-gathering stage. It was changing ecosystems so that more produce was available to man that created gregariousness and the development of civilization. Certain timid animals keep their distance from human aggregations, others adapt to them, even though they remain timid. The fox in the suburbs of London is an example. Plants, being less mobile, have less choice and many species are stamped out. The large human foot with our upright form above it is much heavier on herbage than many people imagine. This has become obvious, and even crucial, in large national parks such as those of North America where blacktop paving is having to be laid in growing quantities to safeguard fragile plant associations in alpine, desert, and forest areas. A spongy woodland floor will show footmarks for years; the tundras of Alaska are crisscrossed by the tracks of caterpillar vehicles and jeeps, and it is doubtful whether some of these tracks will ever disappear. It may be thought that forest trees are immune to the tread of human feet, but not so. The French find it necessary to fence individually the largest oaks in the forest of Fontainebleau because impaction of the ground by picnic parties beneath them eventually kills the trees.

The rapid rise of the human population in India in a climatic situation of wet and dry seasons illustrates the power of human feet to remove vegetation. Village sites are now bare dried mud where three years ago there was secondary jungle. In these situations, where physical building is quite flimsy and sometimes no more than poles, burlap and corrugated iron, it is not definite planned urbanization paving the surface of the earth, but the passage of human feet and of domestic animals. The latter contribute to the general denudation by their own browsing habits and the necessity to cut down branches of trees for their sustenance.

Certain tribes in propitious habitats realize the immediate advantage to be gained by cultivation within the forest edge. The Kikuyu of Kenya are an excellent example: these people have pushed back forest to a dangerous extent, especially in recent years since the tenfold increase in the tribe under the Pax Britannica.

Man altering ecosystems purposely or unconsciously usually simplifies them and makes possible the invasion of plants and animals of earlier stages of succession. Annual and biennial seeds are good examples. The animal world has ready invaders such as the brown rat and the common sparrow, which follow man wherever the wheat plant is grown. This is not the place to explore the subject of exotics, but their influence on habitat is worthy of special study in relation to the changes man has brought about in the biosphere. There exists a limited field for congratulation in pleasant residential suburbs where gardens express beauty and birds come in greater variety than might be found in the countryside. In fact, unconsciously or no, a diversity has been created to which nature has been quick to respond by adding its own contribution.

POLLUTION OF AIR, WATER, AND SOIL

Pollution, the greatest problem of modern times, has been left till last. No longer is it just a matter of England's industrial north, Germany's Ruhr, the U.S.S.R's Yaroslavl and Gorky, as being instances of the polluted areas where money is to be made but where it is better not to live if possible. In the last quarter of a century the whole planet has been polluted by man so that the fat-storing creatures of the remote Antarctic continent, penguins and seals, carry in their fat appreciable quantities of organochlorine compounds, commonly alluded to as pesticides and not used within many hundreds of miles of any part of the Antarctic continent. There is virtually a cessation of reproduction in some species of birds which have not been the objects of dislike but only of pleasure. Man himself carries these impurities in the fat of his own body and is as yet unaware of any effects on him, for good or ill, but most thinking people are concerned about the possible effect of accumulations of these compounds over the years. Some developed countries have regulated the use of pesticides and cut down dosages, but this has meant the economic dumping of large quantities in tropical countries where there are sufficient pests and little control over the use of pesticides. The informed know that to regulate use in some countries but not on a world-wide basis does not prevent pollution of the biosphere. Man is today a world citizen.

Much could also be said on radioactivity. In the recent past the radioactive level has been increased in air, soil, water, and living organisms over large areas of the world. Experimental A and H bomb explosions and some other works with radionuclides represent one of the greatest dangers for all living organisms on earth.

What might be the planetary profit and loss account in relation to the expenditure and production of oxygen? It is thought that the atmospheric figure of 20 percent oxygen has been evolved by the photosynthetic power of plant life and the slow deposition of organic matter in the deep ocean sediments. When a jet plane crossing the Atlantic burns 35 tons of oxygen, is industrial process and combustion dipping seriously into the margin of production over consumption, especially in terms of the rate of destruction of forest growth and other plant life? Oxygen consumption and carbon dioxide production are to some extent linked; the carbon dioxide content of the atmosphere is rising and this could lead eventually to such an increase in the temperature of atmosphere and oceans as to cause a substantial ice melt which would appreciably raise the level of the oceans. Pollution, combustion and destruction are combined here in their ultimate power to change the biosphere.

The deposition of sewage into rivers is of course age-old. Lakes receive this effluent, and eutrophic conditions are being reported from areas as far apart as Lake Erie on the border of Canada and the United States and Lake Baikal in the U.S.S.R. Life in some lakes is being asphyxiated through lack of oxygen and too great a deposit of organic matter. Riverbanks and lakesides are becoming overpopulated. In order to be able to live at high densities of population, all sewage and waste matter must be processed. Man is definitely trying to mend his ways but not fast enough. London's River Thames is improving thanks to civic action, and the sport of trout fishing has saved from pollution many streams that feed the Thames and other rivers in England.

The wreck of the Torrey Canyon focused the attention of everyone on the growing hazard of heavy oil pollution of the sea and the secondary evil of detergents used to combat the oil—which killed more marine life than the oil itself. Responsible oil firms are taking the dangers of oil transportation by sea as a major task of research. Half the world's ocean cargo weight is oil—700 million tons of it last year in 3,218 tankers. Separators of improved design at ports, plastic booms round oil harbors, flocculants for spillage at sea, jelling agents to solidify oil in threatened tankers and new methods of loading and ballasting giant tankers, show a willingness on the part of the oil firms that could well be imitated by those shore-based industries which are potential pollutors of the environment.

Problems of Accelerated Development

Much of what has been said applies to the long, slow changes in past civilizations which have produced the developed nations of today and the undeveloped nations of a generation ago. The rapid advances that characterize developing nations have produced dramatic new kinds of interactions with the environment that were not so evident earlier. Nowhere is the problem of accelerated development and its devastating consequences more evident than in many regions of the tropics and subtropics, where two thirds of humanity now live. Here the relation between man and his environment is changing so rapidly that well-balanced development is seriously endangered and in extreme cases entire populations may be threatened with extinction.

SIDE EFFECTS OF LARGE-SCALE MODIFICATIONS OF TROPICAL ECOSYSTEMS

Owing to the rapid development of ever greater land areas and the widespread use of powerful technological methods, the various tropical biotic communities are today undergoing radical and extensive changes. Deforestation, irrigation, the introduction of exotic plants and animals, the large-scale use of weedkillers, the eradication of certain pathogenic agencies, etc., have transformed the tropical landscape more profoundly in ten years than the traditional agricultural and stock-raising methods in ten centuries. In addition to economic advantages, which are immediate and undeniable, this "reordering" of the tropical ecosystem has sometimes entailed a sudden collapse of the balance, generally going back thousands of years, between man and his environment and given rise to unexpected difficulties.

For example, the introduction of livestock into the savannas of tropical America has promoted the increase of rabies-transmitting haematophagous vampires. Irrigation of the savannas of the Sahel in Africa has caused the spread of bilharzia. Systematic deforestation has frequently brought about transmission to men and domestic animals of certain arboreal viruses, normally confined to the pathogenic cycle of the forest canopy. Some are capable of causing sickness of little gravity to monkeys and tree-dwelling rodents, but they can be much more dangerous to humans: yellow fever, dengue fever, Kyasanur forest disease, etc.

The abandonment of certain traditional foods, both animal and vegetable, or their replacement by other kinds of food more easily produced has often served to aggravate already existing nutritional deficiencies. Even eradication of a number of tropical diseases may give rise to long-term difficulties. Some of the peoples of west Africa appear to be protected against the effects of malaria by certain abnormal haemoglobins which they possess in a heterozygotic state. Once the parasite has been eliminated, however, only the adverse effects of this genetic adaptation to the environment will remain.

BIOLOGICAL, PSYCHOLOGICAL AND SOCIAL CONSEQUENCES OF HASTY AND CHAOTIC URBANIZATION OF THE TROPICS

In Africa, as in Asia and tropical America, the last 20 years have seen the rampant spread of bidonvilles, shantytowns, barrios and favelhas of every description, whose crowded inhabitants are all too often undernourished and illiterate, and have been frequently ruthlessly cut off from their traditional values. The pathology of this situation, which is still little known, is a combination of the effects of malnutrition and poverty, and the physical and psychical effects of manifold stress. It has resulted in the creation of a subproletariat in poor physical condition, which deprives agriculture of a source of manpower but at the same time is unable to offer skilled labor to industry. As regards public health, sanitary conditions in these shantytowns represent a constant menace to the great urban centers of which they are a part.

BIOLOGICAL, PSYCHOLOGICAL AND SOCIAL CONSEQUENCES OF MIGRATIONS

This lure of new cities and industrial centers involves unprecedented population movements which are the cause of a demographic imbalance that is in every way detrimental to the development of many tropical countries. Such migrations are, moreover, frequently selective: they tend to leave the less enterprising groups or individuals behind in the rural areas and to condemn the majority of the better elements to chronic unemployment. Populations biologically well adapted to a special environment (the high plateaus of the Andes, for instance), sometimes find themselves moved by these migrations into areas climatically unsuited to them.

These problems—and there are many—require urgent study by teams of specialists, including ecologists, medical experts, psychosociologists and economists, with a view to finding satisfactory solutions as quickly as possible. Tropical ecosystems are at once so numerous, complex and delicate that nothing could be more detrimental to the interests of the human beings who inhabit them than a pure and simple "transplantation" to the tropics of techniques (and sometimes even of ideas) which have proved their effectiveness in the temperate latitudes. It is not because the airplane has practically done away with distances, or because it is now possible to prevent or cure most tropical diseases, that the fundamental ecological differences between the great biomes have disappeared. To attempt to spread the way of life of the industrialized nations of the temperate zones over the whole planet is dangerously unrealistic. Western norms should be adapted to other environments and cultures and not imported into them just as they are. And this applies not only in the provinces of economics and technology but also as regards food, housing, and clothing.

Human Impacts Leading Toward Maintenance of Environmental Quality

Irrespective of past and present mistakes and of new problems created by accelerated development due to demographic, economic, and social growth, in the past man has already tried to solve these problems of habitat deterioration. Today new ways of thinking and changing technologies give him the desire and the opportunity to build or rebuild an environment which provides esthetic pleasure as well as goods and services.

DELIGHT IN ORGANIC FORMS AND ENVIRONMENTAL DIVERSITY

There is ample reason to believe in the importance of this. There are the great villas of the past, their gardens and the arrangement of fountains and lakes for esthetic pleasure. One calls to mind Imperial China, Ancient Rome, Persia, Renaissance Italy, the architectural gardens of France, and the English private parks. The parks designed by Capability Brown could not be enjoyed with the trees at their maturity. Gardening is becoming enormously popular and municipal authorities are gaining confidence in spending money on their public parks. Increased affluence creates a demand for environmental amenity.

ESTABLISHMENT OF NATIONAL PARKS

This is a major contribution to civilization from conservation-minded people who have a vision of future human aspirations. The best results of conservancy and the study of wild nature have been achieved through the establishment of reserves and national parks with a staff of scientists working permanently in them.

ESTABLISHMENT OF WILDERNESS AND NATURAL AREAS

Action here is prompted by both spiritual and esthetic concern, and biological need. There is a widely diffused element of human feeling that wishes there to be wildernesses in existence even if unattainable. In addition, natural areas fulfill a biological need as reservoirs; their study provides datum lines from which other environments can in turn be studied more intelligently.

CONSERVATION AGRICULTURE

Some tribes of people in limited or constricted habitats have arrived at conservation practices empirically, but the science of conservation in farming and water relations, together with forestry, has come from change of heart after extravagant exploitation. It could be said guardedly that conservation agriculture was fairly firmly established in developed countries, but these same developed countries have made some shocking mistakes in advising the developing countries. This could be stopped if the necessary preliminary studies concerning site potential and limitation and the necessary investments for conservation were made.

INTEREST IN SPORT

Many forms of outdoor sport in developed countries lead to care of the environment. Care of rivers has been mentioned already in the interests of fishing; a diversified agriculture and the maintenance of strategic woodlands and covers has been followed in western countries in the interests of sport. Sometimes sport has removed animals from large areas, but understanding of the ecological place of predators is bringing about much greater tolerance. All in all, farms run by those who like to keep the sporting interest are richer in a variety of plant and animal life than others. Sometimes the sporting farmer progresses to the state of being a naturalist, even enjoying natural life with no sporting interest.

CHANGES IN INDUSTRY

In the United States, agriculture moved from the eastern seaboard to the Middle West. The forests that had never relinquished their hold took over again. Even 60 years ago much wood was being cut for fuel in these eastern forests. Now, oil-fired central heating is fairly general and the woods have grown more beautiful; what is more, the eastern woods have reached a new high density, and have become the suburban forest, the pleasure of the urban communities of the seaboard. Public opinion is now solidly behind the maintenance of these forests.

Likewise, in Europe, the concentration of agriculture on the best soils and the consequent abandonment of marginal lands has opened the way for rural development based on afforestation for soil conservation, forest management for timber production and recreation, the establishment of national parks and, generally speaking, conservation, recreation, and the extensive management of natural resources.

Conclusion

To conclude: does this imply optimism? Ecologists can scarcely afford to be optimists. But a pessimist is a defeatist, and that is no good either. There need not be complete disaster and if man's eyes were open wide enough, worldwide, he could do much toward rehabilitation. The biggest danger of all is man's inability to control the explosive rate of growth of human population. Population control would lessen the all too frequent policy of expediency, bolstered by a technology divorced from the philosophy of science. The scientist as a social entity must eventually establish the necessity for the ecosystem approach to world problems as a safeguard against unbalanced technological action. It is yet to be realized that political guidance and restraint is nothing like so operative in regard to technology as in other major fields of human action.

Ecology and Northern Development
Ian McTaggart Cowan

No attribute of man exceeds in importance his capacity to alter his environment. During the long history of his social evolution from hunter-food-gatherer to twentieth century technological man his capacity for environmental change has been dramatic.

There have been two facets to this change. By direct attack on the vital problems of human mortality man equipped himself to lengthen his life span. At the same time his technological skills and his appetite for energy equipped him, in the industrialized parts of the world, with the means and the incentive to attack the environment in increasingly sophisticated ways and to bend it to his immediate interest. All too frequently the long-term consequences of these actions have been ignored.

There is every likelihood that through the 1970's millions of people will starve as a direct outcome of man's failure to understand the basic ecological parameters from which he cannot escape or, understanding them, his failure to act appropriately.

Population in an ecological sense is a relative concept and even in parts of the Arctic death from starvation has overtaken a distressingly large number of people in the last decade (Vallee 1967); this has been a direct consequence of dramatic man-induced changes in the environment.

While the crisis of numbers haunts the unindustrialized sectors of the human species, advanced peoples are pursuing a course equally at odds with ecological prudence, a course that can only lead to impoverishment and misery to man and catastrophe to a large part of the world's living organisms.

The most obvious symptom of the approaching catastrophe is the rate of our chemical and physical alteration of the air, water and land with waste products. I refer not only to sewage and industrial wastes in our rivers, lakes and inshore marine waters, to noxious gases, aerosol biocides and fission products cast into the air; but also to the cacophony, rapidly increasing in variety and volume as it invades the remaining silent places; to the sprawling chaos that reaches out from many of our cities; to the shack towns where the poor and unfortunate live in filth and distress; to the needless invasion of scarce farmlands by expanding cities; to the defilement and desecration of landscape by strip mines, automobile graveyards, and industrial and military junk; and to so many other activities that reflect a callous disinterest in maintaining a sensitive and effective relationship between man and the living world.

But surely in Alaska, the Yukon, and the Northwest Territories, the last frontiers of the continent, where human populations average about 1 per 150 square miles, these forces cannot be at work. Surely with the mistakes made elsewhere as our lesson, we can proceed to avoid these errors and at least be original in the mistakes we make.

The natural communities of the continent fall roughly into two categories, first the pioneering associations of plants and animals that invade areas where catastrophe has struck by fire, flood or human clearing. These pioneering associations slowly evolve until there arises the second category, an assortment of plants that is relatively self-perpetuating—the climax state of the ecologist. For instance, the fire weed-willow-aspen community proceeds to the spruce forest climax. In each of these associations the classical biological energy cycle is proceeding, with the plants trapping solar energy and taking minerals and water from the soil. All other forms of life live either directly or indirectly upon the plants.

However, the passage from pioneer stage to climax is also a transition from an unstable state to a stable state. In the highly unstable early stages there is rapid change in the total mass of biological material. There is equally rapid change in the species composition. At the other extreme of the time scale there is relatively little change in the amount of living material from year to year, and though the proportions of the constituent species may change, the fluctuations are less extreme. The great surges in numbers of field mice, rabbits, lynx and moose take place generally in the early phase of the arctic-subarctic ecosystems.

Agricultural man survives on his capacity to create and maintain the unstable stages of the ecological cycle. By preference his aim is an annual cycle from plowed field to harvested crop. At the same time he strives to eliminate any competitors for his chosen plants and to assure that he is the only animal user. So he fences out competing animals, he hoes and cultivates and destroys pests with biocide sprays. A prodigious input of energy is needed to maintain the unstable stage; in fact it has been calculated that the energy required to produce 1 calorie of metabolizable energy in an Iowa cornfield is 5 times the yield. In the ecological sense then, man the cultivator is a natural catastrophe.

In the far north of the continent we have watched the impact of our agricultural technological culture, with its strong economic orientation, on that of the indigenous peoples. At its best it has been a dramatic social and biological revolution—too often, however, it has been tragic both in the short-term impact on the indigenes and in its long-term portent even for ourselves. The native peoples were a dynamic element in the balanced ecosystem. By ingenious techniques they lived on the contact zones

between land and water, removing annually some of the biological increase from each. Theirs was not a culture based on surpluses, and storage was limited and local. The causes of mortality were those of the wild living predator, starvation, accident and parasitic disease. Mortality was specially high among the young. Their mortality rate was apparently from 5 to 10 times that experienced by *southern* Canadians today.

There were probably as many people living in arctic North America as the biological resources could support and this capacity was unbelievably low. Evidence arising from recent studies of a human population west of Hudson Bay (Vallee, 1967) suggests a mean carrying capacity of 1 person per 250 square miles. Away from the coast caribou was the major source of food and it was a nomadic species imposing on its predators (wolves and men alike) unpredictable changes in its presence and abundance. The total system was essentially closed in as much as there was neither import nor export by the native peoples. The only exchange occurred through the annual migration from the Arctic of the entire new generation of birds reared on the summer surplus there.

Technological man introduced massive disturbances within the northern ecosystems. Not the least of these was an entirely new, economic strategy of approach to the natural resources. This involved import and export along with novel techniques for the use of energy in man's interest. The advent of the fur economy led to the need for more extensive travel. This in many parts of the North necessitated more dogs. In the Arctic dogs and men share the same food resources; thus more dogs were, in this sense, the ecological equivalent of more men. Lengthy trap lines encouraged the establishment of temporary food caches (caribou) in areas remote from the base village. Provident men would attempt to have a surplus at each site—a surplus that could not be carried over from year to year. This revolution would have been impossible without the accompanying changes that came with the rifle.

We know now that the caribou and musk ox, the only native large ungulates of the high North, were unable to sustain the increased demands on their numbers and the great decline began.

The details of this decline have been studied now for twenty years. We have no accurate figure on the original population of the 600,-000 square miles of range between the Mackenzie River and Hudson Bay, but one million would seem to be a reasonable approximation. Everything we have discovered points to the barren ground caribou as an animal in delicate balance with its environment and highly vulnerable to the vicissitudes of weather during the calving season. Consequently for several successive years there may be a net decline in numbers without the impact of man. The net loss is balanced by periodic *good* years when reproductive input greatly exceeds output. This is a form of production with which we have little experience. In essence it means that if there is to be sustained use of caribou by man for food, the base population of animals must be large enough to permit successive years upon which the breeding herd suffers drawdown without sinking to a level from which the good years cannot restore it. The biological *capital* required in this situation greatly exceeds that wherein an annual surplus can be anticipated, as with our domestic livestock.

In this kind of ecological situation short-term assessment of stock is meaningless; only assessment made over ten-year periods or longer will yield a true picture of the trends in numbers and provide a guide to the possible tolerable use by man. We can only assume that neolithic man entered this equation after the caribou population was fully established. We have no evidence as to the limiting factor operating upon those early men—it may not have been the presence or absence of caribou. We do know, however, that the caribou was over-harvested in the years that followed the advent of economics and the population collapsed in the 1950's to a low point—only 10 to 20 per cent of the primitive herd. At this point it certainly became the limiting factor in human population and many communities and nomadic bands would have perished entirely without the importation of food from *outside.*

The musk ox had preceded the caribou into the status of a resource no longer useful in the indigenous survival of man in the north.

But these are only isolated examples presented to illustrate the concept that in the ecology of the North we are dealing with an environment where the biological ground rules often depart widely from those with which we are familiar.

Our success in establishing permanent settlements in the middle and high North of this continent will depend in large measure upon our effectiveness in managing the natural resources. The resources of this vast area have been considered in detail in two recent studies. Turner (1959) sees little hope that agriculture can be a profitable venture as a base for permanent settlement in the area except in small exceptional localities. Forests, on the other hand, will contribute importantly, and wildlife and fisheries will continue to provide support for native peoples in non-urban areas. Turner is more pessimistic than some would be regarding the long-term contributions of mining to the existence of strong permanent settlement.

In general Dunbar (1962) concurs with these views of the terrestrial resources but is more optimistic about possible yields from the sea.

It is safe to say that we have still to solve the problems of successful use of the living resources of Arctic Canada and Arctic Alaska in the support of human populations much in excess of the primitive ones. In fact our impact over the first three to four centuries has generally reduced the resource base and its productivity.

Following is a brief review of some of the major ecological factors that must be considered in taking decisions on strategy in northern development. A first principle would be that at our stage of social evolution our biological resources should be managed so that their capacity to produce future crops is undiminished.

I shall not deal with the role of forests as timber or as cellulose for manufacture. The elements in this equation are reasonably well known.

Meat Production

There are three potentially important large meat animals of the Arctic: caribou, moose, and musk oxen.

The barren ground caribou is intensely social and migratory. Furthermore its food preferences are met in the climax plant communities, the lichen tundra and the mature taiga or broken forest of spruce and tamarack where lichen is an important element. The tundra proper is relatively fire-resistant but the taiga is highly inflammable and huge areas totalling thousands of square miles have been returned to the disturbed early stages with, in the Northwest Territories, a greatly reduced capacity for caribou production. Experience in Newfoundland, however, suggests that under certain conditions burning does not make tundra range useless to caribou. We do not yet understand the circumstances. Biological succession proceeds much more slowly in the Arctic than farther south and it may well take half a century to restore the tundra lichen-range following serious fire.

The moose, on the other hand, is a creature of the disturbed forests. It flourishes on the abundant second growth of deciduous plants that follow disturbance by fire, flood, avalanche or man (Spencer and Chatelain 1953). Thus as the caribou range has decreased, the moose range has increased, but potential yields do not seem to be equated. Furthermore, the moose is less migratory and thus more dependent on the appropriate mix of environment locally to provide its year round needs. It is also solitary and thus all major elements in the strategy of moose management differ from those of caribou management.

The musk ox in the wild has a low rate of productivity. In general it finds its ecological requirements north of the tree line and extends successfully to the frigid deserts of northermost Ellesmere Island. It is only locally a migrant and except for the low energy yield of its habitat could be an important animal in the production of meat and wool. Captive rearing experiments by Teal in Alaska and Oeming in Alberta (personal communications) confirm a capacity for growth and production almost the equivalent of less specialized breeds of domestic cattle.

The success of the reindeer cultures of equivalent latitudes in Europe and Asia led naturally to attempts to use this semi-domesticated caribou as a substitute for the vanishing wild herds of our Arctic. The history of these trials both in Alaska and Canada have been reported in detail (Krebs 1961, Roberts 1942). At their maximum the Alaskan population numbered some 600,000 animals. Porsild's (1929) study of the range potential of the tundra between the Mackenzie and Coppermine rivers suggested a capacity of about half a million animals. In addition to the trials made on the mainland of Alaska the reindeer was introduced to St. Matthew Island, Nunivak Island and St. Paul and St. George Islands of the Pribiloff group.

The history of those trials, in terms of economic or social impact on the North, can only be summarized as failures. In western Alaska and on some of the islands they have also been most destructive of the natural vegetation. The reasons for failure have been varied but the overriding ones are political bungling and failure of the northern population either indigenous or immigrant to grasp the potential, understand the ecological structures, and make the social adaptations requisite for success.

The island introductions have now become classic textbook models for the population behavior of the herbivore with no predators in an environment with limited food supply. (Scheffer 1951, Klein and Whisenhaut 1958). There was no attempt at management and in ecological terms these introductions can only be regarded as irresponsible acts with far reaching consequences to the biota of the islands.

These failures then are best regarded as socio-political; the ecological potential remains even if now somewhat reduced. On the basis of studies by Porsild (1929), Palmer and Rouse (1945), and the experience in U.S.S.R. (Flerov 1952), it is possible to predict that managed herds of reindeer numbering perhaps a million head could be supported in the Arctic from the Alaskan coast to the Coppermine River and that these could yield as much as 34 million pounds of meat a year. These would in part displace existing caribou herds and arrangements for the continued survival of the wild herds would probably reduce the domesticated potential somewhat. The meat yield from the combination would not be proportionately lowered.

The dispersed characteristic of moose distribution makes this an unsatisfactory animal for the production of a wild-taken harvest for the use of arctic communities. It is almost impossible to take an effective harvest even where the local politicians have overcome their sex bias and arranged for the removal of both sexes. The consequence of under-harvest is local surpluses that eat out the food supply and perish.

Great care should be taken that our quasi religious presence before the shrine of the musk ox does not lead to the same unfortunate consequence as has befallen the St. Matthew Island reindeer. This is of special significance on its island habitats, especially where the musk ox has been introduced to areas not previously inhabited.

It is becoming certain that the burgeoning human population of the world will demand first priority of all plant foods usable directly by man. The loss of energy that results from the conversion of human plant food into the meat of herbivores will not be tolerated. It follows therefore that we should be devoting ourselves to a search among the wild herbivores for species that we can culture as human food on acres, and plant forms, that cannot be used for the primary production of human comestibles. In the U.S.S.R. there are several institutes devoted to reindeer research and at least one at which the domestication of the moose is a primary objective. We, on this continent, have given almost no attention to the various possibilities of adapting the three arctic herbivores for domestic or semi-domestic culture. These might involve animal breeding experiments devoted to changing the physical and behavioural characteristics of the animals as well as attempts to adapt mid-arctic lands to the production of shrub or herb pastures suitable for the culture of these creatures. Such culture may be facilitated by the addition to the diet of low bulk, high energy chemical supplements imported for the purpose.

It would seem to be quite possible, for instance, to develop techniques for converting aspen and willow into pelleted feeds for big game being raised in semi-confinement. Technological and economic changes will be needed before such uses will be practicable. However we should be exploring the possibilities.

The fur resource of the arctic and subarctic areas has been a major contribution to this resource in North America. At the same time I must point to some of the Banks Island white fox trappers as the most affluent practitioners of the trapper's art anywhere in the world. It is a fickle component in the regional economy, however, because of very large fluctuations in the harvestable crop as well as rapid changes in demand based upon the social attitude toward different kinds of fur and the development of fur substitutes. The recent changes in the market for sealskins have made this particularly emphatic. The vagaries of supply can be predicted with reasonable assurance; not so the changes in demand.

The extensive marshlands of such areas as the delta of the Mackenzie River, the delta of the Athabasca, the area adjacent to Old Crow, and other great river deltas of the North have the capacity to provide a sustained harvest of muskrat fur of considerable volume.

The muskrat crop could be increased and stabilized by some relatively simple alterations in the habitat that would deepen the water remaining through the winter in many productive but temporary lakes. At the same time the institution of more refined management regulations, based upon a comprehension of the ecological opportunities and constraints, could also improve the long-term yield.

So far the meat potential of this species has been ignored, probably because of a bias arising from the common name. I can give personal assurance that fresh muskrat properly handled is far superior to rabbit both in flavour and nutritive content. It has enjoyed a limited market elsewhere under a trade name and I

can see an opportunity for the promotion of a specialty market in Canada.

I should not leave the subject of meat production from the arctic ecosystems without referring to the fishery potential. Recent studies by biologists of the Fisheries Research Board of Canada (personal communication) have revealed char stocks as high as 5 lbs. per acre in landlocked waters and stocks of migratory and marine fishes of some 3 lbs. per acre in the western Arctic and ten times that in the eastern Arctic. These stocks may be able to tolerate annual harvests of 20 per cent or higher on a sustained basis. The potential tonnage of high quality protein is immense when economic constraints are overcome.

The Role of the Far North as a Nursery Area for Migratory Birds

So far human activity in the North has had little influence on the hosts of migratory birds that swarm into its vast reaches each spring to make use of the energy surpluses available each summer. Countless millions of birds of some species annually raise their young on the teeming insects and the abundant crop of vegetation, seeds and fruits that characterize the arctic summer.

The majority of these species will remain relatively immune to normal human alteration to the environment. However, to an increasing extent we are searching for oil and minerals over vast regions and our leasing regulations completely ignore the ecology of the areas being assigned for *development*. The Arctic occupies a unique position with respect to many important migratory species of wildlife. Among the rare and vanishing species of birds the very localized and extremely vulnerable nesting grounds of the whooping crane have received wide publicity. No similar concern has been centred on the even rarer Eskimo curlew that may be making its last stand some-where along the lower reaches of the Anderson River. The Tule goose, Ross's goose, of the lower Perry River, the Aleutian Canada goose of Amchitka Island, and the greater snow goose on Bylot Island are just examples of large and valuable species, with small to very small world populations based exclusively upon very circumscribed arctic nesting grounds. A number of other species of Alaska and Northern Canada could be mentioned as almost as vulnerable.

My purpose in naming these few examples is to emphasize the ecological problems they present, and to urge modification of our regulations governing alienation of tracts of arctic land so that concern for proper ecological behaviour becomes an essential condition of the development.

Pesticides and Other Pollutants

The establishment of human communities in the Arctic sometimes leads to the demand for aerial distribution of DDT and other pesticides over adjacent areas. As a former and potential meal for millions of mosquitoes, black flies and tabanids that blight the short arctic summer, I can sympathize with the demand. However, decisions to spray and choice of the chemical should not be taken without adequate consideration of ecological consequences. The appropriate biologists of wildlife services should always be involved in reaching decisions on where, when and how to use eradicants and which ones to apply.

Pruitt has shown that the arctic caribou acquire a heavier load of radioactive fallout products than any other known species. It has been shown also that even on the isolated arctic slope of the Brooks Range the aquatic insects are carrying their quota of DDT. True, no harmful consequences have yet been dem-onstrated for these two situations, but then we have not looked for any. These examples, however, do serve to emphasize that caution must be observed even in such remote habitats. We cannot assume that a species is safe from such contacts with damaging chemicals merely because its nesting ground is remote.

The disposal of the waste products of our culture are a continuing problem in the Arctic. Human sewage is usually dumped into a local river if one is available. In this the north differs little from many more southerly areas. However, rates of decomposition are much slower and we know almost nothing of the impact of such action on arctic rivers. Research is urgent. The beaches of some of our most remote Arctic Islands are littered with plastic bags of human excrement grounded ashore after drifting miles from some northern outpost of our culture!

In the Far North where the number of polluters is small and the cost of better planning correspondingly low, we must face the reality that the assumption of dilution, where biologically active wastes are at issue, is frequently false. The assumption is an attractive one. It seems so obvious that if a relatively small amount of a biologically active toxicant is discharged into a large body of water it will rapidly dilute below the level of toxicity. The evidence from study of radioactive wastes (Woodwell 1968) and from DDT is that the assumption is false and that biological concentration is a frequent sequel. Unfortunately this false assumption is so much part of our philosophy that we accept its corollary, the right to pollute until detailed scientific proof of damage to man is produced. It would be far more valid to adopt here the principle behind the licensing of drugs for use on man—that permission be refused until there is detailed scientific proof that the pollutant will do no damage to the biosystem.

Military and industrial activity in the Far North has shown widespread disregard for the environment. Anyone who has visited the sites has been appalled at the litter that is left behind. In almost every instance government involvement or control was intrinsic and it becomes the responsibility of government to insure a clean-up of each such project in the north.

Mining is one of our most productive uses of the natural resources of the Far North. In most instances it is an activity with high waste residues. In land as heavily glacier-tilled as much of the western Arctic is, the physical wastes of subsurface mining may be relatively unimportant except where they occur in areas of special value in their original state. The discharge of chemical pollutants, however, is a different matter and should be carefully studied and controlled where ecological changes may arise.

The rates of recovery from all forms of exploitation are much slower in the Arctic than in the more southerly areas where we have been forming our procedures. This is a fact of the greatest significance as we pattern our attitudes and behaviour.

The Protection Movement

One of the most significant developments of the last forty years has been the rapid growth in the number of people who look to the natural environment for a large part of their recreation and enjoyment. This segment of our population is not generally antagonistic to economic development and to the use of the natural resources for human enrichment. They realize fully that to a very large extent the many advantages we enjoy stem directly and indirectly from our ingenuity in resource use. They agree that man cannot expand his use or improve his lot without greatly altering many areas of land.

However, these developments do demand that we develop patterns of use that are consistent with the maintenance of natural ecological conditions wherever possible. They also require total proscription in some areas of unique quality or biological importance.

There are many excellent reasons for seeking special immunity from environmental alteration in some areas. One of these is aesthetic. Many people attain a refreshment of the spirit in wildlife and unspoiled wilderness that needs no explanation or justification. A host of others, while not actively participating in the enjoyment of nature in remote places, find their lives richer for the knowledge that the opportunities exist.

Furthermore, the world's store of genetic material may well prove to be exceedingly valuable as we seek to put our new-found knowledge of genetic processes to work. The Arctic includes many species that have acquired unique heritable capacities to survive and prosper under northern climatic circumstances during millions of years of evolution. These qualities are in theory available for our use when we reach the stage in our knowledge that makes it possible. But each genotype lost, before evolution has replaced it, is another step in the degradation of our environment, another potential opportunity irretrievably lost.

The participants in the conservation movement regard the Arctic as of very special interest. It is a part of the continent where we have not yet made the plethora of mistakes that face us with unfortunate consequences farther south. It is a land where vast areas remain essentially as they have always been. It is a land rich in a unique flora, where plants and animals alike have evolved highly specialized attributes of behaviour, anatomy and physiology, much of this still unexplored. Conservationists therefore will increasingly appear as the conscience of our society to demand that the conversion of wild lands and their products into marketable form in the least expensive way is not the sole, nor perhaps the most important goal of this society.

In the northern lands knowledge of the appropriate ecological facts is indispensable if we are quickly to recognize the alternative opportunities for resource development and the constraints within which we must conduct our activities. Here, even more than further south, man will continue to live in the closest contact with the natural environment. His success will depend as much upon the sophistication of his ecological knowledge as upon his technological competence to respond.

REFERENCES

Dunbar, M. J. 1962. The Living Resources of Northern Canada, in: V. W. Bladen, editor, *Canadian Population and Northern Colonization.* Toronto: University of Toronto Press, pp. 124-35.

Flerov, K. K. 1952. Fauna of U.S.S.R.: mammals, vol. 1, No. 2: *Musk Deer and Deer,* pp. 202-28, translated by A. Biron and Z. S. Cole for National Science Foundation.

Klein, D. R. and J. L. Whisenhaut. 1958. Caribou Management Studies/Saint Matthew Island Reindeer—Range Studies. *Job Completion Reports 12* (2a) *Alaska Wildlife Investigations,* 39 pp.

Krebs, C. J. 1961. Population Dynamics of the Mackenzie Delta Reindeer Herd, 1938-1958. *Arctic,* 14:91-100.

Palmer, L. J. and Ch. H. Rouse. 1945. Study of the Alaska Tundra with Reference to its Reactions to Reindeer and Other Grazing. *U. S. Department of the Interior Research Report* 10, 48 pp.

Porsild, A. E. 1929. *Reindeer Grazing in Northwest Canada.* Ottawa: Department of the Interior, 46 pp.

Pruitt, W. O. Jr. 1962. A new "caribou problem". *Beaver,* Outfit 293, winter, pp. 24-25.

Roberts, B. 1942. The Reindeer Industry in Alaska. *Polar Record,* 24:568-72.

Scheffer, V. B. 1951. The Rise and Fall of a Reindeer Herd. *Scientific Monthly,* 43(6):356-62.

Spencer, D. L. and E. F. Chatelain. 1953. Progress in the Management of the Moose of South Central Alaska. *Transactions 18th North American Wildlife Conference,* pp. 539-52.

Turner, D. B. 1959. The Resources Future, in: F. H. Underhill, editor, *The Canadian Northwest: Its Potentialities.* Toronto: University of Toronto Press, pp. 46-89.

Vallee, F. G. 1967. Kabloona and Eskimo. Ottawa: Research Centre for Anthropology, Saint Paul University, 232 pp.

Woodwell, G. M. 1968. Radioactivity and fallout: The model pollution, 16 pp. mimeographed from symposium "Challenge for Survival" 1968. New York Botanical Garden.

The California Desert—Problems and Proposals
Jack F. Wilson

This great California Desert, a 16,000,000 acre national treasure, is in serious trouble. First, let me be more precise about what area I am talking about when I discuss "The California Desert." Since we know of no clear-cut delineation in a physiographic sense we have somewhat arbitrarily defined the desert to be an area that begins at the Colorado River near Yuma, Arizona, then follows the U.S.-Mexico border westward to the escarpment west of Ocotillo; then following northwestward along the backbone of the Santa Rosa and San Jacinto Mountains, along the northern foothills (roughly along the San Bernardino National Forest boundary)—west of Victorville to about Mojave, then generally northward to the northern end of the China Lake Military Reservation, then roughly eastward until one intersects the California-Nevada boundary near Lathrop Wells, Nevada, then down the state line to the Colorado River and the point of beginning. Within this area there is approximately 16,000,000 acres of land—the vast majority of which is in public ownership of some kind. The ownership breaks roughly this way:

National Monuments (2)	1,100,000 acres
Military Reservations (4)	2,900,000 acres
Reclamation W/D (3)	800,000 acres
Indian Lands	200,000 acres
Bureau of Land Management	10,100,000 acres
Private and State	1,100,000 acres
	16,200,000 acres

It might be well to mention at this point that we are embarked on a comprehensive study of the boundary question, and we already know there will be changes. For instance, we did not include the Antelope Valley because of its imminent urbanization, nor did we include the Owen Valley because of primarily administrative considerations. Also, we did not include the Colorado River and its myriad of complications in our original work, yet all or parts of these areas must be considered. This could quite alter the acreage figures presented here. Also, I must emphasize that the word *approximately* should be well considered since much of the desert is as yet unsurveyed in the cadastral sense.

And this brings us to one of the first problems. Each of the major land-holding agencies in the desert, except one, is generally dedicated to a single use. Bureau of Land Management alone in the desert has a multiple-use responsibility by law. Perhaps these concepts might be new to some of you, and there are no black and white distinctions as between multiple use and single-purpose use. But let's look at a couple of these land-holding agencies: for example, the National Park Service has well over 1,000,000

acres in two large national monuments. Their primary mission is defined by law as preservation of scenic, natural and historic values. Hence, they can and do rule out a great many uses, i.e., you cannot mine in Joshua Tree Monument even if you found a valuable deposit of minerals. There is no grazing for domestic livestock in Death Valley, nor do I think you could dig up a paleontological specimen except under the most stringent of controls. You cannot ride a motorcycle off designated trails, and you pay to get into the area, primarily for the recreational values that exist. There is nothing wrong with these restrictions —I simply point out that over 1,000,000 acres is primarily unavailable to many valid uses.

Take the military withdrawals as another case—here nearly 3,000,000 acres are used primarily for military purposes, i.e., weapons testing, rocket ranges, gunnery, and things of this sort. This land is being shot up, torn up, and scarred for reasons probably related to our security, yet there are known mineral deposits, critical wildlife habitat, archaeological sites and many other values that are totally unavailable because of a single-use philosophy. Perhaps this isn't too bad either, because the exclusion of other uses for some 30 years is accidentally providing a relict environment for future observation and study. These may make wonderful preserves for educational purposes.

Reclamation-withdrawn lands are set aside, primarily to further reclamation project activities and are primarily along the Colorado River. These are old withdrawals and consequently preclude mining and some other activities.

So an immediate problem is that nearly 25 per cent of the desert is to one degree or another unavailable to certain activities or uses.

The California Desert until recent years was a relatively untouched expanse larger than the state of West Virginia. Until recently it provided nothing more than a vast expanse to get across as soon as possible and preferably not in the heat of the day. Its riches include a multitude of resources that can be used in many ways for the betterment of man, and it might be well to examine some of these resources. There is no order of importance implied:

1. The Mineral Resources

Generally these fall logically into three classes in the desert. The *leaseable*—potash, phosphate, coal, oil and gas, etc.; the *locatable* —those subject to mining claims, i.e., gold, silver, platinum, titanium, rare earths, etc.; and the *saleable* for common varieties—sand, gravel, building stone, etc. Today there is an already established mineral use economy that exceeds over $160,000,000 annually. One mine produces over 80 per cent of the world's supply of europium—without which the quality of your color T.V. picture would be much less than you now enjoy. Large quantities of iron ore, boron, limestone (cement) and potash come from various places on the desert. Besides there are many known deposits of silver, tungsten, gold, and rare earths in production. This potential—according to our desert studies—should approximate over $300 million dollars a year by 1980 if a manageable system of minerals/land management can be developed.

2. The Wildlife Resource

The California Desert is host to a fascinating array of wildlife including several endemic and native species that are threatened with extinction. In some cases, the habitat for some of these rare endangered species has narrowed to a very few square miles. By today's standards it is estimated—probably conservatively —that 1,900,000 visits were made to the California Desert in pursuit of some type of wildlife-oriented experience, and a price tag of about $20 million annually was established with about 50 per cent being of a non-consumptive nature. While it may sound paradoxical, the California Desert supports fish populations that may range from a few rare and endangered, desert warm-spring species to the huge fisheries of the Salton Sea and the Colorado River.

3. The Vegetative Resources

Over 140 species of flowering plants are endemic to the California Desert only, and conceivably some 5,000,000 acres of desert does support enough, and the right kinds of, vegetation to permit livestock grazing. Besides the normally understood consumptive uses, such as grazing and wildlife habitat, there are some other economic uses of desert flora. Examples are commercial nursery for desert flora; medicinal uses, i.e., digitalis, surgical splints; and decorative (desert holly).

4. The Scientific, Cultural, Historical, and Educational Resources

The answers to many of the unsolved problems in archaeology, anthropology, and paleontology *may* lie in the California Desert. In the process of our California Desert Study over 1,000 archaeological sites were identified. The "Rosetta Stone" that could perhaps unlock the secrets on over 500 known petroglyphs and pictographs is yet to be found, and only very recently an early relative of the Condor was discovered in the desert. The deposits near Barstow now provide the basic correlation for the land mammal age on this continent.

Historically, there are some wonderful opportunities in the desert. Even today there are still some relatively untouched segments of the Old Spanish Trail, the Old Government Road, and historical forts. There are many old mining camps, and the desert is replete with legends of lost gold mines and hidden fortunes.

5. The Recreation Resource

Today a dynamic new use basically anchored in the technological accomplishments of man has literally roared into the desert. Our surveys in 1968 indicated about 5,000,000 visitor days were spent in the desert. While 50 per cent of the use was for camping, most of the campers also enjoyed either a motorcycle or four-wheel drive vehicle. This amazing technology has shrunk the desert, done away with its defenses of remoteness and hazards, and opened it up in all its glory to those who have the desire to use it. The desert also has become a last resort for some kinds of uses. The motorcycle is a case in point—because of its characteristics, i.e., noise, mobility and destructive potential, its use has been run out of private lands—state parks, national forests, flood channels, etc., and now in full force goes to the desert. Still, the most prevalent use for the desert is camping, nature observing, riding, hiking, and other kinds of recreation that seek the attributes of solitude and separation.

By the year 1980 we believe recreational use will swell to about 14,000,000 visitor days without any developments or restrictions, but about 30,000,000 is possible, if there is proper development and direction.

And now we come to the second major problem—nowhere else in the world are so many people in a position, so well equipped, to exert such a dramatic impact on a similar area of undeveloped land. This same accelerating technology that has placed man on the moon has given him the mistaken belief that he can escape dependence upon his environment and has created the probability of monumental environmental mistakes. Of almost equal importance, but really of a causative nature, is the fact that most of the methods devised to use the varied resources of the desert were pursued with only one purpose in mind and without consideration of the impact on the total environment.

Today, right now, we are faced with a range of problems.

1. One of the concepts of desert management is to try to establish utility corridors to provide for the necessary increased service that normal growth of our nation requires, and still retain as much of the environmental value as possible. There are essentially now four major corridors through the desert, to a great extent dictated by military withdrawals. Three of the four are already in heavy use. Now, Southern California Edison Company has applied to utilize the last potential corridor area for 500 KV lines. This corridor runs through Twentynine Palms, Yucca Valley and Morongo Valley, which have over the years developed as recreation and retreat/homesite areas. The people of the area are greatly upset, and are exerting great pressure (the SCE also has a strong case, and have done a commendable job in attempting to blend into the landscape) yet this is a single-purpose, economically-motivated proposal that cannot help but affect one aspect of the environment adversely.

2. Today in the Santa Rosa Mountains behind (west) of the Cathedral City-Palm Springs area, a bulldozer is busily cutting roads and lots into the mountain backdrop of these cities. The net result

is going to be the loss of an extensive area of habitat for the endangered Desert Bighorn Sheep and a scarring of the landscape, probable hazards of flooding, pollution, and costly administrative requirements. This is being done to provide new lots, to provide a cut-off road from the mountain to the desert, and for economic advantages.

3. Very soon a new airport will be built in the desert to accommodate the supersonic aircraft. Yet to be known and assessed are the repercussions of the fantastic sounds produced by these aircraft.

4. Twice in the past few months we've been able to head off "Rock Festivals" in the desert. Can you imagine the effects on a fragile desert area if 200,000 people were to converge on it? The normal daily waste of a crowd this size would require the removal of 40,000 pounds daily, not to mention other damages. Yet, there are some sociological implications that need exploring.

So far, all I've done is to raise or attempt to define the problems. How about the proposals? First off, we recognize two very basic facts:

(a) To date, there has been no real effort to consider the desert in terms of its total environmental contribution to mankind. While there is much knowledge available, there are extensive gaps, and

(b) many of the values of the desert are in danger of immediate loss. (For example, Dr. Gerald Smith of the San Bernardino County Museum estimates that two archaeological sites are being lost each week due to development or destruction.)

Right now we are proposing a two-pronged program for desert management: First, a comprehensive, long-range plan for the desert in the context of the total environment of the southern California area. Second, protection and preservation of the endangered and irreplaceable values while the comprehensive plan is being developed.

The protection and preservation aspects come into form with three main proposals:

1. That a uniformed ranger force be placed in the field as soon as possible. It is not envisioned as a police force, but rather in terms of information and education, and preventive presence.

2. Development of a system of way stations or information centers which are oriented primarily toward education. A knowledgeable user has more chance of being a good user. These will also serve as administrative points.

3. Development of a road and trail network that will permit as many of the uses as can properly be accommodated, and build the kind of sites that are needed to serve this roads/trails system.

Concurrently we already have begun the data gathering and inventory work needed to support the comprehensive plan. The unique feature of this type of planning is that it must be done with the participation of the broadest possible range of cooperation: state, educational institutions, user groups, other federal agencies, and the responsible public must be involved. We hope much of the data can be computerized and that implications and options can be carefully and rapidly assessed.

It is no doubt clear that many of you in the scientific community will need to be contacted because you are among the most important factors, and the real insurance against tragic environmental decisions. Let us cite an example of what I mean. In Imperial County there are some extensive sand dune areas. This area is ideal for the use of dune buggies, and is receiving heavy use. Yet, to the east there are some small oases provided by the run-off from the Chocolate Mountains, probably less than 1,000 acres in total, wherein lies the total known habitat of a spadefoot toad. This little toad has the capability to hibernate for extensive periods—up to two years or more, and

within him may be the clue we need to cope with the extensive time/distance problems of space travel. Without being aware of the spadefoot toad it would be so easy to unknowingly cause his habitat to be destroyed. This is the type of environmental error we fear.

Controlling the Planet's Climate
J. O. Fletcher[1]

As internationally pursued research efforts continue to improve our knowledge of climatic processes and the possibilities of deliberately influencing them, we are also becoming increasingly aware of the disturbing fact that human activity may already be inadvertently and irreversibly doing so. Furthermore, the inadvertent consequences of human activity will increase manyfold in only a few decades, precisely at a time when rapidly growing pressures on world food production make the social consequences of climatic variation ever more serious. The inescapable conclusion is that purposeful management of global climatic resources will eventually become necessary to prevent undesirable changes. That such capabilities could be used to improve existing climatic conditions is obvious.

Recent years have seen an upsurge of research in the fields of weather modification and climate control. Substantial improvements in the accumulation and analysis of environmental data, coupled with a better understanding of the nature and interrelationship of climatic processes, have provided researchers with theoretical insights into how global climate can be modified and what some of the resulting consequences might be. Man already has the technological capability to carry out many climate-influencing schemes, such as the creation of large inland seas, the deflection of ocean currents, the seeding of extensive cloud

or surface areas, and perhaps even the removal of the arctic pack ice.

Still unresolved, however, is the uncertainty about the possible global effects of such large-scale weather modification efforts, which, in addition to bringing about major environmental changes, would give rise to many complex economic, sociological, legal, and political problems.

Let us proceed now to examine more fully the nature of the physical problem, the depth of our present understanding of it, the feasible influencing capabilities available to us, and the prospects for future progress.

The Inadvertent Influencing of Global Climate

Whether human activity has played a significant role in climatic shifts of the past century is a question which cannot yet be answered with any confidence. The complexities of global climate are still too poorly understood to assess the dynamical response of the system to a given change. Some investigators have argued that the effects of man's activities

1. Any views expressed in this paper are those of the author. They should not be interpreted as reflecting the views of his employer, the RAND Corporation, or the official opinion or policy of any of its governmental or private research sponsors.

are already significant, or even dominant, in changing global climate. The influencing factors most frequently suggested are carbon dioxide pollution, particulate pollution (smog and dust), and heat pollution. The physical arguments advanced have to do with the effects of these pollutants on the heat balance of the atmosphere.

Carbon dioxide is one of the three important radiation-absorbing constituents in the atmosphere (the other two being water vapour and ozone). There is no doubt that the carbon dioxide concentration in the atmosphere has been increasing in this century, apparently by some 10-15 per cent, due primarily to the increased combustion of carboniferous fuels.

The physical effect of a greater CO_2 concentration in the atmosphere is to decrease the radiative loss to space. Thus, an increase in CO_2 increases the so-called 'greenhouse effect' and causes global warming.

Some have suggested that the general warming that took place from 1900 until about 1940 was due to just such an increase in the atmospheric content of CO_2. Plass, in 1959, estimated that a warming of 0.5° C during the last century could be attributed to this cause, and this is comparable to the warming that actually did occur.

It is further estimated that, by the year 2000, a further warming of three times this amount could be caused by the increase of CO_2 in the atmosphere. Other estimates have predicted an even greater warming.

Notwithstanding these arguments, the sharp global cooling of the past decade indicates that other, oppositely directed factors are more influential than the increasing atmospheric content of CO_2. For example, Möller (1963) estimates that a 10 per cent change in CO_2 can be counter-balanced by a 3 per cent change in water vapour or by a 1 per cent change in mean cloudiness.

Let it also be noted that the oceans have an enormous capacity to absorb CO_2, this varying according to their temperature with colder oceans being able to store more of the gas. Thus, a warming of the oceans could also be a primary cause of the increase of CO_2 in the atmosphere.

In summary, it appears that, other factors being constant, the CO_2 generated by human activity could bring about important changes of global climate during the next few decades. But other factors, of course, are not constant, and have apparently been more influential than the CO_2 increase in affecting the climate of recent years.

With regard to heat pollution, Budyko (1962, 1966) points out that, although the yearly production of man-made energy on Earth is now only about 1/2,500 of the solar radiation arriving at the Earth's surface which is not returned to space, it could increase to equal the retained solar radiation if compounded annually at 10 per cent for 100 years, or 4 per cent for 200 years. (The present growth rate is about 4 per cent.) From these numbers we may conclude that, sometime during the next century, the problem of heat pollution will become important on a global scale. By then we must be able to compensate for it or face the possibility of a sharp global warming which could, in turn, trigger additional reinforcing transformations such as a melting of the polar ice. But, for the time being, and for the next few decades, the effects of heat pollution will not be sizable enough to exert significant influence on global climate.

One of the most rapidly increasing forms of man-made atmospheric pollution is smog, which embodies all forms of industrial pollution. Bryson (1968) reports a turbidity increase of 30 per cent per decade over Mauna Loa Observatory, which is far from all sources of pollution. This is thus indicative of the general increase. He further argues that a reduced atmospheric transparency, even by only 3-4 per cent, could decrease the global mean temperature by 0.4° C. This is due to the fact that a more turbid atmosphere will reflect back more of the sun's radiation, thus allowing less heat to penetrate through to the Earth.

Bryson believes that the increasing global air pollution, through its effect on the reflectivity of the Earth's atmosphere, is currently the dominant influence on climate and is responsible for the temperature decline of recent decades. Budyko (1968) also attributes climatic changes primarily to the decreased transparency of the atmosphere, caused in the past by volcanic eruptions and in recent decades by man-made pollution. If this interpretation is correct, mankind faces an immediate and urgent need for global climate management, especially in view of the fact that smog production is increasing everywhere at an exponential rate and no means of curbing this increase are in sight.

Curve 4 in Figure 1 shows the observed trends of atmosphere transparency since 1890 and a general correlation with some of the other variations in the global system—such as those of northern hemisphere temperatures

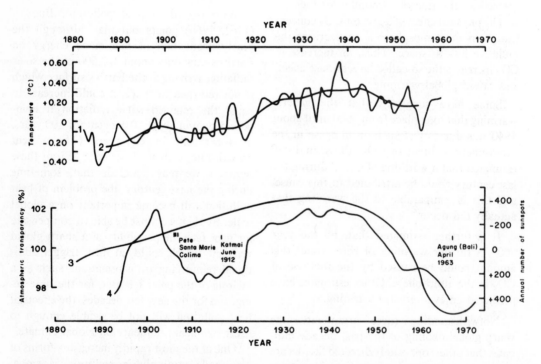

Figure 1. Comparison of annual variations in several climatic factors. In each case the ordinates are the deviations from an annual mean established over a long term. Curves 1 and 2 (from Budyko, 1968) show the mean annual temperature in the northern hemisphere, curve 1 depicting the annual figures and curve 2 being smoothed by taking a ten-year moving average. Curve 3 depicts total number of sun-spots (from Nazarov). Curve 4 (from Budyko, 1968) depicts atmospheric transparency as percentage of a mean. The transparency was determined by measuring direct solar radiation with cloudless sky at several stations in Europe and America. Since atmospheric transparency can be affected by volcanic eruptions the dates of five major eruptions are indicated.

(Curves 1, 2) and sun-spot activity (Curve 3) —can be seen. The sharp decrease in transparency early in this century can be attributed to a series of volcanic eruptions. However, the decrease since 1940 cannot be attributed to this cause although the eruption of Agung in Bali, in 1963 did cause a noticeable world-wide effect. Thus, man-made pollution may have been the most important cause of recent climatic changes.

On the other hand, there also appears to be a connection between solar activity and atmospheric transparency. Curve 3 in Figure 1 shows the trends of sun-spot activity and one can see that much of the recent decrease in atmospheric transparency might be accounted for on this basis. If this is true, a reversal should become apparent during the next decade, when fewer sun-spots are expected.

Still another form of growing pollution, and one whose possible effects have received little study, is the creation of cirrus cloudiness (vapour trails) by the exhaust products of high-flying aircraft. Increased cloudiness of any form tends to increase the reflectivity (albedo) of the Earth and, according to Bryson's calculations, a 1 per cent increase in mean albedo would cool the Earth by 1.6° C. On the other hand, it should be noted that increased cloudiness at high levels greatly reduces radiative loss to space, and this would have a warming effect on the Earth. Thus, the dual effects of more or less cloudiness are great, but the direction of the net influence depends on the type and height of the clouds, and on whether they are in a dark or sunlit region of the Earth.

From the foregoing considerations, we may conclude that man is probably inadvertently influencing global climate at the present time. Certainly several products of man's activity are theoretically influential enough to do so within a few decades. However, there are so many variables and degrees of freedom in the global system that specific cause and effect estimates in this regard are still very uncertain. In order to better understand this uncertainty, let us take a brief look at the dynamic and multifaceted nature of global climate.

The Changing Pattern of Global Climate

The climate of a particular region is determined by a number of relatively static factors such as elevation, latitude, topography, type of surface, and also by the properties of the air which passes over it. The dynamic factor which brings about weather changes is the circulation of the atmosphere, which, in turn, is strongly influenced by the interaction of the ocean/atmosphere system.

Substantial world-wide changes of climate have occurred, even in the course of a few decades, and have been described by many investigators. The data show that the general vigour of the global atmospheric circulation undergoes significant variations, with associated latitudinal shifts of the main wind currents and changes in the nature of their disturbances. Variation in the global atmospheric circulation pattern is the factor which makes possible a coherent interpretation of climatic data from all parts of the Earth.

For example, during the first three decades of this century, the general trend was toward a growing strength of the northern hemisphere circulation, a northward displacement of polar fronts (outer boundaries of cold masses) in both the atmosphere and the ocean, a northward displacement of pack-ice boundaries and cyclone paths (movements of large, rotating wind currents), a weaker development of blocking air masses over the continents, and a pronounced aridity of the south central parts of North America and Eurasia. Conversely, recent decades have exhibited opposite trends:

a weakening circumpolar circulation, south-ward shifts of ice boundaries and cyclone paths, and increased rainfall in the south central parts of the continents.

These trends were underscored in 1968. It was a year in which Icelandic fishermen suffered losses due to the most extensive sea ice in the last half century, while phenomenal wheat yields from the plains of both Asia and North America due to increased rainfall pushed world wheat prices to a 16-year low. In a happier vein, the predicted 1968 famine in India did not occur, with favourable climate and better strains of grain as the important offsetting factors. In the southern hemisphere, the southward displacement of the Chilean rainfall region created severe droughts.

Yet such small variations of climate, though of growing importance to our complex pattern of human activity, are minor compared to the more pronounced variations that have occurred in the relatively recent past. Less than 20,000 years ago, an ice sheet still covered North America and stretched from the Atlantic to the Pacific with a thickness of up to two miles. The last major ice sheet disappeared from Scandinavia only about 8000-7000 B.C., while in North America the ice retreated even later. During the period of ice retreat and somewhat after, rainfall in the Mediterranean area and probably over much of the hemisphere was less than at present, possibly due to cooler oceans. The post-glacial warming culminated in a "climatic optimum" about 4000-2000 B.C., during which world temperatures were 2°-3° C warmer than they are now and rain was much more plentiful in North Africa and the Middle East.

The decline from the warm optimum was abrupt from about 1000 B.C., with cooling continuing to about 400 B.C. This was a period of maximum North African rainfall, which was accompanied by the rapid develop-ment of human activity partly induced by climatic stress. By this time, renewed warming had set in and continued until a "secondary climatic optimum" of A.D. 800-1000, a period characterized by a relatively rainless, warm and storm-free North Atlantic, which made possible the great Viking colonization of Iceland, Greenland and Newfoundland.

The subsequent climatic decline, during which arctic pack ice advanced southward in the North Atlantic, was abrupt from about A.D. 1300, with one partial recovery around 1500. It culminated in the "little ice age" of 1650-1840.

Since about 1840, a new warming trend has predominated and appears to have reached a climax in this century, followed by cooling since about 1940, irregularly at first but more sharply since about 1960. The periods of general warming were accompanied by increasing vigour of the westerly circulation in both hemispheres, bringing a more maritime climate to the continents, a northward displacement of cyclone paths, and a pronounced warming of the Arctic. The recent cooling trend exhibits a reverse pattern: weakened westerly circulation, more variable and southerly cyclone paths, and a colder Arctic.

The pattern of change in the southern hemisphere is more obscure. No reliable index has been found for the strength of the southern hemisphere trade winds and even the indices of mid-latitude westerlies are not adequate. Temperature patterns for the 80 per cent of the southern hemisphere covered by oceans are almost non-existent. Even since the International Geophysical Year (1957-58), year-to-year variations in sea-ice extent in the Southern Ocean are largely unknown. However, the meagre data that are available show that corresponding climatic variations are evident from pole to pole.

The Global 'Climate Machine'

It is increasingly apparent that climatic change can be explained only in terms of the behaviour of the atmosphere and ocean on a global scale. Net heating at low latitudes and net cooling in polar regions forces the motion of the atmosphere, which, in turn, drives the surface circulation of the ocean. On the average, the atmosphere and oceans transport heat vigorously enough to balance the difference in heat loss between equator and poles, with atmospheric motion transforming potential energy into kinetic energy at a rate which balances frictional dissipation.

Climatic variations seem to be associated with variations in the vigour of the whole global circulation, but why the global system varies is still a mystery. It follows that the fundamental problem in the study of climatic change is the development of a quantitative understanding of the general circulation of the atmosphere and, since three-quarters of the heat which forces the atmospheric motion comes by way of the ocean surface, a quantitative understanding of oceanic heat transport and ocean/atmosphere heat exchange is especially vital.

Such an understanding should begin with the planetary distribution of heat loss and gain by the atmosphere and ocean. Fundamental physical laws should then enable us to predict the global distribution of temperature, pressure, motion, water vapour, clouds, and precipitation, together with resulting moisture and heat transports. In practice, this presents enormous difficulties. However, with the development of modern computer technology, rapid progress is being made. Already it is becoming possible to mathematically simulate certain large-scale processes in more detail than we can now observe them in nature.

For further progress in simulating atmospheric dynamics we need a better understanding of the processes of atmosphere. Variations in equatorial heating and polar cooling are poorly understood and have received little study, largely because of the paucity of relevant data.

Nevertheless, it has been discovered that significant year-to-year variations in ocean/atmosphere heat and moisture exchange do occur and that these anomalies are closely related to observed variations in the dynamical behaviour of the atmosphere.

For example, one very influential ocean/atmosphere interaction which is subject to large and sudden anomalies, is associated with the zone of up-welling cold water at the equator. This zone is created by the opposite deflection of warm surface water north and south of the equator in response to the easterly trade winds. In the eastern Pacific, the temperature difference between this up-welling water and the warm waters on either side is normally several degrees and extends for several thousand miles.

During some years, these cold tongues weaken or vanish as the equatorial trade winds wane. Bjerknes (1966) has documented several such cases for the Pacific, showing that the resulting variation of evaporation and subsequent condensation influences the atmospheric circulation of the whole northern hemisphere.

Similar studies for the Indian Ocean have not yet been conducted due to the lack of data, though a 1963-64 expedition found a cold equatorial tongue there nearly 10° C colder than the surrounding waters at 28° C. Yet it seems likely that such processes are associated with the rise of East African rainfall since 1961-63. Indeed, the frequency of such occurrences may be closely connected with the changes in the global system since 1961-63 (Lamb, 1966c).

The interaction of large-scale atmospheric and oceanic circulation in the Indian Ocean is known to vary from year to year. Understanding this interaction is not only necessary for understanding global climate, but has immediate application for forecasting the south-west monsoon, which directly affects the crops and economy of one of the most densely populated areas in the world.

Our present state of knowledge cannot yet explain why the equatorial trade winds wane, though we have some surmises. There is growing evidence that variations of the northern hemisphere circulation may be influenced by variations of the much stronger southern hemisphere circulation, but the basic cause of the planetary variation is still obscure.

Impressive statistical correlations between various indices of climatic change and various indices of solar activity (including sun-spot activity, as in Figure 1) have been presented by many investigators (Fairbridge, 1961; Rubinstein and Polozova, 1966), but no one has yet been able to advance a physically plausible cause-and-effect explanation.

Variations in the quantity of radiation received from the sun (expressed as the 'solar constant') are usually judged to be too small to account for the relatively large observed variations of global climate. Therefore, much attention has been directed towards searching for mechanisms by which upper atmosphere processes, triggered by very small changes in the energy from the sun, can in turn influence much more energetic processes in the lower atmosphere (troposphere). However, a better understanding of ocean/atmosphere interactions may reveal that feedback processes at the surface can amplify the effect of small solar variations to produce large changes in the behaviour of the planetary system. One such 'thermal lever' is the variable extent of ice on the ocean (Fletcher, 1969).

Ocean Ice as a Climate 'Trigger'

The presence of sea ice effectively prevents the transfer of heat from ocean to atmosphere in winter, thus forcing the atmosphere to balance the radiative heat lost to space. For example, in January the mean surface temperature in the central Arctic is about -30° C, while a few feet below the pack ice, the ocean water is near -2° C. The ice and its snow cover are such good insulators that relatively little heat reaches the surface from below. The surface radiates heat to space, and this heat loss simply cools the surface until it is cold enough to drain the needed heat from the atmosphere. The thermal participation of the ocean is greatly suppressed. If the ice were not there, the needed heat would be obtained from the relatively warm ocean.

In summer, on the other hand, an open polar ocean would absorb around 90 per cent of the solar radiation reaching the surface, instead of the 30-40 per cent presently absorbed by the year-round pack ice. Thus, the presence of the ice suppresses heat loss by the ocean in winter and suppresses heat gain by the ocean in summer. For the atmosphere, of course, the reciprocal relation applies; over pack ice, the atmosphere cools more intensely during winter and warms more intensely in summer.

In this way, variations in the extent of ice can amplify the effect of small variations in solar heating. Thus, a decrease in solar radiation causes cooling, which causes ice extension, which in turn cools the atmosphere more, causing further ice growth and stronger thermal gradients. The causes and effects are self-reinforcing, and provide "positive feedback."

How far such a process must go before it triggers other instabilities in the ocean/atmosphere system, such as the sudden variation of equatorial temperature described above, cannot be judged at this time. Clearly, there are many complex feedback processes, both positive and negative, in the ocean/atmosphere "climate machine," and many thresholds beyond which the direction of the feedback can change.

For example, suppose that the warming of the Arctic, which by 1940 had greatly reduced the thickness of the pack ice, had continued? As the ice receded farther in summer and the thinner ice become more fractured in winter, evaporation would have increased, thus increasing the density of the surface waters both by increasing the salt concentration and by cooling; this would tend to decrease the vertical stability of the upper layers of the ocean.

If this process had continued to the point of destroying the present strong stratification of ocean surface layers and inducing deep convection, then refreezing at the surface would have been impossible until the whole water column had cooled to freezing temperature—a process which would take many years at the least. After the whole ocean had cooled to the freezing temperature, additional cooling would have then refrozen the surface, thus re-creating surface stratification and reformation of surface ice, namely, the initial condition. Thus, a 'threshold' exists in each direction: destruction of stratification whose effect is to prevent refreezing, and the eventual depletion of heat content which triggers refreezing.

Budyko (see Fletcher, 1966) has argued that, under present conditions of solar heating, the arctic pack ice would not reform if it were removed. Instead, a new and stable climatic régime would be established in which the Arctic Ocean would remain ice-free.

To answer such questions with more certainty we really need to make a model for the entire planetary circulation under the assumption of an ice-free Arctic Ocean, but as yet this has not been adequately done. However, detailed calculations of zonal temperature distribution at various levels under conditions of an ice-free Arctic have been made by Rakipova (see Fletcher, 1966) using a theoretical model of zonal temperature distribution. According to these calculations, the intensity of atmospheric circulation would decrease, but much more so in the winter than in the summer, so that seasonal contrasts would be much smaller than at present. In high latitudes, poleward atmospheric heat transport would decrease by about 25 per cent during the cold half year, and the Arctic Ocean would remain ice-free.

In summary, it appears that a sufficient warming of global climate would lead to the disappearance of the arctic pack ice, at which time a new and relatively stable climatic régime would be established. Such a régime, while bringing a more temperate climate to the subpolar areas, could make other parts of the world considerably more arid.

Budyko (1968) used a similar empirical approach to estimate the influence on global climate, during planetary cooling, of the interaction of variable solar radiation, changing ice extent, and mean global surface temperatures. For his highly idealized model he concludes that, in the event that mean solar radiation over the earth decreases by 1 per cent, the mean global temperature would drop by 5° C, the cooling being reinforced by an advance of the ice boundary by about 10° of latitude in both hemispheres. Should the solar radiation decrease by 1.5 per cent, the global temperature drop would be 9° C, and the ice advance would be 18° of latitude. If the radiation decrease were more than 1.6 per cent the ice boundary would advance past the 50° latitudes in both hemispheres, and the cooling due to the large ice area would cause continued ice growth until all the oceans were frozen. Once such a condition was established, melting would not occur even with a substantially higher solar radiation intensity.

It should be noted that the empirical dependencies used by Budyko were calculated from northern hemisphere climatic data and he assumes that the southern hemisphere would respond similarly. This assumption probably exaggerates the sensitivity of global climate to solar variations, but Budyko's dramatic conclusions illustrate the necessity of taking such feedback processes into account.

Ice extent is probably the most influential factor capable of quickly transforming the large scale thermal properties of the earth's surface. Thus, understanding the interaction of ice extent, radiation variations and atmospheric circulation is fundamental to understanding global climatic changes.

Possibilities for Deliberately Influencing Global Climate

Theoretical perspectives for modifying global climate by influencing large-scale atmospheric circulation have been discussed by Yudin (1966), who emphasizes that since the energies in nature are so vast compared to man's capabilities, ways must be found to trigger natural instabilities using relatively small energy inputs. He points out that, in theory, it should be possible to influence the velocity of air masses with much less energy than is needed to effect local changes in either atmospheric temperature or pressure. Moreover, in influencing velocity, energy should be applied evenly over a broad area in order to minimize its dissipation.

Yudin then proposes that, following these precepts for the application of energy, emphasis should be placed on identifying critical 'instability points' in the natural development of cyclones. For example, only slight deflections of certain winds are associated with a faster movement of cyclone centres.

These brief criteria clearly identify one difficulty associated with large-scale weather modification, namely that the theoretically most effective approaches involve actions that we do not know how to produce efficiently. On the other hand, various ways of influencing the heat losses and inputs to the atmosphere, although theoretically inefficient from the viewpoint of immediate dynamical consequences, are much more achievable with present technology. It has, for example, already been noted that the creation or dissipation of high cloudiness has an enormous influence on the heat budget of the atmosphere and of the surface. Moreover, under certain conditions, only one kilogram of reagent can seed several square kilometres of cloud surface. It is estimated that it would take only sixty American C-5 aircraft to deliver one kilogram per square kilometre per day over the entire Arctic Basin (10 million square kilometres). Thus, it is a large but not an impossible task to seed such enormous areas.

Assuming that such seeding were effective in creating or dissipating clouds, it is of interest to estimate the effect of such cloud modification on the heat budget of the surface/atmosphere system. It is estimated that the average cloud cover over the Arctic in July decreases the radiative heat loss to space by about 350,000 million calories per square kilometre per day from what it would be without clouds. By comparison, total cloud cover at 500 metres would decrease radiative loss by only one-third as much, whereas total cloud cover at 5,000 metres would decrease radiative loss by three times as much. These numbers demonstrate not only the enormous thermal leverage that might be exercised by influencing mean cloudiness, but also the range of influence that might be possible, depending on cloud type, height, and its influence on the regional heat budget. This conclusion is further underscored by noting that mean monthly values of radiative heat loss at the surface have been observed to vary by more than 100 per cent in different years at some Arctic stations, possibly due to variations in cloudiness.

Similarly, it may be noted that, under certain conditions, influencing the surface reflectivity of arctic pack ice is not beyond the capability of present technology. Since the presence of sea ice severs the intense heat flux from the ocean water to the cold atmosphere, regulating the extent of sea ice is still another possible way of exercising enormous leverage on patterns of thermal forcing of atmospheric motion.

Influencing the temperature of extensive ocean surface areas by changing the courses of certain ocean currents has also been proposed (Rusin and Flit, 1962). These schemes involve large, but not impossible, engineering efforts, some of which are discussed in the next section. The principal difficulty, however, is that the present understanding of ocean dynamics is too rudimentary to reliably predict the effects of such projects and, even if this were possible, the dynamic response of the atmosphere to the new pattern of heating could not be predicted until more realistic simulation models have been developed.

These various examples demonstrate the following essential conclusions:

1. It does appear to be within man's engineering capacity to influence the loss and gain of heat in the atmosphere on a scale that can influence patterns of thermal forcing of atmospheric circulation.

2. Purposeful use of this capability is not yet feasible because present understanding of atmospheric and oceanic dynamics and heat exchange is far too imperfect to predict the outcome of such efforts.

3. Although it would be theoretically more efficient to act directly on the moving atmosphere, engineering techniques for doing so are not presently available.

4. The inadvertent influences of man's activity may eventually lead to catastrophic influences on global climate unless ways can be developed to compensate for undesired effects. Whether the time remaining for bringing this problem under control is a few decades or a century is still an open question.

5. The diversity of thermal processes that can be influenced in the atmosphere, and between the atmosphere and ocean, offers promise that, if global climate is adequately understood, it can be influenced for the purpose of either maximizing climatic resources or avoiding unwanted changes.

Specific Schemes for Climate Modification

Many engineering proposals have been advanced for improving the climatic resources of particular regions. All of these schemes share the common defect that their influence on the global system cannot yet be reliably judged. Some are on a scale that could well influence the global system and possibly even trigger instabilities with far reaching consequences. Sooner or later, some such schemes may be carried out, and it is of interest to consider them in the larger perspective discussed here (Rusin and Flit, 1962).

ICE-FREE ARCTIC OCEAN

The largest scale enterprise that has been discussed is that of transforming the Arctic into an ice-free ocean. As was noted earlier, this has been very carefully studied by the staff of the Main Geophysical Observatory in Leningrad. The central question is the stability of the ensuing global climatic régime. This question cannot be adequately evaluated until global climate simulation models are better developed and suitable simulations performed.

There is also a certain amount of uncertainty in regard to the engineering feasibility of removing the arctic pack ice. It is possible that the capacity of present technology may be sufficient to accomplish this task, but this has not yet been established. Three basic approaches have been proposed: (a) influencing the surface reflectivity of the ice to cause more absorption of solar heat; (b) large-scale modification of Arctic cloud conditions by seeding; and (c) increasing the inflow of warm Atlantic water into the Arctic Ocean.

BERING STRAIT DAM

The Soviet engineer, Borisov, has been the most active proponent of the much-publicized Bering Strait dam. The basic idea is to increase the inflow of warm Atlantic water by stopping or even reversing the present northward flow of colder Pacific water through the Bering Strait. The proposed dam would be 50 miles long and 150 feet high. The net climatic effect of the project, if it were carried out, is still highly uncertain. A good argument can be made that the effect would be less than that of naturally occuring variations in the Atlantic influx.

DEFLECTING THE GULF STREAM

Two kinds of proposals have been discussed, a dam between Florida and Cuba, and weirs extending out from Newfoundland across the Grand Banks to deflect the Labrador current as well as the Gulf Stream. None of these proposals have been supported by detailed engineering studies or reliable estimates of what the resultant effects would be.

DEFLECTING THE KUROSHIO CURRENT

The Pacific Ocean counterpart of the Gulf Stream is the warm Kuroshio Current, a small branch of which enters the Sea of Japan and exits to the Pacific between the Japanese islands. It has been proposed that the narrow mouth of Tatarsk Strait, where a flood tide alternates with an ebb tide, be regulated by a giant one-way 'water valve' to increase the inflow of the warm Kuroshio Current to the Sea of Okhotsk and reduce the winter ice there.

CREATION OF A SIBERIAN SEA

Dams on the Ob, Yenisei and Angara rivers could create a lake east of the Urals that would be almost as large as the Caspian Sea. This lake could be drained southward to the Aral and Caspian Seas, irrigating a region about twice the area of the Caspian Sea. In terms of climatic effects, the presence of a large lake transforms the heat exchange between the surface and atmosphere. Of equal or greater importance in terms of climatic effects, is the land region transformed from desert to growing fields, with accompanying changes in both its reflectivity and evaporation.

CREATION OF AFRICAN SEAS

This is the largest known proposal for creating man-made lakes. If the Congo, which carries some 1,200 cubic kilometres of water per year, were dammed at Stanley Canyon (about 1 mile wide), it would impound an enormous lake (the Congo Sea). The Ubangi, a tributary of the Congo, could then flow to the north-west, joining the Chari and flowing into Lake Chad, which would grow to enormous size (over 1 million square kilometres). This large lake (the Chad Sea) would approximately equal the combined areas of the Baltic Sea, White Sea, Black Sea, and Caspian Sea. The two lakes would cover 10 per cent of the African continent. They could then be drained north across the Sahara, creating an extensive irrigated region, similar to the Nile Valley.

NAWAPA Project

The proposed North American Water and Power Alliance is a smaller scale scheme. It would bring 100 million acre-feet[2] per year of water from Alaska and Canada to be evaporated by irrigation in the western United States and Mexico. The possible climatic effects are highly speculative. For example, would the increased moisture in the air fall out again over the central United States, or would it be transported to some other region?

Prospects for Future Progress

It is convenient to think of progress toward climate control in four stages—observation, understanding, prediction, and control. We must observe *how* nature behaves before we can understand *why,* we must *understand* before we can *predict,* and we must be able to *predict* the outcome before we undertake measures for *control.*

From the foregoing examples it is evident that modern technology is already capable of influencing the global system by altering patterns of thermal forcing, but the consequences of such acts cannot be adequately predicted. The global system is a single, interacting 'heat engine' in which a substantial action anywhere may influence subsequent behaviour everywhere. At present, we do not understand the system well enough to predict this behaviour. Much progress in observation, understanding, and prediction is needed before purposeful climate modification can become feasible, but a more rapid progress can now be anticipated.

In theory, it should be possible to solve the equations which describe the behaviour of the atmosphere and the ocean, given the conditions of thermal forcing and the initial state of the system. Such a quantitative analytical approach was formulated by V. Bjerknes in 1904 and expanded by Richardson in 1922. But, since neither the means to observe the state of the system nor the necessary computational power existed, such an approach had little immediate impact. Recent technological breakthroughs are removing these barriers and we are now entering a period of rapid progress.

As recently as the Second World War, not more than about 20 per cent of the global atmosphere was observed at one time. With the advent of satellite observing systems, some quantities are now observed over the entire planet every day. This observational breakthrough makes possible the surveillance of the entire global system, and the sophistication of the observations that can be made by satellite is rapidly increasing.

Modern computer technology is rapidly overcoming the computational aspects of the problem. Mathematical simulation of the interacting ocean/atmosphere system has already been demonstrated. With computers now being developed that are 500 to 1,000 times faster than existing models, we can reasonably hope that such simulation can be performed in enough detail to reliably evaluate the consequences of specific climate modification acts. With a straightforward means of testing hypotheses, we can expect a surge of new interest in theories of climatic change.

Such simulation capability also provides a means for making long-range forecasts, such as for a season or longer, based on observed and predicted conditions of thermal forcing. This will lead to a shift of emphasis in observing the global system. A short-range forecast can be based largely on the inertial behaviour of the atmosphere, and the 'machine forecasts' of the last decade have basically been of this

2. One acre-foot is the quantity of water which covers 1 acre (0.4 hectare) to the depth of 1 foot (0.3 metre).

type. The needed input data for such a forecast is a detailed description of the initial state, especially the field of motion. Patterns of thermal forcing are too slow-acting to be important in this short-range context.

On the other hand, for a very long time period, we may expect that the mean behaviour of the system will depend primarily on thermal forcing and be relatively independent of the initial dynamical state. It follows that the growing capability for climate simulation and long-range forecasting must also place new emphasis on observing and understanding the processes by which heat is exchanged between the ocean and the atmosphere. Today, we are not yet able to observe the global system in enough detail to know whether or not we are simulating realistic patterns of thermal forcing.

The presently-foreseeable ways by which global climate may be influenced all reduce to changing, in one way or another, the pattern of thermal forcing of atmospheric circulation. Such changes occur naturally for a variety of reasons. Understanding how and why they occur is the key to explaining observed changes of climate and also a necessary step toward being able to evaluate the consequences of man-induced changes.

Climatically important variations in surface characteristics and surface heat exchanges occur naturally and to some extent may be influenced by man. In ocean areas, anomalies of surface temperature occur as a result of the wind-driven oceanic circulation. In land areas, the reflectivity and moisture capacity varies with the extent of vegetation. In ice areas, the reflectivity falls abruptly when melting begins.

Of special importance is the variable extent of ice on the sea, for the presence or absence of ice determines whether the thermal characteristics of the surface will resemble those of land or those of ocean. The climatic significance of this factor can be appreciated by noting that about 12 per cent of the world ocean is ice-covered at some time during the year, but only about 4 per cent is ice-covered during the entire year. That is to say, the thermal behaviour of some 8 per cent of the world ocean area is ocean-like for part of the year and land-like for part of the year, a variable factor of possibly great climatic influence.

Figure 2 shows available observational evidence of this relationship (Fletcher, 1969). It shows that variations in iciness of the Antarctic waters (lower curves) have a high correlation with variations in the character of northern hemisphere atmospheric circulation about five years later. We might surmise that five years is comparable to the circulation time for waters moving from the Southern Ocean to the equator, that variations in the Southern Ocean cause variations in the tropical ocean a few years later, and that these variations in ocean temperatures influence northern hemisphere circulation. However, without more complete observational data, or a realistic simulation model, such a hypothesis cannot be easily tested.

Only in 1968 are ocean temperature patterns and the extent of ice on the seas beginning to be observed on a regular basis. A suitable ocean/atmosphere simulation model will probably be available within five years.

Finally, it may be noted that an understanding of contemporary and future climatic changes can hardly be achieved without understanding the large climatic changes of the more distant past. Defining the patterns of these changes is a way of observing nature's own "climate control experiments." The collection and systematization of palaeoclimatic evidence is a task of great practical importance.

From the foregoing considerations one arrives at a conclusion of great significance,

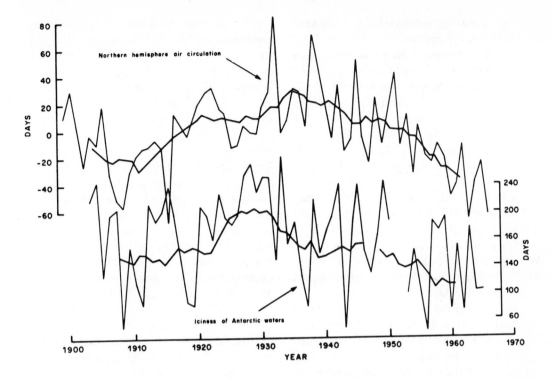

Figure 2. Comparison of iciness of Antarctic waters and atmospheric circulation. Bottom pair of curves (from Schwerdtfeger) shows the number of days per year that Scotia Bay in the Weddell Sea, Antarctica, was closed by ice (fine curve = annual figures; thick curve = figures smoothed by taking a ten-year moving average). Upper pair of curves (from Dzerdzeyevskiy) indicates the number of days per year in northern hemisphere in which there was air circulation of the zonal type (dominant air movement is east-west, as opposed to meridional circulation, which is dominantly north-south), expressed as number of days more or less than a long-term annual mean (fine curve = annual deviations; thick curve = ten-year moving average). Examination of lower and upper smoothed curves suggests that iciness of antarctic waters affects northern hemisphere air circulation patterns five years later.

namely that *we are reaching, or perhaps have already reached, a technological threshold from which progress can be proportional to the investment of effort.* This conclusion, combined with the proposition that sooner or later purposeful climate modification is inevitable, deserves the attention of scientific and government leaders who must organize the needed resources.

International Co-operation

The management of global climatic resources is a problem shared by all nations. So far, international efforts in climatic research have been directed toward *observation* and *understanding,* and co-operation has been good. It is a challenge to political and scientific leadership to preserve this spirit of co-operation as further progress is achieved toward *prediction* and *control*.

In 1961, President John F. Kennedy, in a statement to the United Nations, proposed "further co-operative efforts between all nations in weather prediction and eventually in weather control." In response, on 11 December 1961, the United Nations adopted Resolution 1721, which calls on all of its Member States to join in a co-operative world weather programme.

A first step was taken the following year, when the World Meteorological Organization (WMO) created a special working group to make a proposal in response to this resolution. In 1963, a programme known as World Weather Watch (WWW) took shape under the auspices of WMO.

The goals of the WWW are immediate: to improve the accuracy of weather predictions and extend their usefulness to many new areas.

Most of the Member States, showing awareness of the great potential gains in human well-being promised by improved weather observations and predictions, have participated according to their ability and resources, and have already become actively involved in the World Weather Watch.

On the part of the United States, a national policy was affirmed in 1968 as follows:

"Resolved by the Senate of the United States (The House of Representatives concurring).

"That it is the sense of Congress that the United States should participate in and give full support to the world weather program which included (1) a world weather watch—the development and operation of an international system for the observation of the global atmosphere and the rapid and efficient communication, processing, and analysis of worldwide weather data, and (2) the conduct of a comprehensive program of research for the development of a capability in long-range weather prediction and for the theoretical study and evaluation of inadvertent climate modification and the feasibility of intentional climate modification. . . ."[3]

The ongoing observational programmes emphasize certain typical regions, studying them in great detail and for a limited period, in order to understand the heat exchange processes taking place and their influence on the atmosphere and the ocean. This is especially important in regions which play an important role in the thermal forcing of atmospheric and oceanic circulation, and where large year-to-year variations can occur. In the equatorial heat-source regions, variations in the tropical convergence zone, where rising warm air carries moisture and heat up into the atmosphere, seem to be associated with changing global climate. In the two polar heat-sink regions, variations in extent of ice cover on the ocean also seem to be associated with changing global climate. In all cases, both the causes and the effects of these variations are obscure.

The progress achieved by co-operative international efforts will bring us closer to a realistic capability for managing global climatic resources. Let us hope that the spirit of international co-operation will continue to grow.

SELECTED REFERENCES

Bjerknes, J. 1966. A possible response of the Hadley Circulation to variations of the heat supply from the equatorial Pacific. *Tellus,* vol. XVII, p. 820-9.

Bryson, R. A. 1968. All other factors being constant. *Weatherwise,* vol. 21, no. 2.

Budyko, M. I. 1962. Certain means of climate modification. *Meteorologiia i gidrologiia,* no. 2, p. 3-8.

————. 1968. The effect of solar radiation variations on the climate of the Earth. *Proc. International Radiation Symposium, Bergen, Norway, August 1968.*

————; Drozdov, O. A.; Yudin, M. I., 1966. Influence of economic activity on climate. *Contemporary problems of climatology.* Leningrad.

3. *Congressional Record (Senate),* 1 April 1968.

Fairbridge, R. W. (ed.). 1961. Solar variations, climatic change, and related geophysical problems. *Ann. N. Y. Acad. Sci.,* vol. 95, p. 1-740.

Fletcher, J. O. (ed.). 1966. *Proceedings of the Symposium on the Arctic Heat Budget and Atmospheric Circulation.* Santa Monica, California, The RAND Corporation (RM-5233-NSF.)

————. 1969. *The interaction of variable sea ice extent with global climate.* Santa Monica, California, The RAND Corporation. (RM-5793-NSF.)

Lamb, H. H. 1966a. *The changing climate.* London, Methuen.

————. 1966b. *On climatic variations affecting the far south.* Geneva. (WMO Technical Note, no. 87.)

————. 1966c. Climate in the 1960's. *Geophys. J.,* vol. 132, part 2.

Mitchell, J. M., Jr. 1963. On the world-wide pattern of secular temperature change. *Changes of climate. Proc. Rome Symposium.* Paris, Unesco.

Möller, F. 1963. On the influence of changes in the CO_2 concentration in air on the radiation balance of the earth's surface and on the climate. *J. geophys. Res.,* vol. 68, no. 13.

Rubinstein, E. S.; Polozova, L. G. 1966. *Contemporary climatic variations.* Leningrad, Hydrometeorological Publishing House.

Rusin, N. P.; Flit, L. A. 1962. *Methods of climate control.* Moscow.

Willet, H. C. 1961. Solar climatic relationships, *Ann. N. Y. Acad. Sci.,* vol. 95, art. 1, p. 89-106.

————. 1965. Solar climatic relationships in the light of standardized climatic data. *J. Atmos. Sci.,* vol. 22, p. 120-36.

Yudin, M. I. 1966. The possibilities for influencing large-scale atmospheric processes. *Contemporary problems in climatology.* Leningrad.

Attitudes: The Intangible Environment

Self-awareness involves conception of a relationship between oneself and one's surroundings. Once this is done, the conceptualized relationship becomes, itself, an environmental factor, shaping actions upon, and responses to, the perceived world. Some one and a half million years ago our ape-like ancestors were moving across the vague boundary of being which defines humanness, and one may suppose that self-awareness, as a uniquely human attribute, also came slowly into existence early in that dim past. With the increasing power of technology, and the increasing human population, the intangible world of human thought has become a major force in the biosphere.

Primitive men seem to have identified closely with the natural ecosystems in which they lived. For them, spirits to which respect was due existed in plant and animal, in stone and stream. Hunting was a matter of cooperation between the hunter and the hunted, and a scarcity of game might be the consequence of an offense to the spirit of an individual deer, or mammoth, or mouse. Magic rituals, and effigies, were devised to show respect and invoke cooperation. Gary Snyder, in a little book called "Earth House Hold," has put it this way:

"To hunt means to use your body and senses to the fullest; to strain your consciousness to feel what the deer are thinking today, this moment; to sit still and let your self go into the birds and wind while waiting by a game trail. Hunting magic is intended to bring the game to you—the creature who has heard your song, witnessed your sincerity, and out of compassion comes within your range. Hunting magic is not only aimed at bringing beasts to their death, but to assist in their birth—to promote their fertility. Thus the great Iberian cave paintings are not of hunting alone—but of animals mating and giving birth."

A sense of identity with nature, and of continuity in the cycles of stars and seasons, of death and birth, dominated our ancestors' developing concepts of themselves in relation to their world for most of one and a half million years.

Four or five thousand years are but a fleeting moment in the scale of time with which we have been concerned, but in that moment immediately behind us the sense of identity which the ancient animism provided was swept away, and

233

another, less gentle, mysticism took command. That new mysticism established a duality of man and nature. Nature was a creation for man's benefit, a dominion, to exploit and destroy as need or fancy called. For a while this may have been good, and for a while it mattered little, if at all, that this duality was grossly in error. Unfortunately it has outlived its time, persisting in language, in social institutions, in economics, and even in the minds of many scientists long after the young sciences of evolution and ecology clearly indicated its folly. The Copernican demonstration that the earth is not the center of the universe, and the Darwinian demonstration that man shared a common ancestry with other forms of life, have failed to shake the conviction that man is set apart from, is superior and central to, and is justified in exploiting, the living and non-living world around him. Now, suddenly, the anti-evolutionary, anti-ecological, concept of man as separate-from-ecosystem does matter, and it matters desperately. The myth of duality persisting in the majority of our institutions, in the minds of educated and influential men, and even in our language, subverts our thinking and prejudices our future. It lies behind our operant theories of economics and politics with their irrational commitment to unending growth. Wrong though it may be, it dominates the decision-making processes of our society.

Today, a new and ecologically realistic alternative is needed to supplant the man/nature mythology. Only through the concept of man as part of, rather than apart from, the planetary ecosystem can we avert total disaster. We have all too little time. Yet we must start with the recognition of ourselves as part of nature, and then go on to discover the size and character of the ecological niche in which modern, technological man can find both a free and happy existence for himself, and a viable biosphere.

In those selections which follow, the myth of duality will be more apparent than the ecological reality. I have let it be so because I think it is so in our society. In the context of Lynn White's analysis of the roots of our crisis it should be easy enough to recognize the sources of Buckminster Fuller's faith that "we are functioning in a region of the universe where everything has been made tolerable for our specific technological model of regenerative life," of Edward Teller's definition of progress as the use of nuclear explosions to dig canals, of the AEC's view that adequate ventilation of the Colorado uranium mines is "uneconomic," or Colin Clarke's preoccupation with population as a source of "power." The intangible environment of ideas presented in this chapter may seem a rogues' gallery, but we must see things as they are if we are to change them, if we are to make the right decisions, or even if we are to ask the right questions.

The Historical Roots of Our Ecologic Crisis

Lynn White, Jr.

A conversation with Aldous Huxley not infrequently put one at the receiving end of an unforgettable monologue. About a year before his lamented death he was discoursing on a favorite topic: Man's unnatural treatment of nature and its sad results. To illustrate his point he told how, during the previous summer, he had returned to a little valley in England where he had spent many happy months as a child. Once it had been composed of delightful grassy glades; now it was becoming overgrown with unsightly brush because the rabbits that formerly kept such growth under control had largely succumbed to a disease, myxomatosis, that was deliberately introduced by the local farmers to reduce the rabbits' destruction of crops. Being something of a Philistine, I could be silent no longer, even in the interests of great rhetoric. I interrupted to point out that the rabbit itself had been brought as a domestic animal to England in 1176, presumably to improve the protein diet of the peasantry.

All forms of life modify their contexts. The most spectacular and benign instance is doubtless the coral polyp. By serving its own ends, it has created a vast undersea world favorable to thousands of other kinds of animals and plants. Ever since man became a numerous species he has affected his environment notably. The hypothesis that his fire-drive method of hunting created the world's great grasslands and helped to exterminate the monster mammals of the Pleistocene from much of the globe is plausible, if not proved. For 6 millennia at least, the banks of the lower Nile have been a human artifact rather than the swampy African jungle which nature, apart from man,

would have made it. The Aswan Dam, flooding 5000 square miles, is only the latest stage in a long process. In many regions terracing or irrigation, overgrazing, the cutting of forests by Romans to build ships to fight Carthaginians or by Crusaders to solve the logistics problems of their expeditions, have profoundly changed some ecologies. Observation that the French landscape falls into two basic types, the open fields of the north and the *bocage* of the south and west, inspired Marc Block to undertake his classic study of medieval agricultural methods. Quite unintentionally, changes in human ways often affect nonhuman nature. It has been noted, for example, that the advent of the automobile eliminated huge flocks of sparrows that once fed on the horse manure littering every street.

The history of ecologic change is still so rudimentary that we know little about what really happened, or what the results were. The extinction of the European aurochs as late as 1627 would seem to have been a simple case of overenthusiastic hunting. On more intricate matters it often is impossible to find solid information. For a thousand years or more the Frisians and Hollanders have been pushing back the North Sea, and the process is culminating in our own time in the reclamation of the Zuider Zee. What, if any, species of animals, birds, fish, shore life, or plants have died out in the process? In their epic combat with Neptune have the Netherlanders overlooked ecological values in such a way that the quality of human life in the Netherlands has suffered? I cannot discover that the questions have ever been asked, much less answered.

People, then, have often been a dynamic

element in their own environment, but in the present state of historical scholarship we usually do not know exactly when, where, or with what effects man-induced changes came. As we enter the last third of the 20th century, however, concern for the problem of ecologic backlash is mounting feverishly. Natural science, conceived as the effort to understand the nature of things, had flourished in several eras and among several peoples. Similarly there had been an age-old accumulation of technological skills, sometimes growing rapidly, sometimes slowly. But it was not until about four generations ago that Western Europe and North America arranged a marriage between science and technology, a union of the theoretical and the empirical approaches to our natural environment. The emergence in widespread practice of the Baconion creed that scientific knowledge means technological power over nature can scarcely be dated before about 1850, save in the chemical industries, where it is anticipated in the 18th century. Its acceptance as a normal pattern of action may mark the greatest event in human history since the invention of agriculture, and perhaps in nonhuman terrestrial history as well.

Almost at once the new situation forced the crystallization of the novel concept of ecology; indeed, the word *ecology* first appeared in the English language in 1873. Today, less than a century later, the impact of our race upon the environment has so increased in force that it has changed in essence. When the first cannons were fired, in the early 14th century, they affected ecology by sending workers scrambling to the forests and mountains for more potash, sulfur, iron ore, and charcoal, with some resulting erosion and deforestation. Hydrogen bombs are of a different order: a war fought with them might alter the genetics of all life on this planet. By 1285 London had a smog problem arising from the burning of soft coal, but our present combustion of fossil fuels threatens to change the chemistry of the globe's atmosphere as a whole, with consequences which we are only beginning to guess. With the population explosion, the carcinoma of planless urbanism, the now geological deposits of sewage and garbage, surely no creature other than man has ever managed to foul its nest in such short order.

There are many calls to action, but specific proposals, however worthy as individual items, seem too partial, palliative, negative: ban the bomb, tear down the billboards, give the Hindus contraceptives and tell them to eat their sacred cows. The simplest solution to any suspect change is, of course, to stop it, or, better yet, to revert to a romanticized past: make those ugly gasoline stations look like Anne Hathaway's cottage or (in the Far West) like ghost-town saloons. The "wilderness area" mentality invariably advocates deep-freezing an ecology, whether San Gimignano or the High Sierra, as it was before the first Kleenex was dropped. But neither atavism nor prettification will cope with the ecologic crisis of our time.

What shall we do? No one yet knows. Unless we think about fundamentals, our specific measures may produce new backlashes more serious than those they are designed to remedy.

As a beginning we should try to clarify our thinking by looking, in some historical depth, at the presuppositions that underlie modern technology and science. Science was traditionally aristocratic, speculative, intellectual in intent; technology was lower-case, empirical, action-oriented. The quite sudden fusion of these two, towards the middle of the 19th century, is surely related to the slightly prior and contemporary democratic revolutions which, by reducing social barriers, tended to assert a

functional unity of brain and hand. Our ecologic crisis is the product of an emerging, entirely novel, democratic culture. The issue is
whether a democratized world can survive its
own implications. Presumably we cannot unless we rethink our axioms.

The Western Traditions of
Technology and Science

One thing is so certain that it seems stupid
to verbalize it: both modern technology and
modern science are distinctively *Occidental.*
Our technology has absorbed elements from
all over the world, notably from China; yet
everywhere today, whether in Japan or in Ni-
geria, successful technology is Western. Our
science is the heir to all the sciences of the past,
especially perhaps to the work of the great
Islamic scientists of the Middle Ages, who so
often outdid the ancient Greeks in skill and
perspicacity: al-Rāzi in medicine, for example;
or ibn-al-Haytham in optics; or Omar
Khayyám in mathematics. Indeed, not a few
words of such geniuses seem to have vanished
in the original Arabic and to survive only in
medieval Latin translations that helped to lay
the foundations for later Western developments. Today, around the globe, all significant
science is Western in style and method, whatever the pigmentation or language of the scientists.

A second pair of facts is less well recognized
because they result from quite recent historical
scholarship. The leadership of the West, both
in technology and in science, is far older than
the so-called Scientific Revolution of the 17th
century or the so-called Industrial Revolution
of the 18th century. These terms are in fact
outmoded and obscure the true nature of what
they try to describe—significant stages in two
long and separate developments. By A.D.
1000 at the latest—and perhaps, feebly, as

much as 200 years earlier—the West began to
apply water power to industrial processes
other than milling grain. This was followed in
the late 12th century by the harnessing of wind
power. From simple beginnings, but with remarkable consistency of style, the West rapidly expanded its skills in the development of
power machinery, labor-saving devices, and
automation. Those who doubt should contemplate that most monumental achievement in
the history of automation: the weight-driven
mechanical clock, which appeared in two
forms in the early 14th century. Not in crafts-
manship but in basic technological capacity,
the Latin West of the later Middle Ages far
outstripped its elaborate, sophisticated, and
esthetically magnificent sister cultures, Byzantium and Islam. In 1444 a great Greek ec-
clesiastic, Bessarion, who had gone to Italy,
wrote a letter to a prince in Greece. He is
amazed by the superiority of Western ships,
arms, textiles, glass. But above all he is aston-
ished by the spectacle of waterwheels sawing
timbers and pumping the bellows of blast fur-
naces. Clearly, he had seen nothing of the sort
in the Near East.

By the end of the 15th century the technological superiority of Europe was such that its
small, mutually hostile nations could spill out
over all the rest of the world, conquering, looting, and colonizing. The symbol of this technological superiority is the fact that Portugal,
one of the weakest states of the Occident, was
able to become, and to remain for a century,
mistress of the East Indies. And we must
remember that the technology of Vasco da
Gama and Albuquerque was built by pure em-
piricism, drawing remarkably little support or
inspiration from science.

In the present-day vernacular understanding, modern science is supposed to have begun
in 1543, when both Copernicus and Vesalius
published their great works. It is no deroga-

tion of their accomplishments, however, to point out that such structures as the *Fabrica* and the *De revolutionibus* do not appear overnight. The distinctive Western tradition of science, in fact, began in the late 11th century with a massive movement of translation of Arabic and Greek scientific works into Latin. A few notable books—Theophrastus, for example—escaped the West's avid new appetite for science, but within less than 200 years effectively the entire corpus of Greek and Muslim science was available in Latin, and was being eagerly read and criticized in the new European universities. Out of criticism arose new observation, speculation, and increasing distrust of ancient authorities. By the late 13th century Europe had seized global scientific leadership from the faltering hands of Islam. It would be as absurd to deny the profound originality of Newton, Galileo, or Copernicus as to deny that of the 14th century scholastic scientists like Buridan or Oresme on whose work they built. Before the 11th century, science scarcely existed in the Latin West, even in Roman times. From the 11th century onward, the scientific sector of Occidental culture has increased in a steady crescendo.

Since both our technological and our scientific movements got their start, acquired their character, and achieved world dominance in the Middle Ages, it would seem that we cannot understand their nature or their present impact upon ecology without examining fundamental medieval assumptions and developments.

Medieval View of Man and Nature

Until recently, agriculture has been the chief occupation even in "advanced" societies; hence, any change in methods of tillage has much importance. Early plows, drawn by two oxen, did not normally turn the sod but merely scratched it. Thus, cross-plowing was needed and fields tended to be squarish. In the fairly light soils and semiarid climates of the Near East and Mediterranean, this worked well. But such a plow was inappropriate to the wet climate and often sticky soils of northern Europe. By the latter part of the 7th century after Christ, however, following obscure beginnings, certain northern peasants were using an entirely new kind of plow, equipped with a vertical knife to cut the line of the furrow, a horizontal share to slice under the sod, and a moldboard to turn it over. The friction of this plow with the soil was so great that it normally required not two but eight oxen. It attacked the land with such violence that cross-plowing was not needed, and fields tended to be shaped in long strips.

In the days of the scratch-plow, fields were distributed generally in units capable of supporting a single family. Subsistence farming was the presupposition. But no peasant owned eight oxen: to use the new and more efficient plow, peasants pooled their oxen to form large plow-teams, originally receiving (it would appear) plowed strips in proportion to their contribution. Thus, distribution of land was based no longer on the needs of a family but, rather, on the capacity of a power machine to till the earth. Man's relation to the soil was profoundly changed. Formerly man had been a part of nature; now he was the exploiter of nature. Nowhere else in the world did farmers develop any analogous agricultural implement. Is it coincidence that modern technology, with its ruthlessness toward nature, has so largely been produced by descendants of these peasants of northern Europe?

This same exploitive attitude appears slightly before A.D. 830 in Western illustrated calendars. In older calendars the months were shown as passive personifications. The new

Frankish calendars, which set the style for the Middle Ages, are very different: they show men coercing the world around them—plowing, harvesting, chopping trees, butchering pigs. Man and nature are two things, and man is master.

These novelties seem to be in harmony with larger intellectual patterns. What people do about their ecology depends on what they think about themselves in relation to things around them. Human ecology is deeply conditioned by beliefs about our nature and destiny —that is, by religion. To Western eyes this is very evident in, say, India or Ceylon. It is equally true of ourselves and of our medieval ancestors.

The victory of Christianity over paganism was the greatest psychic revolution in the history of our culture. It has become fashionable today to say that, for better or worse, we live in "the post-Christian age." Certainly the forms of our thinking and language have largely ceased to be Christian, but to my eye the substance often remains amazingly akin to that of the past. Our daily habits of action, for example, are dominated by an implicit faith in perpetual progress which was unknown either to Greco-Roman antiquity or to the Orient. It is rooted in, and is indefensible apart from, Judeo-Christian teleology. The fact that Communists share it merely helps to show what can be demonstrated on many other grounds: that Marxism, like Islam, is a Judeo-Christian heresy. We continue today to live, as we have lived for about 1700 years, very largely in a context of Christian axioms.

What did Christianity tell people about their relations with the environment?

While many of the world's mythologies provide stories of creation, Greco-Roman mythology was singularly incoherent in this respect. Like Aristotle, the intellectuals of the ancient West denied that the visible world had had a beginning. Indeed, the idea of a beginning was impossible in the framework of their cyclical notion of time. In sharp contrast, Christianity inherited from Judaism not only a concept of time as nonrepetitive and linear but also a striking story of creation. By gradual stages a loving and all-powerful God had created light and darkness, the heavenly bodies, the earth and all its plants, animals, birds, and fishes. Finally, God had created Adam and, as an afterthought, Eve to keep man from being lonely. Man named all the animals, thus establishing his dominance over them. God planned all of this explicitly for man's benefit and rule: no item in the physical creation had any purpose save to serve man's purposes. And, although man's body is made of clay, he is not simply part of nature: he is made in God's image.

Especially in its Western form, Christianity is the most anthropocentric religion the world has seen. As early as the 2nd century both Tertullian and Saint Irenaeus of Lyons were insisting that when God shaped Adam he was foreshadowing the image of the incarnate Christ, the Second Adam. Man shares, in great measure, God's transcendence of nature. Christianity, in absolute contrast to ancient paganism and Asia's religions (except, perhaps, Zoroastrianism), not only established a dualism of man and nature but also insisted that it is God's will that man exploit nature for his proper ends.

At the level of the common people this worked out in an interesting way. In Antiquity every tree, every spring, every stream, every hill had its own *genius loci,* its guardian spirit. These spirits were accessible to men; centaurs, fauns, and mermaids show their ambivalence. Before one cut a tree, mined a mountain, or dammed a brook, it was important to placate the spirit in charge of that particular situation, and to keep it placated. By destroying pagan animism, Christianity made it possible to exploit nature in a mood of indifference to the feelings of natural objects.

It is often said that for animism the Church substituted the cult of saints. True; but the cult of saints is functionally quite different from animism. The saint is not *in* natural objects; he may have special shrines, but his citizenship is in heaven. Moreover, a saint is entirely a man; he can be approached in human terms. In addition to saints, Christianity of course also had angels and demons inherited from Judaism and perhaps, at one remove, from Zoroastrianism. But these were all as mobile as the saints themselves. The spirits *in* natural objects, which formerly had protected nature from man, evaporated. Man's effective monopoly on spirit in this world was confirmed, and the old inhibitions to the exploitation of nature crumbled.

When one speaks in such sweeping terms, a note of caution is in order. Christianity is a complex faith, and its consequences differ in differing contexts. What I have said may well apply to the medieval West, where in fact technology made spectacular advances. But Greek East, a highly civilized realm of equal Christian devotion, seems to have produced no marked technological innovation after the late 7th century, when Greek fire was invented. The key to the contrast may perhaps be found in a difference in the tonality of piety and thought which students of comparative theology find between the Greek and the Latin Churches. The Greeks believed that sin was intellectual blindness, and that salvation was found in illumination, orthodoxy—that is clear thinking. The Latins, on the other hand, felt that sin was moral evil, and that salvation was to be found in right conduct. Eastern theology has been intellectualist. Western theology has been voluntarist. The Greek saint contemplates; the Western saint acts. The im-

plications of Christianity for the conquest of nature would emerge more easily in the Western atmosphere.

The Christian dogma of creation, which is found in the first clause of all the Creeds, has another meaning for our comprehension of today's ecological crisis. By revelation, God had given man the Bible, the Book of Scripture. But since God had made nature, nature also must reveal the divine mentality. The religious study of nature for the better understanding of God was known as natural theology. In the early Church, and always in the Greek East, nature was conceived primarily as a symbolic system through which God speaks to men: the ant is a sermon to sluggards; rising flames are the symbol of the soul's aspiration. This view of nature was essentially artistic rather than scientific. While Byzantium preserved and copied great numbers of ancient Greek scientific texts, science as we conceive it could scarcely flourish in such an ambience.

However, in the Latin West by the early 13th century natural theology was following a very different bent. It was ceasing to be the decoding of the physical symbols of God's communication with man and was becoming the effort to understand God's mind by discovering how his creation operates. The rainbow was no longer simply a symbol of hope first sent to Noah after the Deluge: Robert Grosseteste, Friar Roger Bacon, and Theodoric of Freiberg produced startlingly sophisticated work on the optics of the rainbow, but they did it as a venture in religious understanding. From the 13th century onward, up to and including Leibnitz and Newton, every major scientist, in effect, explained his motivations in religious terms. Indeed, if Galileo had not been so expert an amateur theologian he would have got into far less trouble: the professionals resented his intrusion. And Newton seems to have regarded himself more as a theologian than as a scientist. It was not until the late 18th century that the hypothesis of God became unnecessary to many scientists.

It is often hard for the historian to judge, when men explain why they are doing what they want to do, whether they are offering real reasons or merely culturally acceptable reasons. The consistency with which scientists during the long formative centuries of Western science said that the task and the reward of the scientist was "to think God's thoughts after him" leads one to believe that this was their real motivation. If so, then modern Western science was cast in a matrix of Christian theology. The dynamism of religious devotion, shaped by the Judeo-Christian dogma of creation, gave it impetus.

An Alternative Christian View

We would seem to be headed toward conclusions unpalatable to many Christians. Since both *science* and *technology* are blessed words in our contemporary vocabulary, some may be happy at the notions, first, that, viewed historically, modern science is an extrapolation of natural theology and, second, that modern technology is at least partly to be explained as an Occidental, voluntarist realization of the Christian dogma of man's transcendence of, and rightful mastery over, nature. But, as we now recognize, somewhat over a century ago science and technology—hitherto quite separate activities—joined to give mankind powers which, to judge by many of the ecologic effects, are out of control. If so, Christianity bears a huge burden of guilt.

I personally doubt that disastrous ecologic backlash can be avoided simply by applying to our problems more science and more technology. Our science and technology have grown out of Christian attitudes toward man's relation to nature which are almost universally held not only by Christians and neo-Christians but also by those who fondly regard themselves as post-Christians. Despite Copernicus, all the cosmos rotates around our little globe. Despite Darwin, we are *not,* in our hearts, part of the natural process. We are superior to nature, contemptuous of it, willing to use it for our slightest whim. The newly elected governor of California, like myself a churchman but less troubled than I, spoke for the Christian tradition when he said (as is alleged), "when you've seen one redwood tree, you've seen them all." To a Christian a tree can be no more than a physical fact. The whole concept of the sacred grove is alien to Christianity and to the ethos of the West. For nearly 2 millennia Christian missionaries have been chopping down sacred groves, which are idolatrous because they assume spirit in nature.

What we do about ecology depends on our ideas of the man-nature relationship. More science and more technology are not going to get us out of the present ecologic crisis until we find a new religion, or rethink our old one. The beatniks, who are the basic revolutionaries of our time, show a sound instinct in their affinity for Zen Buddhism, which conceives of the man-nature relationship as very nearly the mirror image of the Christian view. Zen, however, is a deeply conditioned by Asian history as Christianity is by the experience of the West, and I am dubious of its viability among us.

Possibly we should ponder the greatest radical in Christian history since Christ: Saint Francis of Assisi. The prime miracle of Saint Francis is the fact that he did not end at the stake, as many of his left-wing followers did. He was so clearly heretical that a General of the Franciscan Order, Saint Bonaventura, a great and perceptive Christian, tried to suppress the early accounts of Franciscanism. The key to an understanding of Francis is his belief in the virtue of humility—not merely for the individual but for man as a species. Francis tried to depose man from his monarchy over creation and set up a democracy of all God's creatures. With him the ant is no longer simply a homily for the lazy, flames a sign of the thrust of the soul toward union with God; now they are Brother Ant and Sister Fire, praising the Creator in their own ways as Brother Man does in his.

Later commentators have said that Francis preached to the birds as a rebuke to men who would not listen. The records do not read so: he urged the little birds to praise God, and in spiritual ecstasy they flapped their wings and chirped rejoicing. Legends of saints, especially the Irish saints, had long told of their dealings with animals but always, I believe, to show their human dominance over creatures. With Francis it is different. The land around Gubbio in the Apennines was being ravaged by a fierce wolf. Saint Francis, says the legend, talked to the wolf and persuaded him of the error of his ways. The wolf repented, died in the odor of sanctity, and was buried in consecrated ground.

What Sir Steven Ruciman calls "the Franciscan doctrine of the animal soul" was quickly stamped out. Quite possibly it was in part inspired, consciously or unconsciously, by the belief in reincarnation held by the Cathar heretics who at that time teemed in Italy and southern France, and who presumably had got it originally from India. It is significant that at just the same moment, about 1200, traces of metempsychosis are found also in western Judaism, in the Provencal *Cabbala*. But Francis held neither to transmigration of souls nor to pantheism. His view of nature and of man rested on a unique sort of pan-psychism of all things animate and inanimate, designed for the glorification of their transcendent Creator, who, in the ultimate gesture of cosmic humility, assumed flesh, lay helpless in a manger, and hung dying on a scaffold.

I am not suggesting that many contemporary Americans who are concerned about our ecologic crisis will be either able or willing to counsel with wolves or exhort birds. However, the present increasing disruption of the global environment is the product of a dynamic technology and science which were originating in the Western medieval world against which Saint Francis was rebelling in so original a way. Their growth cannot be understood historically apart from distinctive attitudes toward nature which are deeply grounded in Christian dogma. The fact that most people do not think of these attitudes as Christian is irrelevant. No new set of basic values has been accepted in our society to displace those of Christianity. Hence we shall continue to have a worsening ecologic crisis until we reject the Christian axiom that nature has no reason for existence save to serve man.

The greatest spiritual revolutionary in Western history, Saint Francis, proposed what he thought was an alternative Christian view of nature and man's relation to it: he tried to substitute the idea of the equality of all creatures, including man, for the idea of man's limitless rule of creation. He failed. Both our present science and our present technology are so tinctured with orthodox Christian arrogance toward nature that no solution for our ecologic crisis can be expected from them alone. Since the roots of our trouble are so largely religious, the remedy must also be essentially religious, whether we call it that or not. We must rethink and refeel our nature and destiny. The profoundly religious, but heretical, sense of the primitive Franciscans for the spiritual autonomy of all parts of nature may point a direction. I propose Francis as a patron saint for ecologists.

The Economics of Pollution
Edwin L. Dale Jr.

Now that environment has become a national concern, it might be well to clean up some of the economic rubbish associated with the subject. There are, alas, a few "iron laws" that cannot be escaped in the effort to reduce the pollution of our air and water, in disposing of solid waste and the like. The laws do not necessarily prevent a clean environment, but there is no hope of obtaining one unless they are understood.

We have all become vaguely aware that there will be a cost—perhaps higher monthly electric bills, perhaps higher taxes, perhaps a few cents or a few dollars more on anything made from steel—if there is a successful and massive effort to have a better environment. But that is only a beginning. There are other problems.

This article will describe the three iron laws that matter. There is no point in hiding the fact that all three are very depressing. The only purpose in adding more depressing information to a world already surfeited with it is a small one: to avoid useless effort based on false premises. A classic example has already arisen in wistful Congressional inquiries into whether we might think of a future with somewhat less electric power, or at least less growth in electric power.

In shorthand, the three laws are:
(1) The law of economic growth.
(2) The law of compound interest.
(3) The law of the mix between public and private spending.

The Law of Economic Growth

Whether we like it or not, and assuming no unusual increase in mass murders or epidemics, the American labor force for the next 20 years is already born and intends to work. It is hard for any of us—myself included—to imagine a deliberate policy to keep a large portion of it unemployed. But that simple fact has enormous consequences.

For more than a century, the average output of each worker for each hour worked has risen between 2 and 3 per cent a year, thanks mainly to new machines, but also to better managerial methods and a more skilled labor force. This increase in what is called *productivity* is by far the most important cause of our gradually rising standard of living—which, pollution aside, nearly all of us have wanted. In simplest terms, each worker can be paid more because he produces more and he consumes more because he earns more. Inflation only increases the numbers and does not change the facts. Machines increase the productivity of an auto worker more than a barber, but both rightly share, through the general rise in real income, in the expansion of productivity in the economy as a whole.

It is difficult to conceive of our society or any other wanting to halt the rise in productivity, or efficiency, which has made real incomes higher for all of us. But even if "we" wanted to, in our kind of society and economy "we" couldn't. The profit motive will almost always propel individual, daily decisions in the direction of higher productivity. A business will always buy a new machine if it will cut costs and increase efficiency—and thank goodness! That is what has made our standard of living—and we do enjoy it—rise.

It is not a matter of enjoying it, however. By any fair test, we are not really affluent; half of our households earn less than $8,500 a year. Apart from redistributing income, which has very real limits, the only way the society can continue to improve the well-being of those who are not affluent—really the majority—is through a continued increase in productivity. Anyone who wants us to go back to the ax, the wooden plow, the horse carriage and the water wheel, is not only living a wholly impossible dream, he is asking for a return to a society in which nearly everybody was poor. We are not talking here about philosophical ideas of happiness, but of what people have proved they want in the way of material things. This society is not about to give up productivity growth. But every increase in productivity adds to output. Now consider the next step:

We can count on the output of the average worker to continue to rise in the years ahead, as it has in the past. Nearly all current forecasts put this rise in productivity much closer to 3 per cent than to 2, and 3 per cent has been about our average in the years since World War II. So without any change in the labor force at all, our national output will go on rising by some 3 per cent a year.

What does output mean?

It means electric power produced—and smoke produced.

It means cans and bottles produced.

It means steel produced—and, unless something is done about it, water and air polluted.

It means paper produced—with the same result as for steel.

And so on and on.

But that is not the end, for there will not be a static labor force. As noted, the force for the next 20 years is already born and it is going to grow year by year (with a caveat, to be described below).

Obviously, we want to offer these people employment opportunity. So, in addition to a 3 per cent productivity growth, there will be an added growth of at least 1 per cent a year in the number of workers. The result is that we are almost "condemned" to a rise in our total output of 4 per cent a year. The only escape, it seems, would be a national decision either to have high unemployment or to try to be less efficient. Both are absurd on their face.

The law of economic growth says, then, that we already know that the national output in 1980 will be, and almost must be, some 50 per cent higher than it is now. President Nixon has said so publicly, and he is right. That is the result of an annual rate of real growth of about 4 per cent, compounded. It is terrifying. If an economy of $900-billion in 1969 produces the pollution and clutter we are all familiar with, what will an economy half again as large produce?

Is there no escape from this law? The answer, essentially, is no. But there is one possible way to mitigate the awesome results. We might reduce the labor input (but, we hope, not the productivity input), without creating mass unemployment.

Each working person has a workday, workweek, workyear and worklife. Any one of them could be reduced by law or otherwise. We could reduce the legal workweek from the present 40 hours. We could add more holidays or lengthen vacations to reduce the workyear. We are already shortening the worklife, without planning it that way: increased participation in higher education has meant later entry into the labor force for many, and retirement plans, including Social Security, have brought about earlier retirement than in the past for others.

If, by chance or by law, the annual manhours of employment are reduced in the years ahead, our output will grow a little less rapidly. This is the only way to cut our economic growth, short of deliberate unemployment or deliberate inefficiency.

There is a cost. It is most easily seen in a union-bargained settlement providing for longer vacations without any cut in annual wages, or a legal reduction in the workweek from 40 to 35 hours, with compulsory overtime payments after that. In each case, more workers must be hired to produce the same output, and if the employer—because of market demand—goes on producing at the same level, wage costs for each unit of output are higher. This is widely recognized. Maybe we would be willing to pay them.

But we cannot guarantee less output. Only if employers produce less—because of the extra cost—would that happen. And in that larger sense, the cost of a reduction of our annual labor input is simply less production per capita because the labor force is idle more of the time.

But less production was the objective of the exercise—the anti-pollution exercise. If we start with the proposition that the growth of production is the underlying cause of pollution, which has merit as a starting point, the only way we can get less growth in production, if we want it, is to have more of our labor force idle more of the time. In that case, we will have more leisure without mass unemployment, as we usually think of the term. Our national output, and our standard of living, will rise less rapidly.

That last idea we may learn to take, if we can cope with the leisure. But under any foreseeable circumstances, our output will still go on rising. With the most optimistic assumptions about a gradual reduction of the workday, workweek, workyear and worklife, we shall undoubtedly have a much higher output in 1980 than we have in 1970. To a man concerned about the environment, it might seem a blessing if our economic growth in the next 10 years could be 2 per cent a year instead of 4 per cent; he cannot hope for zero growth.

The law of economic growth, then, tells us a simple truth: "we" cannot choose to reduce production simply because we have found it to be the cause of a fouled environment. And if we want to reduce the rate of growth of production, the place to look is in our man-hours of work.

The Law of Compound Interest

It is a fair question to ask: Why weren't we bothered about pollution 12 or 15 years ago? In October, 1957, to pick a date, the Soviet Union sent the first earth satellite into orbit. The American economy had just begun a recession that was to send unemployment to 7 per cent of the labor force. The late George Magoffin Humphrey, who had just resigned as Secretary of the Treasury, was warning of what he saw as vast Government spending, at that time $77-billion, and saying it would bring "a depression that would curl your hair." There were plenty of things to think about.

But nobody was worried about pollution. Conservation groups were properly bothered about parts of the wilderness (the Hell's Canyon Dam in Idaho, for example), but that was an entirely different thing. That was an issue of esthetics, not health. Nobody seemed to mention air pollution or waste that might overwhelm the space in which to put it. In a peculiarly sad irony, the late Adlai E. Stevenson had fought and lost an election against Dwight D. Eisenhower in 1956 partially on a "pollution" issue—radiation in the atmosphere from the explosion of atomic weapons.

The question, to repeat: Why didn't we worry about pollution then? The answer is that, relatively speaking, there *was* no pollution. Yes, there were electric power plants then, too. Yes, there were paper mills polluting streams. Yes, there were tin cans and paper and bottles. Some snowflakes, though we didn't know it, were already a bit black, and Pittsburgh got national attention because it tried to do some cleaning up.

But here we come to the law of compound interest. In 1957—*only 13 years ago*—our gross national product was $453-billion. In 1969, in constant dollars, it was $728-billion. That is an increase of nearly $300-billion in tin cans, electric power, automobiles, paper, chemicals and all the rest. It is an increase of 60 per cent.

So what? That was not the result of an unnaturally rapid growth rate, though a bit more rapid than in some periods of our past. The *so what* is this: in the preceding 13 years the growth had been *only $100-billion*. We were the same nation, with the same energy, in those preceding 13 years. We invested and we had a rise both in productivity and in our labor force. But in the first 13 years of this example our output rose $100-billion, and in the second 13 it rose $300-billion.

In the next 13 it will rise more than $500-billion.

That is the law of compound interest. These are not numbers; they are tin cans and smoke and auto exhaust. There is no visible escape from it. Applying the same percentage growth to a larger base every year, we have reached the point where our growth in one year is half the total output of Canada, fully adjusting for inflation. Another dizzying and rather horrifying way of putting it is that the real output of goods and services in the United States has grown as much since 1950 as it grew in the entire period from the landing of the Pilgrims in 1620 up to 1950.

Most investors know the law of compound interest. There is a magic rule, for example, known as the Rule of 72. It says, with mathematical certainty, that money invested at a 7.2 per cent rate of interest, compounded each year, doubles in 10 years. Our G.N.P., happily, does not compound at 7.2 per cent. But it compounds at between 4 and 5 per cent, and it has been compounding. The result is that the

same, routine, full-employment, desirable, nationally wanted, almost unavoidable percentage increase in our national output in 1970 means precisely twice as many extra tin cans, twice as much additional electric power, and so on, as the same rate of growth in 1950. And that is only 20 years ago! We are not doing anything different, or anything awful. We are the same people. Granting approximately the same amount of human carelessness and selfishness, we are the victims solely of the law of compound interest.

The Law of the Mix Between Public and Private Spending

Robert S. McNamara, the eternally energetic and constructive former Secretary of Defense and now president of the World Bank, gave a speech in February about the plight of the poor countries. In the speech he understandably criticized the United States for reducing its foreign aid effort. But in supporting his point he adopted, almost inadvertently, a piece of partly fallacious conventional wisdom:

"Which is ultimately more in the nation's interest: to funnel national resources into an endlessly spiraling consumer economy—in effect, a pursuit of consumer gadgetry with all its senseless by-products of waste and pollution—or to dedicate a more reasonable share of those same resources to improving the fundamental quality of life both at home and abroad?"

Fair enough. It means tax increases, of course, though Mr. McNamara did not say so. This is what the "mix" between public and private spending is all about. But for our purposes the point is different. Let us look more closely at the phrase: ". . . a pursuit of consumer gadgetry with all its senseless by-products of waste and pollution . . ."

As it stands, it is true. Private consumption does create side effects like waste and pollution. But now, assume a Brave New World in which we are all happy to pay higher taxes and reduce our private consumption so that the Government may have more money with which to solve our problems—ranging from poor education to poverty, from crime to inadequate health services. We shall not examine here the issue of whether more Government money solves problems. It is obviously more effective in some areas than in others. But anyway, in our assumption, we are all willing to give the Government more money to solve problems, including pollution.

Now let us see what happens.

*The Government spends the money to reduce pollution. Sewage plants are built. They need steel. They need electric power. They need paperwork. They need workers. The workers get paid, and they consume.

*The Government spends the money on education. New schools are built, which need steel, lumber and electric power. Teachers are hired. They get paid, and they consume. They throw away tin cans.

*The Government spends the money on a better welfare system that treats all poor people alike, whether they work or not. Incomes among the poor rise by some amount between $4-billion and $20-billion, and these people consume. Electric power production rises and appliance and steel production rises and so on and on.

The point is obvious by now. A shifting in our national income or production between "public goods" and "private goods" hardly changes the environment problem at all because it does not reduce total spending, or output, in the economy.

Lest a careful economist raise a valid objection, a slightly technical point must be conceded here. Government spending is done in three categories:

Purchase of goods (tanks, typewriters, sani-

tation trucks and school buildings).

Transfer payments to people outside government (Social Security, veterans' benefits, welfare).

Purchase of services, meaning the services of the people it employs (teachers, policemen, park rangers, tax collectors).

To the extent that a shift to more public spending, through higher taxes and a resulting reduction of private consumption, involves the first two of these categories, the point stands as made: there will be just as much production of steel, tin cans, electric power and toasters as before. To the extent that the higher public spending goes to the third category, employment of more teachers, policemen and the like, there will be slightly less production of goods, even though these people spend their paychecks like everyone else. Essentially what happens in this case is that the society has chosen, through higher taxes, to have more services and fewer goods. If we assume that goods production brings pollution, a society with fewer auto- or steelworkers and more cops will crank out less pollution.

But this remains a relatively minor matter. Hardly anyone who proposes a solution to our problems thinks in terms of vast armies of Government workers. Reforming welfare through the President's new family-assistance plan is the perfect example; this will be a simple expansion of transfer payments. And, for that matter, building more sewage plants will be a purchase of goods. The overriding fact is that we can spend 30 per cent of our G.N.P. for public purposes, as we do now, or 50 per cent, and the G.N.P. will still be there. The law of compound interest will apply, forcing the G.N.P. upward. To the extent that the environment problem is caused by ever-expanding output, the third law says that it will not be essentially changed by altering the mix between private and public spending.

Conclusion

Three nice, depressing laws. They give us a starting point for any rational discussion of the environment problem. Our output is going to go on growing and growing under any conceivable set of choices we make.

But the starting point does not mean despair. It simply means that trying to solve the problem by reducing output, or the growth of output, is waste of time and energy. It won't and can't work.

How is the problem solved then? The purpose here is not, and has not been to solve any problems. It has been to try to head off useless solutions. But a few things can be said:

There is, first, technology itself. The very energy and inventiveness that gave us this rising output—and got us to the moon—can do things about pollution. A fascinating case is the sulphur dioxide put into the air by coal-burning electric power plants. A very strong argument can be made that under any foreseeable circumstances we will have to burn more and more coal to produce the needed growth of electric power. And the ground does not yield much low-sulphur coal. Thus, somebody is going to have to have the incentive to develop a way to get the sulphur out before it leaves the smokestack; and if this costs the utilities money, the regulatory commissions are going to have to allow that cost to be passed along in electric bills.

Next, there is the related idea—being increasingly explored by economists, regulators and some legislators—of making antipollution part of the price-profit-incentive system. In simplest terms, this would involve charging a fee for every unit of pollutant discharged, with meters used to determine the amount. There would be an economic incentive to stop or reduce pollution, possibly backed up with the threat to close down the plant if the meter readings go above a specified level. The company—say a paper company—would be faced with both a carrot and a stick.

There is also the simple use of the police power, as with poisonous drugs or, lately, D.D.T. It is the "thou shalt not" power: automobiles can emit no more than such-and-such an amount of this or that chemical through the exhaust pipe. Once again, if the engineers cannot find a way out, the car simply cannot legally be sold. There will be, and should be, all sorts of debate "at the margin" —whether the higher cost of the different or improved engine is worth the extra reduction of pollution. The argument exists now over D.D.T.; there are clearly costs, as well as benefits, in stopping its use. But the "thou shalt not" power exists.

Finally, there are many possibilities for using a part of our public spending for environmental purposes. Sewage plants are the obvious case. President Nixon has proposed a big expansion of the current level of spending for these plants, though not as much as many interested in clean water—including Senator Edmund Muskie—would like to see.

In this case, and only in this case, a greater effort at curing polluting must be at the expense of some other Government program unless we pay higher taxes. It is proper to point out here the subtle dimensions of the issue. There are all sorts of possible gimmicks, like tax rebates for antipollution devices for industry and federally guaranteed state and local bonds. One way or another, spending more for pollution abatement will mean spending that much less for something else, and the something else could mean housing or medical services. Every local sewage plant bond sold means that much less investment money available for mortgages, for example.

A final reflection is perhaps in order, though it is almost banal. Our rising G.N.P.

gives us the "resources" to do the antipollution job. These resources include rising Government receipts. Our technology, which has given us the rising G.N.P., might find the way out of one pollution problem after another—and they are all different.

But, in the end, we cannot be sure that the job will be done. Growth of total output and output per capita will continue. The long-term relief is perfectly obvious: *fewer "capita."* That sort of "solution" might help, in our country, by about 1990. If we survive until then, the law of compound interest will be much less horrifying if the population is 220 million instead of 250 million.

A Corollary to the Dismal Theorem
Conrad Istock

Deleterious environmental alterations press widely and diversely upon the modern world. Expressions of concern and suggestions for remedying environmental ills have begun to appear in most sectors of our society and from this widespread concern one can adopt an easy optimism—an optimism which ignores the fundamental contradiction between modern economics and ecology.

Specific deleterious or unpleasant environmental conditions cannot be discussed singly except when a narrow technological solution for a specific local malaise is sought. The integrative aspects of human physiology, human societies, and ecosystems cast the discussion into a single broad theoretical framework. This frame of reference is complicated and not generally admitted.

Beginning to Perceive Ancient Limits

The evolution of the unique features of man's anatomy, physiology, and sociality reach back through hundreds of thousands of generations. And it is clear that less than a thousand generations of civilization can have produced only little genetic adaptation of man to his modern living conditions.

While we often dwell upon the remarkable biological and social accouterments of man's genetic heritage, we seldom give equivalent attention to the restrictions it has placed upon us. We may admit that there are certain common and annoying weaknesses in the human frame and that the sense organs disadvantage us with respect to certain wave lengths, tastes, or smells. But how many have noticed, for example, that the mind can only count six or seven objects at a single glance and that man's immediate sustainable and generalizable concern is usually with a very small group of people? Recent evidence indicates that early man probably lived in quite small groups (30-50 individuals) with low absolute population density (5-10 square miles per person) (Washburn and DeVore, 1961). It is not difficult to conclude that selection so molded the human nervous system that a major intellectual effort is required to consider populations as large as a whole city, while it is easy to sustain thought during hours of planning for a family, gang, or fraternity. If man did evolve largely in a social context of small semi-isolated bands, should we really expect the human nervous system to be otherwise with regard to the contemplation of large groups? I have chosen this plausible evolutionary deduction because, if true, it illuminates many seemingly intractable difficulties in national and international affairs. Further, this speculation causes one to wonder what other restrictions we have carried through our meteoric social expansion into life in an industrial age. Surely all our physiological and psychological tolerances predate civilization and form a collective heritage of genetic constraints—constraints which men wise in technology and cognizant of self-interest would seek to understand and accommodate.

Economic Theory in Practice

In discussing present technological societies Galbraith and Randhawa (1967) have emphasized the following two conclusions:

1) Large industry is today collectively free of a market economy and in fact has substituted economic planning and highly effective control over supply, demand, and capital. This is a condition essential to highly technical enterprises, and modern capitalist and socialist economies are very much alike in this respect.

2) Even in nonsocialist countries there is tight cooperation between governments and corporations. In the United States this is made particularly so by the continuous exchange of leadership between corporations and government, and because the U.S. Government is such a large and dependable buyer. Thus socialist countries, with their blatant planning of market and production, and capitalist economies, which accomplish this planning partly by de facto arrangements, converge on one pattern. This Galbraith has called the industrial state (Galbraith and Randhawa, 1967). Such industrial states today occupy most of the northern hemisphere.

The practical economic theory by which all industrial nations proceed requires steadily increasing production and consumption. This expectation of growth has until now been satisfied partly by increasing per-capita consumption and partly by increasing population. It is debatable whether increasing consumption can long continue without some concomitant population growth.

Consider the following example. The U.S. population doubles about every 60 years. Annual American automobile production doubles about every 20 years. At present, American manufacturers produce a new car for every 25 American people of all ages each year. With the given rates of steady increase in population and automobile production, it would require 130 years to reach the utopia of one *new* car per head per year. If the population instead remained constant at its present size while production increase continued, we could make it to one car per head in 90 years. In either case, the minimum driving age will have to be sharply lowered quite soon if we are to make use of so much shiny transportation.

The American industrial system as a whole has a doubling time of 13-15 years, suggesting that great economic strains would arise if the population suddenly ceased to grow. The real situation is, of course, complicated by the origin of new industries and novel desires to consume, and the partial openness of the American market to foreign trade.

Still, one might expect that as a natural process the rate of expansion in production would come to approximate the rate of population growth with sporadic and temporary margins created by increases in per-capita consumption. If this were the case, and population growth went to zero, we would surely require a different practical theory of economics. To the contrary, it seems more likely that the motivating forces resulting from planned industrial expansion and advertising will actually be able to provoke population growth for some time into the future. The rest of the increase in production will be taken up by increases in real per-capita consumption and by governmental buying. From a businessman's view the increasing consumption of military goods by government during war or the arms race is roughly and powerfully equivalent to a sustained expansion in per-capita consumption of "luxury" items. Thus, if population, real per-capita consumption, and governmental buying can all be kept increasing, the practical theory of economics will stand vindicated.

The important conclusion at this point is that industrial societies have come, to a great extent, to serve the esoteric goals of industry rather than to apply technology to the amelioration of conditions of existence stemming from man's essentially fixed and ancient genetic background. The industrial society expects an expanding population and does not in any effective, collective manner worry over the deleterious environmental effects attendant on industrial growth.[1] It is important to note that even without my ecologically tainted viewpoints Galbraith was led to some deeply personal worries:

If we continue to believe that the goals of the industrial system—the expansion of output, the companion increase in consumption, technological advance, the public images that sustain it—are coordinate with life, then all of our lives will be in the service of these goals. What is consistent with these ends we shall have or be allowed; all else will be off limits.

Ecological Views

Populations of most species of plants, microbes, and animals fluctuate within definable limits in nature. The constraining of increase and decrease in local ensembles of species engenders stability in the entire ecosystem and its processes of mineral cycling, gaseous exchange, energy capture and transfer, and other by-processes of geographically fixed natural assemblages (e.g., the retention of runoff water and climatic amelioration). By almost any definition of ecological dominance man

qualifies as the dominant species in most terrestrial, freshwater, and shallow marine ecosystems.

Human populations through technology are sometimes free of the nearly ubiquitous existence of population limitation and regulation in nature. Thus, for a part of humanity, technology has for the past century managed to expand the means of subsistence faster than population growth, leading some in that part to doubt the validity of Malthus' (1798) dismal theorem.[2] Malthus' theorm states that because population can grow geometrically and resources only arithmetically, population generally outruns resources, thus fomenting human misery. While Malthus was incorrect locally and the arithmetic condition for resources has not generally held, his generalization has been qualitatively and globally true since he set it forth a century and a half ago.

Malthus was an astute ecologist. He understood, at least intuitively, the universal processes of population control and restriction operating in nature. Modern ecology justifies such intuition and as a result of such control the gross physiological processes of our planet

1. From Galbraith one concludes, further, that the industrial state with its *sine qua non,* the modern corporation, is an almost naturally developed system for the making of decisions. Once a business achieves the status of large corporation it will automatically have achieved an impersonality which leaves no individual or small group truly responsible for its good or ill deeds. All the important decisions leading to corporate activity emerge from the technical, hierarchial deliberations of many individuals. No one individual, wherever placed in the hierarchy, generally has sufficient information to counter such conglomerate decisions. When environmental degradation by a large corporation is revealed, it is usually meaningless to rail against the president or board of directors as if they were "the enemy." These men seldom really led the corporation into its misdeeds, and they do not possess the technical knowledge to rectify environmental or social problems. They may, however, under criticism display emotional insensitivities akin to the natural collective motivations and "economic realities" of the corporation itself.
2. I first heard this expression used by Marston Bates and imagine it was original with him.

mediated by plants, animals, and microbes all seem to tend toward stability and equilibrium. We have no theory of general ecological expansion on a short enough time scale to match the practical theory of an expanding economy.

The Fundamental Contradiction

Thus there arises a contradiction between the practice of economics in industrial states and the requirements for stable ecosystems. The two are probably at odds at every point of contact, with human population increase a concomitant of almost every difficulty. So general is the conflict between human economics and the ecology of this planet, and so pervasive, powerful, and unswerving the dynamic of the industrial state that it is impossible to imagine a suitable industrial ecology under prevailing economic theory. We can, however, identify some of the signs which would indicate a massive international effort to turn technology to the construction of a physiologically and socially healthy environment.

Suggested Signs of Hope

If most of the following existed or were generally sought after there would be some license for optimism.

1) Virtually complete international agreement on the means and pace by which human population growth rates throughout the world will be brought to zero and in some cases made to become temporarily negative.

2) Identification and study of the natural ecological subdivisions of the planet and effective preservation of these to at least the extent required for indefinite global ecological stability.

3) International cooperation in the identification and study of the most natural economic subunits of the globe and planning for the fu-

ture development and optimal interaction of such units.

4) Complete world-wide disarmament.

5) Technological achievement of absolute control of all quantitatively and biologically meaningful gaseous, liquid, solid, and thermal wastes including elimination of almost all release of radioactive wastes into the environment because of the absence of any meaningful biological thresholds.

6) Cycling, reuse as raw materials, or mineralization of virtually all manufactured products and sound management of renewable resources. This would shortly mean application of an equilibrium theory of economics.

7) Intense study of the roots of disruptive human behavior.

8) Increased analysis of the genetics of modern and aboriginal human populations and development of both population and individual genetic therapy.

9) Planning for the effective utilization and distribution of wealth consonant with diverse ethnic and esthetic values.

10) Design and construction of rational cities with effective mass transit systems, public educational, health, and recreational facilities, communications, energy sources, and other human services.

11) Clear recognition that the above categories encompass the most difficult problems man has faced and that the lag time from concept through experiment to production in the solution of such problems probably approaches 30 to 50 years—a research and development time which only the most idealistic planner could bring himself to call a challenge and which would instantly reduce most businessmen to insanity.

A Corollary

One cannot dismiss these suggested signs as outrageous utopianism. They are what must happen—should have been happening—if we are to avoid ever-increasing and socially disruptive calamity. They reflect the present size of our dilemma. That most of these do not exist as clear signs means that man has been busy for the past 20 years solving a good many of the easy and less relevant problems. Science and technology have been applied to easily quantified problems—ones for which the physics, chemistry, and biology were well understood ahead of time. Many of today's problems are indirectly the result of this short-sighted use of technology, while tougher problems remain unsolved.

In industrial societies we seem to be rapidly substituting indiscriminate or egalitarian population effects for older, more discriminating ones. By indiscriminate effects I mean environmental features which do not differentially influence the birth and death probabilities of individuals as a function of population density. Rather, such effects tend to treat all individuals alike and are not capable of regulating and stabilizing population size. Because of its complex structure, modern society can poison, explode, infect, stress, and thus disrupt itself far more effectively and widely than hunger or disease ever could in the past. Here then is a corollary to the dismal theorem. When a society through industrialization relieves itself of ancient limiting conditions one by one in order to satisfy a theory of expanding economy and population, it ultimately creates novel and varied local limiting effects which are largely indiscriminate and nonregulative. Concurrently, in the climb into technology an elaborate and vital social order with many interdependencies must be built. *When indiscriminate or weakly discriminating environmental effects, fostered by dense population and high social interdependence, exceed human physiological or psychological tolerances or basic requirements for ecological sta-*

bility, most individuals will simultaneously experience detrimental effects as the social order begins to crumble.

I submit that there is virtually no citizen of a modern industrial state whose very survival is not dependent on the intact social order—a social order that will undergo disruption with the first epidemic to exceed all hope of medical care in several cities simultaneously; or when several of our cities become uninhabitable through toxic or inadequate water supplies, clinically toxic air pollutants, local carbon dioxide elevation, and oxygen depression; or when large numbers of people break psychologically under stresses induced by inequalities in the distribution of wealth. We know in part what these scenes will look like. The early subclinical forms of each of them surround us and we have no defense so massive as the problem.

We have no need for prophets of environmental doom. The storm clouds gathering over Western technological civilization are, figuratively and literally, visible.

ACKNOWLEDGMENTS

In working out and writing the above synthesis I have been aided by the comments of Uzi Nur, Jerram Brown, Peter Frank, Joseph Connell, William Murdoch, and Seth Reice.

REFERENCES

Galbraith, J. K., and M. S. Randhawa. 1967. *The New Industrial State.* Houghton-Mifflin Company, Boston.

Malthus, T. R. 1798. *An Essay on the Principle of Population.* Johnson, London.

Washburn, S. L., and I. DeVore. 1961. Social behavior of baboons and early man. In: *Social Life of Early Man.* S. L. Washburn (ed.). Aldine Publishing Co., Chicago.

Can a Progressive Be a Conservationist?
Edward Teller

The question of pollution is serious. We must do something, and that is one of the very few things on which practically everybody in the United States, young or old, Democrat or Republican, is agreed. It is important to do something about clean air and clean water. At the same time, any good cause can be exaggerated. There are those who worry about a slight rise in temperature and call it thermal pollution. I, for one, am for thermal pollution in the winter.

I like to think about myself as a progressive; and I like to imagine that the best possible state of the world is not the present state, not even the past state. With imagination, with work, with restraint, with proper thoughtfulness, we can make sure that the future will be better than the present and the past. On generalities it is easy to agree. Specifics raise more problems, and so it is in the case of "nuclear ecology." The use of nuclear energy, particularly the use of nuclear explosives for peaceful purposes, can produce harmful effects, and of these there is one—radioactive contamination—which already has received a great deal of attention and which seems to receive increasing attention at the present time.

I would like to discuss radioactive contamination and also the questions: What can Plowshare, the constructive use of nuclear explosives, do for all of us? In what manner can it make life more abundant? And can it make our environment cleaner? Radioactive contamination is frightening; its horrors have excellent spokesmen, such as Linus Pauling. It is disquieting because the danger of radiation itself is something of which our senses give no warning, and at the same time, the damage that has been discussed does exist. Radioactivity is also frightening by association. It is connected with the atom—with nuclear explosions—and this association of ideas is, of course, closer in the case of Plowshare than in other cases. Radioactivity is, furthermore, frightening in a very practical sense because radiation can be detected so easily, not by our senses, but by our instruments. While you have seen nothing of it, felt nothing of it, somebody can come along with a Geiger counter and show you: it clicks, get scared! What can be more frightening in this day and age than a piece of apparatus that clicks in an ominous manner?

There are also clear and simple reasons why radioactivity need not be frightening when proper care is exercised. This is so, in the first place, precisely because of these clicking pieces of apparatus. What can be easily discovered by a simple counter can be easily guarded against. But there is more than that in it; life is full of danger. In fact, I can think of nothing as dangerous as life. And, in life, I am really scared of micro-organisms and of chemistry. I am scared of these because two micro-organisms which appear similar might be very different in their effect upon us. Two similar molecules may differ so greatly in their effects that one is a useful medicine and the other is a deadly poison. At the same time the actual chemical difference consists in a rather small rearrangement of the constituent parts of the molecule. It is very difficult for chemists to predict which will act in which way, and it takes a long and arduous research enterprise to find out whether a certain kind of sugar substitute is indeed dangerous or not.

With radiation, it is very different. In good approximation one can say that all radiation acts alike. If, in a certain cell a certain amount of high energy radiation is deposited, the effect is similar whether the energy has been deposited by alpha particles, by beta particles, by gamma particles, by cosmic rays, by neutrons, or by any other hard radiation. Possible harmful effects of radiation can therefore be evaluated much more simply in a much more complete and reliable manner. All of us, including our ancestors who lived in the trees, those who went around on four legs, those who swam in the oceans, and even those who were in the monocellular state, all of them have been exposed to some radiation. On the average, this radiation was constant ever since the human race existed on Earth. The present guidelines which regulate the safety of radiation prescribe that we should add no more than 100 per cent to the natural background or more precisely to the radiation which we get anyway. This limit is to be applied separately to each tissue in the human body.

It is an interesting footnote that people who are foolhardy enough to live in Colorado—who spend their lives closer to the sky, the source of cosmic rays and who are in greater proximity of uranium deposits—are exposed to almost as much radiation as the US Atomic Energy Commission permits. Many thousands of people in Brazil and in the Kerala Province of India are exposed, and have been exposed for centuries, to 10 times as much as the maximum permissible dose. Maybe they suffer from it, but they are backward people and they suffer from being backward much more. Thus their additional suffering on account of radiation has not, as yet, been verified.

I am a progressive; I believe in progress. I am also a conservationist, and I believe in clean air and clean water. I also know that only in case of radioactive contamination have we demanded that the pollutant be essentially no more than the natural amount to which all of us would have been exposed in the original pristine natural state—if, indeed, anybody can tell me precisely what the word "natural" means. Imagine that similarly stringent conditions should be imposed on all materials which cause pollution. Automobiles would, of course, be banned. The same would hold for our power plants, except the hydroelectric plants and the nuclear plants. Our industry would be reduced to a negligible fraction of what it is today, and I am also a little uncertain what would happen to the production of iron.

All this should at least in part be considered in a serious manner. One may hope that in the foreseeable future most of the electricity will be produced by nuclear reactors. Electricity may become cheap enough to be used for space heating, thus eliminating further contamination due to burning of coal and oil. It is not impossible that electricity will play a greater role for providing motive power in the transportation of people, and one might even hope that with cheap electricity aluminum that can be cheaply produced in a clean way by electrical processes will replace iron to some extent. In all of these ways contributions can be made to greater cleanliness although the transition will necessarily take a long time if it is to be executed at a reasonable cost. At the same time, it should be remembered that the needed growth of nuclear energy consumption will be accompanied by some release of radioactivity. These releases have been held to minimal values and standards similar to those prevailing at present can be enforced with proper and continuous care. Thus we can fight pollution wherever it becomes serious—provided we do not introduce unnecessarily rigid limitations on radioactive releases, but are satisfied with the standards which are based on the long experience of all human existence and on the even longer history of the living world.

The fight against pollution has entered into a particularly popular phase, but thoughts on pollution are older. In 1954 Otto Frisch, one of the discoverers of fission, wrote a short parody on the safety measures connected with nuclear reactors. He pretended that in the year 4995 the uranium and thorium mines from the Earth and Moon mining systems were near exhaustion and wrote: "The recent discovery of coal (black, fossilized plant remains) in a number of places offers an interesting alternative to the production of power from fission. . . . The power potentialities depend on the fact that coal can be readily oxidized, with the production of a high temperature and an energy of about 0.0000001 megawatt day per gramme. . . ."

Further on, he remarks: "The main health hazard is attached to the gaseous waste products. They contain not only carbon monoxide and sulphur dioxide (both highly toxic) but also a number of carcinogenic compounds such as phenanthrene and others. To discharge those into the air is impossible; it would cause the tolerance level to be exceeded for several miles around the reactor."

These words sound to me a little more appropriate than some recent discussions. But let us return to the particular role that the constructive uses of nuclear explosives can play in improving our ecology. In order to clean up our continent, in order to keep the civilized world free of dangerous contamination, we not only should tolerate Plowshare—we need it. Plowshare can make positive contributions to cleanliness; and the words "nuclear ecology" should not mean that the use of Plowshare must be further restricted, it should mean that Plowshare must be expanded to fulfill its

proper purposes. It is true that if we should not exercise careful control we would be in trouble; but no big-scale enterprise has ever been carried out with as much assurance to human health, to human life and cleanliness as the atomic energy enterprises.

Where are the positive uses? We are engaged in producing more natural gas. The amount of the natural gas in the USA which we now have available will keep us supplied at the present rate of usage for less than 15 years. Yet, natural gas is valuable, because it contains practically no sulphur and, therefore, it is one of the cleanest fuels that we can use. Los Angeles has to use it almost exclusively. New York would like to use it; but because there is not enough of it, New York must burn more sulphur-containing fuels. At the time of any extended period of atmospheric inversion in New York (when released contaminants stay near their source of origin), statistics show that there are several hundred excess deaths in New York hospitals. I cannot say that this is proven, but according to the figures I have seen, the correlation is impressive.

There is as much natural gas again in the United States as the amount I have already mentioned, but this additional 100 per cent is contained inside tight rock formations from which we cannot extract the gas in an economical manner. By exploding a nuclear device a few thousand feet underground one can loosen up the rock formations and produce the appropriate rubble chimneys. Thus we have an excellent chance to make these additional amounts of gas available. We have actually performed an experiment in breaking up rock and stimulating gas production. The name of the experiment was "Gasbuggy" and it was performed near the "Four Corners" where the states of New Mexico, Arizona, Utah and Colorado meet. A nuclear explosive was buried 4000 feet under the surface in a gas-bearing formation.

This experiment was carried out on 10 December 1967, and it was a success in that it produced more gas and gas carrying less radioactivity than we had expected. It is very clear that the nuclear explosion did not cause the end of the world, not even in a strictly local sense. In the meantime an attempt is being made in Colorado for the massive exploitation of gas production by the same principle. It is too early to say whether this will succeed in the near future. In the long run I have little doubt of success.

Another similar example is the following: Gas is consumed in our big cities and the demand is much greater when there is a cold spell. Gas is brought in by pipelines. We need storage space for this gas, and on the eastern seaboard all the natural storage space has been exhausted. We could produce lots more storage space by nuclear explosions. Similar storage places could be used in any location where gas is brought in by ships in a refrigerated form. The storage space could be established below the sea bottom on the continental shelf and could be brought into population centres by relatively short pipelines.

In connection with the general power economy, I should mention one more example. It is a 10-year-old dream, and I hardly believed it 10 years ago. My good friend at the University of California, Los Angeles, Dr. George Kennedy, was its first apostle; my colleagues in the Livermore branch of the Lawrence Radiation Laboratory have been converted by him. In many places there is a lot of usable heat underground; in Italy, in New Zealand, even in California, some of the accessible heat has been used as a power source. But this amounts to a dribble. Deep underground, much geothermal heat can be found. Logically

enough, there seems to be a lot of it around extinct volcanoes, some of which are located all along the Pacific coast in the northwest part of the United States. A nuclear explosion could be used not so much to produce heat but to open up this geothermal reservoir. The explosion would produce a rubble cone, expose a lot of surface of the hot rock so that we can pump water down, convert it into steam and use it. All this could be done in a closed cycle, and hopefully no radioactivity need escape. We might get a source of energy as clean as power from a waterfall.

I would like to mention just one more example connected with ecology. I have already mentioned that we can make holes deep underground, open spaces, or partly open spaces. These spaces can be produced cheaply in many locations and we can use them for waste disposal. We can remove dirt from the biosphere. We must dispose of the by-products which we throw away each day. The most inaccessible place, the place best isolated and yet reached when needed, could be established a couple of thousand feet right under the surface of the Earth. Places could be selected where no ground water would get to them, and where the dirt would never bother us again—at least not for the next few million years.

The word ecology has a somewhat flexible meaning. One might be justified in including the question of how to improve the conditions of life for people in general. There are many people in the world and there will be more; a good life for the billions is, in the end, our great problem, and water is one of the most important necessities for a good life. Plowshare can be used to divert streams to store water, even possibly to distill water underground using, again, geothermal heat. If there is a real way to make the desert bloom, I think that Plowshare is the best candidate to accomplish this goal.

It is also worthwhile to reflect on a general class of necessities of our modern civilization: raw materials. By using new methods of shattering the rocks deep underground and of employing liquid extraction methods, some important raw materials could be obtained. In the case of copper, a method utilizing weak acid solution has been worked out. In other cases it will take more research to find proper methods. Decades ago, scarcity of raw materials was considered the main reason for international conflict and eventually for war. Today this problem has been displaced in people's minds by the danger of nuclear weapons. One should not forget, however, that nuclear explosives might be used to supply human needs and thereby decrease the reasons for international conflict.

Along similar lines, one may remember the increasing role that big oil tankers play in the cheap distribution of the necessities of modern life. Many new ships have a displacement of 300,000 tons and ships of 700,000 tons are on the drawing board. Compared to these carriers, one may consider the *Queen Mary* as an oversized canoe. These carriers may become important not only to transport oil, but to haul iron ore from the abundant deposits of northwest Australia to all corners of the world. The difficulty is that neither our canals nor our harbours can accommodate these new monstrous ships. Plowshare may be the best means by which to create the appropriate waterways and the necessary harbour facilities.

By a number of simultaneous explosives placed in a row, one can produce an elongated crater whose rim would rise above the waves and would also serve as a windbreak. The entrance of this artificial crater could be adjacent to water of a depth of little more than 100 feet, thus permitting the biggest ships

to enter. If a causeway to the shore is needed it would probably be located in shallow water and might be constructed by conventional means. Ambitious plans for canals through which big ships could pass have been discussed. One extensive discussion is centred around a new canal connecting the Atlantic and Pacific oceans. Thus it is possible that by shipping, and in directly by the use of Plowshare, the economy of the world can become more united and more efficient.

Returning to the question of the association between Plowshare and nuclear weapons, there is no doubt that this "guilt by association" does influence, consciously or otherwise, the feelings of many people. Nuclear explosions killed people—women and children—at the end of the Second World War. They are now considered as weapons of terror; they must be banned and not be used for anything.

I want to remind you of a remarkable historic parallel, one that is known but perhaps not sufficiently recognized. I refer to a horrible ancient weapon, a weapon more than 1000 years old, the "Greek Fire," a mixture considered mysterious because it caught fire on contact with water; it was this "Greek Fire," an invention of the eastern Roman Empire, which turned back the first Arab invasion and which saved the eastern half of the Roman Empire and kept it safe for hundreds of years. The weapon was effective; it also was considered horrible. It was outlawed and this limitation stuck. Constantinople lost its defences; in the end it fell.

The "Greek Fire" also happened to be the first really impressive use of chemical energy in human affairs, the first big step beyond the most primitive, and the most important control of fire itself. I suspect that suppression of "Greek Fire"—the fact that "Greek Fire" was not only outlawed but kept secret—delayed the industrial development of the world by almost a millenium. If the discovery of the "Greek Fire" would have evoked more interest and less horror, more openness and less secrecy, the dark ages may have been avoided.

Progress cannot be and will not be stopped, and I know that Plowshare will proceed. Whether it will proceed as rapidly as it should, whether it will proceed in the United States or in some other part of the world—these are most important questions of detail that we should consider with some care. Today those conservationists who have become reactionary, who are opposed to all progress, who seem to believe that everything that is good lies in the past, may bring about another dark age. I hope that they will not succeed.

Confronting the A. E. C.
John W. Gofman

In 1940, I came to the University of California to work for a Ph.D. For my thesis I chose a problem suggested by Glen Seaborg. That work became part of what was the Manhattan Project. I stayed with it until 1944, when I left to complete medical school, which I had begun before going to Berkeley. So I know something about the early developments of the modern atomic era.

In 1947, after medical school, I returned to the University of California at Berkeley. I was teaching medical physics and doing research, in part under Atomic Energy Commission auspices, and was involved in studying the origins of heart disease. In many of these research projects, I collaborated with Dr. Arthur R. Tamplin. We were associates both in teaching and in carrying out research on the problems of atomic energy and its hazards.

In 1953, the Lawrence Radiation Laboratory was divided into two labs. To the Berkeley laboratory was added a new branch at Livermore. For the most part, Livermore was created for Edward Teller to devise hydrogen bombs with the approval and blessing of its director, Ernest Lawrence. Because I was a close friend of Ernest Lawrence's and because he thought there were several hazardous features connected with the weapons work being done at Livermore he asked me to come to Livermore two days a week to watch over the people involved. They were handling radioactive hydrogen, tritium, and plutonium in very large amounts, and doing so under rather urgent schedules, preparing for tests in the Pacific. So for two days a week I went to Livermore to observe the people there. I continued my observations for four years. In 1957, I asked to be relieved of the responsibility.

Some six years later, I received a call from Dr. John Foster, whom I had come to know at Livermore. He asked me to see him at the lab. Dr. Foster said that he had an interesting request from the Atomic Energy Commission. The A. E. C. wanted to know if the Livermore laboratory would undertake a long-range program to evaluate the impact of its activities—weapons testing, the peaceful uses of nuclear explosives; in fact, every aspect of atomic energy—on man and the biosphere. I thought it was strange that in 1963, something like eighteen years after the A. E. C. had been formed, with some eighteen or twenty laboratories around the country involved in biomedical aspects of radiation and radioactivity, it wanted to set up still another laboratory to consider the impact of radiation on man.

But there was a definite need for it. The background for it was set in 1961, after the voluntary moratorium, when the Russians resumed testing. Then, during the 1961-62 period, President Kennedy ordered a resumption of American testing to take place in the Pacific and in Nevada. As a result of some tests made above ground in Nevada in 1962, the state of Utah was hit hard with radioactivity —the levels of radio-iodine found in milk were quite high, for instance, and that of course has an effect on children's thyroids. The tests had been smaller than some conducted in the nineteen-fifties and the actual dose to thyroids was not as great in the 1962 tests as those that resulted from earlier ones. Even so, the A. E. C. found itself in trouble. The agency decided that by setting up a biomedical laboratory at Livermore—with biologists working closely with the people who were making nuclear explosives and firing them off—it might in some way be able to avoid a recurrence of the Utah incident.

It was obvious that research was needed. In spite of all the other laboratories in the country under A. E. C. auspices that were investigating one or another aspect of the hazards of radioactivity, no one in any one of those laboratories and no one in the A. E. C. ever asked the question: If we do this weapons test, what will it mean? How will it affect health? If we use nuclear explosives for making harbors or canals or getting at underground resources, what will be the cost to society?

Dr. Tamplin and I and a few others decided we would consider going into this job of evaluating the impact of radioactivity on man and the various activities within the atomic energy field—provided that we were able to conduct the research in our own way, had long-term support, and were given professional independence. We said, in effect, that we would investigate the problems, but whether the results were favorable or unfavorable to atomic energy programs, we would make the results available on an open, unclassified basis to both the scientific community and the public, because these are matters of vital concern to everyone. This was agreed to and the program was scheduled. Three months later, however, the Nuclear Test Ban Treaty, calling off tests in the atmosphere, was signed. Thereupon the A. E. C. seemed to think the problem of fallout was solved. We had already committed ourselves to doing the research and went to work, even though our budget was cut to a third of what was originally planned.

Our problem, as we saw it, was clear enough. We set out to develop an integrated system of calculating the extent over space and time of radiation and radioactivity from any kind of nuclear source—an explosion, a reactor, any radio-isotope source. Dr. Tamplin took it upon himself to develop a thoroughgoing system to enable us to make exactly this kind of prediction, tracing and measuring the effect of, say, the radioactivity from an underground nuclear explosion traveling through food chains to man. The other half of the problem was crucial; we wanted to know, after we had knowledge of the measure of radioaction man would receive, what its effects would be.

Dr. Tamplin's findings showed that the standards established by the Federal Radiation Council, detailing the allowable levels of human exposure to radiation, were useless. The regulations which state how much a nuclear reactor could release do not take into consideration the increased concentration of radioactivity as it works through the food chains; the radioactivity in water, for instance, might become five thousand times as concentrated in a fish.

The A. E. C. was extremely disturbed by the airing of these facts. It contended that the Commission could rely on the fact that there was a primary standard, not on how much radiation a reactor could release, but on how much was tolerable for humans. Even if the standards establishing what level of radiation reactor or power companies may release are wrong, the people are protected from exposure to any more than a certain amount set by the Federal Radiation Council. Though there is no scientific basis for accepting the Council's guidelines as safe, the A. E. C. consistently, whenever questioned about those standards, claims that the standards represent a harmless amount of radiation, or an amount of radiation to the individual so small it would hardly be noted when detected. If pushed a little harder, the Commission retreats to the position that everything one does in life entails a risk. The view that the benefits the atom can bring far outweigh the risks was taken into account in setting the standards of tolerance. Indeed, one of the most grievous errors of government management was placing in the hands of the Federal Radiation Council the authority not only to evaluate the risks but to evaluate the benefits and set standards accordingly.

One such risk-benefit ratio involves the question of ventilating the Colorado Plateau uranium mines a little more in order to bring a halt to the current epidemic of lung cancer among the miners. It is estimated that over a hundred miners are already dead of lung cancer from radiation, and that out of a few thousand miners the ultimate toll will be a thousand.

When this problem first came up, the A. E. C. and the Joint Committee on Atomic Energy disavowed all responsibility for setting standards for mines; it was traditionally part of the Labor Department, the Bureau of Mines. Nevertheless, hearings were held and Willard Wirtz, then Secretary of Labor, recommended lower levels. Before he left office, Secretary Wirtz directed that the levels in those mines must go down; the directive is supposed to take effect later this year. Just recently, the Federal Radiation Council and

the A. E. C. awarded a two-hundred-thousand-dollar contract to the Arthur D. Little Corporation to evaluate the economic impact on the uranium industry if the levels are pushed down to those required in the directive. So, having once abdicated from responsibility, they now want to assume it, because it is compromising the industry they are trying to promote.

Even when it was pointed out that the conditions in the mines may bring about a repetition of the European experience in Joachimsthal, Czechoslovakia (where, over a period of years, seventy per cent of the miners—five thousand men—died of lung cancer), the Federal Radiation Council said we must remember that the mines are important to the economy of Colorado, the national defense needs uranium, and the country is going to need atomic electricity.

All are non sequiturs with respect to why the miners should be dying, but comprise the strange ways of risk-benefit calculations. The benefit to the electric power industry is cheap uranium. The risk is death to the miners.

Sooner or later we have to come to grips with a reliable estimate of what the tolerance levels the Federal Radiation Council has set actually mean. Evidence concerning the effects of radiation continues to come in from Hiroshima and Nagasaki, and the follow-up period is getting longer. In the early period, leukemia showed up, but as the years passed a series of new cancers began to appear in excessive numbers.

From Britain have come reports on some fourteen thousand men who had radiation treatments for arthritis of the spine, and what the effects were after five, ten, and fifteen years. An analysis of the data from a Nova Scotia tuberculosis sanatarium showed that women who have received repeated fluoroscopic examinations to the chest (averaging a hundred and fifty examinations in the course of their treatment) had, in the fifteen years following, twenty-four times as much breast cancer as the women who had not been fluoroscoped. Dr. Karl Morgan, an eminent health physicist at Oak Ridge, Tennessee, and a member of the International Commission on Radiological Protection, estimates that the current use of fluoroscopy and x-ray in the United States, giving only a third or a tenth as good medical information as one would want, is responsible for somewhere between twenty-five and a hundred thousand deaths each year.

All the evidence that we have examined has led us to believe, with as good assurance as most laws of biology or chemistry provide, that all forms of cancer can be induced by radiation. (Spontaneous cancer, of course, occurs for reasons no one knows, but our studies concentrated on those induced by radiation exposure.) Not only are all forms of cancer induced by radiation, but the dose that it takes to double the incidence of a particular cancer remains, within a factor or two, the same for other types of cancer. That is, if there is one cancer that is very rare—maybe only one case per year in a million people—the dose necessary to double it to two cases turns out to be about the same dose it takes to double a cancer that is a hundred cases a year to two hundred.

From data on pregnant women who had been irradiated in utero for diagnostic purposes, it was found that even one diagnostic x-ray, or a couple, can give a fifty-per cent increase in the incidence of childhood cancers and leukemia in the first ten years of the child's life. Another of our conclusions (from these data and some others), was that a child is roughly ten times as sensitive as an adult to induction of cancer by radiation.

So, too, with animals. Doses given all at once to young animals result in a great incidence of cancer. If an animal receives the same

dosage spread out over a longer period, when the animal is older and therefore less sensitive, the result is less cancer. On that erroneous basis, the A. E. C. has contended that fractionating the dose over an interval provides the necessary protection; since the effects of all peaceful nuclear energy activity will be spread out in this way, there will be no danger to human beings.

Our own estimation was that if everyone in the United States were to get the dose that the Federal Radiation Council allows, it would mean at least one extra cancer or leukemia for every twenty that occur normally. Put another way, it would mean sixteen thousand additional, unnecessary cancers per year in the United States. That was a conservative estimate. We knew it was conservative when we presented testimony before the Joint Committee on Atomic Energy last January; we raised it to thirty-two thousand because we had evidence to support it.

The A. E. C. said our data were wrong— even after we pointed out that our data were in agreement with the highly respected International Commission on Radiological Protection—and said that much of the data concerning cancers came from people who have had appreciable dosages. Somewhat below that dose, they said, there must be a dose that is safe.

The International Commission and all responsible biologists, however, have long pointed out that that attitude shows no real regard for public health. A safe threshold can never be assumed unless there is proof for it. All the evidence shows no suggestion whatever for a threshold.

In spite of the claims the A. E. C. makes for the safety of the standards currently in force, there may now be some hope that the standards will be revised. Earlier this year, at the request of Senator Edmund Muskie, we submitted our findings to his Subcommittee on Air and Water Pollution. Senator Muskie brought our findings to the attention of Robert Finch, Secretary of Health, Education and Welfare, who is also chairman of the Federal Radiation Council. The Council, while not agreeing with every premise and conclusion of our findings, did agree that no safe threshold of radiation exists, that every amount of radiation produces its commensurate amount of cancer. Mr. Finch has ordered a complete review of the standards, with a view to resetting them if needed, for the exposure of the population at large.

Clean Air and Future Energy
E. F. Schumacher

The purpose of the Des Voeux Memorial Lectures, I have been told, is to put the special subject of this National Society into a wider perspective, and this has been accomplished most admirably by previous Des Voeux Lecturers who managed, in one way or another, to talk about clean air while at the same time delivering a learned discourse on their own special subject.

I have set myself a similar task, to talk about my own subject, which is economics, but to study in particular how economics relates to the kind of thing in which the National Society for Clean Air is concerned, namely, the protection and conservation of the environment in which we live. I very much welcome the opportunity of doing this because I have been worried for some time about certain misunderstandings about the nature and relevance of economics, that seem to me to exist not only among laymen but also among many of the economists themselves, misunderstandings inimical to conservation. When the economist delivers a verdict to the effect that this or that activity is "uneconomic," two important and closely related questions arise: first, what does this verdict mean? And, second, is the verdict conclusive in the sense that practical action can reasonably be based on it?

Going back into history we may recall that when there was talk about founding a professorship for political economy at Oxford some 150 years ago, many people were by no means happy about the prospect. Edward Copleston, the great Provost of Oriel College, did not want to admit into the University's curriculum a science "so prone to usurp the rest;" even Henry Drummond Esq. of Albury Park, who endowed the professorship in 1825, felt it necessary to make it clear that he expected the University to keep the "new study in its proper place."[1] The first professor, Nassau Senior, was certainly not to be kept in an *inferior* place. Immediately, in his inaugural lecture, he predicted that the new science "will rank in public estimation among the first of moral sciences in interest and in utility" and claimed that "the pursuit of wealth . . . is, to the mass of mankind, the great source of moral improvement."[2] Not all economists, to be sure, have staked their claims quite so high. John Stuart Mill (1806-1873) looked upon political economy "not as a thing by itself, but as a fragment of a greater whole; a branch of Social Philosophy, so interlinked with all the other branches that its conclusions, even in its own peculiar province, are only true conditionally, subject to interference and counteraction from causes not directly within its scope."[3] And the great John Maynard Keynes, some 80 years later, admonished us not to "overestimate the importance of the economic problem, or sacrifice to its supposed necessities other matters of greater and more permanent significance."[4]

Such voices, however, are but seldom heard today. It is hardly an exaggeration to say that, with increasing affluence, economics has moved into the very centre of public concern, and economic performance, economic growth, economic expansion, and so forth have become the abiding interest, if not the obsession, of all modern societies. In the current vocabulary of condemnation there are few words as final and conclusive as the word "uneconomic." If an activity has been branded as uneconomic, its right to existence is not merely questioned but energetically denied. Anything that is found to be an impediment to economic growth is a shameful thing, and if people cling to it, they are thought of as either saboteurs or fools. Call a thing immoral or ugly, soul-destroying or a degradation of man, a peril to the peace of the world or to the well-being of future generations; as long as you have not shown it to be "uneconomic" you have not really questioned its right to exist, grow and prosper.

But what does it *mean* when we say something is uneconomic? I am not asking what most people mean when they say this; because that is clear enough. They simply mean that it is like an illness: you are better off without it. The economist is supposed to be able to diagnose the illness and then, with luck and skill, remove it. Admittedly, economists often disagree among each other about the diagnosis and, even more frequently, about the cure; but that merely proves that the subject matter is uncommonly difficult and economists, like other humans, are fallible.

No, I am asking what *it* means, *what sort of meaning the method of economists actually produces.* And the answer to this question cannot be in doubt: something is uneconomic when it fails to earn an adequate profit in terms of money. The method of economics does not, and cannot, produce any other meaning. Numerous attempts have been made to obscure this fact, and they have caused a very great deal of confusion; but the fact remains. Society, or a group or individual within society, may decide to hang on to an activity or asset *for non-economic reasons*—social, aesthetic, moral, or political—but this does in no way alter their *uneconomic character.* The judgment of economics, in other words, is an extremely *fragmentary* judgment; out of the large number of aspects which in real life have to be seen and judged together before a decision can be taken, economics supplies only one —whether a thing yields a money profit *to those who undertake it* or not.

Do not overlook the words "to those who undertake it." It is a great error to assume, for instance, that the methodology of economics is normally applied to determine whether an activity carried on by a group within society yields a profit to society as a whole. Even nationalized industries are not considered from this more comprehensive point of view. Every one of them is given a financial target—which is, in fact, an obligation[5]—and is expected to pursue this target without regard to any damage it might be inflicting on other parts of the economy. In fact, the prevailing creed, held with equal fervour by all political parties, is that the common good will necessarily be maximized if everybody, every industry and trade, whether nationalized or not, strives to earn an acceptable "return" on the capital employed. Not even Adam Smith had a more implicit faith in the "hidden hand" to ensure that "what is good for General Motors is good for the United States."

However that may be, about the *fragmentary* nature of the judgments of economics there can be no doubt whatever. Even within the narrow compass of the economic calculus, these judgments are necessarily and *methodically* narrow. For one thing, they give vastly more weight to the short than to the long term, because in the long term, as Keynes put it with cheerful brutality, we are all dead. And then, secondly, they are based on a definition of cost which excludes all "free goods," that is to say, the entire God-given environment, except for those parts of it that have been privately appropriated. This means that an activity can be economic although it plays hell with the environment, and that a competing activity, if at some cost it protects and conserves the environment, will be uneconomic.

Economics, moreover, deals with goods in accordance with their market value and not in accordance with what they really are. The same rules and criteria are applied to primary goods, which man has to win from nature, and secondary goods, which pre-suppose the existence of primary goods and are manufactured from them; and among primary goods no distinction is made between renewable and non-renewable goods, although from many points of view this is the most vital distinction of all. All goods are treated the same, because the point of view is fundamentally that of private profit making, and this means that it is inherent in the methodology of economics *to ignore man's dependence on the natural world.*

It is not surprising, therefore, that the idea of conservation has no home in economics. It is obviously an uneconomic idea, an impediment to the maximization of immediate profits. You may have noticed this in connection with the discovery of gas in the North Sea. In ministerial speeches and leading articles in *The Times* it was announced as an obvious and unquestionable truth that any failure to exploit this new power resource with the utmost speed and at the highest possible rate would be grossly uneconomic.

Now, there would be no need to enlarge on these points, if everyone were aware of the extreme narrowness of the base on which such judgments are built. Nor would there be any cause for criticism, for it is the acknowledged right of any specialist to specialize as narrowly as he wishes. The trouble is, however, that the words "economic" and "uneconomic" (as we have observed already) have acquired an infinitely wider meaning than they can legitimately claim: they are taken as almost synonymous with good and bad, or useful and useless.

It is a remarkable fact: the gloomy forebodings of the Oxford dons 150 years ago have come true—economics is indeed a science "so prone to usurp the rest." In spite of its palpable and obvious narrowness, it has been enthroned as universal judge. In spite of its specialization on private profit, verdicts are taken as equally applicable to the public interest. In spite of its concentration on the short term, which may be sufficient for the purposes of private persons, its doctrines are being applied to the affairs of nations whose life-spans are counted in centuries, if not millennia. Instead of using economics as a useful, if narrowly specialized, tool, modern society has embraced it as its primary religion, thereby laying itself open to dangers of an unprecedented kind.

It is obvious that the idea of conservation is more than ever in need of support, as the tempestuous advances of science and technology multiply the hazards. But as I said before, it is an uneconomic idea and has therefore no acknowledged place in a society under the dictatorship of economics. When it is occasionally introduced into the discussion, it tends to be treated not merely as a stranger but as an undesirable alien, probably dishonest and almost certainly immoral. In the past, when religion taught men to look upon Nature as God's handiwork, the idea of conservation was too self-evident to require special emphasis. But now that the religion of economics lends respectability to man's inborn envy and greed and Nature is looked upon as man's quarry to be used and abused without let or hindrance, what could be more important than an explicit theory of conservation? We teach our children that science and technology are the instruments for man's battle with nature, but forget to warn them that, being himself a part of nature, man could easily be on the losing side.

Modern economic thinking, as I have said, is peculiarly unable to consider the long term and to appreciate man's dependence on the natural world. It is therefore peculiarly defenceless against forces which produce a gradual and cumulative deterioration in the environment. Take the phenomenon of urbanization. It can be assumed that no one moves from the countryside into the city unless he expects to gain a more or less immediate personal advantage therefrom. His move, therefore, is economic, and any measure to inhibit the move would be uneconomic. In particular, to make it worthwhile for him to stay in agriculture by means of tariffs or subsidies, would be grossly uneconomic. That it is done nonetheless is attributed to the irrationality of political pressures. But what about the irrationality of cities with millions of inhabitants? What about the cost, frustration, congestion and ill health of the modern monster city? Yes, indeed, these are problems to be looked at, but (we are told) they do not invalidate the doctrine that subsidized farming is grossly uneconomic.

It is not surprising, therefore, that all around us the most appalling malpractices and malformations are growing up, the growth of which is not being inhibited, because to do so would be uneconomic. Something like an ex-

plosion has to occur before warning voices are listened to, the voices of people who had been ridiculed for years and years as nostalgic, reactionary, unpractical and starry-eyed. No one would apply these epithets today to those who for so many years had raised their voices against the heedless economism which has turned all large American cities into seedbeds of riots and civil war. Now that it is almost too late, popular comments are outspoken enough: "Throughout the U.S., the big cities are scarred by slums, hobbled by inadequate mass transportation, starved for sufficient finances, torn by racial strife, half-choked by polluted air." And yet: "The nation's urban population is expected to double by the beginning of the next century."[6] You might be tempted to ask, Why? The answer would come back: Because it would be uneconomic to attempt to resettle the rural areas. The American economist, John Kenneth Galbraith, has brilliantly shown how the conventional wisdom of economics produces the absurdity of "private opulence and public squalor."[7]

Other changes, equally destructive or even more so, are going on all around us, but they must not be talked about because to do so might cause alarm and even impede economic growth. All the same, we cannot claim that we have not been warned. For instance, in spite of enormous advances in medicine, on which we do not fail to congratulate ourselves, there is a relentless advance in the frequency of chronic illness. The U.S. Public Health Service states that "About 40.9 per cent of persons living in the United States were reported to have one or more chronic conditions. While some of these conditions were relatively minor, others were serious conditions such as heart disease, diabetes, or mental illness."[8] "We are exchanging health for mere survival," writes Lewis Herber in his comprehensive and invaluable book on *Our Synthetic Environ-*

ment. "We have begun to measure man's bio-logical achievements, not in terms of his ability to live a vigorous, physically untroubled life, but in terms of his ability to preserve his mere existence in an increasingly distorted environment. Today, survival often entails ill health and rapid physical degeneration."[9] Even the achievements in prolonging life are not impressive, except for the very young. In America, the life expectancy of a white male aged 45 years has increased by only 2.9 years since 1900, and that of a 65-year-old man by only 1.2 years. Considering the enormous economic advances in America since the beginning of this century, these results are astonishingly small. "Almost any improvement in social conditions or medical techniques," comments Lewis Herber, "would have rescued large numbers of people from premature death and added substantially to their life span ... Nevertheless, most of the increase in longevity is due to the fact that more children survive the diseases of infancy and adolescence today than two generations ago. What this means, in effect, is that if it weren't for the extraordinary medical advances and great improvements of the material conditions of life, today's adult might well have a much shorter life span than his grandparents had. This is a remarkable indication of failure."[10] At the same time, the expenditure on medical services in the United States now amounts to some 50,000 million dollars a year, or about *five dollars a week* for every man, woman, and child, on average.

It is not my purpose to investigate the causes of this extraordinary development. It is well known that the infectious diseases, which were the principal causes of death in 1900, have been reduced almost to the vanishing point, but that deaths from the so-called degenerative diseases have greatly increased, particularly deaths from cancer, heart disease, and diabetes, involving increasing numbers of children and young adults. "Many individuals seem to be succumbing to degenerative diseases long before they reach the prime of life. Not only is cancer a leading cause of death in childhood and youth, but. . . . many American males between 20 and 30 years of age are on the brink of a major cardiac disease. . . . If diseases of this kind represent the normal deterioration of the body, then human biology is taking a patently abnormal turn. A large number of people are breaking down prematurely."[11] Deaths from infectious diseases are now so low that further medical advances in this field cannot have a large impact; yet the growth of the degenerative diseases continues. The time may not be far off when death rates overall start rising in the most "advanced" countries. The real costs of a deteriorating environment are heavy indeed.

Developments of this kind are invariably the result of imbalance and disharmony. In the blind pursuit of immediate monetary gains modern man has not only divorced himself from nature by an excessive and hurtful degree of urbanization, he has also abandoned the idea of living in harmony with the myriad forms of plant and animal life on which his own survival depends; he has developed chemical substances which are unknown to nature and do not fit into her immensely complex system of checks and balances; many of them are extremely toxic, but he nonetheless applies them or discharges them into the environment, as if they would be out of action when they had fulfilled their specific purpose or could no longer be seen.

The religion of economics, at the same time, promotes an idolatry of rapid change, unaffected by the elementary truism that a change which is not an unquestionable improvement is a doubtful blessing. The burden of proof is placed on those who take the "ecological viewpoint:" unless *they* can produce evidence of marked injury to man, the change will proceed. Common sense, on the contrary, would suggest that the burden of proof should lie on the man who wants to introduce a change; *he* has to demonstrate that there *cannot* be any damaging consequences. But this would take too much time, and would therefore be uneconomic. Ecology, indeed, ought to be a compulsory subject for all economists, whether professionals or laymen, as this might serve to restore at least a modicum of balance. For ecology holds:

"that an environmental setting developed over millions of years must be considered to have some merit. Anything so complicated as a planet, inhabited by more than a million and a half species of plants and animals, all of them living together in a more or less balanced equilibrium in which they continuously use and re-use the same molecules of the soil and air, cannot be improved by aimless and uninformed tinkering. All changes in a complex mechanism involve some risk and should be undertaken only after careful study of all the facts available. Changes should be made on a small scale first so as to provide a test before they are widely applied. When information is incomplete, changes should stay close to the natural processes which have in their favour the indisputable evidence of having supported life for a very long time."[12]

Of all the changes introduced by man into the household of nature, large-scale nuclear fission is undoubtedly the most dangerous and profound. As a result, ionizing radiation has become the most serious agent of pollution of the environment and the greatest threat to man's survival on earth. The attention of the layman, not surprisingly, has been captured by the atom bomb, although there is at least a chance that it may never be used again. The danger to humanity created by the so-called peaceful uses of atomic energy is hardly ever mentioned. There could indeed be no clearer example of the prevailing dictatorship of economics. Whether to build conventional power stations, based on coal or oil, or nuclear sta-

tions, is being decided solely on economic grounds, with perhaps a small element of regard for the "social consequences" that might arise from an over-speedy curtailment of the coal industry. But that nuclear stations represent an incredible, incomparable, and unique hazard for human life does not enter any calculation and is never mentioned. People whose business it is to judge hazards, the insurance companies, are not prepared to insure nuclear power stations anywhere in the world for third party risk, with the result that special legislation has had to be passed whereby the State accepts all liabilities.[13] Yet, insured or not, the hazard remains, and such is the thraldom of the religion of economics that the only question that appears to interest either governments or the public is whether "it pays."

It is not as if there were any lack of authoritative voices to warn us. The effects of alpha, beta, and gamma rays on living tissues are perfectly well known: the radiation particles are like bullets tearing into an organism, and the damage they do depends primarily on the dosage and the type of cells they hit.[14] As long ago as 1927, the American biologist, H. J. Muller, published his famous paper on genetic mutations produced by x-ray bombardment,[15] and since the early 'thirties the genetic hazard of exposure to ionizing radiation has been recognized also by non-geneticists.[16] It is clear that here is a hazard with a hitherto unexperienced "dimension," endangering not only those who might be directly affected by this radiation but their offspring as well for all future generations.

A new "dimension" of hazard is given also by the fact that while man now can—and does—create radioactive elements, there is nothing he can do to reduce their radioactivity once he has created them. No chemical reaction, no physical interference, only the passage of time reduces the intensity of radiation once it has been set going. Carbon-14 has a half-life of 5,900 years, which means that it takes nearly six thousand years for its radioactivity to decline to one-half of what it was before. The half-life of strontium-90 is 28 years. But whatever the length of the half-life, some radiation continues almost indefinitely, and there is nothing that can be done about it, except to try and put the radioactive substance into a safe place.

But what is a safe place, let us say, for the enormous amounts of radioactive waste products created by nuclear reactors? No place on earth can be shown to be safe. It was thought at one time that these wastes could be safely dumped into the deepest parts of the oceans, on the assumption that no life could subsist at such depths.[17] But this has since been disproved by Soviet deep-sea exploration. Wherever there is life, radioactive substances are absorbed into the biological cycle. Within hours of depositing these materials in water, the great bulk of them can be found in living organisms. Plankton, algae, and many sea animals have the power of concentrating these substances by a factor of 1,000 and in some cases even a million. As one organism feeds on another, the radioactive materials climb up the ladder of life and find their way back to man.[18]

No international agreement has yet been reached on waste disposal. The conference of the International Atomic Energy Organization at Monaco, 16th to 21st November, 1959, ended in disagreement, mainly on account of the violent objections raised by the majority of countries against the American and British practice of disposal into the oceans.[19] "High level" wastes continue to be dumped into the sea, while large quantities of so-called "intermediate" and "low-level" wastes are discharged into rivers or directly into the ground. An A.E.C. report observes laconically that the liquid wastes "work their way slowly into

ground water, leaving all or part (sic!) of their radioactivity held either chemically or physically in the soil."[20]

The most massive wastes are, of course, the nuclear reactors themselves after they have become unserviceable. There is a lot of discussion on the trivial economic question of whether they will last for 20, 25 or 30 years. No one discusses the humanly vital point that they cannot be dismantled and cannot be shifted but have to be left standing where they are, probably for centuries, perhaps for thousands of years, an active menace to all life, silently leaking radioactivity into air, water and soil. No one has considered the number and location of these satanic mills which will relentlessly accumulate in these crowded islands, so that, after a generation or two, there will be no habitation in Britain outside the "sphere of influence" of one or more of them. Earthquakes, of course, are not supposed to happen, nor wars, nor civil disturbances, nor riots like those that infested American cities. Disused nuclear power stations will stand as unsightly monuments to unquiet man's assumption that nothing but tranquility, from now on, stretches before him, or else—that the future counts as nothing compared with the slightest economic gain now.

Meanwhile, a number of authorities are engaged in defining "maximum permissible concentrations" (MPC's) and "maximum permissible levels" (MPL's) for various radioactive elements. The MPC purports to define the quantity of given radioactive substance that the human body can be allowed to accumulate. But it is known that *any* accumulation produces biological damage. "Since we don't know that these effects can be completely recovered from," observes the U. S. Naval Radiological Laboratory, "we have to fall back on an arbitrary decision about how much we will put up with; i.e. what is 'acceptable' or 'permissible'—not a scientific finding, but an administrative decision."[21] We can hardly be surprised when men of outstanding intelligence and integrity, like Albert Schweitzer, refuse to accept such administrative decisions with equanimity: "Who has given them the right to do this? Who is even entitled to give such a permission?"[22] The history of these decisions is, to say the least, disquieting. The British Medical Research Council noted some 12 years ago that:

"The maximum permissible level of strontium-90 in the human skeleton, accepted by the International Commission on Radiological Protection, corresponds to 1,000 micro-micro-curies per gramme of calcium (=1,000 S.U.). But this is the maximum permissible level for adults in special occupations and is not suitable for application to the population as a whole or to the children with their greater sensitivity to radiation."[23]

A little later, the MPC for strontium-90, as far as the general population was concerned, was reduced by 90 per cent and then by another third, to 67 S.U. Meanwhile, the MPC for workers in nuclear plants was raised to 2,000 S.U.[24]

We must be careful, however, not to get lost in the jungle of controversy that has grown up in this field. The point is that very serious hazards have already been created by the "peaceful uses of atomic energy," affecting not merely the people alive today but all future generations, although so far nuclear energy is being used only on a statistically insignificant scale. The real development is yet to come, on a scale which few people are capable of imagining. If this is really going to happen, there will be a continuous traffic of radioactive substances from the "hot" chemical plants to the nuclear stations and back again; from the stations to waste processing plants; and from there to disposal sites. The slightest accident, whether during transport or production, can cause a major catastrophe; and the radiation levels throughout the world will rise relentlessly from generation to generation. Unless all living geneticists are in error, there will be an equally relentless, though no doubt somewhat delayed, increase in the number of harmful mutations. K. Z. Morgan, of the Oak Ridge Laboratory, emphasizes that the damage can be very subtle, a deterioration of all kinds of organic qualities, such as mobility, fertility, and the efficiency of sensory organs. "If a small dose has any effect at all at any stage in the life cycle of an organism, then chronic radiation at this level can be more damaging than a single massive dose ... Finally, stress and changes in mutation rates may be produced even when there is no immediately obvious effect on survival of irradiated individuals."[25]

Leading geneticists have given their warnings that everything possible should be done to avoid an increase in mutation rates;[26] leading medical men have insisted that the future of nuclear energy must depend primarily on researches into radiation biology which are as yet still totally incomplete;[27] leading physicists have suggested that "measures much less heroic than building. ... nuclear reactors" should be tried to solve the problem of future energy supplies—a problem which is in no way acute at present;[28] and leading students of strategic and political problems, at the same time, have warned us that there is really no hope of preventing the proliferation of the atom bomb, if there is a spread of plutonium capacity, such as was "spectacularly launched by President Eisenhower in his 'atoms for peace' proposals of 8th December, 1953."[29]

Yet all these weighty opinions play no part in the debate on whether we should go immediately for a large "second nuclear programme" or stick a bit longer to the conventional fuels which, whatever may be said for or against them, do not involve us in entirely novel and admittedly incalculable risks. None of them are even mentioned: the whole argument, which may vitally affect the very future of the human race, is conducted exclusively in terms of immediate economic advantage, as if two rag and bone merchants were trying to agree on a quantity discount.

I wonder how Dr. Des Voeux would have reacted to such an absurdly improbable situation. What, after all, is the fouling of air with smoke compared with the pollution of air, water, and soil, with ionizing radiation? Not that I wish in any way to belittle the evils of conventional air and water pollution; but we must recognize "dimensional differences" when we encounter them; radioactive pollution is an evil of incomparably greater "dimension" than anything mankind has known before. One might even ask; what is the point of insisting on clean air, if the air is laden with radioactive particles? And even if the air could be protected, what is the point of it, if soil and water are being poisoned?

Even an economist might well ask: what is the point of economic progress, a so-called higher standard of living, when the earth, the only earth we have, is being contaminated by substances which may cause malformations in our children or grandchildren? Have we learned nothing from the thalidomide tragedy? Can we deal with matters of such a basic character by means of bland assurances or official admonitions that "in the absence of proof that (this or that innovation) is in any way deleterious, it would be the height of irresponsibility to raise a public alarm?"[30] Can we deal with them simply on the basis of a short-term profitability calculation?

"It might be thought," wrote Leonard Beaton, "that all the resources of those who fear the spread of nuclear weapons would have been devoted to heading off these developments for as long as possible. The United States, the Soviet Union and Britain might be expected to have spent large sums of money trying to prove that conventional fuels, for example, had been underrated as a source of power. . . . In fact. . . . the efforts which have followed must stand as one of the most inexplicable political fantasies in history. Only a social psychologist could hope to explain why the possessors of the most terrible weapons in history have sought to spread the necessary industry to produce them . . . Fortunately, . . . power reactors are still fairly scarce."[31] In fact, a prominent American nuclear physicist, A. W. Weinberg, has given some sort of explanation: "There is," he says, "an understandable drive on the part of men of good will to build up the positive aspects of nuclear energy simply because the negative aspects are so distressing." But he also adds the warning that "there are very compelling personal reasons why atomic scientists sound optimistic when writing about their impact on world affairs. Each of us must justify to himself his preoccupation with instruments of nuclear destruction (and even we reactor people are only slightly less beset with such guilt than are our weaponeering colleagues)."[32]

Our instinct of self-preservation, one should have thought, would make us immune to the blandishments of guilt-ridden scientific optimism or the unproved promises of pecuniary advantages. "It is not too late at this point for us to reconsider old decisions and make new ones," says a recent American commentator. "For the moment at least, the choice is available."[33] Once many more centres of radioactivity have been created, there will be no more choice, whether we can cope with the hazards or not.

It is clear that certain scientific and technological advances of the last 30 years have produced, and are continuing to produce, hazards of an altogether intolerable kind. At the

Fourth National Cancer Conference in America in September, 1960, Lester Breslow of the California State Department of Public Health reported that tens of thousands of trout n western hatcheries suddenly acquired liver ncers, and continues thus:

echnological changes affecting man's environ ient are being introduced at such a rapid rate and with so little control that it is a wonder man has thus far escaped the type of cancer epidemic occurring this year among the trout."[34]

To mention these things, no doubt, means laying oneself open to the charge of being against science, technology, and progress. Let me therefore, in conclusion, add a few words about future scientific research. Man cannot live without science and technology any more than he can live against nature. What needs the most careful consideration, however, is the *direction* of scientific research. We cannot leave this to the scientists alone. As Einstein himself said,[35] "almost all scientists are economically completely dependent" and "the number of scientists who possess a sense of social responsibility is so small" that they cannot determine the direction of research. The latter dictum applies, no doubt, to all specialists, and the task therefore falls to the intelligent layman, to people like those who form the National Society for Clean Air and other, similar societies concerned with *Conservation*. They must work on public opinion, so that the politicians, depending on public opinion, will free themselves from the thraldom of economism and attend to the things that really matter. What matters, as I said, is the *direction* of research, that the direction should be towards non-violence rather than violence; towards a harmonious co-operation with nature rather than a warfare against nature; towards the noiseless, low-energy, elegant and economical solutions of our immature sciences.

The continuation of scientific advance in the direction of ever increasing violence, culminating in nuclear fission and moving on to nuclear fusion, is a prospect of terror threatening the abolition of man. Yet it is not written in the stars that this must be the direction. There is also a life-giving and life-enhancing possibility, the conscious exploration and cultivation of all relatively non-violent, harmonious, organic methods of co-operating with that enormous, wonderful, incomprehensible system of God-given nature, of which we are a part and which we certainly have not made ourselves.

NOTES

1. *Cf.* A. Dwight Culler, *The Imperial Intellect,* Yale University Press, 1955, p. 250.
2. Nassau Senior, *An Introductory Lecture on Political Economy,* delivered before the University of Oxford on the 6th of December, 1826, London, 1827, pp. 1 and 12.
3. John Stuart Mill, *Autobiography,* 1924 ed., pp. 165-6.
4. John Maynard Keynes, *Essays in Persuasion,* London, 1933, p. 373.
5. *Cf. The Financial and Economic Obligations of the Nationalised Industries,* Command 1337, H.M.S.O., London, 1961.
6. *Time,* Aug. 26, 1966, p. 11.
7. John Kenneth Galbraith, *The Affluent Society,* Penguin, 1962, p. 211.
8. U.S. Public Health Service, Health Statistics from the *U.S. National Health Survey,* Series C, No. 5, Washington, D.C., 1961, p. 1.
9. Lewis Herber, *Our Synthetic Environment,* London, 1963, pp. 198-9.
10. *Ibid.* pp. 197-98.
11. *Ibid.* pp. 8-9; *Medical Tribune,* Vol. 2, No. 29 (1961), p. 16.
12. Ralph and Mildred Buchsbaum, *Basic Ecology,* Pittsburgh 1957, p. 20 (Quoted by Herber, *op. cit.*).
13. *Cf.* C. T. Highton, "Die Haftüng für Strahlenschäden in Grossbritannien," *Die Atomwirtschaft, Zeitschrift für wirtschaftliche Fragen de Kernumwandlung,* 1959, p. 539.
14. *Cf.* Jack Schubert and Ralph Lapp: *Radiation: What it Is and How it Affects You,* New York,

1957. Also Hans Marquardt and Gerhard Schubert, *Die Strahlengefahrdung des Menschen durch Atomenergie,* Hamburg, 1959. And Volume XI of *Proceedings* of the International Conference on the Peaceful Uses of Atomic Energy, Geneva, 1955; and Volume XXII of *Proceedings* of the Second United Nations International Conference on the Peaceful Uses of Atomic Energy, Geneva, 1958.

15. *Cf.* H. J. Muller, "Changing Genes: Their Effects on Evolution," in *Bulletin of the Atomic Scientists,* Chicago, 1947.

16. *Cf.* Statement by G. Failla, Hearings before the Special Subcommittee on Radiation, of the Joint Committee on Atomic Energy, 86th Congress of the United States, May 5-8, 1959, *Fallout from Nuclear Weapons.* Washington, D.C. 1959, Vol. II, p. 1577.

17. R. Revelle and M. B. Schaefer, "Oceanic Research Needed for Safe Disposal of Radioactive Wastes at Sea"; and V. G. Bogorov and E. M. Kreps, "Concerning the Possibility of Disposing of Radioactive Waste in Ocean Trenches." Both in Vol. XVIII of *Proceedings,* Geneva Conference 1958 (see Note 14 above).

18. *Ibid.,* B. H. Ketchum and V. T. Bowen, "Biological Factors determining the Distribution of Radioisotopes in the Sea," pp. 429-33.

19. Conference Report by H. W. Levi, in *Die Atomwirtschaft,* 1960, pp. 57 *et. seq.*

20. U.S. Atomic Energy Commission, Annual Report to Congress, Washington D.C., 1960, p. 344.

21. U.S. Naval Radiological Defence Laboratory Statement, in *Selected Materials on Radiation Protection Criteria and Standards, their Basis and Use,* p. 464. (Quoted by Herber, *op. cit.*)

22. Albert Schweitzer, *Friede oder Atomkrieg,* 1958, p. 13.

23. British Medical Research Council, *The Hazards to Man of Nuclear and Allied Radiation,* p. 68. (Quoted by Herber, *op. cit.*)

24. Lewis Herber, *op. cit.,* p. 181.

25. K. Z. Morgan, "Summary and Evaluation Environmental Factors that must be considered in the Disposal of Radioactive Wastes," in *Industrial Radioactive Disposal,* Vol. 3, p. 2378 (Quoted by Herber, *op. cit.* p. 193).

26. *Cf.* H. Marquardt, "Natürliche und künstliche Erbänderungen," *Probleme der Mutationsforschung,* Hamburg, 1957, p. 161.

27. *Cf.* Schubert and Lapp, *op. cit.*

28. A. M. Weinberg, "Today's Revolution," in *Bulletin of the Atomic Scientists,* Chicago, 1956.

29. Leonard Beaton, *Must the Bomb Spread?* Penguin Books in association with the Institute of Strategic Studies, London, 1966, p. 88.

30. W. O. Caster, "From Bomb to Man," in *Fallout,* ed., by John M. Fowler, New York, 1960, pp. 48-9.

31. *Op. cit.,* pp. 88-90.

32. *Op. cit.,* pp. 299 & 302.

33 Walter Schneir, "The Atom's Poisonous Garbage," in *Reporter,* March 17, 1960, p. 22 (Quoted by Herber, *op. cit.,* p. 194.)

34. Lewis Herber, *op. cit.,* p. 152.

35. Albert Einstein, *On Peace,* ed. by O. Nathan and H. Norden, with a Preface by Bertrand Russell, New York, 1960, pp. 455-6.

World Power and Population
Colin Clark

There are many obvious things which we take for granted, of which, however, it is necessary to remind ourselves occasionally; not only in order to obtain a better understanding of our situation; but also because what is obviously true today has an awkward way of ceasing to be true tomorrow.

Let us begin with one of these obvious statements. The United States is a world power. What precisely do we mean by this? It means that statesmen all over the world, planning their countries' foreign policies, have to take into account what the United States may think or do; indeed further, that the United States is

in a position to influence, by methods ranging from tactful suggestions to open threats, events anywhere in the world. (It is of course by no means implied that it is either wise or expedient for the United States always so to intervene; or that past decisions to intervene, or not to intervene, have always been soundly judged.) In this sense, it is clear, the United States is a world power, and Mexico is not—important though Mexico's foreign policy decisions may be to her immediate neighbors.

It is another well-known fact that the United States now has a population of 202 million, and Mexico under 50 million. The facts of world power and of population are inescapably connected. Americans sometimes like to think that it is their advanced industrial technology, rather than the size of their population, which makes them a world power. Switzerland has an industrial technology comparable with that of the United States; Soviet Russia comparable with that of Mexico. Soviet Russia is a world power, and Switzerland is not. Some people have yielded to the superficial theory that, in these days of nuclear and other advanced weaponry, size of population has ceased to matter. Indeed, it probably matters more than before, as the historical table [Fig. 1] will show. The technical ability to produce advanced weapons is only one element in a country's ability to become a world power. Equally necessary are the very large economic resources needed to produce and maintain them. The size of a country's population also plays an important part in its ability to exert economic, political or cultural influence in the world; or indeed, to construct adequate defenses or retaliatory measures against nuclear warfare or, if the worse comes to the worst, to recover from its consequences.

Looking at the world situation in four succeeding centuries, it is indeed possible to specify the "ante" (if one may use a card

1670 (Minimum 1 million)		1770 (Minimum 6 million)	
Turkey	25	Russia	33
France	20	Turkey	30
Russia	16	France	24
Spain-Portugal	7	Austria	20
England	6	U.K.	14
Poland	5	Spain	9
Netherlands	1	Prussia	6
1870 (Minimum 27 million)		1970 (Minimum 100 million)	
Russia	78	China	670
Germany	41	India	550
Austria-Hungary	40	USSR	245
U.S.A.	39	U.S.A.	205
France	36	Indonesia	120
U.K.	31	Pakistan	110
Italy	27	Japan	100

Figure 1. Population required for world power status (millions)

player's word) required for entering the grim contest of world politics.

It is true that many idealistic Americans, up to 1917, thought that it was possible and desirable, while being militarily prepared, to stay outside the "world power game." But later generations have concluded that a powerful country may owe a real duty to the rest of the world to enter world politics on certain occasions; or indeed that those who think that they can stay out of the world power game are in some danger of becoming subjugated to those who actively play it.

The reader will see that this excludes countries which have largely or completely isolated themselves from dealings with other countries, such as seventeenth-century India, China or Japan; or are ruled by another country, as was eighteenth- and nineteenth-century India; or are in a state of internal political confusion, as were nineteenth-century China and Turkey.

It is true that seventeenth-century military equipment was vastly simpler than that of the present century. But it must be remembered also that the economic ability to produce it was relatively even more retarded. So we get

the paradoxical result that in the seventeenth century two nations of small population, economically advanced by the standards of the time, Sweden and the Netherlands, were able to create, respectively, the most powerful army and navy of their times, and to be leading world powers. (The Dutch indeed were then wealthy enough to be able to obtain as large an army as they wanted by hiring Germans at sixpence a day.) The Swedish Empire, which covered a large part of Northern Europe, was effectively brought to an end by Peter the Great of Russia in the early eighteenth century. The Dutch had to retreat from New York, and lost their position as a world power, when confronted by a temporary Anglo-French alliance in the time of Charles II.

Increasing the Ante

The table for 1970 is admittedly somewhat forward-looking. There may be statesmen now who, in planning their foreign policy, do not think that they need take into account what China, Indonesia or Japan might think about it. But they will be imprudent if they do not do so.

It cannot be denied, however, that the "ante" is increasing rapidly, at least fourfold per century. We really should be asking ourselves what the position will be in a hundred, or even in fifty, years' time.

Population changes, in spite of all the superficial talk about "population explosion," do not come suddenly. A demographic change, whether in an upward or downward direction, takes more than a century to work out all its consequences; and those who recommend or try to bring about such changes should bear in mind that they are carrying some responsibility for events at least a century into the future. The present size of the American population has been achieved, in the superficial sense of the word, by the present generation of Americans, and by those of the recent past. But the decisions which really made the difference were those of the nineteenth-century immigrants who had the courage to leave their homelands, and of the hardy Midwestern pioneers who brought up large families under austere circumstances.

From the point of view of becoming a world power it is of importance to build up a population, and to preserve it in a political union. If it had not been for the hundreds of thousands of Americans who fought and died to preserve the Union in 1861-65, or had they been unsuccessful, there can be no doubt that the political face of the world now would be immeasurably different from what it is—and, almost certainly, for the worse. It may be that in a generation's time (though there is hardly any sign of it at present) Latin Americans, Africans, even Southeast Asians may have succeeded in welding what are now weak and quarreling states into genuine political unions, which will become world powers. Within our grasp in the past two decades—though we have in fact failed to grasp it—has been the shining opportunity of creating a real political union of Western Europe, which could have played a vital part in helping to preserve a peaceful and civilized world. Responsibility for this failure will be placed by future historians, I think, upon inaction by British governments between 1945 and 1957, under Attlee, Churchill and Eden. These were the critical years during which, with British participation, a Western European political union could have been easily achieved; and it is permissible to speculate that, under these circumstances, France would have resisted the temptation to rock the boat, to which she succumbed under de Gaulle.

We cannot speculate any further on these political possibilities. But we can to some extent foresee demographic trends; and the conclusions are disquieting.

In the first place, the reader may have noticed that the population given for China in the table is considerably lower than is generally supposed. The United States Government's official advisers on this subject estimated the Chinese population at just over 600 million for 1953 (somewhat higher than the Chinese census of that date); but subsequently they have accepted unquestioningly the Chinese claim that population has been growing 2 per cent per year or more, bringing it to 850 million by 1970. It was the same group of advisers who persistently overestimated the population of Soviet Russia in the 1950s and they are almost certainly wrong again. It is true that information about Chinese population, fertility and mortality, is extremely scanty, scattered and indirect. But some scientific analysis is possible. During the troubled period of the 1940s, population had been declining. For 1953, I estimate it at 560 million, a little below that given by the census. The rate of growth never reached anything like that officially claimed of 2.2 per cent per year. During the comparatively undisturbed 1950s, I estimated it at 1.1 per cent per year. Agricultural chaos and near-famine prevailed after "The Year of the Great Leap Forward" in 1958, and the rate of population growth probably fell to .05 per cent per year or less. There was some recovery of agriculture during the mid-1960s, due in part to some restoration of small private plots. In 1966 further chaos supervened in the form of the Cultural Revolution. The estimate of 670 million for 1970 is indeed probably on the high side. If China's future has in store a prolonged spell of ordered government and agricultural improvement, we will then see a rapid increase in population, at 2 per cent per year or even more. But it has not happened yet.

Below Replacement Level

Meanwhile, if present rates of growth persist, India's population will overtake that of China in the early 1980s.

Of the other world powers mentioned in the table, we know that population is growing at over 2 per cent per year in Pakistan. Independent critics consider that the recent census of Indonesia was reasonably accurate; and there are clear signs of agricultural recovery there. Population growth in Indonesia can be estimated at over 2 per cent per annum.

In Japan, on the other hand, while the number of births still exceed the number of deaths (present deaths being on the average those of the considerably smaller generation born some sixty years ago), the number of births now taking place falls considerably short of those required to replace the present generation of parents. Unless this situation is radically and permanently altered, the Japanese population must ultimately age and decline.

But what of Soviet Russia and the United States? There has been a decline in the size of family in Soviet Russia, and throughout Eastern Europe (with the exception of Rumania). In Hungary and Bulgaria the number of births is below replacement level. Though material for complete analysis is lacking, the same is now almost certainly true of Soviet Russia also. Though the process is starting later than in Japan, the consequence is none the less inevitable.

'Net Reproduction Rate'

But what of the United States? It is generally known that a heavy fall in births began in 1961, and indeed the social and propagandist pressures for its continuance are being maintained with unabated force.

There are many people who reassure themselves with the shallow fallacy that, so long as the number of births exceeds the number of deaths, they need not seriously concern themselves about a check to population growth.

They fail to realize the elementary fact that the deaths now occurring (with a few exceptions) represent the passing of the generation born sixty to seventy years ago, when the United States had a very much smaller population. The critical question is the extent to which the number of children now being born suffices to replace, or enlarge the numbers of their parents of the present generation. The indications now are that within a year or so American births will fail to replace the parents, and that the United States also, within a measurable time, will enter the cycle of population decline.

To measure, from information contemporarily available, whether the number of children being born suffices to replace their parents is, in fact, a complex mathematical problem. (The problem is easy enough when one is dealing with the past generation, when the total offspring born to the generation can readily be counted; but it is very much harder when analyzing the fertility of the generation which is still in the process of reproduction.) America has many highly skilled demographers. But, without exception, they are fervently committed to the advocacy of population restriction and they all seem unable or unwilling to measure or to criticize the consequences of the reduction of births which has already taken place.

An attempt may be made to describe in non-mathematical language the various successive methods which have been devised for making, from contemporary information, a true estimate of fertility. The reader with some familiarity with this subject must, however, be warned that methods believed until recently to be reliable are now known to be unreliable, and have been replaced; demographic science has been advancing rapidly.

It is easy to see that the crude birthrate, or number of births per thousand of population, does not tell you very much. The first refine- ment, devised in the nineteenth century, was to measure the annual number of births per head of the female population aged between fifteen and 45, the generation actually concerned in reproduction. This method was further refined in the 1930s by the calculation of "net reproduction rates." This ingenious method took account of the number of contemporary births to mothers under twenty, to mothers aged twenty to 24, etc. and then added them up to obtain, as it were, a snapshot picture of the reproduction of a generation, but based on a single year's evidence. Allowance was made for the proportion of children (now very small) who were expected to die before reaching maturity, to give an estimate, called the "net reproduction rate," of the extent to which a generation was failing to maintain, maintaining or enlarging itself.

The Prospects

The defect in this method is that it implicitly assumes, as the discerning reader may have observed, that ages at marriage do not significantly change. One effect of the war years in the 1940s was that there was an acceleration of marriage in many countries, and the tendency to the reduction of the age of marriage has persisted, for other reasons. When the age of marriage is changing, the method of net reproduction rates completely breaks down.

The next step forward was taken by the American demographer, Whelpton, who introduced into his methods of calculation the sensible assumption that parents, having had a child, were less likely to have another in the immediate future. He divided all the women of reproductive age into "cohorts" i.e., the generation born in each succeeding year, and then analyzed their reproductive performance, taking all their past offspring into account, to measure the extent to which they were replacing themselves. Whelpton's method made it clear that it was the cohorts born between 1900 and 1910 who were the least reproductive —and this is not all to be blamed on the Great Depression, because most of them reached marriageable age during the 1920s. The cohorts born during the two subsequent decades showed considerably higher reproductivity.

America: Falling Behind

But Whelpton's method has two drawbacks. The first is that his curves require extrapolation, a difficult mathematical procedure at best, and always likely to lead to error. The second and more serious objection is that it conceals an underlying assumption that families know in advance how many children they are going to have and how they are going to be spaced, and that therefore the observation of their record over a limited number of years makes it possible to predict their final totals. While this is true for most times, it breaks down in a period of violent change of reproductivity, such as has occurred since 1961.

About the same time, a quite different method was pioneered by Karmal in Australia, namely the method of "marriage-duration specific fertilities." This takes account of another obvious fact, that most children are born in the early years of marriage. If the duration of the parents' marriage is registered at the time of the child's birth (not all countries do this), and the number of marriages is also recorded, it is possible to sum the specific fertilities for each year of duration of marriage. These can then be extrapolated, without serious risk of error, to predict the expected total offspring of the marriages of recent years. Unfortunately the American system of registration does not lend itself to this method. Fur-

thermore, American statistics of marriages are incomplete; and in any case made much more difficult to handle by the high proportion of divorces and re-marriages, as compared with other countries.

The best method now available for analyzing contemporary information in order to estimate the final reproductivity of a generation was devised in 1953 by Henry, of the Institut National des Etudes Démographiques in Paris, the method of "probability of family enlargement." Unlike Whelpton's cohort method, this method does not make the assumption that parents plan their offspring and their spacing in advance, but rather the opposite, namely that they change their minds drastically from year to year. It is a method of considerable mathematical elegance. In nonmathematical language it can best be described as a method of analyzing succeeding dropouts. Out of each generation, in the first place, a certain small proportion will die before reaching maturity and these can be counted. At the next stage we need not, if we do not wish to, or if we do not have the information, calculate those who drop out of the race without marrying. We make our next count for those who fail to have even one child—less than 10 per cent. Of those who have had one child, however, some 20-30 per cent or more drop out without having a second child. These in their turn show a 35-40 per cent drop-out without having a third child. After the third child the drop-out proportion remains curiously constant for subsequent children; but it must be remembered that we now are dealing with a greatly reduced number of parents, and the contribution of these few large families to the total of reproduction is not very great.

The application of Henry's method required some detailed tables showing the distribution over past years of first births which might be expected to produce a second birth this year, of second births likely to produce a third birth, etc. The official tabulations of births by order unfortunately are always considerably delayed, and it is not therefore possible to make complete calculations fully up to date. Use is therefore made of the recognized demographic device of taking "standard rates" of the numbers and time distribution of first births expected to the rising generation, of second births expected to those who have already had a first, etc., and then of comparing actual births with those predicted from the "standard rates." To bring it up to date, this method requires a limited extrapolation, unlikely to lead to serious error, of the numbers of recent births of given order. (See Fig. 2.)

Unless there is a very drastic reversal of trend, the number of American births will very shortly fall below replacement rate. And this has happened in less than a decade from the time when they were providing for a one-third increase in the size of the generation.

The advent of oral contraceptives may have played a part in this extraordinary movement. But other forces must have been at work. In the first place, oral contraceptives were not abundant in 1961. Secondly, a marked downward movement began at almost exactly the same time in many other countries, including Soviet Russia where, so far as is known, oral contraceptives are not available.

1959	32
1960	33
1961	32
1962	27
1963	25
1964	22
1965	14
1966	10
1967	8
1968	2

Figure 2. Percentage by which current U.S. number of births exceeds number required to replace parental generation.

In France and England, however, the downward movement started later, in 1965, and has been more moderate, and births in those countries are still 15-20 per cent above replacement rate.

We conclude by looking back at our world power table. It is of course always possible that powerful political confederations will be formed in time in Europe, Latin America, Africa and Southeast Asia, to create a new balance of power in the world. But at present, it does not look probable, on the face of it. As far as we can see it now, the prospects for fifty years hence are of a world in which both the United States and Soviet Russia have fallen out of the race, a world dominated by the Asian countries, with India and China in the lead, and Pakistan and Indonesia as the runners-up.

SECOND EDITION / The Politics of Population
Aldous Huxley

In politics, the central and fundamental problem is the problem of power. Who is to exercise power? And by what means, by what authority, with what purpose in view, and under what controls? Yes, under what controls?

For, as history has made it abundantly clear, to possess power is *ipso facto* to be tempted to abuse it. In mere self-preservation we must create and maintain institutions that make it difficult for the powerful to be led into those temptations which, succumbed to, transform them into tyrants at home and imperialists abroad.

For this purpose what kinds of institutions are effective? And, having created them, how then can we guarantee them against obsolescence? Circumstances change, and, as they change, the old, the once so admirably effective devices for controlling power cease to be adequate. What then? Specifically, when advancing science and acceleratingly progressive technology alter man's long-established relationship with the planet on which he lives, revolutionize his societies, and at the same time equip his rulers with new and immensely more powerful instruments of domination, what ought we to do? What *can* we do?

Very briefly let us review the situation in the light of present facts and hazard a few guesses about the future.

On the biological level, advancing science and technology have set going a revolutionary process that seems to be destined for the next century at least, perhaps for much longer, to exercise a decisive influence upon the destinies of all human societies and their individual members. In the course of the last fifty years extremely effective methods for lowering the prevailing rates of infant and adult mortality were developed by Western scientists. These methods were very simple and could be applied with the expenditure of very little money by very small numbers of not very highly trained technicians. For these reasons, and because everyone regards life as intrinsically good and death as intrinsically bad, they were in fact applied on a worldwide scale. The results were spectacular. In the past, high birthrates were balanced by high death rates. Thanks to science, death rates have been halved but, except in the most highly industrialized, contraceptive-using countries, birthrates remain as high as ever. An enormous and

accelerating increase in human numbers has been the inevitable consequence.

At the beginning of the Christian era, so demographers assure us, our planet supported a human population of about two hundred and fifty millions. When the Pilgrim fathers stepped ashore, the figure had risen to about five hundred millions. We see, then, that in the relatively recent past it took sixteen hundred years for the human species to double its numbers. Today world population stands at three thousand millions. By the year 2000, unless something appallingly bad or miraculously good should happen in the interval, six thousand millions of us will be sitting down to breakfast every morning. In a word, twelve times as many people are destined to double their numbers in one-fortieth of the time.

This is not the whole story. In many areas of the world human numbers are increasing at a rate much higher than the average for the whole species. In India, for example, the rate of increase is now 2.3 per cent per annum. By 1990 its four hundred and fifty million inhabitants will have become nine hundred million inhabitants. A comparable rate of increase will raise the population of China to the billion mark by 1980. In Ceylon, in Egypt, in many of the countries of South and Central America, human numbers are increasing at an annual rate of three per cent. The result will be a doubling of their present populations in about twenty-three years.

On the social, political, and economic levels, what is likely to happen in an underdeveloped country whose people double themselves in a single generation, or even less? An underdeveloped society is a society without adequate capital resources (for capital is what is left over after primary needs have been satisfied, and in underdeveloped countries most people never satisfy their primary needs); a society without a sufficient force of trained

teachers, administrators, and technicians; a society with few or no industries and few or no developed sources of industrial power; a society, finally, with enormous arrears to be made good in food production, education, road building, housing, and sanitation. A quarter of a century from now, when there will be twice as many of them as there are today, what is the likelihood that the members of such a society will be better fed, housed, clothed, and schooled than at present? And what are the chances in such a society for the maintenance, if they already exist, or the creation, if they do not exist, of democratic institutions?

Mr. Eugene Black, the former president of the World Bank, once expressed the opinion that it would be extremely difficult, perhaps even impossible, for an underdeveloped country with a very rapid rate of population increase to achieve full industrialization. All its resources, he pointed out, would be absorbed year by year in the task of supplying, or not quite supplying, the primary needs of its new members. Merely to stand still, to maintain its current subhumanly inadequate standard of living, will require hard work and the expenditure of all the nation's available capital. Available capital may be increased by loans and gifts from abroad, but in a world where the industrialized nations are involved in power politics and an increasingly expensive armament race there will never be enough foreign aid to make much difference. And even if the loans and gifts to underdeveloped countries were to be substantially increased, any resulting gains would be largely nullified by the uncontrolled population explosion.

The situation of these nations with such rapidly increasing populations reminds one of Lewis Carroll's parable in *Through the Looking Glass,* where Alice and the Red Queen start running at full speed and run for a long time until Alice is completely out of breath. When they stop, Alice is amazed to see that they are still at their starting point. In the looking glass world, if you wish to retain your present position, you must run as fast as you can. If you wish to get ahead, you must run at least twice as fast as you can.

If Mr. Black is correct (and there are plenty of economists and demographers who share his opinion), the outlook for most of the world's newly independent and economically non-viable nations is gloomy indeed. To those that have shall be given. Within the next ten or twenty years, if war can be avoided, poverty will almost have disappeared from the highly industrialized and contraceptive-using societies of the West. Meanwhile, in the underdeveloped and uncontrolled breeding societies of Asia, Africa, and Latin America, the condition of the masses (twice as numerous, a generation from now, as they are today) will have become no better and may even be decidedly worse than it is at present. Such a decline is foreshadowed by current statistics of the Food and Agriculture Organization of the United Nations. In some underdeveloped regions of the world, we are told, people are somewhat less adequately fed, clothed, and housed than were their parents and grandparents thirty and forty years ago. And what of elementary education? UNESCO provided an answer. Since the end of World War II heroic efforts have been made to teach the whole world how to read. The population explosion has largely stultified these efforts. The absolute number of illiterates is greater now than at any time.

The contraceptive revolution which, thanks to advancing science and technology, has made it possible for the highly developed societies of the West to offset the consequences of death control by a planned control of births, has had as yet no effect upon the family life of people in underdeveloped countries. This is not surprising. Death control, as I have already remarked, is easy, cheap, and can be carried out by a small force of technicians. Birth control, on the other hand, is rather expensive, involves the whole adult population, and demands of those who practice it a good deal of forethought and directed will power. To persuade hundreds of millions of men and women to abandon their tradition-hallowed views of sexual morality, then to distribute and teach them to make use of contraceptive devices or fertility-controlling drugs—this is a huge and difficult task, so huge and so difficult that it seems very unlikely that it can be successfully carried out, within a sufficiently short space of time, in any of the countries where control of the birthrate is most urgently needed.

Extreme Poverty Now Breeds the Expectation that Desires Will Soon be Satisfied

Extreme poverty, when combined with ignorance, breeds that lack of desire for better things which has been called "wantlessness"— the resigned acceptance of a subhuman lot. But extreme poverty, when it is combined with the knowledge that some societies are affluent, breeds envious desires and the expectation that these desires must of necessity, and very soon, be satisfied. By means of the mass media (those easily exportable products of advancing science and technology) some knowledge of what life is like in affluent societies has been widely disseminated throughout the world's underdeveloped regions. But, alas, the science and technology which have given the industrial West its cars, refrigerators, and contraceptives have given the people of Asia, Africa, and Latin America only movies and radio broadcasts, which they are too simple-minded to be able to criticize, together with a population explosion, which they are still too

poor and too tradition-bound to be able to control through deliberate family planning.

In the context of a three, or even of a mere two per cent annual increase in numbers, high expectations are foredoomed to disappointment. From disappointment, through resentful frustration, to widespread social unrest, the road is short. Shorter still is the road from social unrest, through chaos, to dictatorship, possibly of the Communist Party, more probably of generals and colonels. It would seem, then, that for two-thirds of the human race now suffering from the consequences of uncontrolled breeding in a context of industrial backwardness, poverty, and illiteracy, the prospects for democracy, during the next ten or twenty years, are very poor.

From underdeveloped societies and the probable political consequences of their explosive increase in numbers, we now pass to the prospect for democracy in the fully industrialized, contraceptive-using societies of Europe and North America.

It used to be assumed that political freedom was a necessary precondition of scientific research. Ideological dogmatism and dictatorial institutions were supposed to be incompatible with the open-mindedness and the freedom of experimental action, in the absence of which discovery and invention are impossible. Recent history has proved these comforting assumptions to be completely unfounded. It was under Stalin that Russian scientists developed the A-bomb and, a few years later, the H-bomb. And it is under a more-than-Stalinist dictatorship that Chinese scientists are now in the process of performing the same feat.

Another disquieting lesson of recent history is that, in a developing society, science and technology can be used exclusively for the enhancement of military power, not at all for the benefit of the masses. Russia has demonstrated, and China is now doing its best to demonstrate, that poverty and primitive conditions of life for the overwhelming majority of the population are perfectly compatible with the wholesale production of the most advanced and sophisticated military hardware. Indeed, it is by deliberately imposing poverty on the masses that the rulers of developing industrial nations are able to create the capital necessary for building an armament industry and maintaining a well-equipped army with which to play their parts in the suicidal game of international power politics.

We see, then, that democratic institutions and libertarian traditions are not at all necessary to the progress of science and technology, and that such progress does not of itself make for human betterment at home and peace abroad. Only where democratic institutions already exist, only where the masses can vote their rulers out of office and so compel them to pay attention to the popular will, are science and technology used for the benefit of the majority as well as for increasing the power of the state. Most human beings prefer peace to war, and practically all of them would rather be alive than dead. But in every part of the world men and women have been brought up to regard nationalism as axiomatic and war between nations as something cosmically ordained by the Nature of Things. Prisoners of their culture, the masses, even when they are free to vote, are inhibited by the fundamental postulates of the frame of reference within which they do their thinking and their feeling from decreeing an end to the collective paranoia that governs international relations. As for the world's ruling minorities, by the very fact of their power they are chained even more closely to the current system of ideas and the prevailing political customs; for this reason they are even less capable than their subjects of expressing the simple human preference for life and peace.

Some day, let us hope, rulers and ruled will break out of the cultural prison in which they are now confined. Some day. And may that day come soon! For, thanks to our rapidly advancing science and technology, we have very little time at our disposal. The river of change flows ever faster, and somewhere downstream, perhaps only a few years ahead, we shall come to the rapids, shall hear, louder and ever louder, the roaring of a cataract.

Modern war is a product of advancing science and technology. Conversely, advancing science and technology are products of modern war. It was in order to wage war more effectively that first the United States, then Britain and the U.S.S.R., financed the crash programs that resulted so quickly in the harnessing of atomic forces. Again, it was primarily for military purposes that the techniques of automation, which are now in the process of revolutionizing industrial production and the whole system of administrative and bureaucratic control, were initially developed. "During World War II," writes Mr. John Diebold, "the theory and use of feedback was studied in great detail by a number of scientists both in this country and in Britain. The introduction of rapidly moving aircraft very quickly made traditional gunlaying techniques of anti-aircraft warfare obsolete. As a result, a large part of scientific manpower in this country was directed toward the development of self-regulating devices and systems to control our military equipment. It is out of this work that the technology of automation as we understand it today has developed."

The headlong rapidity with which scientific and technological changes, with all their disturbing consequences in the fields of politics and social relations, are taking place is due in large measure to the fact that, both in the U.S.A. and the U.S.S.R., research in pure and applied science is lavishly financed by military planners whose first concern is in the development of bigger and better weapons in the shortest possible time. In the frantic effort, on one side of the Iron Curtain, to keep up with the Joneses—on the other, to keep up with the Ivanovs—these military planners spend gigantic sums on research and development. The military revolution advances under forced draft, and as it goes forward it initiates an uninterrupted succession of industrial, social, and political revolutions. It is against this background of chronic upheaval that the members of a species, biologically and historically adapted to a slowly changing environment, must now live out their bewildered lives.

Old-fashioned war was incompatible, while it was being waged, with democracy. Nuclear war, if it is ever waged, will prove in all likelihood to be incompatible with civilization, perhaps with human survival. Meanwhile, what of all the preparations for nuclear war? If certain physicists and military planners had their way, democracy, where it exists, would be replaced by a system of regimentation centered upon the bomb shelter. The entire population would have to be systematically drilled in the ticklish operation of going underground at a moment's notice, systematically exercised in the art of living troglodytically under conditions resembling those in the hold of an eighteenth century slave ship. The notion fills most of us with horror. But if we fail to break out of the ideological prison of our nationalistic and militaristic culture, we may find ourselves compelled by the military consequences of our science and technology to descend into the steel and concrete dungeons of total and totalitarian civil defense.

In the past, one of the most effective guarantees of liberty was governmental inefficiency. The spirit of tyranny was always willing, but its technical and organizational flesh was weak. Today the flesh is as strong as the spirit. Governmental organization is a fine art, based upon scientific principles and disposing of marvelously efficient equipment. Fifty years

ago an armed revolution still had some chance of success. In the context of modern weaponry a popular uprising is foredoomed. Crowds armed with rifles and homemade grenades are no match for tanks. And it is not only to its armament that a modern government owes its overwhelming power. It also possesses the strength of superior knowledge derived from its communication systems, its stores of accumulated data, its batteries of computers, its network of inspection and administration.

Where democratic institutions exist and the masses can vote their rulers out of office, the enormous powers with which science, technology, and the arts of organization have endowed the ruling minority are used with discretion and a decent regard for civil and political liberty. Where the masses can exercise no control over their rulers, these powers are used without compunction to enforce ideological orthodoxy and to strengthen the dictatorial state. The nature of science and technology is such that it is peculiarly easy for a dictatorial government to use them for its own anti-democratic purposes. Well-financed, equipped, and organized, an astonishingly small number of scientists and technologists can achieve prodigious results. The crash program that produced the A-bomb and ushered in a new historical era was planned and directed by some four thousand theoreticians, experimenters and engineers. To parody the words of Winston Churchill, never have so many been so completely at the mercy of so few.

Throughout the nineteenth century the state was relatively feeble, and its interest in, and influence upon, scientific research were negligible. In our day the state is everywhere exceedingly powerful and a lavish patron of basic and ad-hoc research. In western Europe and North America the relations between the state and its scientists on the one hand and individual citizens, professional organizations, and industrial, commercial, and educational institutions on the other are fairly satisfactory. Advancing science, the population explosion, the armament race, and the steady increase and centralization of political and economic power are still compatible, in countries that have a libertarian tradition, with democratic forms of government. To maintain this compatibility in a rapidly changing world, bearing less and less resemblance to the world in which these democratic institutions were developed —this, quite obviously, is going to be increasingly difficult.

A rapid and accelerating population increase that will nullify the best efforts of underdeveloped societies to better their lot and will keep two-thirds of the human race in a condition of misery in anarchy or of misery under dictatorship, and the intensive preparations for a new kind of war that, if it breaks out, may bring irretrievable ruin to the one-third of the human race now living prosperously in highly industrialized societies—these are the two main threats to democracy now confronting us. Can these threats be eliminated? Or, if not eliminated, at least reduced?

My own view is that only by shifting our collective attention from the merely political to the basic biological aspects of the human situation can we hope to mitigate and shorten the time of troubles into which, it would seem, we are now moving. We cannot do without politics; but we can no longer afford to indulge in bad, unrealistic politics. To work for the survival of the species as a whole and for the actualization in the greatest possible number of individual men and women of their potentialities for good will, intelligence, and creativity—this, in the world of today, is good, realistic politics. To cultivate the religion of idolatrous nationalism, to subordinate the interests of a single national state and its ruling minority—in the context of the population explosion, missiles, and atomic warheads, this is bad and thoroughly unrealistic politics. Unfortunately, it is to bad and unrealistic politics that our rulers are now committed.

Ecology is the science of the mutual relations of organisms with their environment and with one another. Only when we get it into our collective head that the basic problem confronting twentieth-century man is an ecological problem will our politics improve and become realistic. How does the human race propose to survive and, if possible, improve the lot and the intrinsic quality of its individual members? Do we propose to live on this planet in symbiotic harmony with our environment? Or, preferring to be wantonly stupid, shall we choose to live like murderous and suicidal parasites that kill their host and so destroy themselves?

Committing that sin of overweening bumptiousness which the Greeks called *hubris,* we behave as though we were not members of earth's ecological community, as though we were privileged and, in some sort, supernatural beings and could throw our weight around like gods. But in fact we are, among other things, animals—emergent parts of the natural order. If our politicians were realists, they would think rather less about missiles and the problem of landing astronauts on the moon, rather more about hunger and moral squalor and the problem of enabling three billion men, women, and children, who will soon be six billions, to lead a tolerably human existence without, in the process, ruining and befouling their planetary environment.

Animals have no souls; therefore, according to the most authoritative Christian theologians, they may be treated as though they were things. The truth, as we are now beginning to realize, is that even things ought not to be treated as *mere* things. They should be treated as though they were parts of a vast living organism. "Do as you would be done by." The Golden Rule applies to our dealings with nature no less than to our dealings with our fellowmen. If we hope to be well treated by nature, we must stop talking about "mere things" and start treating our planet with intelligence and consideration.

Power politics in the context of nationalism raises problems that, except by war, are practically insoluble. The problems of ecology, on the other hand, admit of a rational solution and can be tackled without the arousal of those violent passions always associated with dogmatic ideology and nationalistic idolatry. There may be arguments about the best way of raising wheat in a cold climate or of reforesting a denuded mountain. But such arguments never lead to organized slaughter. Organized slaughter is the result of arguments about such questions as: Which is the best nation? The best religion? The best political theory? The best form of government? Why are other people so stupid and wicked? Why can't they see how good and intelligent *we* are? Why do they resist our beneficent efforts to bring them under our control and make them like ourselves?

To questions of this kind the final answer has always been war. "War," said Clausewitz, "is not merely a political act but also a political instrument, a continuation of political relationships, a carrying out of the same by other means." This was true enough in the eighteen-thirties, when Clausewitz published his famous treatise, and it continued to be true until 1945. Now, obviously, nuclear weapons, long-range rockets, nerve gases, bacterial aerosols, and the laser (that highly promising addition to the world's military arsenals) have given the lie to Clausewitz. All-out war with modern weapons is no longer a continuation of previous policy; it is a complete and irreversible break with previous policy.

Power politics, nationalism, and dogmatic ideology are luxuries that the human race can no longer afford. Nor, as a species, can we afford the luxury of ignoring man's ecological situation. By shifting our attention from the now completely irrelevant and anachronistic politics of nationalism and military power to the problems of the human species and the still inchoate politics of human ecology we shall be killing two birds with one stone—reducing the threat of sudden destruction by scientific war and at the same time reducing the threat of more gradual biological disaster.

The beginnings of ecological politics are to be found in the special services of the United Nations Organization. UNESCO, the Food and Agriculture Organization, the World Health Organization, the various technical-aid services—all these are, partially or completely, concerned with the ecological problems of the human species. In a world where political problems are thought of and worked upon within a frame of reference whose coördinates are nationalism and military power, these ecology-oriented organizations are regarded as peripheral. If the problems of humanity could be thought about and acted upon within a frame of reference that has survival for the species, the well-being of individuals, and the actualization of man's desirable potentialities as its coördinates, these peripheral organizations would become central. The subordinate politics of survival, happiness, and personal fulfillment would take the place now occupied by the politics of power, ideology, nationalistic idolatry, and unrelieved misery.

In the process of reaching this kind of politics we shall find, no doubt, that we have done something, in President Wilson's prematurely optimistic words, "to make the world safe for democracy."

Who Is Really Uprooting This Country?
Josephine W. Johnson

The wild cherry leaves are waxy green and bronze in this spring morning. The rain is falling over the buckeye pyramids, over the lime-green flowers and the dark-green leaves. The woods are flooded with flowers. The fern-cut leaves of dutchman-breeches and squirrel corn, the white hearts of their flowers, shake in the rain.

When the sun comes out it is as though a veil of silver ice sheathed every living thing. A green-white ice where the slow rain stayed on every leaf and shoot it touched. And as far as the eye can see white glittering spiderwebs are flung on the hillside, and long strands, pearl-hung, wave from the beech twigs, invisibly anchored somewhere far out in space.

These are only a few hours out of a spring woods that will unfold, flower, change with every day. The buds of buckeye have that waxy rose of petalled sap, and then a burst of green leaves and the rose petals falling. The great oaks and maples and beeches will begin to leaf, begin their giant breathing, the soundless respiration and purifying to which our own lives are tied.

This is indeed a beautiful country, one thinks. A land to be loyal to. One's soul expands. But down beyond this valley, beyond the circle of hills that ring this wild greenness, plans are being made to destroy it. Anarchic, sweeping, enormous and expensive plans. Who is making these plans?

A half mile away in the early spring morning, the great orange school buses begin to roll, gathering up the children, weighted down with books, carrying them to their new and beautiful buildings on old pasture lands. Among the thousands of things these children will hear are words about a *Man Without a Country:* "Breathes there a man with soul so dead, who never to himself has said, this is my own, my native land?"

From these words should flow love, patriotism, and law into the veins of the growing child. But busily behind the edifice, under cover of the mighty music of the organ sound, a vast throng of people are working night and day, destroying all they still call their native land.

Who are these people? Who are the destroyers? Breathes there a man. . . . Try and breathe. Who pollutes the air? Who fouls the rivers? Who cuts down the trees, builds houses on the stripped hillsides? Who poisons the sheep, shoots the deer, oils the beaches, dams and rivers, dries up the swamps, concretes the countryside? Who bulldozes homes, builds missile sites, pours poison wastes underground, poison gas overground, slabs over mountain tops, rocks the earth with explosions, scars the earth with strip mines?

Who is doing this? Who is responsible for this anarchy and ruin? Is it the revolutionaries, the black militants, the draft refusers? Is it the college students, the pacifists, the hippies? Who is taking our country away from us before our eyes?

It is the well-dressed, law-abiding, patriotic and upright citizens who are taking our country away from us. In the name of saving us, protecting us and civilizing us, statesmen and generals, scientists and engineers, businessmen and Congressmen, are making us into a people without a country, dead souls and exiles. And we are paying to do it.

In the name of saving and protecting us, the

Pentagon has become the symbol of the greatest power on earth today. There it sits, a terrible mass of concrete. Its power penetrates into every single life. It is in the air we breathe, the water we drink. Because of its insatiable demands we are drained and polluted. Nothing in the world is like this concrete monster. It is like the great god Moloch into which the children were thrown as sacrifice. It is our greatest unnatural disaster.

We are dying of preconceptions, outworn rules, decaying flags, venomous religions, and sentimentalities. We need a new world. We've wrenched up all the old roots. The old men have no roots. They don't know it. They just go on talking and flailing away and falling down on the young with their tons of dead weight and their power. For the power is still there in their life-in-death. But the roots are dead, and the land is poisoned for miles around.

After the first silent spring will come a short and suffocating summer, then asphalt autumn, and in the end, winter. Cold, clean, orderly, concrete winter. Winter forever. And then we will have nothing to fear anymore, nothing to be protected from, nothing to be protected for, nothing at all, in fact, that we or anyone else will want to call our own.

UPI

UPI

Prospects for the Future: Population

At 55 persons per square mile, population density in the United States is approximately equivalent to the mean density of population on the earth as a whole. That density is 160 times as great as was the density of the aboriginal United States population at the time of colonization less than 400 years ago. It is 14,000 times as great as the density that prevailed in stable Eskimo populations in the North American arctic.

Projections call for doubling of the mean population density of the earth as a whole by, or shortly after, the year 2000. In the increasingly prevalent urban concentrations of today, human population densities reach 200,000 per square mile. The proportion of individuals living in such concentrations is expected to increase considerably faster than the growth of the population.

There is ample evidence that despite technological palliatives, unrelieved exposure to present urban densities is at variance with the biological nature of human beings. Social disruption, and decreasing quality and expectancy of life, originating in the tensions of urban existence, are synergistic with associated rapid and massive increases in pollution-related mortality and morbidity. On an ever greater scale these internal stresses are associated with chronic and growing resource shortages and local and planet-wide deterioration of the ecosystem. All recent and current experience supports the conclusion that these problems will increase henceforth at a proportionately faster rate than will population density.

The complex of phenomena confirms a fundamental point which must be accepted. After 10,000 years of accelerating growth, the human population of Earth has reached (and exceeded) the number that can be supported if the biosphere is to continue. No further expansion of the ecological niche of man is possible, even to accommodate the present excess.

The situation is wholly revolutionary in both human and evolutionary history. No single species, to our knowledge, has ever threatened the continued existence of the biosphere. And in previous human history there has always been the possibility that the human ecological niche might be further expanded by a new innovation in technology. In addition, there has always been some room for geographical expansion. These possibilities no longer exist.

Mankind has reached this crucial turning point so swiftly that reproductive capacity, mores, social systems, economics, and systems of government remain adapted only to the continuing and accelerating growth of the past. Inflexibility is built in. Reproductive capacities and instincts change only in evolutionary time. The minds of those in a position to influence events are frozen in the growth tradition. The basic assumptions which present-day leaders acquired twenty or thirty years ago are no longer relevant!

The problems of the cities, of war and political unrest, of pollution and of resource shortage have common roots in overpopulation. Efforts to reduce the emission and accumulation of pollutants, to recycle resources, to increase food production, and to minimize the stressfulness and psychic deprivation of urban environments are treatments of symptoms only. They are, simultaneously, vitally necessary and dangerous. Technological bandages on the lesions of population growth, whatever temporary ease they may give us, are dangerous in so far as they encourage us to evade treatment of the underlying disease. We must use them as a means of surviving crises, of relieving pressure, and of gaining time, but we cannot allow them to become addictive, or use them to drug ourselves into complacency.

The essential first step to survival of both man and the biosphere is identification of the carrying capacity of our ultimate ecological niche. This step is not being taken. To my knowledge, there has been no truly comprehensive effort to agree upon an acceptable quality of human existence, and to determine the level of population at which that quality can be indefinitely maintained. Nor does such a step seem to be contemplated. The guesstimate made by Hulett in the first of the following selections is probably as good as any presently available (carrying capacity of one billion at U. S. standards).

Once a carrying capacity has been identified and plans made to adjust it in the light of future knowledge and conditions, the question of means arises. What is the *optimum* way, if there is one, to bring current population growth to a halt, and then control reproduction in such a way that population size will drop to the desired level? Is there, in fact, any *practical* way? How much time do we have in which to make such a fantastic transition? The time will be determined by how soon we begin, by our success in keeping pollution below lethal levels, and our willingness and ability to stretch supplies of non-renewable resources while avoiding the destruction of those which are renewable. The time available will also be dependent on our ability to devise social, economic, and political strategems to withstand the severe current and future tensions without major war or social collapse. Yet, at this time, the only step with which governments are seriously concerned is the evaluation of the resource base (to be reviewed in the next chapter).

Given the absence of the radical steps which are necessary, the prospects for future population trends are approximately as outlined by the United Nations projections: increments of between 53 and 97 percent in world population, between 68 and 120 percent in population in the underdeveloped countries, and between 20 and 46 percent in the developed countries within the next thirty years. Accepting Hulett's estimate of carrying capacity, we will have approximately 6 billion *excess* people in the year 2000 if we have not been blessed with visitations from the apocalyptic horsemen. By then we will have consumed a vast amount of our vital resources, and we will have multiplied the tensions and toxins of the population many fold. Will we have fatally damaged the biosphere? Or will man and the biosphere die before that time? The honest answer is that we do not know with any reasonable degree of assurance.

The readings in this chapter leave us with about as black a prospect for effective population control as can be imagined. Techniques to prevent conception are necessary. So is the back-up possibility of abortion on request. Motivation is even more necessary, however, and the evidence is that no government, the United Nations included, has shown the ability to face the issue squarely, even to the extent of studying the goals and possible means of genuine population control. In fact, as the case of Japan seems to illustrate, government *reaction* may be expected when restraint by the people themselves begins to be effective.

With government leadership imbedded in the irrelevant past the only hope seems to lie in the public at large. It is here that the few bright spots lie. Young adults have shown a sure instinct in grasping the environmental issue, in campaigning for elimination of restrictions on abortion and voluntary sterilization, and developing the womens' liberation movement. Pushed from below, the anachronism-encrusted mass of society has at least quivered. Such public pressure is probably the only hope of effective action. If the attitudes, and the institutions, nurtured in 10,000 years of population increase are to be overthrown, the revolution will have to start with those who are as yet uncommitted to the fantasy of eternal expansion.

Let us suppose, then, that a revolution of the young is successful. Assume that the most drastic of possible commitments to solving the problem of human overpopulation were made. If, for example, all new families formed in the United States from 1975 onwards produced an average of only two children (the goal of "Zero Population Growth"), the United States population would not peak until 75 years later. The continued growth would be the consequence of the high proportion of young in the population. That proportion is higher in the underdeveloped countries and their growth would therefore be even more persistent. Consider a still more drastic and improbable event: total worldwide moratorium on births, beginning in 1970 and lasting for thirty years. In the year 2000, even allowing for a "generous" death rate, the world population would still be close to two billion, or twice the presumed planetary carrying capacity at North American standards of consumption.

These extravagant examples cannot be viewed as possibilities, but they may drive home the point that, even with genuine and total worldwide commitment to stabilization and eventual reduction of population, effective action will take many decades, even centuries. Short of mass mortality, the peak population will certainly approach or exceed 10 billion before any decline is evident.

Optimum World Population
H. R. Hulett

No hunter of the age of fable
Had need to buckle in his belt
More game than he was ever able
To take ran wild upon the veldt.
Each night with roast he stocked his table
Then procreated on the pelt,
And that is how, of course, there came
At last to be more men than game.
  A. D. Hope, *Texas Quarterly, Summer 1962*

The population explosion is now widely recognized as a world-wide problem. Many have attempted to calculate the ultimate supportable population of the world, with resulting numbers ranging from a few to many times the current population of about 3.5 billion (Schmidt, 1965). There is no reason to believe that such a population is optimum, unless it is assumed that it is innately desirable to have as many people as possible. On the other hand, there are many reasons why such a population would not be optimum—reasons concerned with such aspects of the quality of life as pollution, loss of open space and wildlife, overcrowding, and lowered individual allotments of food and other raw materials. It is probably impossible to come to any quantitative conclusions as to the most desirable value for any of these parameters. However, the average U. S. citizen would certainly assume that the amount and variety of food and other raw materials available to him are not greater than optimum. The ratio of current world production of these materials to current average American consumption then can be used as a rough indication of the upper limit of optimum world population at present production rates. This optimum population can increase no more rapidly than the production of these essential raw materials can increase.

Food

Food is probably the most useful indicator of supportable population. Even at current levels, which over most of the world entail much lower quantity and quality than American diets, there is not enough food in many areas to provide an adequate diet for the present population (F.A.O., 1968). The picture is much darker when the American diet is used as a reference standard. We each purchased about 3200 kcal/day in 1966, probably corresponding to about 2600 kcal/day consumed (USDA, 1968). Of this 3200 kcal/day about one-third came from meat, milk, eggs, and other animal products, with the remainder, about 2100 kcal/day or 7.5×10^5 kcal/yr, directly from plants. The animal products required 750 x 10^9 feed units (corn equivalent pounds) or about 6.3×10^6 kcal/person/yr, so total requirement was slightly over 7×10^6 kcal/person/yr—about six times what it would have been on a strictly vegetarian diet.

The increased primary calorie requirements because of use of animal products are not strictly comparable to calories available from cropland, since much of the land used by animals for grazing is not suitable for cropland,

and its produce would be wasted if it were not grazed. Some 1 billion acres are used for grazing in the United States, but its caloric yield is only equivalent to about 200 million cropland acres (Landsberg et al., 1963). There is an additional 16 million cropland acres equivalent from grazing on previously harvested fields, but about 66 million acres of cropland are included in the grazed area, giving a caloric output equivalent to that from about 150 million cropland acres from grazing areas not suitable to crops. There are a total of 465 million acres presently suitable for crops. Thus the increased yield from grazing is about one-third.

There is more grazing and cropland in the United States relative to total area than in most countries, but the ratio of the two is not far from average (Schmidt, 1965). However, let us assume optimistically that caloric output throughout the world from grazing land is equal to that from present cropland. The production of plant crops in the world is about 1.2 x 10^6 kcal/person/yr or about 4.2 x 10^{15} kcal/yr (Schmidt, 1965). If this is doubled to include grazing by domestic animals, the total production would be less than 10^{16} kcal/yr, enough to provide for only about 1.2 billion people at American food standards. Other methods of calculation, based on present protein consumption in various areas, together with the efficiency of conversion of plant food by individual animal species, give very similar results. These figures show the impossibility of lifting the rest of the world to our dietary standards without a several-fold increase in world food production or a massive reduction in population.

In addition to food from the land, man utilizes some of the food in the ocean. However, the amount is relatively small — less than 1% of the total food calories and less than 4% of the total protein in the global diet (Schmidt, 1965). It may be possible to increase the harvest from the oceans in the future, but at present it has only a minimal effect on available food.

Other Renewable Resources

The other major renewable resources are lumber and other forest products such as paper. Here again the picture is dim. On a world basis, forest reserves are probably shrinking. Total worldwide wood production in 1965 was 2.0 x 10^9 m^3. Of this about 0.33 x 10^9 m^3 were used in the United States (USDA, 1968; UN Statistical Yearbook, 1969). If the present cut could be maintained, it would supply a world population of a little over one billion people at the U.S. level of consumption. The energy content in the wood used, estimated at 0.65 g/cm^3 of carbohydrate equivalent, is about 4 x 10^6 cal/yr for each U.S. inhabitant, a large fraction of that utilized for food. If the energy in the wasted roots, branches, and leaves is included, the total is undoubtedly more than that used for food.

Of course the basic limitation on renewable resources is the photosynthetic process. Man tries to convert solar energy into various plant and animal products. Unfortunately, he is not yet able to attain high efficiency in the conversion. When irrigation water "makes the desert bloom," or when large-scale fertilizer applications remove natural limitations on plant food, man increases photosynthetic production. However, in many cases where agricultural crops are substituted for natural ecosystems, total photosynthesis is decreased because man keeps certain plants out of available ecological niches where they would utilize sunlight which is otherwise wasted. In addition, we take land out of production for cities and highways, vast areas of the earth have been made deserts by man's activities, and environmental pollution has drastically reduced plant growth in some areas. On balance, photosynthesis has been reduced by man.

Recent estimates of net photosynthetic production on land include about 10^{10} tons of carbon fixed/year (Lieth, 1963), 2.25 x 10^{10} tons (Wassick, 1968), and 2.2 to 3.2 x 10^{10} tons (Vallentyne, 1966). These figures correspond to between 2.5 and 10 x 10^{10} tons of dry plant material as carbohydrate. With a caloric value for such material of about 4 kcal/g, this is a total conversion of about 1 to 4 x 10^{17} kcal of solar energy into available plant energy. If all 3.5 billion inhabitants of the world utilized this energy for food and forest products at the same rate as U.S. citizens, from 10 to 40% of all material photosynthesized on land would be used for humans. When it is realized that most of the solar energy goes into roots, stalks, leaves, and other materials which are currently unusable by man, it is obvious that either the efficiency of photosynthesis will have to be greatly increased or some other energy source will have to be developed if the growing world population is to adopt U.S. patterns of using renewable resources.

Nonrenewable Resources

The picture is even bleaker in terms of present production and use of many nonrenewable resources (UN Statistical Yearbook, 1969). World production of energy in 1967 was equivalent to about 5.8 billion tons of coal, that of the United States was equivalent to almost 2 billion tons. Thus, fewer than 600 million people could have used energy at the rate we did. In theory, coal, oil, and gas can be considered renewable resources since they are derived ultimately from photosynthesis. If all the material currently photosynthesized on land were burned, it would provide energy at the U.S. rates of consumption for about one to four billion people (and, of course, there would be nothing left for food). Steel consumption of the world in 1967 was

443 million tons, in the United States, about 128 million. About 700 million could have used steel at this rate. Fertilizer use in the world was 44 million tons, in the United States, about 10 million. About 900 million people could have used fertilizer at our rate. Aluminum production was 7.5 million tons, U.S. production, almost 3 million. Thus, only 500 million people could have used aluminum at the rate we did. Practically all other mineral resources show similar ratios. The world's present industrial complex is sufficient to provide fewer than a billion people with the U.S. standard of affluence. Production of all these substances can be increased, but the increases will be slow because of the heavy capital investment required. In addition, of course, the increased production is accompanied by more rapid depletion of mineral reserves, and, in many cases, by increased environmental pollution, certainly not a component of optimum conditions.

In all the areas treated, it appears that on the order of a billion people is the maximum population supportable by the present agricultural and industrial system (and the present work force) of the world at U.S. levels of affluence.

It would obviously be very difficult to produce food and raw materials at the present rate with the smaller work force consistent with a world population of about a billion people; therefore, this number is, if anything, too large to be self-supporting at U.S. affluence levels. As our technology, knowledge, and industrial and agricultural systems expand so can the optimum population, although if might be more desirable either to channel the increased production into an increased standard of living or to reduce the depletion rate of our nonrenewable resources rather than simply to produce more bodies. The difference between one billion people and the present world population is an indication of the magnitude of the problem caused by the population explosion.

REFERENCES

F.A.O.: Third World Food Survey, Freedom from Hunger Campaign, Basic Study No. 1, 1963; *Mon. Bull. Agri. Econ. Statist.,* 17: 1 (No. 5), 1968.

Landsberg, H. H., L. L. Fischman, and I. L. Fisher. 1963. *Resources in America's Future.* The Johns Hopkins Press, Baltimore, Md.

Leith, H. 1963. The role of vegetation in the carbon dioxide content of the atmosphere, *J. Geophys. Res.,* 68: 3887-3898. (1.5 x 10^{10} tons C. exchanged, with 30% lost on short time basis through respiration and direct exchange).

Schmidt, W. R. 1965. The planetary food potential, *Ann. N. Y. Acad. Sci.,* 118: 647-718.

United Nations Statistical Yearbook 1968. Statistical Office of the U.N., Department of Social & Economic Affairs, New York, 1969.

U.S. Dept. of Agriculture. 1968. *Agricultural Statistics.* Govt. Printing Office, Washington, D.C.

Vallentyne, J. R. 1966. New primary production and photosynthetic efficiency in the biosphere, p. 309. In: *Primary Productivity in Aquatic Environment,* C. R. Goldman (ed.), University of California Press.

Wassink, E. C. 1968. Light energy conversion in photosynthesis and growth of plants, p. 53-66. In: *Functioning of Terrestrial Ecosystems at the Primary Production Level,* F. E. Eckhardt (ed.), UNESCO.

Population Policy: Will Current Programs Succeed?
Kingsley Davis

Throughout history the growth of population has been identified with prosperity and strength. If today an increasing number of nations are seeking to curb rapid population growth by reducing their birth rates, they must be driven to do so by an urgent crisis. My purpose here is not to discuss the crisis itself but rather to assess the present and prospective measures used to meet it. Most observers are surprised by the swiftness with which concern over the population problem has turned from intellectual analysis and debate to policy and action. Such action is a welcome relief from the long opposition, or timidity, which seemed to block forever any governmental attempt to restrain population growth, but relief that "at last something is being done" is no guarantee that what is being done is adequate. On the face of it, one could hardly expect such a fundamental reorientation to be quickly and successfully implemented. I therefore propose to review the nature and (as I see them) limitations of the present policies and to suggest lines of possible improvement.

The Nature of Current Policies

With more than 30 nations now trying or planning to reduce population growth and with numerous private and international organizations helping, the degree of unanimity as to the kind of measures needed is impressive. The consensus can be summed up in the phrase "family planning." President Johnson declared in 1965 that the United States will "assist family planning programs in nations which request such help." The Prime Minister of India said a year later, "We must press forward with family planning. This is a programme of the highest importance." The Republic of Singapore created in 1966 the Singapore Family Planning and Population

Board "to initiate and undertake population control programmes."[1]

As is well known, "family planning" is a euphemism for contraception. The family-planning approach to population limitation, therefore, concentrates on providing new and efficient contraceptives on a national basis through mass programs under public health auspices. The nature of these programs is shown by the following enthusiastic report from the Population Council:[2]

"No single year has seen so many forward steps in population control as 1965. Effective national programs have at last emerged, international organizations have decided to become engaged, a new contraceptive has proved its value in mass application, . . . and surveys have confirmed a popular desire for family limitation. . . . An accounting of notable events must begin with Korea and Taiwan. . . .

"Taiwan's program is not yet two years old, and already it has inserted one IUD *intrauterine device* for every 4-6 target women (those who are not pregnant, lactating, already sterile, already using contraceptives effectively, or desirous of more children). Korea has done almost as well . . . has put 2,200 full-time workers into the field, . . . has reached operational levels for a network of IUD quotas, supply lines, local manufacture of contraceptives, training of hundreds of M.D.'s and nurses, and mass propaganda . . . "

Here one can see the implication that "population control" is being achieved through the dissemination of new contraceptives, and the fact that the "target women" exclude those who want more children. One can also note the technological emphasis and the medical orientation.

What is wrong with such programs? The answer is, "Nothing at all, if they work." Whether or not they work depends on what they are expected to do as well as on how they try to do it. Let us discuss the goal first, then the means.

Goals

Curiously, it is hard to find in the popula-

tion-policy movement any explicit discussion of long-range goals. By implication the policies seem to promise a great deal. This is shown by the use of expressions like *population control* and *population planning* (as in the passages quoted). It is also shown by the characteristic style of reasoning. Expositions of current policy usually start off by lamenting the speed and the consequences of runaway population growth. This growth, it is then stated, must be curbed—by pursuing a vigorous family-planning program. That family planning can solve the problem of population growth seems to be taken as self-evident.

For instance, the much-heralded statement by 12 heads of state, issued by Secretary-General U Thant on 10 December 1966 (a statement initiated by John D. Rockefeller III, Chairman of the Board of the Population Council), devotes half its space to discussing the harmfulness of population growth and the other half to recommending family planning.[3] A more succinct example of the typical reasoning is given in the Provisional Scheme for a Nationwide Family Planning Programme in Ceylon:[4]

"The population of Ceylon is fast increasing . . . [The] figures reveal that a serious situation will be created within a few years. In order to cope with it a Family Planning programme on a nationwide scale should be launched by the Government."

The promised goal—to limit population growth so as to solve population problems—is a large order. One would expect it to be carefully analyzed, but it is left imprecise and taken for granted, as is the way in which family planning will achieve it.

When the terms *population control* and *population planning* are used, as they frequently are, as synonyms for current family-planning programs, they are misleading. Technically, they would mean deliberate influence over all attributes of a population, including its age-sex structure, geographical distribution, racial composition, genetic quality, and total size. No government attempts such full control. By tacit understanding, current population policies are concerned with only the *growth* and *size* of populations. These attributes, however, result from the death rate and migration as well as from the birth rate; their control would require deliberate influence over the factors giving rise to all three determinants. Actually, current policies labeled population control do not deal with mortality and migration but deal only with the birth input. This is why another term, *fertility control,* is frequently used to describe current policies. But, as I show below, family planning (and hence current policy) does not undertake to influence most of the determinants of human reproduction. Thus the programs should not be referred to as population control or planning, because they do not attempt to influence the factors responsible for the attributes of human populations, taken generally; nor should they be called fertility control, because they do not try to affect most of the determinants of reproductive performance.

The ambiguity does not stop here, however. When one speaks of controlling population size, any inquiring person naturally asks, What is "control"? Who is to control whom? Precisely what population size, or what rate of population growth, is to be achieved? Do the policies aim to produce a growth rate that is nil, one that is very slight, or one that is like that of the industrial nations? Unless such questions are dealt with and clarified, it is impossible to evaluate current population policies.

The actual programs seem to be aiming simply to achieve a reduction in the birth rate. Success is therefore interpreted as the accomplishment of such a reduction on the assumption that the reduction will lessen population growth. In those rare cases where a specific demographic aim is stated, the goal is said to be a short-run decline within a given period. The Pakistan plan adopted in 1966 (5, p.889) aims to reduce the birth rate from 50 to 40 per thousand by 1970; the Indian plan[6] aims to reduce the rate from 40 to 25 "as soon as possible;" and the Korean aim[7] is to cut population growth from 2.9 to 1.2 percent by 1980. A significant feature of such stated aims is the rapid population growth they would permit. Under conditions of modern mortality, a crude birth rate of 25 to 30 per thousand will represent such a multiplication of people as to make use of the term *population control* ironic. A rate of increase of 1.2 percent per year would allow South Korea's already dense population to double in less than 60 years.

One can of course defend the programs by saying that the present goals and measures are merely interim ones. A start must be made somewhere. But we do not find this answer in the population-policy literature. Such a defense, if convincing, would require a presentation of the *next* steps, and these are not considered. One suspects that the entire question of goals is instinctively left vague because thorough limitation of population growth would run counter to national and group aspirations. A consideration of hypothetical goals throws further light on the matter.

Industrialized nations as the model. Since current policies are confined to family planning, their maximum demographic effect would be to give the underdeveloped countries the same level of reproductive performance that the industrial nations now have. The latter, long oriented toward family planning, provides a good yardstick for determining what the availability of contraceptives can do to population growth. Indeed, they provide more than a yardstick; they are actually the model which inspired the present population policies.

What does this goal mean in practice? Among the advanced nations there is considerable diversity in the level of fertility.[8] At one extreme are countries such as New Zealand, with an average gross reproduction rate (GRR) of 1.91 during the period 1960-64; at the other extreme are countries such as Hungary, with a rate of 0.91 during the same period. To a considerable extent, however, such divergencies are matters of timing. The birth rates of most industrial nations have shown, since about 1940, a wave-like movement, with no secular trend. The average level of reproduction during this long period has been high enough to give these countries, with their low mortality, an extremely rapid population growth. If this level is maintained, their population will double in just over 50 years—a rate higher than that of world population growth at any time prior to 1950, at which time the growth in numbers of human beings was already considered fantastic. The advanced nations are suffering acutely from the effects of rapid population growth in combination with the production of ever more goods per person.[9] A rising share of their supposedly high per capita income, which itself draws increasingly upon the resources of the underdeveloped countries (who fall farther behind in relative economic position), is spent simply to meet the costs, and alleviate the nuisances, of the unrelenting production of more and more goods by more people. Such facts indicate that the industrial nations provide neither a suitable demographic model for the nonindustrial peoples to follow nor the leadership to plan and organize effective population-control policies for them.

Zero population growth as a goal. Most discussions of the population crisis lead logically to zero population growth as the ultimate goal, because *any* growth rate, if continued, will eventually use up the earth. Yet hardly ever do arguments for population policy consider such a goal, and current policies do not dream of it. Why not? The answer is evidently that zero population growth is unacceptable to most nations and to most religious and ethnic communities. To argue for this goal would be to alienate possible support for action programs.

Goal peculiarities inherent in family planning. Turning to the actual measures taken, we see that the very use of family planning as the means for implementing population policy poses serious but unacknowledged limits on the intended reduction in fertility. The family-planning movement, clearly devoted to the improvement and dissemination of contraceptive devices, states again and again that its purpose is that of enabling couples to have the number of children they want. "The opportunity to decide the number and spacing of children is a basic human right," say the 12 heads of state in the United Nations declaration. The 1965 Turkish Law Concerning Population Planning declares:[10]

Article 1. Population Planning means that individuals can have as many children as they wish, whenever they want to. This can be ensured through preventive measures taken against pregnancy. . . .

Logically, it does not make sense to use *family* planning to provide *national* population control or planning. The "planning" in family planning is that of each separate couple. The only control they exercise is control over the size of *their* family. Obviously, couples do not plan the size of the nation's population, any more than they plan the growth of the national income or the form of the highway network. There is no reason to expect that the millions of decisions about family size made by couples in their own interest will automatically control population for the benefit of society. On the contrary, there are good reasons to think they will not do so. At most, family planning can reduce reproduction to the extent that unwanted births exceed wanted births. In industrial countries the balance is often negative—that is, people have fewer children as a rule than they would like to have. In underdeveloped countries the reverse is normally true, but the elimination of unwanted births would still leave an extremely high rate of multiplication.

Actually, the family-planning movement does not pursue even the limited goals it professes. It does not fully empower couples to have only the number of offspring they want because it either condemns or disregards certain tabooed but nevertheless effective means to this goal. One of its tenets is that "there shall be freedom of choice of method so that individuals can choose in accordance with the dictates of their consciences,"[11] but in practice this amounts to limiting the individual's choice, because the "conscience" dictating the method is usually not his but that of religious and governmental officials. Moreover, not every individual may choose: even the so-called recommended methods are ordinarily not offered to single women, or not all are offered to women professing a given religious faith.

Thus, despite its emphasis on technology, current policy does not utilize all available means of contraception, much less all birth-control measures. The Indian government wasted valuable years in the early stages of its population-control program by experimenting exclusively with the "rhythm" method, long after this technique had been demonstrated to be one of the least effective. A greater limitation on means is the exclusive emphasis on contraception itself. Induced abortion, for example, is one of the surest means of controlling reproduction, and one that has been proven capable of reducing birth rates rapidly. It seems peculiarly suited to the threshold stage of a population-control program—the stage when new conditions of life first make large families disadvantageous. It was the principal factor in the halving of the Japanese birth rate, a major factor in the declines in birth rate of East-European satellite countries after legalization of abortions in the early 1950's, and an important factor in the reduction of fertility in industrializing nations from 1870 to the 1930's.[12] Today, according to *Studies in Family Planning*,[13] "abortion is probably the foremost method of birth control throughout Latin America." Yet this method is rejected in nearly all national and international population-control programs. American foreign aid is used to help *stop* abortion.[14] The United Nations excludes abortion from family planning, and in fact justifies the latter by presenting it as a means of combating abortion.[15] Studies of abortion are being made in Latin America under the presumed auspices of population-control groups, not with the intention of legalizing it and thus making it safe, cheap, available, and hence more effective for population control, but with the avowed purpose of reducing it.[16]

Although few would prefer abortion to efficient contraception (other things being equal), the fact is that both permit a woman to control the size of her family. The main drawbacks to abortion arise from its illegality. When performed, as a legal procedure, by a skilled physician, it is safer than childbirth. It does not compete with contraception but serves as a backstop when the latter fails or when contraceptive devices or information are not available. As contraception becomes customary, the incidence of abortion recedes even without its being banned. If, therefore, abortions enable women to have only the number of children they want, and if family planners do not advocate—in fact decry—legalization of abortion, they are to that extent denying the central tenet of their own movement. The irony of anti-abortionism in family-planning circles is seen particularly in hair-splitting arguments over whether or not some contraceptive agent (for example, the IUD) is in reality an abortifacient. A Mexican leader in family planning writes:[17]

One of the chief objectives of our program in Mexico is to prevent abortions. If we could be sure that the mode of action (of the IUD) was not interference with nidation, we could easily use the method in Mexico.

The questions of sterilization and unnatural forms of sexual intercourse usually meet with similar silent treatment or disapproval, although nobody doubts the effectiveness of these measures in avoiding conception. Sterilization has proved popular in Puerto Rico and has had some vogue in India (where the new health minister hopes to make it compulsory for those with a certain number of children), but in both these areas it has been for the most part ignored or condemned by the family-planning movement.

On the side of goals, then, we see that a family-planning orientation limits the aims of current population policy. Despite reference to "population control" and "fertility control," which presumably mean determination of demographic results by and for the nation as a whole, the movement gives control only to couples, and does this only if they use "respectable" contraceptives.

The Neglect of Motivation

By sanctifying the doctrine that each woman should have the number of children she wants, and by assuming that if she has only that number this will automatically curb population growth to the necessary degree, the leaders of current policies escape the necessity of asking why women desire so many children and how this desire can be influenced (18, p.

41; 19). Instead, they claim that satisfactory motivation is shown by the popular desire (shown by opinion surveys in all countries) to have the means of family limitation, and that therefore the problem is one of inventing and distributing the best possible contraceptive devices. Overlooked is the fact that a desire for availability of contraceptives is compatible with *high* fertility.

Given the best of means, there remain the questions of how many children couples want and of whether this is the requisite number from the standpoint of population size. That it is not is indicated by continued rapid population growth in industrial countries, and by the very surveys showing that people want contraception—for these show, too, that people also want numerous children.

The family planners do not ignore motivation. They are forever talking about "attitudes" and "needs." But they pose the issue in terms of the "acceptance" of birth control devices. At the most naive level, they assume that lack of acceptance is a function of the contraceptive device itself. This reduces the motive problem to a technological question. The task of population control then becomes simply the invention of a device that *will* be acceptable.[20] The plastic IUD is acclaimed because once in place, it does not depend on repeated *acceptance* by women, and thus it "solves" the problem of motivation.[21]

But suppose a woman does not want to use *any* contraceptive until after she has had four children. This is the type of question that is seldom raised in the family-planning literature. In that literature, wanting a specific number of children is taken as complete motivation, for it implies a wish to control the size of one's family. The problem woman, from the standpoint of family planners, is the one who wants "as many as come," or "as many as God sends." Her attitude is construed as due to ignorance and "cultural values" and the policy deemed necessary to change it is "education." No compulsion can be used, because the movement is committed to free choice, but movie strips, posters, comic books, public lectures, interviews, and discussions are in order. These supply information and supposedly change values by discounting superstitions and showing that unrestrained procreation is harmful to both mother and children. The effort is considered successful when the woman decides she wants only a certain number of children and uses an effective contraceptive.

In viewing negative attitudes toward birth control as due to ignorance, apathy, and outworn tradition, and "mass-communication" as the solution to the motivation problem,[22] family planners tend to ignore the power and complexity of social life. If it were admitted that the creation and care of new human beings is socially motivated, like other forms of behavior, by being a part of the system of rewards and punishments that is built into human relationships, and thus is bound up with the individual's economic and personal interests, it would be apparent that the social structure and economy must be changed before a deliberate reduction in the birth rate can be achieved. As it is, reliance on family planning allows people to feel that "something is being done about the population problem" without the need for painful social changes.

Designation of population control as a medical or public health task leads to a similar evasion. This categorization assures popular support because it puts population policy in the hands of respected medical personnel, but, by the same token, it gives responsibility for leadership to people who think in terms of clinics and patients, of pills and IUD's, and who bring to the handling of economic and social phenomena a self-confident naiveté. The study of social organization is a technical field; an action program based on intuition is no more apt to succeed in the control of human beings than it is in the area of bacterial or viral control. Moreover, to alter a social system, by deliberate policy, so as to regulate births in accord with the demands of the collective welfare, would require political power, and this is not likely to inhere in public health officials, nurses, midwives, and social workers. To entrust population policy to them is "to take action," but not dangerous "effective action."

Similarly, the Janus-faced position on birth-control technology represents an escape from the necessity, and onus, of grappling with the social and economic determinants of reproductive behavior. On the one side, the rejection or avoidance of religiously tabooed but otherwise effective means of birth prevention enables the family-planning movement to avoid official condemnation. On the other side, an intense preoccupation with contraceptive technology (apart from the tabooed means) also helps the family planners to avoid censure. By implying that the only need is the invention and distribution of effective contraceptive devices, they allay fears, on the part of religious and governmental officials, that fundamental changes in social organization are contemplated. Changes basic enough to affect motivation for having children would be changes in the structure of the family, in the position of women, and in the sexual mores. Far from proposing such radicalism, spokesmen for family planning frequently state their purpose as "protection" of the family—that is, closer observance of family norms. In addition, by concentrating on *new* and *scientific* contraceptives, the movement escapes taboos attached to old ones (the Pope will hardly authorize the condom, but may sanction the pill) and allows family planning to be regarded as a branch of medicine: over-population becomes a disease, to be treated by a pill or a coil.

We thus see that the inadequacy of current population policies with respect to motivation is inherent in their overwhelmingly family-planning character. Since family planning is by definition private planning, it eschews any societal control over motivation. It merely furnishes the means, and, among possible means, only the most respectable. Its leaders, in avoiding social complexities and seeking official favor, are obviously activated not solely by expediency but also by their own sentiments as members of society and by their background as persons attracted to the family-planning movement. Unacquainted for the most part with technical economics, sociology, and demography, they tend honestly and instinctively to believe that something they vaguely call population control can be achieved by making better contraceptives available.

The Evidence of Ineffectiveness

If this characterization is accurate, we can conclude that current programs will not enable a government to control population size. In countries where couples have numerous offspring that they do not want, such programs may possibly accelerate a birth-rate decline that would occur anyway, but the conditions that cause births to be wanted or unwanted are beyond the control of family planning alone as its population policy.

This conclusion is confirmed by demographic facts. As I have noted above, the widespread use of family planning in industrial countries has not given their governments control over the birth rate. In backward countries today, taken as a whole, birth rates are rising, not falling; in those with population policies, there is no indication that the government is controlling the rate of reproduction. The main "successes" cited in the well-publicized policy literature are cases where a large number of contraceptives have been distributed or where the program has been accompanied by some decline in the birth rate. Popular enthusiasm

for family planning is found mainly in the cities, or in advanced countries such as Japan and Taiwan, where the people would adopt contraception in any case, program or no program. It is difficult to prove that present population policies have even speeded up a lowering of the birth rate (the least that could have been expected), much less that they have provided national "fertility control."

Let us next briefly review the facts concerning the level and trend of population in underdeveloped nations generally, in order to understand the magnitude of the task of genuine control.

Rising Birth Rates in Underdeveloped Countries

In ten Latin-American countries, between 1940 and 1959,[23] the average birth rates (age-standardized), as estimated by our research office at the University of California, rose as follows: 1940-44, 43.4 annual births per 1000 population; 1945-49, 44.6; 1950-54, 46.4; 1955-59, 47.7.

In another study made in our office, in which estimating methods derived from the theory of quasi-stable populations were used, the recent trend was found to be upward in 27 underdeveloped countries, downward in six, and unchanged in one.[24] Some of the rises have been substantial, and most have occurred where the birth rate was already extremely high. For instance, the gross reproduction rate rose in Jamaica from 1.8 per thousand in 1947 to 2.7 in 1960; among the natives of Fiji, from 2.0 in 1951 to 2.4 in 1964; and in Albania, from 3.0 in the period 1950-54 to 3.4 in 1960.

The general rise in fertility in backward regions is evidently not due to failure of population-control efforts, because most of the countries either have no such effort or have programs too new to show much effect. Instead, the rise is due, ironically, to the very circumstance that brought on the population crisis in the first place—to improved health and lowered mortality. Better health increases the probability that a woman will conceive and retain the fetus to term; lowered mortality raises the probability of widowhood during that age.[25] The significance of the general rise in fertility, in the context of this discussion, is that it is giving would-be population planners a harder task than many of them realize. Some of the upward pressure on birth rates is independent of what couples do about family planning, for it arises from the fact that, with lowered mortality, there are simply more couples.

Underdeveloped Countries with Population Policies

In discussions of population policy there is often confusion as to which cases are relevant. Japan, for instance, has been widely praised for the effectiveness of its measures, but it is a very advanced industrial nation and, besides, its government policy has little or nothing to do with the decline in the birth rate, except unintentionally. It therefore offers no test of population policy under peasant-agrarian conditions. Another case of questionable relevance is that of Taiwan, because Taiwan is sufficiently developed to be placed in the urban-industrial class of nations. However, since Taiwan is offered as the main showpiece by the sponsors of current policies in underdeveloped areas, and since the data are excellent, it merits examination.

Taiwan is acclaimed as a showpiece because it has responded favorably to a highly organized program for distributing up-to-date contraceptives and has also had a rapidly dropping birth rate. Some observers have carelessly attributed the decline in the birth rate—from 50.0 in 1951 to 32.7 in 1965—to the family-planning campaign,[26] but the campaign began

TABLE 1

Decline in Taiwan's fertility rate, 1951 through 1966.

Year	Registered births per 1000 women aged 15-49	Change in rate (percent)*
1951	211	
1952	198	−5.6
1953	194	−2.2
1954	193	−0.5
1955	197	+2.1
1956	196	−0.4
1957	182	−7.1
1958	185	+1.3
1959	184	−0.1
1960	180	−2.5
1961	77	−1.5
1962	174	−1.5
1963	170	−2.6
1964	162	−4.9
1965	152	−6.0
1966	149	−2.1

*The percentages were calculated on unrounded figures. Source of data through 1965, *Taiwan Demographic Fact Book* (1964, 1965); for 1966, *Monthly Bulletin of Population Registration Statistics of Taiwan* (1966, 1967).

only in 1963 and could have affected only the end of the trend. Rather, the decline represents a response to modernization similar to that made by all countries that have become industrialized.[27] By 1950 over half of Taiwan's population was urban, and by 1964 nearly two-thirds were urban, with 29 percent of the population living in cities of 100,000 or more. The pace of economic development has been extremely rapid. Between 1951 and 1963, per capita income increased by 4.05 percent per year. Yet the island is closely packed, having 870 persons per square mile (a population density higher than that of Belgium). The combination of fast economic growth and rapid population increase in limited space has put parents of large families at a relative disadvantage and has created a brisk demand for abortions and contraceptives. Thus the favorable response to the current campaign to encourage use of the IUD is not a good example of what birth-control technology can do for a genuinely backward country. In fact, when the program was started, one reason for expecting receptivity was that the island was already on its way to modernization and family planning.[28]

At most, the recent family-planning campaign—which reached significant proportions only in 1964, when some 46,000 IUD's were inserted (in 1965 the number was 99,253, and in 1966, 11,242) (29; 30, p. 45)—could have caused the increase observable after 1963 in the rate of decline. Between 1951 and 1963 the average drop in the birth rate per 1000 women (see Table 1) was 1.73 percent per year; in the period 1964-66 it was 4.35 percent. But one hesitates to assign all of the acceleration in decline since 1963 to the family-planning campaign. The rapid economic development has been precisely of a type likely to accelerate a drop in reproduction. The rise in manufacturing has been much greater than the rise in either agriculture or construction. The agricultural labor force has thus been squeezed, and migration to the cities has skyrocketed.[31] Since housing has not kept pace, urban families have had to restrict reproduction in order to take advantage of career opportunities and avoid domestic inconvenience. Such conditions have historically tended to accelerate a decline in birth rate. The most rapid decline came late in the United States (1921-33) and in Japan (1947-55). A plot of the Japanese and Taiwanese birth rates (Fig. 1) shows marked similarity of the two curves, despite a difference in level. All told, one should not attribute all of the post-1963 acceleration in the decline of Taiwan's birth rate to the family-planning campaign.

The main evidence that *some* of this acceleration is due to the campaign comes from the fact that Taichung, the city in which the family-planning effort was first concentrated, showed subsequently a much faster drop in fertility than other cities (30, p.69; 32). But the campaign has not reached throughout the island. By the end of 1966 only 260,745 women

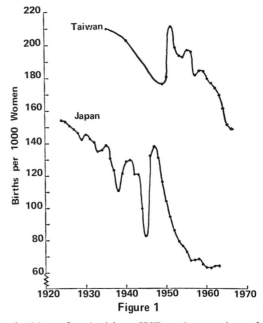

Figure 1

had been fitted with an IUD under auspices of the campaign, whereas the women of reproductive age on the island numbered 2.86 million. Most of the reduction in fertility has therefore been a matter of individual initiative. To some extent the campaign may be simply substituting sponsored (and cheaper) services for those that would otherwise come through private and commercial channels. An island-wide survey in 1964 showed that over 150,000 women were already using the traditional Ota ring (a metallic intrauterine device popular in Japan); almost as many had been sterilized; about 40,000 were using foam tablets; some 50,000 admitted to having had at least one abortion; and many were using other methods of birth control (30, pp. 18; 31).

The important question, however, is not whether the present campaign is somewhat hastening the downward trend in the birth rate but whether, even if it is, it will provide population control for the nation. Actually, the campaign is not designed to provide such control and shows no sign of doing so. It takes for granted existing reproductive goals. Its aim is "to integrate, through education and information, the idea of family limitation *within the existing attitudes, values, and goals* of the people" [30, p.8 (italics mine)]. Its target is *married* women who do not want any more children; it ignores girls not yet married, and woman married and wanting more children.

With such an approach, what is the maximum impact possible? It is the difference between the number of children women have been having and the number they want to have. A study in 1957 found a median figure of 3.75 for the number of children wanted by women aged 15 to 29 in Taipei, Taiwan's largest city; the corresponding figure for women from a satellite town was 3.93; for women from a fishing village, 4.90; and for women from a farming village, 5.03. Over 60 percent of the women in Taipei and over 90 percent of those in the farming village wanted 4 or more children.[33] In a sample of wives aged 25 to 29 in Taichung, a city of over 300,000, Freedman and his co-workers found the average number of children wanted was 4; only 9 percent wanted less than 3, 20 percent wanted 5 or more.[34] If, therefore, Taiwanese women used contraceptives that were 100-percent effective and had the number of children they desire, they would have about 4.5 each. The goal of the family-planning effort would be achieved. In the past the Taiwanese woman who married and lived through the reproductive period had, on the average, approximately 6.5 children; thus a figure of 4.5 would represent a substantial decline in fertility. Since mortality would continue to decline, the population growth rate would decline somewhat less than individual reproduction would. With 4.5 births per woman and a life expectancy of 70 years, the rate of natural increase would be close to 3 percent per year.[35]

In the future, Taiwanese views concerning reproduction will doubtless change, in response to social change and economic modernization. But how far will they change? A good indication is the number of children desired by couples in an already modernized country long oriented toward family planning. In the United States in 1966, an average of 3.4 children was considered ideal by white women aged 21 or over.[36] This average number of births would give Taiwan, with only a slight decrease in mortality, a long-run rate of natural increase of 1.7 percent per year and doubling of population in 41 years.

Detailed data confirm the interpretation that Taiwanese women are in the process of shifting from a "peasant-agrarian" to an "industrial" level of reproduction. They are, in typical fashion, cutting off higher-order births at age 30 and beyond.[37] Among young wives, fertility has risen, not fallen. In sum, the widely acclaimed family-planning program in Taiwan may, at most, have somewhat speeded the later phase of fertility decline which would have occurred anyway because of modernization.

Moving down the scale of modernization, to countries most in need of population control, one finds the family-planning approach even more inadequate. In South Korea, second only to Taiwan in the frequency with which it is cited as a model of current policy, a recent birth-rate decline of unknown extent is assumed by leaders to be due overwhelmingly to the government's family-planning program. However, it is just as plausible to say that the net effect of government involvement in population control has been, so far, to delay rather than hasten a decline in reproduction made inevitable by social and economic changes. Although the government is advocating vasectomies and providing IUD's and pills, it refuses to legalize abortions, despite the rapid rise in the rate of illegal abortions and despite the fact that, in a recent survey, 72 percent of the people who stated an opinion favored legalization. Also, the program is presented in the context of maternal and child health; it thus emphasizes motherhood and the family rather than alternative roles for women. Much is made of the fact that opinion surveys show an overwhelming majority of Koreans (89 percent in 1965) favoring contraception (38, p.27), but this means only that Koreans are like other people in wishing to have the means to get what they want. Unfortunately, they want sizable families: "The records indicate that the program appeals mainly to women in the 30-39 year age bracket who have four or more children, including at least two sons . . . " (38, p.25).

In areas less developed than Korea the degree of acceptance of contraception tends to be disappointing, especially among the rural majority. Faced with this discouragement, the leaders of current policy, instead of reexamining their assumptions, tend to redouble their effort to find a contraceptive that will appeal to the most illiterate peasant, forgetting that he wants a good-sized family. In the rural Punjab, for example, "a disturbing feature . . . is that the females start to seek advice and adopt family planning techniques at the tag end of their reproductive period."[39] Among 5196 woman coming to rural Pujabi family-planning centers, 38 percent were over 35 years old, 67 percent over 30. These women had married early, nearly a third of them before the age of 15;[40] some 14 percent had eight or more *living* children when they reached the clinic, 51 percent six or more.

A survey in Tunisia showed that 68 percent of the married couples were willing to use birth-control measures, but the average number of children they considered ideal was 4.3.[41] The corresponding averages for a village in eastern Java, a village near New Delhi, and

a village in Mysore were 4.3, 4.0, and 4.2, respectively.[42,43] In the cities of these regions women are more ready to accept birth control and they want fewer children than village women do, but the number they consider desirable is still wholly unsatisfactory from the standpoint of population control. In an urban family-planning center in Tunisia, more than 600 of 900 women accepting contraceptives had four living children already.[44] In Bangalore, a city of nearly a million at the time (1952), the number of offspring desired by married women was 3.7 on the average; by married men, 4.1.[43] In the metropolitan area of San Salvador (350,000 inhabitants) a 1964 survey[45] showed the number desired by women of reproductive age to be 3.0, and in seven other capital cities of Latin America the number ranged from 2.7 to 4.2. If the women in the cities of underdeveloped countries used birth-control measures with 100-percent efficiency, they still would have enough babies to expand city populations senselessly, quite apart from the added contribution of rural-urban migration. In many of the cities the difference between actual and ideal number of children is not great; for instance, in the seven Latin-American capitals mentioned above, the ideal was 3.4 whereas the actual births per women in the age range 35 to 39 was 3.7.[46] Bombay City has had birth-control clinics for many years, yet its birth rate (standardized for age, sex, and marital distribution) is still 34 per 1000 inhabitants and is tending to rise rather than fall. Although this rate is about 13 percent lower than that for India generally, it has been about that much lower since at least 1951.[47]

Is Family Planning the "First Step" in Population Control?

To acknowledge that family planning does not achieve population control is not to impugn its value for other purposes. Freeing women from the need to have more children than they want is of great benefit to them and their children and to society at large. My argument is therefore directed not against family-planning programs as such but against the assumption that they are an effective means of controlling population growth.

But what difference does it make? Why not go along for awhile with family planning as an initial approach to the problem of population control? The answer is that any policy on which millions of dollars are being spent should be designed to achieve the goal it purports to achieve. If it is only a first step, it should be so labeled, and its connection with the next step (and the nature of that next step) should be carefully examined. In the present case, since no "next step" seems ever to be mentioned, the question arises: Is reliance on family planning in fact a basis for dangerous postponement of effective steps? To continue to offer a remedy as a cure long after it has been shown merely to ameliorate the disease is either quackery or wishful thinking, and it thrives most where the need is greatest. Today the desire to solve the population problem is so intense that we are all ready to embrace any "action program" that promises relief. But postponement of effective measures allows the situation to worsen.

Unfortunately, the issue is confused by a matter of semantics. "Family *planning*" and "fertility *control*" suggest that reproduction is being regulated according to some rational plan. And so it is, but only from the standpoint of the individual couple, not from that of the community. What is rational in the light of a couple's situation may be totally irrational from the standpoint of society's welfare.

The need for societal regulation of individual behavior is readily recognized in other spheres—those of explosives, dangerous drugs, public property, natural resources. But in the sphere of reproduction, complete in-

dividual initiative is generally favored even by those liberal intellectuals who, in other spheres, most favor economic and social planning. Social reformers who would not hesitate to force all owners of rental property to rent to anyone who can pay, or to force all workers in an industry to join a union, balk at any suggestion that couples be permitted to have only a certain number of offspring. Invariably they interpret societal control of reproduction as meaning direct police supervision of individual behavior. Put the word *compulsory* in from of any term describing a means of limiting births—*compulsory sterilization, compulsory abortion, compulsory contraception*—and you guarantee violent opposition. Fortunately, such direct controls need not be invoked, but conservatives and radicals alike overlook this in their blind opposition to the idea of collective determination of a society's birth rate.

That the exclusive emphasis on family planning in current population policies is not a "first step" but an escape from the real issue is suggested by two facts. (1) No country has taken the "next step." The industrialized countries have had family planning for half a century without acquiring control over either the birth rate or population increase. (2) Support and encouragement of research on population policy other than family planning is negligible. It is precisely this blocking of alternative thinking and experimentation that makes the emphasis on family planning a major obstacle to population control. The need is not to abandon family-planning programs but to put equal or greater resources into other approaches.

New Directions in Population Policy

In thinking about other approaches, one can start with known facts. In the past, all surviving societies had institutional incentives for marriage, procreation, and child care which were powerful enough to keep the birth rate equal to or in excess of a high death rate. Despite the drop in death rates during the last century and a half, the incentives tended to remain intact because the social structure (especially in regard to the family) changed little. At most, particularly in industrial societies, children became less productive and more expensive.[48] In present-day agrarian societies, where the drop in death rate has been more recent, precipitate, and independent of social change,[49] motivation for having children has changed little. Here, even more than in industrialized nations, the family has kept on producing abundant offspring, even though only a fraction of these children are now needed.

If excessive population growth is to be prevented, the obvious requirement is somehow to impose restraints on the family. However, because family roles are reinforced by society's system of rewards, punishments, sentiments, and norms, any proposal to demote the family is viewed as a threat by conservatives and liberals alike, and certainly by people with enough social responsibility to work for population control. One is charged with trying to "abolish" the family, but what is required is selective restructuring of the family in relation to the rest of society.

The lines of such restructuring are suggested by two existing limitations on fertility. (1) Nearly all societies succeed in drastically discouraging reproduction among unmarried women. (2) Advanced societies unintentionally reduce reproduction among married women when conditions worsen in such a way as to penalize childbearing more severely than it was penalized before. In both cases the causes are motivational and economic rather than technological.

It follows that population-control policy can de-emphasize the family in two ways: (1) by keeping present controls over illegitimate childbirth yet making the most of factors that lead people to postpone or avoid marriage, and (2) by instituting conditions that motivate those who do marry to keep their families small.

Postponement of Marriage

Since the female reproductive span is short and generally more fecund in its first than in its second half, postponement of marriage to ages beyond 20 tends biologically to reduce births. Sociologically, it gives women time to get a better education, acquire interests unrelated to the family, and develop a cautious attitude toward pregnancy.[50] Individuals who have not married by the time they are in their late twenties often do not marry at all. For these reasons, for the world as a whole, the average age at marriage for women is negatively associated with the birth rate: a rising age at marriage is a frequent cause of declining fertility during the middle phase of the demographic transition; and, in the late phase, the "baby boom" is usually associated with a return to younger marriages.

Any suggestion that age at marriage be raised as a part of population policy is usually met with the argument that "even if a law were passed, it would not be obeyed." Interestingly, this objection implies that the only way to control the age at marriage is by direct legislation, but other factors govern the actual age. Roman Catholic countries generally follow canon law in stipulating 12 years as the minimum *legal* age at which girls may marry, but the actual average age at marriage in these countries (at least in Europe) is characteristically more like 25 to 28 years. The actual age is determined, not by law, but by social and economic conditions. In agrarian societies, postponement of marriage (when postponement occurs) is apparently caused by difficulties in meeting the economic prerequisites for matrimony, as stipulated by custom and opinion. In industrial societies it is caused by housing shortages, unemployment, the requirement for overseas military service, high costs of education, and inadequacy of consumer services. Since almost no research has been devoted to the subject, it is difficult to assess the relative weight of the factors that govern the age at marriage.

Encouraging Limitation of Births within Marriage

As a means of encouraging the limitation of reproduction within marriage, as well as postponement of marriage, a greater rewarding of nonfamilial than of familial roles would probably help. A simple way of accomplishing this would be to allow economic advantages to accrue to the single as opposed to the married individual, and to the small as opposed to the large family. For instance, the government could pay people to permit themselves to be sterilized;[51] all costs of abortion could be paid by the government; a substantial fee could be charged for a marriage license; a "child-tax"[52] could be levied; and there could be a requirement that illegitimate pregnancies be aborted. Less sensationally, governments could simply reverse some existing policies that encourage childbearing. They could, for example, cease taxing single persons more than married ones; stop giving parents special tax exemptions; abandon income-tax policy that discriminates against couples when the wife works; reduce paid maternity leaves; reduce family allowances;[53] stop awarding public housing on the basis of family size; stop granting fellowships and other educational aids (including special allowances for wives and children) to married students; cease outlawing abortions and sterilizations; and relax rules that allow use of harmless contraceptives only with medical permission. Some of these

policy reversals would be beneficial in other than demographic respects and some would be harmful unless special precautions were taken. The aim would be to reduce the number, not the quality, of the next generation.

A closely related method of de-emphasizing the family would be modification of the complementarity of the roles of men and women. Men are now able to participate in the wider world yet enjoy the satisfaction of having several children because the housework and childcare fall mainly on their wives. Women are impelled to seek this role by their idealized view of marriage and motherhood and by either the scarcity of alternative roles or the difficulty of combining them with family roles. To change this situation women could be required to work outside the home, or compelled by circumstances to do so. If, at the same time, women were paid as well as men and given equal educational and occupational opportunities, and if social life were organized around the place of work rather than around the home or neighborhood, many women would develop interests that would compete with family interests. Approximately this policy is now followed in several Communist countries, and even the less developed of these currently have extremely low birth rates.[54]

That inclusion of women in the labor force has a negative effect on reproduction is indicated by regional comparison (18, p.1195; 55). But in most countries the wife's employment is subordinate, economically and emotionally, to her family role, and is readily sacrificed for the latter. No society has restructured both the occupational system and the domestic establishment to the point of permanently modifying the old division of labor by sex.

In any deliberate effort to control the birth rate along these lines, a government has two powerful instruments—its command over economic planning and its authority (real or potential) over education. The first determines (as far as policy can) the economic conditions and circumstances affecting the lives of all citizens; the second provides the knowledge and attitudes necessary to implement the plans. The economic system largely determines who shall work, what can be bought, what rearing children will cost, how much individuals can spend. The schools define family roles and develop vocational and recreational interests; they could, if it were desired, redefine the sex roles, develop interests that transcend the home, and transmit realistic (as opposed to moralistic) knowledge concerning marriage, sexual behavior, and population problems. When the problem is viewed in this light, it is clear that the ministries of economics and education, not the ministry of health, should be the source of population policy.

The Dilemma of Population Policy

It should now be apparent why, despite strong anxiety over runaway population growth, the actual programs purporting to control it are limited to family planning and are therefore ineffective. (1) The goal of zero, or even slight, population growth is one that nations and groups find difficult to accept. (2) The measures that would be required to implement such a goal, though not so revolutionary as a Brave New World or a Communist Utopia, nevertheless tend to offend most people reared in existing societies. As a consequence, the goal of so-called population control is implicit and vague; the method is only family planning. This method, far from deemphasizing the family, is familistic. One of its stated goals is that of helping sterile couples to *have* children. It stresses parental aspirations and responsibilities. It goes along with most aspects of conventional morality, such as condemnation of abortion, disapproval of

premarital intercourse, respect for religious teachings and cultural taboos, and obeisance to medical and clerical authority. It deflects hostility by refusing to recommend any change other than the one it stands for: availability of contraceptives.

The things that make family planning acceptable are the very things that make it ineffective for population control. By stressing the right of parents to have the number of children they want, it evades the basic question of population policy, which is how to give societies the number of children they need. By offering only the means for *couples* to control fertility, it neglects the means for societies to do so.

Because of the predominantly pro-family character of existing societies, individual interest ordinarily leads to the production of enough offspring to constitute rapid population growth under conditions of low mortality. Childless or single-child homes are considered indicative of personal failure, whereas having three to five living children gives a family a sense of continuity and substantiality.[56]

Given the existing desire to have moderate-sized rather than small families, the only countries in which fertility has been reduced to match reduction in mortality are advanced ones temporarily experiencing worsened economic conditions. In Sweden, for instance, the net reproduction rate (NRR) has been below replacement for 34 years (1930-63), if the period is taken as a whole, but this is because of the economic depression. The average replacement rate was below unity (NRR = 0.81) for the period 1930-42, but from 1942 through 1963 it was above unity (NRR = 1.08). Hardships that seem particularly conducive to deliberate lowering of the birth rate are (in managed economies) scarcity of housing and other consumer goods despite full employment, and required high participation of women in the labor force, or (in freer econo-mies) a great deal of unemployment and eco-nomic insecurity. When conditions are good, any nation tends to have a growing popula-tion.

It follows that, in countries where con-traception is used, a realistic proposal for a government policy of lowering the birth rate reads like a catalogue of horrors: squeeze con-sumers through taxation and inflation; make housing very scarce by limiting construction; force wives and mothers to work outside the home to offset the inadequacy of male wages, yet provide few childcare facilities; encourage migration to the city by paying low wages in the country and providing few rural jobs; in-crease congestion in cities by starving the transit system; increase personal insecurity by encouraging conditions that produce unem-ployment and by haphazard political arrests. No government will institute such hardships simply for the purpose of controlling popula-tion growth. Clearly, therefore, the task of contemporary population policy is to develop attractive substitutes for family interests, so as to avoid having to turn to hardship as a correc-tive. The specific measures required for de-veloping such substitutes are not easy to deter-mine in the absence of research on the ques-tion.

In short, the world's population problem cannot be solved by pretense and wishful thinking. The unthinking identification of family planning with population control is an ostrich-like approach in that it permits people to hide from themselves the enormity and un-conventionality of the task. There is no reason to abandon family-planning programs; con-traception is a valuable technological instru-ment. But such programs must be supple-mented with equal or greater investments in research and experimentation to determine the required socioeconomic measures.

REFERENCES AND NOTES

1. *Studies in Family Planning, No. 16* (1967).
2. *Ibid., No. 9* (1966), p.1.
3. The statement is given in *Studies in Family Planning (1,* p.1), and in *Population Bull.* 23, 6 (1967).
4. The statement is quoted in *Studies in Family Planning (1,* p.2).
5. *Hearings on S. 1676, U.S. Senate, Subcommit-tee on Foreign Aid Expenditures, 89th Con-gress, Second Session, April 7, 8, 11* (1966), pt. 4.
6. B. L. Raina, in *Family Planning and Popula-tion Programs,* B. Berelson, R. K. Anderson, O. Harkavy, G. Maier, W. P. Mauldin, S. G. Segal, Eds. (Univ. of Chicago Press, Chicago, 1966).
7. D. Kirk, *Ann. Amer. Acad. Polit. Soc. Sci.* 369, 53 (1967).
8. As used by English-speaking demographers, the word *fertility* designates actual reproduc-tion performance, not a theoretical capacity.
9. K. Davis, *Rotarian* 94, 10 (1959); *Health Educ. Monographs* 9, 2 (1960); L. Day and A. Day, *Too Many Americans* (Houghton Mifflin, Boston, 1964); R. A. Piddington, *Li-mits of Mankind* (Wright, Bristol, England, 1956).
10. *Official Gazette* (15 Apr. 1965); quoted in *Studies in Family Planning* (1, p. 7).
11. J. W. Gardner, Secretary of Health, Education, and Welfare, "Memorandum to Heads of Op-erating Agencies" (Jan. 1966), reproduced in *Hearings on S. 1676* (5), p. 783.
12. C. Tietze, *Demography* 1, 119 (1964) *J. Chronic Diseases* 18, 1161 (1964); M. Mura-matsu, *Milbank Mem. Fund Quart.* 38, 153 (1960); K. Davis, *Population Index* 29, 345 (1963); R. Armijo and T. Monreal, *J. Sex Res.* 1964, 143 (1964); Proceedings World Popula-tion Conference, Belgrade, 1965; Proceedings International Planned Parenthood Federation.
13. *Studies in Family Planning, No. 4* (1964), p.3.
14. D. Bell (then administrator for Agency for In-ternational Development), in *Hearings on S. 1676* (5), p.862.
15. *Asian Population Conference* (United Nations, New York, 1964), p.30.
16. R. Armijo and T. Monreal, in *Components of Population Change in Latin America* (Mil-bank Fund, New York, 1965), p. 272; E. Rice-

Wray, *Amer. J. Public Health* 54, 313 (1964).

17. E. Rice-Wray, in "Intra-Uterine Contraceptive Devices," *Excerpta Med. Intern. Congr. Ser. No. 54* (1962), p. 135.

18. J. Blake, in *Public Health and Population Change*, M. C. Sheps and J. C. Ridley, Eds. (Univ. of Pittsburgh Press, Pittsburgh, 1965).

19. J. Blake and K. Davis, *Amer. Behavioral Scientist*, 5, 24 (1963).

20. See "Panel discussion on comparative acceptability of different methods of contraception," in *Research in Family Planning*, C. V. Kiser, Ed. (Princeton Univ. Press, Princeton, 1962), pp. 373-86.

21. "From the point of view of the woman concerned, the whole problem of continuing motivation disappears, . . . " [D. Kirk, in *Population Dynamics*, M. Muramatsu and P. A. Harper, Eds. (Johns Hopkins Press, Baltimore, 1965)].

22. "For influencing family size norms, certainly the examples and statements of public figures are of great significance . . . also . . . use of mass-communication methods which help to legitimize the small-family style, to provoke conversation, and to establish a vocabulary for discussion of family planning." [M. W. Freymann, in *Population Dynamics*, M. Muramatsu and P. A. Harper, Eds. (Johns Hopkins Press, Baltimore, 1965)].

23. O. A. Collver, *Birth Rates in Latin America* (International Population and Urban Research, Berkeley, Calif., 1965), pp. 27-28; the ten countries were Colombia, Costa Rica, El Salvador, Equador, Guatemala, Honduras, Mexico, Panama, Peru, and Venezuela.

24. J. R. Rele, *Fertility Analysis through Extension of Stable Population Concepts*. (International Population and Urban Research, Berkeley, Calif., 1967).

25. J. C. Ridley, M. C. Sheps, J. W. Lingner, J. A. Menken, *Milbank Mem. Fund Quart.* 45, 77 (1967); E. Arriaga, unpublished paper.

26. "South Korea and Taiwan appear successfully to have checked population growth by the use of intrauterine contraceptive devices" [U. Borell, *Hearings on S. 1676* (5), p. 556].

27. K. Davis, *Population Index* 29, 345 (1963).

28. R. Freedman, *ibid.* 31, 421 (1965).

29. Before 1964 the Family Planning Association had given advice to fewer than 60,000 wives in 10 years and a Pre-Pregnancy Health Program

had reached some 10,000, and, in the current campaign, 3650 IUD's were inserted in 1965, in a total population of 2 1/2 million women of reproductive age. See *Studies in Family Planning, No. 19* (1967), p.4, and R. Freedman *et. al., Population Studies* 16, 231 (1963).

30. R. W. Gillespie, *Family Planning on Taiwan* (Population Council, Taichung. 1965).

31. During the period 1950-60 the ratio of growth of the city to growth of the noncity population was 5:3; during the period 1960-64 the ratio was 5:2; these ratios are based on data of Shaohsing Chen, *J. Sociol. Taiwan* 1, 74 (1963) and data in the United Nations *Demographic Yearbooks.*

32. R. Freedman, *Population Index* 31, 434 (1965). Taichung's rate of decline in 1963-64 was roughly double the average in four other cities, whereas just prior to the campaign its rate of decline had been much less than theirs.

33. S. H. Chen, *J. Soc. Sci. Taipei* 13, 72 (1963).

34. R. Freedman *et al., Population Studies* 16, 227 (1963); *ibid.,* p.232.

35. In 1964 the life expectancy at birth was already 66 years in Taiwan, as compared to 70 for the United States.

36. J. Blake, *Eugenics Quart.* 14, 68 (1967).

37. Women accepting IUD's in the family-planning program are typically 30 to 34 years old and have already had four children. [*Studies in Family Planning No. 19* (1967). p. 5].

38. Y. K. Cha, in *Family Planning and Population Programs*, B. Berelson *et al.*, Eds. (Univ. of Chicago Press, Chicago, 1966).

39. H. S. Ayalvi and S. S. Johl, *J. Family Welfare* 12, 60 (1965).

40. Sixty percent of the women had borne their first child before age 19. Early marriage is strongly supported by public opinion. Of couples polled in the Punjab, 48 percent said that girls *should* marry before age 16, and 94 percent said they should marry before age 20 (H. S. Ayalvi and S. S. Johl, *ibid.,* p.57). A study of 2380 couples in 60 villages of Uttar Pradesh found that the women had consummated their marriage at an average age of 14.6 years [J. R. Rele, *Population Studies* 15, 268 (1962)].

41. J. Morsa, in *Family Planning and Population Programs*, B. Berelson *et al.*, Eds. (Univ. of Chicago Press, Chicago, 1966).

42. H. Gille and R. J. Pardoko, *ibid.,* p.515; S. N. Agarwala, *Med. Dig. Bombay* 4, 653 (1961).

43. *Mysore Population Study* (United Nations, New York, 1961), p.140.
44. A. Daly, in *Family Planning and Population Programs,* B. Berelson *et al.,* Eds. (Univ. of Chicago Press, Chicago, 1966).
45. C. J. Goméz, paper presented at the World Population Conference, Belgrade, 1965.
46. C. Miro, in *Family Planning and Population Programs,* B. Berelson *et al.,* Eds. (Univ. of Chicago Press, Chicago, 1966).
47. *Demographic Training and Research Centre (India) Newsletter* 20, 4 (Aug. 1966).
48. K. Davis, *Population Index* 29, 345 (1963). For economic and sociological theory of motivation for having children, see J. Blake [Univ. of California (Berkeley)], in preparation.
49. K. Davis, *Amer. Economic Rev.* 46, 305 (1956); *Sci. Amer.* 209, 68 (1963).
50. J. Blake, *World Population Conference [Belgrade, 1965]* (United Nations, New York, 1967), vol. 2, pp. 132-36.
51. S. Enke, *Rev. Economics Statistics* 42, 175 (1960); _____ *Econ. Develop. Cult. Change*

8, 339 (1960); _____, *ibid.* 10, 427 (1962); A. O. Krueger and L. A. Sjaastad, *ibid.,* p. 423.
52. T. J. Samuel, *J. Family Welfare India* 13, 12 (1966).
53. Sixty-two countries, including 27 in Europe, give cash payments to people for having children [U. S. Social Security Administration, *Social Security Programs throughout the World, 1967* (Government Printing Office, Washington, D.C., 1967), pp. xxvii-xxviii].
54. Average gross reproduction rates in the early 1960's were as follows: Hungary, 0.91; Bulgaria, 1.09; Romania, 1.15; Yugoslavia, 1.32.
55. O. A. Collver and E. Langlois, *Econ. Develop. Cult. Change* 10, 367 (1962); J. Weeks, [Univ. of California (Berkeley)], unpublished paper.
56. Roman Catholic textbooks condemn the "small" family (one with fewer than four children) as being abnormal [J. Blake, *Population Studies* 20, 27 (1966)].
57. Judith Blake's critical readings and discussions have greatly helped in the preparation of this article.

Japan: A Crowded Nation Wants to Boost Its Birthrate

Philip M. Boffey

Japan is the most crowded nation in the world. It has 102 million people—half as many as the United States—all crammed into a string of narrow islands that are smaller in total area than Montana. Moreover, 85 percent of Japan's territory is mountainous—a scenic splendor but ill-suited for habitation —so the huge population is actually squeezed into a series of narrow valleys and coastal plains. Japan far exceeds any other country in population density per inhabitable area. As of 1968, Japan had 1333 inhabitants per square kilometer of cultivable land, compared with 565 for runner-up Holland.

The resulting congestion seems unbelievable to many Westerners. Farmland is so scarce that one finds crops growing everywhere—up

the sides of steep hills, in the narrow alleys between adjacent railroad tracks, even at the front stoop, where one ordinarily expects to find a lawn. In the cities, and even in rural villages, tiny houses are jammed side by side, with little or no yard space and barely enough room to walk between. Living is so close that privacy is difficult. As one Japanese physician expresses it: "It's a standing joke among us that you can always tell what a neighbor is cooking. If you can't smell it, you can hear the conversation."

Thus it came as a shock to many Westerners last summer when Prime Minister Eisaku Sato publicly advocated an *increase* in Japan's birthrate. Sato's statement, made in a speech to Japanese newspaper editors, seemed to mark a major reversal of Japan's population policy. For the past two decades, Japan has struggled to curb its population growth, and to a large extent it has succeeded. But now, the Prime Minister indicated, the population control effort may have gone too far. Sato noted that Japan's birthrate had fallen below the average for other advanced nations, and he said the government would strive to bring it back up to that average level. Thus, while other world leaders are struggling to curb the widely feared "population explosion," Japan seems to have embarked on a somewhat contrary course.

The Prime Minister's remarks caused great consternation in family planning circles in Japan, for even at the current rate of expansion, Japan's population is expected to rise to 131 million by early next century before starting to decline. Takuma Terao, an economist who is chairman of the Family Planning Federation of Japan, told *Science:* "I am entirely against the idea of raising the birthrate. Japan already has too large a population." Similarly, Minoru Muramatsu, one of Japan's leading authorities on the public health aspects of population growth, said in an interview: "In terms of space, Japan already has too many people. If you live in Tokyo, all you can find is a place to eat and a place to earn money. There is no green, no trees. I don't feel that people are living a very human life."

A High-Level Recommendation

Yet Sato's statement was no irrational, off-the-cuff remark by an uninformed politician. It was based on some cautiously worded recommendations made by the Population Problems Inquiry Council, a cabinet-level advisory group which includes some of Japan's leading demographers. Moreover, the recommendations are aimed at alleviating some potentially serious economic and social problems that are related, at least in part, to Japan's success at curbing its population growth. One such problem is a worsening labor shortage that threatens to undermine Japan's "economic miracle;" another is an increasing number of elderly people who will have to be cared for somehow, particularly now that Japan's traditional descendant family system, in which the younger generations cared for the older, is breaking up.

This article will make no attempt to prescribe what Japan's population policy should be, for the Japanese, one of the world's most highly educated and industrious peoples, are certainly capable of deciding for themselves what sort of future environment they want. But the Japanese situation is worth examining in some detail because the same problems—and the same political and economic pressures—may well arise in this country as the population growth here is brought under tighter control.

Japan has undeniably achieved remarkable success at controlling its birthrate. In the early 1920's, the birthrate stood above 36 per 1000 population, but then it declined moderately and steadily, a phenomenon that usually accompanies the transition from an agricultural to an industrial society. The rate fell as low as 26.6 in the late 1930s before the trend was reversed by the pronatal policy of Japan's military leaders. After the Second World War the rate soared back up as Japan experienced the normal "baby boom" that occurs when soldiers and overseas civilians return home. The birthrate reached 34.3 in 1947 (an intermediate level by world standards) and stayed above 33 in 1948 and 1949, before beginning the precipitous drop that has brought Japan such praise for its "population miracle." By 1957 Japan's birthrate had fallen to 17.2, a historically unprecedented drop of 50 percent in just 10 years. The decline seems especially sharp when measured from the peak of the postwar baby boom, but even compared with pre-war trends, the reduction is considered significant.

What was the secret of Japan's success? Interestingly enough, many Japanese demographers describe the achievement as largely "spontaneous," in the sense that the Japanese people, faced with near-starvation economic conditions after the war, concluded on their own that they should limit the number of children. The news media and women's magazines issued dire warnings, particularly at the height of the baby boom, about the bleak future faced by a nation with too many mouths and a war-ravaged economy, and the highly literate Japanese population obviously got the message. The national government unquestionably helped the population control effort, chiefly by reversing its pronatal policy of the war period. A national Eugenic Protection Law, passed in 1948 and subsequently amended, removed the previous obstacles to birth control, abortion, and sterilization. But many Japanese experts believe the government was always at least one step behind what the people were already doing. One reason for the Eugenic Protection Law, for example, was that so many women were obtaining illegal abortions that the government decided it should protect their health by legalizing the procedure. "The government had no definite policy to bring about population control," says Toshio Kuroda, chief of the migration research division at the government's Institute for Population Problems. "It just happened under the very extraordinary situation after the war. Ten years later people looked back and said we were successful at controlling our population. But no expert in Japan predicted it would happen."

The chief method for curbing the birthrate was induced abortion. The Japanese do not seem to have the strong religious scruples against "taking a life" that have hobbled efforts to increase the use of abortion in this country. Indeed, during the 18th and 19th cen-turies Japanese peasants often resorted to infanticide to get rid of unwanted children at times of crop failure. Today, abortions are legally obtainable for a number of health and economic reasons. In practice, they are said to be obtainable almost at will. The vast majority of abortions are performed by private physicians within the first 3 months of pregnancy, and most of these take place without overnight admission to a hospital or another medical facility. The operations are quite inexpensive, costing an average of $10 to $15, according to one estimate published in 1967. Health insurance benefits often bring the out-of-pocket cost down much further—sometimes even below $1.

Abortions Declining

The number of officially reported abortions (which is believed to represent about half the total number of abortions) reached a high of 1.17 million in 1955 but has since declined to 757,000 in 1968, largely because of government efforts to encourage contraception as an alternative to abortion. In the early 1950's, according to studies by the Institute for Population Problems, abortion accounted for roughly 70 percent of the decline in Japan's fertility while family planning accounted for 30 percent, but in recent years the percentages have been reversed.

The percentage of couples practicing contraception in Japan seems to be somewhat lower than the figure for comparable populations elsewhere. A 1965 survey indicated that about 67 percent of all Japanese couples either had practiced or were then practicing contraception, compared with perhaps 80 to 90 percent for Great Britain and for the white population of the United States. The most popular contraceptive methods have consistently been the condom and the "safe period" or a combination of both. The Japanese make little use of the "pill" or the intrauterine device (IUD), which are mainstays of the population control effort elsewhere, largely because conservative medical opinion in Japan believes it is not wholesome to introduce foreign materials into a healthy body. The government officially prohibits the insertion of IUD's and the sale of oral contraceptives, and while there are large loopholes in these laws, few Japanese use either of the methods.

Japan's success at curbing its population growth is believed to have contributed significantly to the fantastic economic boom that has propelled Japan's gross national product to third rank in the world. If Japan had not curbed its birthrate so sharply, some analysts say, then a sizable portion of the nation's capital resources would have been used to support new additions to the population and would not have been available for economic recovery and industrial investment. Yet the curbing of population growth has not been an unmixed blessing. As conditions have changed in recent years, industry has increasingly complained about a labor shortage, particularly a shortage of young laborers.

I found considerable disagreement as to whether Japan is really suffering from a labor shortage and, if so, what should be done about it. The age composition of the Japanese population has changed considerably over the past decade or two. There has been a sharp decrease, both absolute and relative, in the population of children below the age of 15, and a sharp increase, both absolute and relative, in the population over 65. Meanwhile, the working age population, from 15 to 64, has continued to increase, but at a slower and slower rate. The average annual increase in the working age population exceeded 1 million for the 1965-70 period, but it will drop to 620,000 for the next 5 years and will become negative by the end of the century. When viewed against

the needs of a rapidly expanding economy, the labor pool appears to be shrinking.

"The labor supply has changed rather remarkably from surplus to shortage," says Saburo Okita, director of the Japan Economic Research Center and a member of the Population Problems Inquiry Council. "There is already a shortage of young workers, and while there is still some surplus of middle-aged workers and women, many of us predict there will be a serious labor shortage in the coming years." Some Japanese economists contend that a decline in West Germany's economic growth rate in the late 1950's was caused primarily by a drop in the growth rate for Germany's labor population, and they suggest that Japan's "economic miracle" may be stalled by the same problem.

Seeking Cheap Labor?

Yet Takuma Terao, the economist who heads the Family Planning Federation of Japan, offers a much different analysis. "The industrialists say the labor shortage is very severe," he says. "But I say what is deficient is young labor, which is very cheap. So all we can say is that we lack cheap labor, only that." Terao and most other experts agree that the chief factor behind the shortage of young labor has not been the low birthrates, but rather the great growth in the number of young people who now go on to high school or college instead of beginning work at an early age. Terao believes it would be "rash to raise fertility" simply to assure more laborers. He believes it would be more sensible for Japan to "rationalize" its traditionally inefficient business enterprises so as to gain greater labor productivity. "We already have an abundance of laborers," he says, "but they are not well utilized."

The Population Problems Inquiry Council —the cabinet-level advisory group whose recommendations provided the basis for Prime Minister Sato's remarks—took a middle-of-the-road position. The council, which is made up of some 40 public and private members, including academics and business and labor leaders, was asked in April 1967 to study the implications of Japan's low birthrate. Last August a subcommittee of the council issued an interim report on its findings; a final report is due this year. According to Kuroda, who sat on the council, the interim report represents a "compromise between those who are worried about a labor shortage and those who think Japan is already too populated." The report is said to have been drafted by Minoru Tachi, an eminent demographer who heads the government's Institute of Population Problems. An unofficial English translation was prepared by the U. S. State Department.

The report, if read carefully, does not seem especially earth-shaking. It notes that Japan's population, by some measures, is no longer replacing itself; it warns that this is causing certain problems, and it recommends that Japan seek to achieve a "stationary" population in terms of both total size and age distribution. The report makes no mention of what the ideal population for Japan should be, and as far as I could tell from talking to two members of the council—namely, Kuroda and Okita—there was little discussion of optimum population size. Instead, the report focused its attention on indicators that measure the changing growth rate and age composition of the Japanese population.

The report expressed particular concern over trends in the net reproduction rate, a measure of the extent to which the female population of child-bearing age is reproducing itself with female babies. If the net reproduction rate is 1, the population will potentially become stationary one generation later. If the rate exceeds 1, the population will continu-

ously increase, and if it falls below 1, the population is expected to begin to decrease one generation later. Japan's rate is currently the lowest in the world except for some East European Communist bloc nations. It has remained slightly below 1 almost every year since 1956, generally ranging between 0.9 and 1.

A Rare Occurence

The report states that while the rate has occasionally dipped below 1 in other countries, "it is very rare for such a situation to continue for more than ten years." (The net reproduction rate for the United States was 1.2 in 1967 and has not dropped below 1 since the 1930's.) The report suggests that Japan's population reproductivity is now "too low," and while it acknowledges that "a high population increase rate cannot be welcomed," it nevertheless believes it would be "desirable" for the net reproduction rate to return to 1 "in the near future" in order to ease the "severe changes in population composition by age."

But the report is very careful not to suggest any direct intervention by the government, such as subsidies to support additional children. Instead, the report simply urges the government to improve social conditions so that Japanese couples will spontaneously decide to have more children. The report also recommends that Japan improve its old-age welfare system and increase the productivity of its labor system.

The report gives no hint as to how its recommendations would affect the size of Japan's population, but there is no question that the population will continue to rise substantially. Government estimates for Japan's population in the year 2025 range from a minimum of 129 million (if the net reproduction rate remains below 1) to a maximum of 152 million, with the median projection being 140 million.

The report is so cautiously written that even such critics as Muramatsu acknowledge there is "nothing really wrong with it if you read the text very carefully." After all, who can object to the government improving social conditions? But opponents of the report are upset that mass media stressed the need for more births and largely ignored the question of social improvement. Some also feel the government was premature in its announcement, since they believe the net reproduction rate is already heading back toward 1, or even higher, without any encouragement. Other critics fear the government will eventually decide to intervene in a very direct way to encourage more births, and some even fear that the government's action was partly motivated by a desire to grow soldiers for a future large army.

At bottom, the disagreement is one of priorities. Those who regard economic expansion as the greatest good want more bodies to man the assembly lines. Those who are worried about overcrowding are willing to sacrifice some economic growth in return for more living space. The question of how much living space is desirable, however, is a knotty one. My own reaction to Japan was to be appalled at the overcrowding. But there is some evidence that the Japanese have grown accustomed to their close living conditions and actually even like them. Ichiro Kawasaki, a former Japanese diplomat, has written that the massive stone buildings of the West "overwhelm" Japanese travelers, and they soon "begin to miss the light wooden structures and small landscape gardens to which they have so long been accustomed." Similarly, Maramatsu, who spent several years at Johns Hopkins University and who frequently travels abroad, laments that many Japanese have no idea what he is talking about when he extols the "spacious way of living" in other countries. "For generations," he says, "many of our

people have been living under the same conditions, so they don't question whether it is wrong or right."

Such differing attitudes toward space needs make it difficult for the experts in one advanced nation to suggest the best population policy for another advanced nation. Such differences also make it difficult to visualize how much of a burden population growth in any one country would really impose on future generations in that country. Perhaps future generations will enjoy living shoulder to shoulder.

The Lesson of Japan's Experience

Japan's decision to boost its birthrate slightly may have an impact and significance beyond its own borders. Some family planning advocates fear Japan's action may throw a monkey wrench in worldwide efforts to curb population growth by somehow downgrading the importance of birth control. Others fear Japan demonstrates that radical population control can never succeed, for the minute a nation reaches the point where its population is apt to level off and then decline, various pressures—political, economic, and nationalistic—build up to reverse the trend. Both views are probably too apocalyptic, for Japan is merely trying to boost its net reproduction rate by a modest amount until it returns to 1 —a level that is considered the desirable goal by planners in many other countries. Some U. S. experts, for example, have called for a stationary population and a zero rate of growth, and that is precisely what Japan is seeking.

The real significance of Japan's experience may be that it underlines the costs involved in achieving population control. Some experts in this country, such as Ansley J. Coale, director of the Office of Population Research at Princeton University, have pointed out that a stationary population and a zero growth rate have unfavorable as well as advantageous effects. Coale suggests, for example, that a stationary population "is not likely to be receptive to change and indeed would have a strong tendency towards nostalgia and conservatism." He also suggests that such a society would no longer offer "a reasonable expectation of advancement in authority with age," since there would be essentially the same number of 50-year-olds as 20-year-olds. Zero growth is unquestionably desirable at some point before crowding becomes painful, but in the current rush to jump on the population control bandwagon, it is well to remember that population control is not an unmixed blessing. There are costs involved, and someone will have to pay them.

Prospects for the Future: Resources

It is clear that stabilization and subsequent reduction of the human population are very distant achievements, even on the basis of wildly optimistic assumptions. It is also clear that no effective steps have yet been taken toward such goals.

The human population that exists today, and the unavoidable immediate increase, can be supported only on the basis of an immense and highly technical industrial establishment, which consumes resources at an enormous rate. Industrial and post-industrial societies have proved themselves prodigal in their use of raw materials. There is little cause to anticipate voluntary austerity on the part of the industrialized nations, either in order to extend supplies of critical resources, or to achieve a more equitable distribution to mankind as a whole. On the contrary, the industrialized nations appear bent on increasing their already disproportionate consumption, while the underdeveloped nations are just as desperately bent on matching the rate at which their more fortunate neighbors are dissipating the finite resources of Earth. The realities of politics and present-day economics, like those of population growth, seem to insure that the ruthless gutting of resources will continue and accelerate.

What, in the face of unrestrained looting, are the resource prospects? What will be the environmental costs of the orgy?

Mineral resources are fundamental to all other resource problems. It is perhaps significant that no comprehensive survey of the mineral situation was available for inclusion in this anthology. Yet it appears from available references that even if a zero population growth were instantaneously attained and the present unconscionable gap between the haves and the have-nots were allowed to persist, thus fixing exploitation at its present rate, known supplies and foreseeable discoveries of at least ten major minerals will be exhausted by the year 2000. Included in the list are the most useful, least polluting, and most economically advantageous energy sources. Known and anticipated discoveries of natural gas will be exhausted on a worldwide basis in less than 15 years, assuming no increase in present rates of use! It is ironic that we hear serious proposals for conversion of automobiles to liquified natural gas as an anti-pollution measure!

Despite the finite nature of known reserves and predictable discoveries of usable deposits of many critical minerals, the majority of writers on the subject are afflicted with a curious optimism, a kind of fairy-godmother-itis. The fairy godmother currently in favor is technology, particularly the technology of nuclear energy, the philosopher's stone of today's alchemy. Assisted by various other lesser technologies, nuclear technology is supposed to make practical the exploitation of ever less accessible, ever less concentrated, mineral deposits. It will not just convert base metals into gold. It is supposed to convert anything into anything, given only the necessary economic demand generated by an ever increasing gross national product.

Externalized, in this optimism, is the fact that as ore grade and accessibility decrease, the energy cost, and the environmental costs in terms of pollution, erosion, and loss of arable land, wildlife habitat, livable space, and aesthetic values increase exponentially. Furthermore, the premise of inexhaustible, economically available nuclear energy, which is the key to the whole resource fantasy, is dubious. The supply of currently usable nuclear fuel (Uranium 235) is adequate for less than 20 years at current rates of use (Preston Cloud, in the *Texas Quarterly,* Vol. 11, no. 2, page 123). Those who believe nuclear energy is the philosopher's stone place their faith in the "breeder reactor" which can make use of the abundant Uranium 238, or in the fusion reaction. We seem to be making little progress with the fusion process, and if the breeder reactor is ever successfully developed, the U-238 supply is only adequate for up to 3000 years, which is, after all, less than the duration of the ancient Egyptian civilization.

Beyond the fact that neither nuclear fusion nor the breeder reactor have been proven operable, there are some other considerations. It appears that nuclear power plants have a limited life span (25 years?) before unavoidable contamination converts them to inoperable and deadly mausoleums. The problem of disposal of radioactive wastes has not yet been solved, nor has the problem of thermal pollution, nor that of biological concentration of even minimal levels of radioactive effluents. Even if the technical problems of nuclear energy are solved to the point where the economic costs, and the risks of catastrophic nuclear accidents are acceptably low, the environmental costs may be prohibitive. Without adequate energy supplies the fairy godmother outlook for mineral resources becomes wholly untenable and critical shortages will appear within ten years.

Fresh water is a critical mineral resource which has long been in inadequate supply in certain regions. It is reusable and renewable, but human use is reaching the level where the hydrologic cycle does not renew supplies at the rate demanded on either a worldwide or regional basis. The response of our technological establishment to the rising demand has generated projects which involve spiralling environmental and human costs. The Aswan dam has destroyed a major ocean fishery and caused an epidemic of virulent disease, yet far more ambitious and potentially destructive plans are underway. In Siberia three major arctic rivers are being diverted southwards into central Asia. The climatic consequences may include the development of desert or semi-desert conditions along the entire northern coast of the Mediterranean Sea from Italy eastwards. The proposed NAWAPA plan in Canada has the potential for similar climatic changes. Other projects, underway or in advanced planning stages, will drain all the rivers of the northern California coast dry before they reach the sea. These projects, and others like them, are generated by rapidly increasing domestic, industrial, and agricultural demands for water. They have the potential for enormous environmental damage in the form of diversion or elimination of nutrient discharge to coastal ocean waters, extermination of vertebrate and invertebrate populations of whole river and estuarine systems, regional or worldwide changes in patterns of precipitation and temperature, and consequent disruption and extirpation of major ecological communities on a heretofore unimaginable scale within time spans of one or two decades. The pressure for such water projects is so great that potential consequences may be given little or no consideration.

It is difficult to believe that the pattern evident in our utilization of water will not be repeated with respect to one mineral resource after another in the face of increasing demand and diminishing supply. It is pleasanter to listen to optimists than to realists, or pessimists. Despite the current rise in environmental concern it seems improbable that today's political and economic establishments are capable of resisting the pressure to exploit resources at the maximum rate without regard to future needs or consequences. The voices of those who believe in an infinite earth, and a magical technology, are likely to prevail.

As is the case with the mineral resources, there are many who maintain that future food needs can be met. It appears that they may be on firmer ground in the short run, but these optimists also arrive at their rosy view by ignoring or overlooking certain peripheral problems and environmental costs. The recent advent of the "miracle" grains, for example, has been publicized without consideration of numerous limitations, attendent economic problems, and vastly increased requirements for fertilizers and pesticides.

If the weather is favorable, recent agricultural advances may make it possible to meet caloric and protein needs at the present, or slightly improved, levels through 1985. This "optimistic" view is predicated upon political stability, continued economic growth and industrial development in the underdeveloped countries, and increasing availability of mineral resources and energy supplies. It also assumes increasing financing for international aid programs (funds for this purpose are in fact decreasing). Finally, it is dependent on a ten-fold increase in total world fertilizer use and a six-fold increase in pesticide use. The pesticides in question will be those now on the market, including the persistent chlorinated hydrocarbons.

There have been attempts to take stock of worldwide food and mineral resources. Generally one of these areas has been evaluated with little or no reference to the other. Even if a highly accurate combined assessment were available, however, it would be of little use without a system for continuously monitoring supply and demand, anticipating shortages and taking remedial action, and rationing use on an international basis. An international political organization with the power to impose austerity on the developed nations in order to distribute the limited resources of "Spaceship Earth" wisely and fairly is scarely imaginable tomorrow, or in 30 years.

The combined prospects of population growth and resource exhaustion appear to leave no ground for doubt. Barring a last-minute inter-stellar rescue effort by some super-civilization from outer space, we are irrevocably committed to a mass human die-off within the next 20 or 30 years. Rescue would be nice, but the problem has been obscured by too much science fiction already. The question is likely to be whether man, or even the ecosystem, can survive the imminent and apparently inevitable catastrophe.

The Future of World Energy
Sir Harold Hartley

Looking back at the gradual evolution of civilization, Man's use of energy has played an essential part at each stage, both in meeting his domestic needs and comfort and in making possible the winning of raw materials, their processing and transportation. For millenia Man had relied on wind and water power until 200 years ago Watt's steam engine quickened the great expansion of the Industrial Revolution, based for a century on coal. Later, in the last hundred years, oil, natural gas, and now nuclear energy have partly replaced it. It was the local availability of coal that determined so largely the emergence of western Europe and North America as the first great centres of industry, but now industrial nations are drawing on all the sources of primary energy.

The statistics of energy production and consumption in the J series of UN statistical papers, *World Energy Supplies,* enable us to take a comprehensive view of the changing pattern of the energy industries of the world since 1950 when conditions were stabilized after the Second World War. Table 1 shows the steady increase in the world's consumption

of primary energy, in terms of the coal equivalent of each fuel, and the changes in the proportion contributed by the various sources. Although liquid and gaseous fuels have largely supplied and indeed created the rising demands for energy, the actual consumption of coal has continued to rise, due mainly to its increased use in the Communist countries. The percentage contribution of hydro power has hardly varied although its actual contribution has doubled.

The different sources of primary energy are distributed very irregularly over the Earth's surface. Ninety per cent of the proved reserves of hard coal are in North America, the Soviet bloc, Germany and the UK, 90 per cent of the proved reserves of soft coal are in Germany, Australia and the USSR, and 80 per cent of the proved reserves of petroleum are in the Middle East and the Soviet bloc. Table 2 shows the different patterns of consumption of primary energy by seven industrialized countries in 1966 and the proportion of energy they import, due mainly to their lack of resources of oil, the USSR being the only net exporter. This

TABLE 1
World consumption of primary energy
(Million metric tons of coal equivalent)

Year	Solid	Liquid	Gas	Hydro	Nuclear	Total
1951	1626	705	318	200	0	2849
percentage	57.1	24.7	11.2	7.0	0	
1956	1882	1028	427	228	0	3565
percentage	52.8	28.8	12.0	6.4	0	
1961	2035	1394	666	289	0	4384
percentage	46.4	31.8	15.2	6.6	0	
1966	2296	2075	1013	373	13	5770
percentage	39.8	36.0	17.6	6.5	0.2	

TABLE 2
Percentage consumption of primary energy
and percentage imported in 1966

	Solid	Liquid	Gas	Hydro(1)	Nuclear(1)	Imports
UK	61.0	35.2	0.4	0.6	2.8	36.2
France	38.3	43.8	4.6	12.9	0.4	53.3
Germany	52.6	41.8	2.0	3.6	0.0	35.6
Italy	10.6	62.2	10.2	16.0(2)	1.3	76.2
USA	23.1	38.8	34.1	4.0	0.1	8.0
USSR	47.1	28.2	20.9	3.8	unknown	−12.9
Japan	31.3	52.6	1.3	14.7	0.1	67.7

1. Hydro and nuclear electricity converted at current average rate of thermal efficiency (32.4 per cent).
2. Includes 1.0 per cent geothermal.

TABLE 3
1966

	Percentage of world population	Percentage of world energy consumption	Per capita energy consumption (metric t.c.e.)	Average annual per cent increase in per capita consumption
North America	6.5	37	9.44	2.0
West Europe	10.3	20	3.14	2.5
Communist countries	32.3	30	1.51	5.6
Sub total	49.1	87	2.90	3.3
Rest of the world	50.9	13	0.44	4.2

mal-distribution of energy reserves poses a difficult problem for many of the developing countries which suffer from a shortage of indigenous fuels.

Aurelio Peccei, in his recent book *The Chasm Ahead,* has emphasized the urgency of the world problem presented by the growing gap between the economies of the richer and the poorer countries. Their average personal income is roughly proportional to their individual consumption of energy and Table 3, which summarizes the consumption of energy by three main groups of nations in 1966, re-

veals the immense gap between their energy consumption and that of the rest of the world. Forty-nine per cent of the world's population consume 87 per cent of its commercial energy, and if three other advanced countries—South Africa, Australia and Japan—are included, there remains only 7-1/2 per cent of the commercial energy to be divided between 48 per cent of the world's population in the less fortunate countries. One result is that they burn wood and the agricultural wastes and dung that are so badly needed as fertilizers. Clearly supplies of energy must take a high priority in the plans to lessen the gap between the richer and poorer countries. But the gap is so large that it is unrealistic to think of equating conditions in the poorer to the lavish pattern of consumption of the richer, and aid must be planned to meet their major needs, employment, nutrition and water.

The contrast between them I can illustrate by what I saw in 1957 in the north-west province of Pakistan and in the Snowy Mountains of Australia. Warsak, the site of the new power station on the Cabul River, was like a human ant-heap with many hundreds of men and women carrying small loads of earth and stones on their heads to make the dam. The station is now providing electricity and irrigation water to bring some thousands of acres of land into cultivation and provide a new means of livelihood to the local tribes. When I asked the chief minister what he would do with the electricity he said, "Light their houses. More light, less murder," and he might have added, "fewer babies." In the Snowy Mountains the water of the Snowy River is being impounded by great dams and led by tunnels through the mountains from east to west and then dropped through power stations with an ultimate capacity of 3740 megawatts and distributed to provide two million acre-feet of water to irrigate great tracts of land needing only water to make them fertile. So the objectives of the two projects were the same, but at Snowy I saw few men and a great assembly of machines of various types. In the richer countries energy is being used lavishly in mechanization and automatic control to replace highly paid labour, while in the poorer, energy must be directed to meeting their most pressing needs by developing their indigenous resources and creating employment. As an example, the discovery of natural gas at Sui in West Pakistan enabled a power station to be built which supplies current to the tube wells in the Rechna Doab and to the pumps in the drainage canals, which have rescued a great area of cultivated land from destruction by salination.

After this brief summary of the recent patterns of energy consumption, I turn now to what I know from experience is the hazardous task of trying to forecast the future. We are in a period of rapid change and without statistics based on the operating experience of the next 10 or 15 years it becomes almost a guessing competition as great developments are in hand. The consumption of electricity in recent years has been rising steadily at about 8 per cent compound, as Figure 1 illustrates. No doubt in the richer countries the rate of increase may fall but my guess is that this will be balanced by an increase in the poorer countries, helped by international aid, so urgently needed. At the same time the annual rate of growth of the world's consumption of primary energy, expressed in coal equivalents, has been 5-1/4 per cent compound, due partly to the growth of per capita consumption and partly to the increase in population. The convention of coal equivalents is not altogether satisfactory as it neglects the differences in the efficiencies with which the various fuels can be used. However, it is the best method we have and any attempt to separate the amounts of fuel used in generating electricity from those used for other purposes would be increasingly invalidated by the technology of "total energy"

in which the heat losses in generation are harnessed to useful purposes.

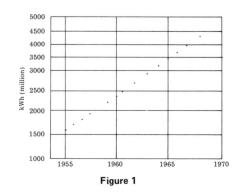

Figure 1

If these two rates of growth continue the primary energy consumption in the year 2000 will be about 30000×10^6 metric tons of coal equivalent and the consumption of electricity will be about $50000 \times 10^9 kWh$. A rough estimate indicates that 30 per cent of the primary energy in four of the largest industrialized nations in 1967 was used in generating electricity. By the year 2000, on this extrapolation, one half of the primary energy would be so used. This seems a reasonable progression as electricity takes over more and more of the present fuel functions. The amount of carbonaceous fuels then available for other purposes will depend on the proportions of electricity generated by nuclear reactors and hydro. This brings me to the major hazard of estimating the growth of nuclear power when there are several competing systems of nuclear reactors and there are serious engineering problems associated with the breeder reactor, which, if it is an economic success, will effect such a great economy in the consumption of uranium. Otherwise there are doubts about the reserves of uranium ore. We lack also long-term data of the cost of operating advanced reactors. From the projected rates of construction of nuclear power stations I estimate that by 1985 they will generate 20 to 25 per cent of the world's electricity. Philip Sporn has estimated that by the year 2000 half of the electricity in the USA will come from nuclear stations. For the world as a whole, considering the vast capital expenditure involved and the local need for sophisticated engineering skills, my guess is that about 40 per cent of its electricity will come from nuclear reactors. This may well err on the conservative side. However, assuming that nuclear power and hydro will together supply half the electricity, this would leave three-quarters of the primary energy input to be supplied by carbonaceous fuels. Hydro has two advantages: it is renewed by Nature and it also provides irrigation water to help to solve the world's food problem. It runs the risk of the silting up of storage reservoirs unless steps are taken to reduce soil erosion based on ecological studies, as was done in the Tennessee Valley and in the Snowy Mountains of Australia.

On the above assumptions about 22,000 million metric tons of coal equivalent of carbonaceous fuels would be available by the year 2000 to generate their share of electricity and for other purposes. In forecasting the probable amount of each we must take account of the present trends, the proved reserves and possible changes in types of demand. The proved reserves reported by the World Energy Conference in 1968 are shown in Table 4.

There are thus ample reserves for the consumption forecast for the next 30 years, with the possible exception of natural gas, and it is significant that a considerable effort in the USA is now devoted to the gasification of coal in case it should be needed.

My guess is that the amounts of fuel consumed in 2000 will be roughly as shown in Table 5.

Opinions will differ but these figures will provide data for discussion.

The rapidly growing consumption of energy will demand every effort to increase efficiency, both in its production, distribution and use. There will be a great incentive to develop means of storage to spread the electricity load and to avoid waste of energy. The concept of total energy is not new. In 1925 at the Beckton gas works a back-pressure turbine gave us

TABLE 4

	Proved reserves of fuels in 1968	Mean annual consumption 1961-2000
	10^9 metric tons of coal equivalent	10^6 metric tons of coal equivalent
Hard coal	400	3100
Brown coal and lignite	180	
Peat	90	small
Oil shales and tar sands	178	4300
Petroleum	72	
Natural gas	40	3100

TABLE 5
Estimated consumption of primary energy in 2000 AD

	10^6 metric tons of coal equivalent	Percentage share of total	Average increase since 1961 (per cent per annum)
Solid	4500	15	2.1
Liquid	9500	32	5.0
Gas	8000	27	6.6
Nuclear & hydro	8000	27	8.9
Total	30000	100	5.1

some of the cheapest electricity and process steam in the country. There is a wide field of application for the principle of total energy, for example in the concept of super-power stations, serving the electrochemical industry and using waste heat as process steam or to desalinate sea water. There are also the possibilities of storing energy in various ways, thermal storage, storage batteries and pumped storage, to utilize off-peak power. The energy pattern of the future will depend on technological developments and on the availability and the costs of different types of fuel, but the role of electricity, the most sophisticated form of energy, will continue to increase.

To sum up: the energy industries of the world are now passing through a phase of rapid transition. With several competing systems of nuclear reactors, the dependence for supplies of cheap uranium on the economic success of the breeder reactor, the absence of long-term operating costs of advanced reactors, the possibilities of nuclear fusion and the doubts about MHD, it might not unreasonably be called an experimental phase. Hence nations will be well advised to maintain in being those energy industries that show promise of potential production, as in the long run all may be needed. However, some may regard this view as just an old man's caution.

I am most grateful to J. S. Huggins for compiling the statistical tables, which are the basis of this paper.

Non-Fuel Mineral Resources in the Next Century
T. S. Lovering

The total volume of commercial mineral deposits is an insignificant fraction of 1 per cent of the earth's crust and each deposit represents some accident of geology in the remote past. It must be exploited where it occurs. Each has its limits, however, and if worked long enough must sooner or later be exhausted. No second crop will materialize; rich mineral deposits are a nation's most valuable but ephemeral possession—its quick assets. Continued extraction of ore leads to increasing costs as the material mined comes from greater and greater depths, but sometimes improved technology makes possible continuation or renewal of work in deposits that would otherwise have been shut down because of competition with deposits more favorably situated or where cost per unit of output was appreciably less.

Demand for mineral products comes chiefly from the chemical and manufacturing industries and from agriculture. The constant arrival of large quantities of raw materials is essential to maintaining output and continuing output is vital to a healthy industrial economy. An adequate supply is more easily assured if the mineral deposits are under domestic control and within the national boundaries of the country where they are to be consumed. The greater the dependence on foreign sources the greater the risk of industrial stress caused by foreign institutional action such as price rises due to cartel action, trade barriers set up by governments, or by adverse military action.

Satisfactory substitutes for most raw materials exist if the price of the substitute is not a consideration. Some minerals, however, have unique properties for which there is no satisfactory substitute and many others are essential to successful commercial competition in the world markets. The minerals that are currently most essential to civilization are probably coal, iron, copper, aluminum, petroleum, and the fertilizer minerals. Only a very few industrial countries have adequate internal sources of all these minerals to supply their current industrial needs and none have reserves adequate for the next century using foreseeable technology. Although it is possible in times of distress, when free access to world sources is cut off, to work deposits that otherwise are noncommercial, it must be remembered that low-grade deposits require much time and capital to develop. Economic chaos can result if foreign sources of supply are denied to a country that has allowed itself to become dependent on them. And dependent all industrial nations are, entirely or in large part, on foreign sources of supply for some or most of the essential metals demanded by their industries. This dependence grows ever greater with the years.

That sources of supply will inevitably shift is evident not only from the hard geological facts but also from a review of historical events. Three thousand years ago the Middle East was the center of the iron mining industry of the ancient world. For several hundred years ancient Greece was the center of lead and silver mining of the western civilized world. For a long time Germany was the leading producer of lead, zinc, and silver in medieval Europe, but Belgium became the chief producer of zinc at the beginning of the industrial era. In the nineteenth century Great Britain was successively the world's foremost producer of lead, of copper, of tin, of iron, of coal, and during that period she was the wealthiest nation in the world; from 1700 to 1850 the United Kingdom mined 50 per cent of the world's lead, from 1820 to 1840 she produced

45 per cent of the world's copper, from 1850 to 1890 she increased her iron production from one-third to one-half of the entire world output. At the turn of the century, Russia was the leading producer of petroleum in the world and during the latter part of the nineteenth century she was the foremost gold producer—until the discovery of the great Transvaal gold deposits in South Africa. Most of the manganese ore mined until after the second World War, came from a comparatively small area in southern Russia. Now large new deposits in Africa, Australia, and Brazil threaten the dominance of the U.S.S.R. and have cut deeply into the manganese market.

Hewett (1929) made a penetrating analysis of the normal history of mineral exploitation in a nation and recognized five stages between discovery and exhaustion. In the forty years since his study, exploitation of mineral resources has strengthened his conclusions that discovery is followed successively by flush production, proliferation of smelters, decline in production, cessation of export metal, growing dependence on imports, and ultimately complete dependence on imported metals.

Between 1905 and 1938 more metal was produced than had been consumed in the entire history of the world prior to 1905; a similar amount of metal was produced and consumed in a far shorter interval after the start of World War II. In the meanwhile the world population has more than doubled so that per capita consumption has not increased appreciably even in industrial countries. The reserves of known and undiscovered ore deposits of commercial grade are finite and diminishing, whereas the demand for metals is growing at an exponential rate; it is clear that exhaustion of deposits of currently commercial grade is inevitable. The cost of mining deposits of lower grade involves increased costs for capital, labor energy, and transportation; decreased costs of energy from the ultimate development of breeder reactors will only cut energy costs a relatively small amount because their major cost is that of invested capital.

It would seem that grave concern may be justified over mineral supplies during the lifetime of those now living. There is, however, an optimism among mineral economists now (1968) that may be shortsighted. Future mineral resources depend on the cost of supplying market places, and this in turn is a function of relative locations, cost per unit of mine output, market price, the market location, and any institutional incentives or restrictions placed on operations. The current but questionable assumptions of the majority of mineral economists as given in The Paley Report, the book *Scarcity and Growth* (Barnett and Chandler, 1963) and *Natural Resources for U.S. Growth* (Landsberg, 1964) may be summarized as follows:

1. Technology for the past fifty years has steadily made increasing amounts of raw material available at lower costs per unit and, therefore, will continue to do so into the "foreseeable future;"

2. As the grade of a mineral deposit decreases arithmetically the reserves increase geometrically;

3. Non-renewable resources are therefore inexhaustible;

4. Scarcity can always be prevented by a rise in price of the raw materials and

5. Since the cost of raw materials is only a fraction of the final cost, a material rise in price for any raw material will have an insignificant effect on the price of manufactured items and on the general economy;

6. Any industrial nation will have adequate access to deposits throughout the world;

7. There will be only insignificant institutional restraints on access;

8. The United States and other industrial nations must have an "ever-expanding economy"—the GNP (Gross National Product) increasing continuously at an annual rate of about 4 per cent because their economic well-being requires it;

9. The population of western industrial nations will continue to increase at a rate of about 1.5 per cent per year for several generations; and finally,

10. The under-developed nations will achieve a per capita income comparable to that currently enjoyed by the United States within one or two generations.

All these assumptions are debatable; some are based more on rhetoric than reason, some seem a sort of struthionic optimism unrelated to physical factors, some are simply wrong. If accepted as guides to future policy they may lead to a lethal complacency as to natural resources.

The basic assumptions underlying the current optimism concerning the adequacy of mineral resources stems in large part from Lasky's classic paper (Lasky, 1948, 1950) in which he analyzed the relations between grade and reserves of ore in eight copper porphyries of the Southwest. The optimism generated by the current interpretation of Lasky's work has been greatly increased by Barnett and Morris (1963) whose book *Scarcity and Growth* and summary article "The Myth of Our Vanishing Resources" (Barnett, 1967) formulate clearly, specifically, and persuasively the basic position and the happy philosophy of most mineral economists. They reformulate the Ricardian hypothesis of scarcity as follows (p. 249): "The character of the resource base presents man with a never ending stream of problems;—that man will face a series of particular scarcities as the result of growth, is a foregone conclusion; that these problems will impose general scarcity as shown by increasing costs per unit output is not a legitimate corollary." According to them (p. 230), "technological progress is automatic and self-reproducing in modern economies." They state that "natural resource building blocks are now to a large extent atoms and molecules . . . units of mass and energy, and the problem thus is one of manipulating the available stores of iron, magnesium, aluminum, carbon, and oxygen atoms—even electrons—so that we obtain those resources required by industry." The exuberant optimism reflected in such a belief leads Barnett and Morris to conclude: "The progress of growth generates antidotes to a general increase in resource scarcity" (p. 290), and to believe also that the sea is a continually augmenting store of resource of all kinds, ready at hand to supply man bountifully for all future time. Barnett and Morris state (p. 249) that principles "that have clear relevance in a Ricardian world where today's depletion curtails tomorrow's production, have little if any relevance in a progressive world." This will be heartening news to all mine owners!

Barnett and Morris looked at the net decrease in unit costs for the mineral industry and certain specific mineral commodities over a period of some seventy-five years and concluded that the improvement in efficiency caused by improved technology had resulted in a continuing net decrease in the cost per unit of mineral extracted; furthermore, that this increased efficiency would continue into the indefinite future. Within the mineral industry, the increased mechanization that accompanied the increased production of coal and oil, resulted in current costs of one-third to one-fourth those of 1920; this Barnett interprets as showing the opposite of increased scarcity or the lessening of economic quality. The curves for lead and zinc, however, show no such change and rather an increase during this period. For copper and for iron, the trend towards lower unit cost flattens greatly between 1940 and 1960 and as shown in their graph for timber, the trend for many years has been the opposite of that which Barnett postu-

lates for minerals. Conservationists, who are unduly belabored by Barnett (1967) might make something of this!

Changing costs in mineral extraction reflect the relative efficiency of operation as opposed to the increasing costs of capital, labor, and energy. The efficiency in operations must approach an irreducible minimum as maximum mechanization is achieved; it would require major innovations not yet in sight to start another marked downward slope in the curve representing price times grade. The present trend of the ratio of copper grade to price (in constant dollars) is not encouraging (Figure 1) but the discovery rate is.

Lasky's analysis of the relation of reserves to tonnage and average grade of ore produced in copper porphyries (Lasky, 1950) resulted in his stating the principle that is now known as the "arithmetic-geometric ratio" (or simply the A/G ratio): The reserves of ore increase geometrically as the average grade mined decreases arithmetically, or as Lasky (1951) expressed it

$$\text{Grade} = K_1 - K_2 \log \text{tonnage}.$$

The problem of exhaustibility thus need cause no concern according to current (1967-68) economic philosophy because over a long period of time the decreased grade that is now mined, shows that reserves have increased or at least been maintained. The past half century refutes the gloomy forecasts made in 1912 which indicated that the United States would run out of many major resources long before now. From this unequivocal fact, some very dubious deductions stem.

In referring to his equation and the curve it generates, Lasky (1948) said, "It fits the porphyry coppers . . . and apparently also other deposits of similar type in which small quantities of ore minerals are scattered through great volumes of shattered rock." According to Lasky then, "It may be stated as a general principle that in many mineral deposits in which there is a gradation from relatively rich to relatively lean material, the tonnage increases at a constant geometric rate as the grade decreases." A typical curve for a porphyry copper plots as a straight line on semilogarithmic paper for which the decrements in grade are represented by the arithmetically spaced vertical lines and the increments in tonnage are scaled by logarithmic ordinate lines. The straight line plot represents the cumulative production or tonnage of a porphyry copper for decreasing average grade of the *total tonnage extracted* as ever leaner ore is mined. *This constraint is a very important one*—but has apparently been misunderstood by many economists who use this analysis as a springboard from which to leap into a sea of optimism.

Lasky averaged the characteristics of the eight major porphyry copper deposits in the United States (as of 1950) and notes that for such an average deposit, there would be somewhat more than 600 million tons of ore averaging about 0.6 per cent of copper when it had been *mined out—down to and including* " *zero cut-off grade.*" In this average deposit, the tonnage increases at a rate of about 18 per cent for each unit decrease in grade of 0.1 per cent. Most of the copper is contained in the 175 million tons having a copper content between 0.5 and 0.9 per cent. Lasky's analysis is a major contribution to our concepts of grade and tonnage for the type of deposit that he considers. It should be noted, however, that the curves rigidly defined the situation over a range of grade of about 1.5 per cent, but that even for porphyry coppers they cannot express the relations for higher grades of ore and for the very lean rocks which should be represented by the extensions of the curve to the left and to the right respectively. For his average copper porphyry deposit which contains sixty million tons of ore averaging 2 per cent copper

and six hundred million tons of ore averaging 0.6 per cent of copper, Lasky's mathematical expression is "Grade = 12.9 - 1.4 log tonnage;" from this we deduce that Zero grade = 12.9 - 1.4 X 9.2 and as 9.2 = log (1.58 X 10^9). Zero grade is reached at 1.58 X 10^9 tons, *which is impossible!* The clarke (average abundance of an element) of many igneous rocks is 0.004 to 0.010 per cent copper; for Lasky's average porphyry copper deposit in question, an average grade of 0.3 per cent copper corresponds to a tonnage of one billion tons—not an unreasonable figure if the cut-off grade is held close to the average grade mined. Somewhere between a grade of a few tenths of a per cent and a hundredth of a per cent, the tonnage must increase astronomically but there is no geologic reason why the curve should maintain its slope or change smoothly from the one calculated by Lasky's formula into the curve that might express the change in copper content of the various rock units found in the crust of the earth. It is more than likely that the cumulative curve would first flatten and then rise precipitously. At the other end of the curve, consider what happens for ore minerals. The formula shows a "tonnage" of one milligram of ore having a grade of 24.5 per cent copper or one ton (out of sixty million tons averaging 2 per cent copper) having a grade of 12.8 per cent copper. It is obvious that the curve generated by the Lasky formula departs widely from reality between 0.01 per cent and 0.3 per cent on the one side and for any masses of ore containing substantial amounts of the common copper ore minerals.

Most ores are deposited over an appreciable time interval by complex processes. The large mass of mineralized intensely fractured rock that marks deposits of the porphyry copper type does have its limits, and beyond these limits the metal content drops off sharply.

Most such bodies show a history of repeated fracturing and mineralization; in some several stages of metallization are present, in others only one stage of metallization but several stages of barren alteration may be represented. Where the fractured rock has been enriched by successive waves of metallizing solutions, each wave is apt to be localized in somewhat restricted masses of refractured rock; these localized blocks of better ore result in a stepwise change in grade for the deposit as a whole, but such a change is not reflected in the product of mining which represents a predetermined mixture. Where a very low grade protore—less than 0.10 Cu—has been enriched by weathering processes, a shallow mass of ore may carry ten times as much metal as the protore. In such porphyry copper deposits there is an abrupt transition from ore grade to unmineable rock. Any attempt to use an A/G curve for them is obviously absurd. Lowering the grade by decrements of 0.10 per cent would increase the total tonnages of ore by only a small fraction until the grade of the protore was reached, when an enormous increase might take place in reserves of protore, though not necessarily in total tonnage of contained copper.

Even though the porphyry copper type deposits share many characteristics in common, each individual deposit is unique. It is clear that generalizations based on a few biased samples will not apply to the entire population; the extension of these generalizations to deposits of entirely different origin and geologic habitat is not only unwarranted, unscientific, and illogical, it is also downright dangerous in its psychological effects.

In his cautious first statement of the problem, Lasky says that the curve generated by his equation is meaningful for many types of deposits, but he adds the proviso "the geological evidence permitting." It is this reservation that has been neglected. There is absolutely no

geological reason for concluding that the so-called arithmetic-geometric ratio holds for ore deposits in general. Most especially is it an error to believe there must be undiscovered low grade deposits of astronomical tonnage to bridge the gap between known commercial ore and the millions of cubic miles of crustal rocks that have measurable trace amounts of the various metals in them. The closer the mineable grade approaches the clarke of an element for major rocks, the more probable is the existence of an exponential ratio of tonnage to grades far below that of commercial ore.

The geologic processes that operate to give us mineable ores include huge tonnages provided by *sedimentation;* this process will cause dilution as well as concentration in detrital deposits and there are, of course, far larger volumes of sediments containing metals in the parts per million range—except for iron, aluminum, and magnesium—than there are that contain ore metals in per cent amounts. In sedimentary rocks, however, the A/G ratio may well hold far down towards the clarke of certain metals. The areas where the greatest concentrations of a mineral occur in sedimentary precipitates, whether by evaporation, inorganic chemical reactions, or biogenic activity, are much more limited in time and space than in precipitation areas of less intense chemical selectivity contaminated both by other precipitates and by detrital components. A special kind of mechanical precipitates would include some magmatic segregations where a desired mineral constituent has concentrated in a liquid magma as an igneous sediment at high temperatures. Many ores of this type show abrupt gradations and others show gradual changes in concentration, especially when followed along the layers that contain the segregation itself. For some segregation deposits the A/G ratio will hold through a tenfold change in grade, but for many segregations, there is an abrupt change from ore deposit to barren rock.

Both for igneous segregations and for precipitates from aqueous solutions in fractured or chemically reactive rocks such as limestone and dolomite, there is no geologic reason to expect geometric increase in tonnage with decrease in grade beyond certain well defined limits which vary not only with type of deposits but with the individual deposit considered. Nearly all these deposits have an outer margin of mineralization controlled by igneous contacts or by fractures or by hostrock where the grade decreases sharply.

Fracture controlled deposits range from those with abrupt transitions between high and low grade ores such as are found in the typical "fissure" veins characteristic of the western United States to the disseminated ores of the porphyry copper type where the A/G ratios have been established for changes in grade of approximately one order of magnitude. Ore deposits the majority of which do not show the A/G ratio would include those of mercury, gold, silver, tungsten, lead, zinc, antimony, beryllium, tantalum, niobium, and the rare earth elements.

Weathering, the precursor of sedimentation, may result in widespread gradational deposits derived from rock protores of huge tonnage; both residual enrichment through leaching, and secondary enrichment through precipitation of elements at depth form important ore bodies having large tonnages. Some of these deposits show abrupt changes in grade from worthless material to valuable ore, as in the upper part of most sulfide deposits; others show a gradual enrichment, as with many lateritic nickel ores.

In many important types of mineral deposits, all available evidence indicates a paucity of the low-grade material essential to the concept of the arithmetic-geometric ratios. For lead-zinc replacement deposits in carbonate rocks, this zone is commonly but a few feet wide, and limestone carrying 20 or 30 ppm may be within arm's reach of a huge ore body having a grade ten thousand times that of the countryrock (Morris and Lovering, 1952). Lowering the grade of a typical large high-grade lead manta ore body from 20 per cent to 10 per cent would not increase the total quantity of reserves greatly and lowering the grade from 10 per cent to 1 per cent would increase the reserves only as barren rock was added to the ore to bring its average down to the proposed low cut-off value.

Many other geologic types of ore deposits scattered over the surface of our planet have characteristics in common. Bonanza epithermal silver and gold ores are characterized by rich ores that commonly bottom at shallow depths and have relatively sharp boundaries with their wallrocks. Most mercury deposits and antimony deposits fall in this epithermal class. For the vast majority of such deposits there is little hope of a geometric increase in tonnage with arithmetic decrease in grade, but we should expect to find an exceptional deposit occasionally somewhere in the world that shows promise of a geometric (but finite) increase in ore reserves with decreasing grade.

Production of several of the metal vitamins essential to the life of industrial giants is concentrated in a few major deposits contained in very small areas of the world, but minor production may come from a large number of small intermittently worked deposits. Mercury belongs in this class. It is worthy of note that the clarke of mercury is 0.00004 per cent or 400 ppb; this is four orders of magnitude less than in ore, which commonly contains from 0.2 to 0.5 per cent mercury. A similar range exists for other important industrial elements such as tungsten (clarke less than 2 ppm), tantalum, silver, tin, vanadium, molybdenum, and others. Several metals fall in an intermediate class between the iron-aluminum group and the mercury-tungsten group; these would include copper (ore grade approximately 0.4 to 0.8 per cent and having a clarke in basic igneous rocks of 87 ppm, although in some igneous rock it averages several hundred ppm, only an order of magnitude below the present cut-off grade in some porphyry coppers). Cobalt, nickel, and vanadium have similar abundance ratios of clarke to ore grades. Both zinc and lead approach this group but currently the ratio of ore grade to clarke is distinctly higher than in this latter trio.

It has been optimistically said (Brown, Bonner, and Weir, p. 91, 1957) that with the advent of cheap nuclear energy, common rock —granite—would become "ore" and supply unlimited quantities of all the metals needed by industry but even the breeder reactor is not expected to make energy costs appreciably less than current costs of cheap hydroelectric or geothermal power (2.5 mills per kilowatt hour). Surprisingly enough, many men unfamiliar with the mineral industry believe that the beneficent gods of Technology are about to open the cornucopia of granite and sea, flooding industry with any and all metals desired. Unfortunately cheap energy little reduces the total costs—chiefly made up of capital and labor—required for mining and processing rock. The enormous quantities of unusable waste produced for each unit of metal also are more easily disposed of on a blueprint than in the field.

The difference in physical and chemical form of the compounds containing the metals in common rock would require development of a new and complex technology to extract them, and the unit costs of labor and capital (and even cheap energy!) could be orders of magnitude above those of the present. For at least another century or so metals will come from ores that have metal concentrations well above the clarkes of metals in rocks, with only few exceptions such as that represented by the

magnetic black sands concentrated in the U.S.S.R.

Even where the arithmetic-geometric ratio holds, each deposit will approach zero grade at a very finite tonnage. This does not mean that large tonnages of very low grade currently noncommercial copper-bearing rock will not be developed ultimately in the porphyry type copper deposits. Indeed, with a moderate increase in price and substantial decrease in grade, there would seem to be ample copper within the Western Hemisphere to supply the needs of North America for at least another fifty years. Currently the outlook for adequate supplies of molybdenum also invites cautious optimism.

Few if any mining geologists see any reason to expect semi-infinite volumes of copper-bearing rock containing about 0.1 per cent copper; most especially they do not by analogy with porphyry type deposits assume any A/G increase in tonnage for ore bodies of the many other metals that are totally different in their geologic habitat and genesis. In spite of the excellent start on a continuing inventory of United States and world ore reserves made by the Geological Survey under Lasky's guidance and by the U.S. Bureau of Mines in the early fifties, far too little has been done in this field with geologic guidance since, and we desperately need such studies of a wide variety of ore deposits, guided by geologic insight and much field study.

Technology and Mineral Resources

It may be true that in the future technology and science will always provide answers to our problems but it is also true that much time, money, and effort will be required as grade diminishes, mineralogy changes, and entirely different types of deposits are exploited. The widespread belief that technology is continually lowering the unit costs while allowing us to work lower grade deposits is belied by the trends revealed in the copper industry as shown in Figure 1. Here the continuing change in average grade of copper as mined is shown, ranging from a high of 2.12 per cent to the current low of about 0.70 per cent; the tonnage produced yearly shows a general upward trend as would be expected. Assuming that the price of copper represents the summation of costs plus profit and that the costs of capital, labor and energy comprise the total costs, it then follows that the product of grade times price should give us an index representing the contribution of technology in lowering the cost per unit produced. The three-year moving average of this "grade times price" is shown in the same illustration and exhibits the

expected sharp down slope from the early 1920s to the end of World War II; during the past twenty years, however, this type curve is nearly horizontal. Improving technology was almost compensating for the ever decreasing grade of ore mined and the increasing costs of capital and labor. Of especial interest is the plot for corrected price representing zero profit; the average price is reduced by the percentage of the total sales represented by net earnings of the major copper companies each year since 1954 to indicate the break-even price at which they might have operated. (Figures taken from annual review of 500 major U.S. corporations in *Fortune* Magazine.)

When this break-even price is used and multiplied by the current average grade of cop-

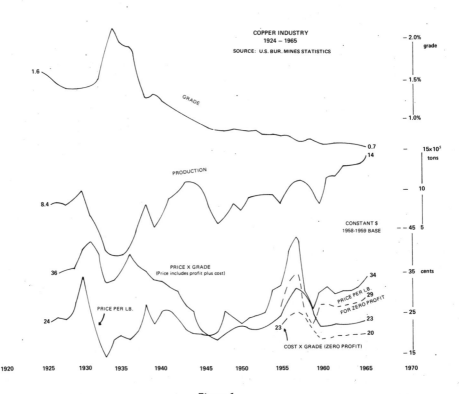

Figure 1.

per mined, the line plots with a perceptible upward trend showing that current technology is not quite keeping pace with the increased costs of extraction; contrary to the Barnett school of thought, *unit costs are not* declining. The major costs of extraction are those of labor and capital, so that a decline in the cost of energy from 3 to 2.5 mills per kilowatt hour will not greatly affect the cost of extraction. To maintain the current dividend rate the major porphyry copper producers will need a continued price increase or will have to devise an entirely different and appreciably more efficient way of extracting copper from the ores. A rough approximation for zero profit is price times grade must equal 0.20. To maintain present production at the average grade of copper mined in the United States and to allow the net earnings comparable to those currently had by the major copper companies, price times grade (in constant dollars 1958-59 base) would have to be 0.24. To maintain present returns on investment, assuming average grade of copper ore mined times the price must be 0.24, an average grade of 0.5 per cent (average grade in 1966 was 0.7 per cent cu.) would require a price in constant dollars (1958-59 base) of forty-eight cents per pound in the United States. For this type of deposit, it seems probable that an increase in price, if guaranteed and maintained, would bring out a corresponding increase in production. This is not true, however, of some other metals. Some elements occur in minor amounts in ores exploited chiefly for some other metal; molybdenum is a valuable by-product of the concentrating processes used for getting copper from porphyry coppers. Tin and tungsten are worthwhile by-products from some molybdenum deposits that have all the characteristics of the copper porphyries except that copper is present in only trace amounts.

Most of the production of silver, gold, and antimony in the United States comes as by-products from the mining of other ores and such deposits can form a limited source of additional by-product metals for a sufficient increase in price of the by-product metal, but at the risk of flooding and depressing the market for the major constituents. When this happens, most such deposits would become uneconomic to operate.

It is often suggested that successful development of the breeder reactor will bring unlimited quantities of cheap (almost free!) power and revolutionize mining and other industries. The cost of nuclear fuel for the breeder reactor will indeed be negligible, but the cost of the large capital investment, of power transmission, of waste disposal, and of operation, combine to bring the price per kilowatt hour to essentially that of cheap steam-coal electric generating plants. Breeder reactors will be a wonderful asset to industrial nations not because they provide power at very low cost but because they may provide desperately needed power when fossil fuels are depleted.

Mining costs always increase with depth, and for those deposits which have an A/G ratio that extends down to low grades and huge tonnages, technology will inevitably cease to produce larger tonnages at lower unit costs; the trend that is currently evident in open pit copper mines manifests itself also in other types of deposits. To maintain production would require a gradually rising price for the domestic product, or increased dependence on foreign sources and new discoveries. For types of deposits other than those characterized by an A/G ratio, the increase in price may not be gradual, but instead may accelerate rapidly to where the metal will price itself out of the market except for the most unusual and essential uses. In other words, a real scarcity will develop as measured by the change in price.

If the change in grade with time is gradual

as with copper, the change in price (in constant dollars) should also be gradual. The assumption that a similar slow change in price will hold for other mineral industries is not borne out by either geologic theory or current mineral statistics. A 20 per cent increase in the price of copper, if maintained, would certainly stimulate activity in the copper industry and increase the production more than 20 per cent within a year or so; for deposits that characteristically lack the Lasky grade/tonnage relation but instead show an abrupt transition between ore and country rock, no such increase in production will result.

Rise in price may well increase production temporarily, hastening producers toward exhaustion, stimulating marginal and submarginal mines into a flurry of activity, with the result that industrial nations increasingly are vulnerable to the vagaries of outside institutional action. For some commodities, however, a substantial rise in price does not bring about any commensurate production of a new material either at home or abroad though it may make available temporarily ore from marginal producers and sufficient secondary or hoarded metal (a stockpile may be regarded as a "hoard") to satisfy demand—as witness the supply-price behavior of mercury for the past few decades.

As a pointed illustration of the impropriety of generalizing about all mineral deposits from data on a special kind of deposit the two basic assumptions for copper are considered in relation to mercury: (1) geometric increase in tonnage accompanies an arithmetic decrease in grade; (2) a moderate increase in price brings a large increase in production. Bearing significantly on the A/G ratio as applied to mercury, we have the figures resulting from an intensive exploration effort by the Bureau of Mines and the Geological Survey during World War II. More than 330 examinations of mercury oc-

currences were made and forty-three deposits were explored. Commercial ore was developed on thirty-eight of them; 370,000 tons of ore averaging 0.8 per cent (16.2 pounds per ton); 1,220,000 tons averaging 0.125 per cent (2.5 pounds per ton), and only 285,000 tons averaging 0.08 per cent or 1.6 pounds per ton. This is equivalent to 2,960 tons of mercury in the 0.8 per cent ore, 1,525 tons of mercury in the 0.12 per cent ore, and only 228 tons of mercury in the 0.008 per cent ore. To sum up the relations, an arithmetic decrease in grade from 0.8 per cent to 0.125 per cent (and actually a sixfold change) resulted in a geometric increase in tonnage of threefold *but the total mercury in the larger tonnage was only one-half* that in the smaller high-grade tonnage. A further decrease of 0.004 per cent in grade resulted in additional tonnage containing only one-tenth as much mercury as that in the high-grade ore.

The U.S. Bureau of Mines Yearbook figures for mercury show the lowest output in twenty-five years for the United States was 5,000 flasks (of seventy-six pounds each) in 1950 when the price was approximately $90 per flask, which was the average price from 1946 to 1950, during which time the output fell steadily from 25,000 to 5,000 flasks. From 1951 to 1953, the price rose rapidly to an average of $200 per flask and in 1954, the United States government guaranteed a minimum price of $225 a flask for a three-year period—a 250 per cent increase in the 1946-50 price! United States production climbed to nearly 12,000 flasks by the end of 1953, about half the 1946 production. Although the price after 1956 fluctuated on both sides of $200 per flask, this is not far from the average that obtained until a few years ago. During this period the production from the United States mercury mines reached a high of 38,000 flasks in 1958 but from then on dropped steadily until it reached a low of

14,000 flasks in 1964. In that year, however, the price of mercury again more than doubled, reaching an all-time high of about $800 per flask in the latter part of 1965, resulting in an average price of over $500 a flask. The substantial increase in price—a sixfold increase in twenty-one years—resulted in another increase in domestic production which reached 22,000 flasks in 1965, somewhat less than was produced in 1947 when the average price of mercury was only $83 per flask. It is worthy of note that at $1,500 per flask the price of mercury would approximate the coinage value of silver at $1.35 per troy ounce.

Mercury resembles silver in many of its occurrences but most silver now produced comes as a by-product of mining complex lead ores. The high-grade silver mines are depleted just as most of the high-grade mercury mines have been depleted. Very little mercury is recovered as a by-product from complex ores so that this source will contribute only an insignificant amount of mercury in the future whereas silver will continue to increase as the production of the complex ores continues to increase. The average consumption of mercury in the United States has increased at a rate of 3 per cent per year for twenty years, equivalent to a doubling in consumption every twenty-three years. But in the same period *the price of mercury has increased more than 500 per cent!* Presumably a price of more than $1,000 per flask might maintain world production of mercury for fifty years or more but the source of this mercury would be increasingly concentrated in a few large deposits such as those of Spain and Italy where cartel action rather than costs per unit output determine market price.

Distribution of Future Demand and Supply

Hans Landsberg (1964, p. 6) in his scholarly study of the sources of natural resources

for the United States during the next forty years, has based his projections on the following assumptions: "Three basic assumptions built in from the start were: continuing gains in technology, improvements in political and social arrangements, and a reasonably free flow of world trade . . . and that there will be neither a large scale war nor a widespread economic depression like that of the early 1930's." There is little evidence in the history of the past half century to justify such an optimistic outlook, but there is some encouragement if one believes, with the writer, that history does *not* repeat itself. Perhaps Landsberg's assumptions will be justified by the next few decades, but only if attention is given to their alternatives, and if policy is devised to prevent the unpleasant consequences of failure to achieve conditions implicit in the assumptions.

Before considering future supplies of the non-fuel minerals, it is necessary to consider potential future demands. The demand for products in a modern world is a function of culture and population, of the relation of the growth rates of GNP and specific populations as well as their particular type of culture (Blackett, 1967). At present cultures all over the world show a marked convergent trend toward a materialistic western way of life which demands an increasing standard of living for the average person; such an increased standard is quickly reflected in the per capita income of the population.

The average income in the United States in 1965 was something over $2,500 per capita. In India the average income then was about $80 per capita, and the average income from some underdeveloped countries was even less. For several years the GNP in the United States has grown at a rate of about 4 per cent per annum while the rate of population growth has been about 1.5 per cent per annum; this results in a per capita income growth rate of approximately 2.5 per cent per year. Projected into the future this would suggest a per capita income increase from $2,500 in 1965 to $5,000 in 1993 and to $10,000 in the year 2021. Meanwhile if the effective fertility rate of 2.5 per cent per year for India were maintained, its population would double every twenty-eight years. The per capita income increase in India, if governed by the present ratio of GNP to population increase, would be far slower; if the GNP were to increase at 4 per cent per annum but the population continued to increase at 2.5 per cent, the actual increase in per capita income would be at the rate of 1.5 per cent per annum. The current per capita income of about $80 would reach $160 in the year 2012 and $320 in the year 2060. Such figures cannot be dismissed as unrealistic unless we also dismiss the basic premise of an "ever expanding economy" at 4 per cent and a population increasing at a constant rate as also unrealistic. Even if population growth slows down substantially in the underdeveloped countries and their GNP is greatly increased, it may be many generations before per capita income can approximate that currently enjoyed by the United States. In the meantime the disparity between incomes will not only be maintained but greatly increased unless a drastic and dramatic change in current trends takes place.

During the early years of industrialization, however, the rate of increase of GNP in backward countries where outside financial assistance is available may be phenomenal. The investment of excess Japanese capital in Korea together with a campaign to control family size has resulted in a steady decline in the birth rate (about 0.1 per cent per year decrease) and an increase in the GNP of approximately 8 per cent per year from 1964-67.

A few demographers (Bogue, 1965) believe that the present population explosion will soon be regarded as an anachronism and that populations will stabilize themselves shortly. This hopeful attitude is not shared by the majority of demographers and is not even considered by economists. The dramatic decline in birth rate in Japan during the fifties following a year of intensive propaganda and appropriate legislation startled the western world. The steady but slow decline in birth rate in the United States since the introduction of modern (1963) contraceptives is also worthy of note. If the industrial and political leaders of a nation become convinced that it is helpful to the national economy and the strength of the nation as a whole to slow down the rate of effective fertility, this can happen within a surprisingly short time.

The demand for goods whether in small amounts per capita for rapidly increasing population or in large amounts per capita for a static population will strain industrial capacity and mineral productivity for the next century. Per capita consumption in the industrial nations first shows exponential growth, usually for a few decades, and then flattens to a merely arithmetic increase usually at a relatively low figure. A nearly static population in underdeveloped countries would certainly result in a tremendous surge in the demand for the nonrenewable resources of the world. For this reason it would seem the current estimates of the United States and world consumption during the rest of the twentieth century as given in Landsberg's study (1964) are too conservative. All available evidence indicates that demand for minerals will increase at an exponential rate for at least fifty or seventy-five years before it begins to level off—as it must do eventually if for no other reason than that the supply is finite.

If events in South Korea and Japan foreshadow those in other Asiatic nations, the demand for minerals and other resources clearly will increase far faster than the rate of 1.5 per cent per year suggested by projection of current fertility rates and an assumed 4 per cent per year increase in GNP. The world's demand (i.e., production) for industrial metals has been growing at a rate of more than 6 per cent per annum for nearly a decade. It is most unlikely that all underdeveloped nations will want a completely western type culture, but enough of them have already indicated their desire to become industrialized to justify the conclusion that major demands will be made on mineral resources far into the twenty-first century.

Industry requires an increasing tonnage and variety of mineral raw materials. Many that are deemed essential to modern industry have understudies that can play their part adequately, but technology has found no satisfactory substitute for some. The locations of mineral raw materials are fixed by geologic accident in the remote past and many potential sources of supply lie far from the centers of consumption.

In considering future mineral resources certain factors stand out. Because all known individual deposits will be exhausted sooner or later if present patterns of use are maintained, future production depends on the continued discovery of new deposits. Currently discovery techniques seem to be developing rapidly and many new deposits have been found by the application of geology, geophysics and geochemistry. Future sources of ore will be of two types; many known noncommercial deposits will become ore through technical innovations, future availability of cheaper transportation, or a rise in price. The second and larger group includes future sources that are now unknown but will be discovered by exploitation. These deposits will be chiefly in remote or underdeveloped areas if they crop out at the surface, and will be found more and more by a combination of geology, geophysics and geochemistry. Such deposits are expensive to find and will require well financed companies or government supported groups because of the high cost of exploration and development far from supply centers. A second type of unknown ore bodies includes those that do not come to the surface—blind ore bodies. To search successfully for them is even more expensive than to find exposed ore bodies in remote areas. Much money will be spent in preliminary exploration and reconnaissance for every drill hole that zeros in on a commercial ore body. It might be said that the cost of discovery and the cost of development are proportional to the depth below the surface of the ore body.

The kind and locale of blind ore bodies can be guessed by a study of the metalogenetic provinces of the world and their geology. At present our geologic theories are inadequate to say with confidence where to drill for blind ore bodies and a substantial amount of adequately financed well planned research should be devoted to establishing sound theories of ore genesis.

Since the number of deposits is finite, the lead time necessary for successful search increases as the number of shallow deposits found grows larger. Both time and money must be allocated for an inventory of known resources and for planning programs and program revision.

The amount of metal consumed in one generation at the current rate of increase in consumption approximates all the metal that has been used previously. The entire metal production of the world prior to World War II is about that which has been consumed since the beginning of that unhappy event. Ore deposits have been sought most actively ever since the industrial era began and the major industrial powers have a similar history of mineral exploitation and exhaustion. The Middle East, Greece, Spain, England, Belgium, France, Germany, Sweden, Mexico, Peru, and the United States have all had their day as the world's foremost producer of one or several metals only to become plagued by either exhaustion or continuing decline.

As individual nations use up the cheap supplies in their own countries first, they inevitably become more and more dependent on foreign sources for most of their raw materials. At present all industrial nations except possibly the U.S.S.R. are net importers of most of the metals or the ores used by them. This dependence on foreign sources will almost certainly grow far greater for the United States during the next generation except possibly for molybdenum, magnesium, copper, and the fertilizer minerals potash, sulphur, and phosphate. Increasing dependence on foreign sources inevitably brings increased vulnerability to military, political, or economic action. Some of the metals most vital to the economic well-being of industrial nations are in areas of political instability or lie in the Communist countries with which many nations are currently at odds. Most of the known reserves of tungsten and antimony lie in Communist lands, as well as a large part of the world's manganese, nickel, chromium, and platinum. The future and present sources of manganese for Europe and North America are mostly in Africa; of tin, in Southeast Asia; of aluminum ore, in various underdeveloped tropical countries. If the present dichotomy in the world economy persists, the non-Communist nations must develop a technology that will insure a viable economy which is independent of Communist control of vital resources, or else exploit Communist scarcities in those commodities of which they have an inexplorable surplus. It would seem profitable to further the policy of "co-existence."

Discovery is required that will develop reserves at an exponential rate until a static population is achieved together with a constant per capita demand for metals. The lead time from discovery to production of a major deposit averages three to five years or more

and a similar time is usually required for the exploration involved in its discovery. It is apparent then that mineral reserves should be available for a minimum of ten years future production on the basis of present demand and that for those metals which show increasing demand, the reserves must increase even faster.

The resources ready at hand are those first mined, and hence future resources will come from less industrialized countries which, however, will be making maximum efforts to become industrialized. The increased vulnerability of industrial nations in both peace and war suggests that the assumption of equitable access will be affected in the future as it has in the past by institutional restraints and the counter measures which will seem in the best interests of the countries concerned. National attitude, economic advantages, military advantages, and current leaders, will all change with time. Their interplay will require constant vigilance on the part of the industrial nations if they are to maintain a healthy economy. In order to plan for the decades ahead which may be needed to give sufficient lead time for economic and technical adjustments, national and international watchdog groups of experts will be required to warn of impending critical resource situations and to recommend the most desirable remedies. For such groups industrial countries should begin at once to develop continuing national inventories of reserves both at home and abroad. Unfortunately the results of studies guided primarily by mineral economists have led to a "cornucopian" concept of mineral resources that allows policy makers to ignore potential mineral shortages. We are rapidly approaching a time when indifferences can be disastrous.

The main escape hatch for scarcity is technical advance along a broad front, as noted by Landsberg (p. 240), and this will depend on extensive and effective programs of research and development in science, engineering, economics, and management. The more immediate the promise of practical results the greater the willingness of private industry to finance the necessary research. Much of the required research, however, would be of the more exploratory kind, the results of which cannot be guaranteed, but from which eventually answers of crucial importance will come. Planning and prosecution of such research should be a continuing function of the government of major industrial nations. There is a desperate need for integrated national and international resources policies that harmonize or adjudicate the needs of the many special segments of an economy that utilizes natural resources. We especially need research on blind bodies and the factors that are critical to their implacement, research on the genesis of ore bodies. This should lead to new discovery techniques which must be tested by the drill. Exploration should emphasize metals in short supply or which have diminishing reserves. The over-all policy concerning mineral resources should be in the hands of a continuing group of legislative officers working in close cooperation with a full time watchdog group of experts from business, government and the educational field. With such a group functioning effectively, stresses caused by utilization of exhaustible foreign and domestic ore sources should be minimized; without it recurrent or persistent shortages will occur in the supply of some metals. Periods of needlessly high prices for some mineral products will alternate with prices so low that many potential sources of ore will be lost, and certain it is that many disastrous political errors will be made through lack of appreciation of the future importance of trade between areas of industry and potential mineral raw materials.

At present we are living in an epoch of localized affluence such as has recurred throughout historical time when new treasures of metal or mineral were discovered. In the past mineral deposits have led to invasion, conquest, and wealth, for a comparative few. Now, more happily, the utilization of mineral resources of not only the industrial countries but also of some of the underdeveloped countries has led to affluence for many people rather than a select few, but it is well to realize that mineral deposits are still the "quick assets" of the country that possesses them and that ultimately these resources will be exhausted or will decrease their contribution to a fraction of their flush production. To insure a more equitable distribution of end results of the exploitation of mineral wealth, is one of the foremost but rarely stated objectives of successful, modern economic systems. During the next century this will not be achieved merely by recycling metal as scrap, nor by processing dozens of cubic kilometers of common rock to supply the metal needs of each major industrial nation. When the time comes for living in a society dependent on scrap for "high grade" and on common rocks for "commercial ore," the "affluent society" will be much overworked to maintain a standard of living equal to that of a century ago. The foreseeable exhaustion of ores of some metals and the continually decreasing grade of most ore deposits now used show that it is no small prudence to provide ample lead time for technology to work out such answers as it can and to allow the economy time to make the necessary adjustments to changing mineral supplies.

BIBLIOGRAPHY

Barnett, H. J. and Morris, Chandler, 1963. Scarcity and Growth. Johns Hopkins Press, Baltimore, Maryland.

Barnett, H. J., 1967. The Myth of Our Vanishing Resources. Transactions Social Science and Modern Society, June, p. 7-10.

Blackett, P. M. S., 1967. The Ever Widening Gap. Science Feb. Vol. 155, No. 3765, p. 959-964.

Bogue, D. J., 1965. The prospects for world population control. Community and Family Study Center, University of Chicago.

Brown, Harrison; Bonner, James; and Weir, John; 1957. The next hundred years. Viking Press, New York, 193 p.

Bureau of Mines Staff, 1965. Minerals Year Book, Vol. 1, Metals and Minerals. U. S. Government Printing Office, Washington.

————, 1965. Mercury potential of the United States. U.S. Bureau of Mines, I. C. 8252, U. S. Government Printing Office, 376 p.

————, 1956. Mineral Facts and Problems. U.S. Government Printing Office, Washington, D. C., 1042 p.

Landsberg, H. M., 1964. Natural Resources for U. S. Growth. The Johns Hopkins Press, Baltimore, Maryland, 260 p.

Lasky, S. G., 1948. Mineral Resources Appraisal by the U. S. Geological Survey. Colorado School of Mines Quarterly, Golden, Colorado, p. 1-27.

Lasky, S. G., 1950. How Tonnage Grade Relations Help Predict Ore Reserves. Engineering and Mining Journal, Vol. 151, No. 4, p. 81-86.

Lasky, S. G., 1951. Mineral Industry Futures. Engineering and Mining Journal, Vol. 152, No. 8, p. 60-63.

Lasky, S. G., 1955. Mineral Industry Futures Can Be Predicted II. Engineering and Mining Journal, Vol. 155, No. 9.

McMahon, A. D., 1965. Copper, A Material Survey. IC. Bureau of Mines Information Cir. 8225, U. S. Bureau of Mines.

Morris, H. T., and Lovering, T. S., 1952. Supergene and hydrothermal dispersion of heavy metals in wall rocks near ore bodies, Tintic district, Utah: Econ. Geol., v. 47, p. 685-716.

Staffs, Bureau of Mines and U. S. Geological Survey, 1948. Mineral Resources of the United States. Public Affairs Press, Washington, D. C. 212 p.

Dimensions of the World Food Crisis
H. F. Robinson

At the September, 1967, Plenary Session, the Honorable Henry M. Jackson, Senator from the State of Washington, spoke on the topic, "Public Policy and Environmental Administration," and began with relating a fable written by James C. Rettie, which seems most appropriate to begin our discussions of the world food problem. I will not give that fable in detail but will relate only its general theme. As the story goes, a film had been made to represent a record of the earth's life history during the past 750 million years. The film, scaled in time and length to the total time span of the earth's history, required one full year to show and with the life span of an individual man taking only 30 seconds. The complexities of man in his interaction with environment were pointed out in the fable which ended with the statement and a question saying, "Man has just arrived on this earth. How long will he stay?"

We have heard this question raised in connection with ecological discussions and the emphasis placed on control of the environment with the plea to remove the pollution from the air and from the water and to restore these two to their once desirable characteristics. The same question regarding man and his ability to exist on this planet, coupled with fears of his possible extinction, may be more appropriate when the balance of the population growth with the available food supply is considered.

Seriousness of the Problem

The problem of feeding the people of the world is as old as civilization itself. The crisis between population growth and food supply has been variously predicted for the past 200 years, and I suppose we would say that the only new feature of the problem is its dimensions. Possibly another new and distressing feature of the problem is the disproportionality in the economic development, population density, and food supply throughout the world. This is really basic to the total problem. It is most important that we be reminded of our fortunate position in this nation as compared to that in the less developed countries. We must be prepared to assume our obligations and responsibilities to insure a proper diet and a decent existence for the less fortunate. The following excerpt from a recent issue of the Eli Lilly Company News Letter gives emphasis to our economic advantages and minority in numbers:

"If all the people in the world could be reduced proportionately into a theoretical town of 1,000 people, the picture would look something like this: In this town there would be 60 Americans, with the remainder of the world represented by 940 persons. This is the proportion of the population of the United States to the population of the world, 60 to 940. The 60 Americans would have half the income of the entire town with the other 940 dividing the other half. About 350 of these would be practicing Communists, and 370 others would be under Communistic domination. White people would total 303, with 697 being non-white. The 60 Americans would have 15 times as many possessions per person as all the rest of the world. The Americans would produce 60 percent of the town's food supply although they eat 72 percent above the maximum food requirements. They would either eat most of what they grow or store it for their own future use at an enormous cost. Since most of the 940 non-Americans in the town would be hungry most of the time, it would create ill feelings toward the 60 Americans, who would appear to be enormously rich and fed to the point of sheer disbelief by the great majority of the townspeople. The Americans would also have a disproportionate share of the electric power, fuel, steel and general equipment. Of the 940 non-Americans, 200 would have malaria, cholera, typhus, and malnutrition. None of the 60 Americans would get these diseases or probably ever be worried about them."

There is reason for skepticism about the extent of assistance with the problems of food supply and economic development of the less

fortunate by the more affluent nations. There is no doubt that people are more knowledgeable and have greater interest in solutions to problems of poverty, hunger, and malnutrition of the less fortunate than ever before. I do not know whether this concern is one of genuine interest in a better life for all mankind or whether it is based on fear. Unless effective steps are taken to improve economic conditions among the less fortunate, revolutionary tendencies and insecurity of governments may lead to a general chaotic condition throughout the world. It was this latter condition to which Dr. James Bonner, biologist at the California Institute of Technology, referred recently in a speech when he stated:

"We will, I suspect, begin to regard the starving population of the underdeveloped nations as a race or species apart, people totally different from us as indeed they will be. They are just animals, we will say, and a serious reservoir of disease. The inevitable culmination of the two cultures will be that the one culture (the rich) will devour the other."

The PSAC Study

The deteriorating situation of the developing countries with respect to their food supply and balancing of the available food with the rapidly expanding populations prompted President Johnson's request to the President's Science Advisory Committee in his Food for Freedom message on February 10, 1966, to do the following:

Search out ways to:

1. ". . . develop inexpensive, high-quality, synthetic dietary supplements. . . ."
2. ". . . improve the quality and nutritional content of food crops."
3. ". . . apply all of the resources and technology to increasing food production . . ."

The study was organized around 14 subpanels whose membership consisted of a total of approximately 100 representatives from industry, universities, private foundations, and the Federal government. A wide range of the major issues were represented by the subpanels: population growth, nutrition, plant and animal production, fertilizers, pesticides, machinery, marketing and economic aspects of the food problem, and financial requirements needed to provide reasonable solutions. The report, including over 1200 pages, was prepared in three volumes at the conclusion of the study.

When the panel began its work, there was no general agreement as to the dimensions of the task and the approaches that should be taken in the study. As we began to examine the seriousness of the problem and the scope of the study, as well as to set objectives in examining food production in the developing nations, the importance and magnitude of the total economic development in relation to foreign technical assistance became increasingly apparent. The conception of the assignment was set forth in the following paragraph:

"The situation as the Panel now views it, after nearly a year of study, is that hunger and malnutrition are not primary 'diseases' of the last half of the twentieth century. Rather, along with the so-called population explosion, they are symptoms of a deeper malady—lagging economic development of the countries of Latin America, Asia and Africa, in which nearly two-thirds of the people of the earth now live."

The seriousness of the world food problem and the predictions with respect to its most crucial hours have been described and discussed by many, and it is not the intention here to analyze these predictions as to their validity and accuracy. It is clear that if present trends continue, the size and severity of the problem will become increasingly worse. The Panel surmised that within 10 to 15 years, when the serious consequences are readily apparent, it may then be too late to provide workable solutions.

The encouraging reports of increased production from various parts of the world due to the use of improved technology, including improved varieties, additional fertilizers, and modern cultural practices, have given new hope, at least on a temporary basis, and have further convinced us that certain programs can contribute toward the solution of the problem. However, the seriousness of the situation and the impending hunger resulting from the imbalance of available food with the population growth continues.

General Conclusions

The conclusions reached by the PSAC Panel in its report are as appropriate at this moment as at the time they were written and should serve as a guide in considering the global strategy that will be involved. These conclusions are as follows:

1) The scale, superiority, and duration of the world food problem is so great that a massive, long-range, innovative effort unprecedented in human history will be required to master it.

2) The solution of the problem, which will exist after about 1985, demands that programs of population control be initiated now. For the immediate future, the food supply is critical.

3) Food supply is directly related to agricultural development and, in turn, agricultural development and overall economic development are critically interdependent in hungry countries.

4) A strategy for attacking the world food problem will, of necessity, encompass the entire foreign economic assistance of the United States in concert with other developed countries, voluntary institutions, and international organizations.

There are two very basic factors in the food problem faced by the developing nations of the world. The problem can be stated clearly and succinctly, but it is often so obvious that we fail to comprehend the magnitude and scope.

These factors are: (1) Population growth is occurring at a more rapid rate than the food to sustain the people, and (2) Increase in food production continues to lag behind expectations in many of the developing nations.

Another basic consideration in development of the report was the time period over which the findings were expected to be generally applicable and most appropriate in terms of solutions that were to be suggested. It was decided that the time period of two decades, namely 1965-1985, would likely be the most critical period of the problem to be experienced. One of two conditions will probably become apparent by the end of this 20-year span in the developing nations with which we are chiefly concerned: (1) Either the population growth will have been brought under control and a balance will have been developed between the population and the food supply, or (2) Some nations, possibly many, will have passed the crest and will be accelerating on a declining grade of malnutrition, economic deterioration, and political instability to the point where no reasonable solution to the problem can be found.

Requirements for Solution

Total economic development is required in these less developed nations where the food supply is critical and population growth most rapid if there is to be meaningful solutions for these less fortunate people in the world. It is obvious to all of those closely involved in technical assistance that the task is much more complicated now than was visualized at the time President Truman set forth his Point Four Program in the 1940's. We now know that the task involves building an entirely new economy of which agricultural development is only a part. Political stability and the general attitude by the recipient governments as well as the will and determination of the people to

work for a higher standard of living are of critical importance. The disruptions, constraints, and disturbing political conditions with which many of these nations try our patience, and the frustrations we meet in working with the governments, are often so great as to threaten our withdrawal from any future involvement. However, we know that overcoming these difficulties will continue to require long periods of tolerance, and we must exercise the most astute management to avoid crises that may disrupt all of our foreign technical assistance programs. We cannot leave these responsibilities to others because of the frustrations and difficulties that may be encountered.

The PSAC Panel on World Food Supply indicated a strong belief for agricultural development and economic growth in the developing nations of the world as an absolute requirement if the people were to avoid starvation. We in the academic world must contribute in every possible and appropriate matter in these activities.

The Panel was in general agreement that the food problem would not be solved from our continuing the shipment of food abroad. We were most critical of that part of the foreign aid program which has relied on food aid to overcome the hunger problems in those nations experiencing the imbalance between food and population. The governments of those recipient nations have contributed to this problem, and many of them are continuing to impede domestic developments by appealing to outside sources for food aid rather than increase their own local production. However, shipments of food required to prevent starvation in a few cases where there are major problems with drought and other difficulties that have prevented normal production of the crops can be justified.

It is known that some government leaders in the nations receiving food aid have taken special advantage of this program to either discourage, or not encourage, increased agricultural production. It has been considered economically desirable to try and provide the deficit in food from outside sources through food aid rather than furnish financial aid and other resources required to encourage agricultural productivity within the developing nation. Every possible means should be used to identify those who are exploiting for personal gain the assistance being provided from developed nations.

In accumulating data on total food requirements for the world, areas considered representative of major developing nations of the world where the food problem is either most urgent now or expected to become of increasing seriousness were selected for attention. These areas included India and Pakistan in the Far East and Brazil in South America. Little special attention was given to Africa since it is expected that problems of food supply in all of the African nations will become serious, but the critical period is expected to be beyond the time to which this study was focused.

One of the first requirements in considering solutions to the food problem of the future is to seek reliable estimates of the magnitude of the population within the range of time to which the study is intended to apply. The World Food Report by the President's Science Advisory Panel projected growths to year 2000, but, as indicated, basic emphasis was given to the period between now and 1985. In predicting population growth, two estimates were derived for each point in time. One was based on continued growth at the present rate and the second, designated "low estimate," was based on the assumption that family planning programs during the next two decades might lead to a progressive decrease to 30% in the probability that a woman of a given age would bear a child (fertility rate).

Projections indicated that the population would rise from the present possibly 3.6 billion people in the world to between 4.65 billion (low estimate) or 5.03 billion by 1985. By the year 2000, the population at its present rate of growth would be approximately 7.15 billion, compared to 6 billion from use of the low estimate, which gives the maximum amount of optimism to the effectiveness of family planning programs.

If populations continue to grow at the present rate, caloric requirements are expected to increase until, by 1985, 52% more calories will be required on a world-wide basis than at present. We would need 43% higher calories by the end of the next two decades if the population could be controlled to the extent indicated in the low estimates. Even the most effective family planning programs, which could be expected to be implemented in the rapidly growing nations, will provide only minor curbs on the populations to be fed during the immediate years ahead. By the year 1985, we are predicting more than 5 billion with normal population growth. The optimistic family planning program could result in only about 385 million fewer people. Thus, we recognize the urgency of the problem. While our family planning programs are expected to have long-term effect, they will have relatively little impact on the food needs of the world during the next two or three decades.

In seeking the food that is required to meet present deficiencies and provide for the population growth of the future, we gave consideration to all known sources. This included the food from the sea, bacteria, petroleum, as well as synthetic and all traditional sources. We came to the definite conclusion that there is *no panacea to this global problem*. Assuming that the marine sources of food might be greatly increased, possibly expanding to 10 to 20 times the present food from the sea, it is not expected that this source will have a major impact on the solution of the problem. The same can be said for other unique sources. Although we do not offer these as potential reservoirs expected to meet the major demands, research should be continued and expanded on all new and novel sources of food in order to achieve greater productivity in the future. All possible avenues of food production will be needed and must be exploited in the years ahead.

If the world food problems are to be solved and catastrophic conditions prevented that are likely to result from population growth, then the increased production of food in the less developed nations must come from improved basic agricultural practices and a generally better economic condition within those nations. This will include the increased use of proper fertilizers, pesticides, improved varieties of seeds, additional machinery appropriate to the various conditions within these nations, the proper use of available water, and most emphatically—generally improved economic conditions. Parallel with these needs, attention must be given to transportation. These nations must develop roads needed to move food from farms to the markets and to enable the farmer to move his requirements from markets back to his producing lands. Markets, both domestic and foreign, must be improved and emphasis given to preservation of the food being produced, marketed, and utilized by the people. Arrangements must be made to provide the farmer with the necessary incentive to increase his production. If these modern inputs, which are generally known by all, cannot be supplied to those nations needing increased food supply, then all other efforts to reach meaningful solutions are likely to fail.

Land

In considering requirements for increasing food on a world-wide basis, first attention was given to the question of availability of land. While this study did reveal approximately double the amount of potentially arable land ever reported in any previous study, the requirements for clearing, draining, irrigating, and other needs for putting it into production are likely to prevent its availability for crop and animal production during the most critical period of this problem.

The conclusion is obvious—we must give major attention to increasing the productivity of the presently arable land and not expect to increase the food supply by expanding onto lands not previously arable. This is especially true in Asia and in Europe. There is also relatively little additional land to be cultivated in the Soviet Union. Further, the great potential for increased productivity of the available land is in the tropics where yields are not only generally low but the requirements for additional food are the greatest. More than half of the potentially arable land, which amounts to more than 4 billion acres, is in the tropics, and about one-sixth is in the humid tropics. The difficulties in increasing productivity in this part of the world are largely associated with a lack of research results that are applicable to this geographical area. Increased attention should be given to tropical soils and climates in order to provide for manifold increases in the productivity of the lands in the tropical and sub-tropical areas.

Manufactured Inputs

The subpanel concerned with fertilizers estimated that approximately 61 million metric tons of plant nutrients would be required, in addition to the present roughly 6 million metric tons now being applied, in the developing nations if their food production is to be increased 100% above present levels. Such fertilizer use would necessitate a capital outlay in the neighborhood of $30.5 billion for the production of raw material, mixing plants, distribution, and other requirements. Even the 25%

increase in food yields, which we urge should be provided by the early 1970's, would require a total additional investment cost of approximately $2.5 billion.

The fertilizer component needed for increasing food production, while considered to be the most important component, is not the only one necessary for good crop production. It is only with adequate water, improved seeds, appropriate machinery, and adequate cultivation that the farmer can be expected to realize optimum returns from increased fertilizer applications.

Recent emphasis that has been placed on fertilizer requirements has led to some difficulties in depressing prices for fertilizers on the world market. Many of the major fertilizer manufacturing industries have attempted to respond to the request for greatly increased use of fertilizers by providing the materials to the people of less developed nations. There is not an overproduction of needed fertilizers, but a lack of their distribution. The developing nations are unable to purchase the needed fertilizers with the financial resources they have at their disposal. This apparent surplus of fertilizers due to last year's increased production must not be considered as indicative that we have met the needs of this important component in the food production program. Rather, it is an indication of the complexity of the problem and of the need to take all possible measures to increase economic development in those nations that need to purchase the manufactured inputs required for increasing food supplies.

Varieties, Pesticides, Machinery, and Irrigation

Effective breeding programs must be developed within the nations needing improved crops. The national programs should be ap-

plied because their principle objective should be the production of high yielding strains of adapted and desired crops. Emphasis may also be placed on producing crops, particularly cereals, with improved nutritional characteristics. Following the procedures developed in basic research in the United States, genes for high quality protein may be incorporated into crops, such as corn, that are developed within a developing nation's breeding program.

The importance of improved crop varieties in solving the world food problem is indicated in the higher yields of rice from the International Rice Institute in the Philippines and the high-yielding, short, stiff stalk wheats from the International Maize and Wheat Center in Mexico.

However, shipping in new varieties and strains from research centers abroad is not the final answer. Lack of complete adaptation to climates, the need for quality to satisfy the appetites and customs of the people, and resistance to diseases and insects peculiar to the area are some important factors dictating that adaptive research be conducted to provide crop varieties and strains specifically for the needs of a country.

The use of pesticides will have to be increased at least sixfold if the insects and diseases are to be controlled in the developing nations. Adequate farm machinery must provide at least one-half horsepower per hectare for the absolute minimum power needs to achieve a reasonable level of agricultural productivity. The need for new irrigation projects to yield the required water was recognized in those nations where moisture supply is a problem. Many nations can provide the needed water from underground wells, while others will have to resort to major impounding projects in order to obtain an adequate supply for crop production. Irrigation cannot be fully effective unless the total program of improved practices by the various farming programs is rigidly followed.

Education and Research

Those who have participated in technical assistance in the developing nations have emphasized from the outset the importance of education and research appropriate to the needs of the people of these less developed nations. The tremendous magnitude of the problem assumes even greater dimensions when we consider the time scale which must be used in solving it and the great difficulties encountered because of lack of education. Traditional procedures followed by the Western nations must meet the needs of the less developed areas if we are to make rapid progress in the immediate future. Approaches must be used that will educate people of all ages. Ways must be found to communicate with the people if the recommendations made are to be adopted on a wide scale and at a rapid pace. This pace is needed to achieve economic and social development. Technical assistance will have to move at a more rapid rate than the normal educational procedures. Universities of developed nations will have to lead in the innovations needed in the educational processes. Individuals will be expected to work with people of the hungry nations to provide effective communication as one of the most important components of the educational program.

The rate of progress toward improving the educational development of the people is quite slow. Vice-President Humphrey in an address of February 9, 1968, entitled "Nation Building and Peace Building" stated: "Since 1960, despite enormous investments in education, world illiteracy has grown by some 200 million people. Of 373 million children in developing nations, 115 million (or 30%) are in school

and about 250 million (or 70%) are not in school."

Not only education itself but the type of training needed among the people must be considered. Efforts must be redirected if the food requirements are to be met in the future. Vocational education, particularly education relating to agricultural careers, must be emphasized. Statistics indicate that only about 4 to 5% of the some 85,000 foreign students studying in the United States are pursuing academic programs that will lead to careers in agriculture and closely related fields. Dr. Roger Revelle, reporting on the special study on "Education in India," found that less than 10,000 agricultural graduates were among the 273,000 who had degrees from universities as of 1961. The manpower requirements are likely to be at least ten- to twentyfold above the presently available number and will tax to capacity the institutional facilities of not only the developed but the developing nations.

Research requirements are in many ways as great and even more demanding than those in education. Errors in approaches and judgment have been made in attempting to help the needy nations. Suggested programs that are essentially a transfer of knowledge of one country to another without the needed adapted research to insure success of the technology in the new environment do not help. The PSAC Report is specific in its recommendations that the misconceptions of the "know-how"-"show-how" approach must be erased. Developing nations cannot be "shown how" until the "know how" is available and specific for the needs of the social structure, the physical environment, the economic conditions, and general agricultural development of the nation. Adaptive research is needed, basic research should be carried out by the affluent nations until the economic development of the needy nations has reached a point where they can afford to allocate their resources to these problems.

Economic Implications and Financial Requirements

The Panel in its analysis of the overall problem indicated required compound growth rates during the next two decades for the developing countries would probably be in aggregate about as indicated below:

	Per cent	
	Present	Required
Increase in Food Demand	3.0	4.0
Increase in Food Production	2.7	4.0
Increase in Gross National Income . . .	4.5	5.5

Such growth rates would require massive efforts and progress which have never before been achieved in any 20-year period. Only Mexico and Taiwan are at the present realizing growth rates of the order considered necessary.

The total financial requirements for technical assistance is staggering and the likelihood of our meeting the needs is most discouraging when we consider the resources being allocated by the developed and more affluent nations. Although the Panel had serious doubts that the financial requirements could be met, they did decide that it was still economically possible when considered on a multi-lateral basis. These needs were approximated to be in the range of $12 billion above the 1965 base of capital investments expended in the some 70 developing nations having food problems. This is the estimated present extent of the increase required to achieve the 4% annual growth rate in food demand and supply. These annual expenditures will be expected to increase to some $25 billion by about

1985. The report concluded on this point with the following statement: "to achieve such a feat would require capital and technical involvement of developing and developed nations alike on a scale unparalleled in peace time history of man."

The attitude of the present Congress toward providing the technical assistance needed is demonstrated in the foreign-aid bill, which is at the lowest level in modern history. Unless the financial needs are made available at a much greater rate in the immediate future than in the past, not only will the requirements of the future be greatly amplified but there is a possibility that the rapidly deteriorating conditions of many of the hungry nations will have proceeded to the point where mass starvation and general chaos cannot be avoided.

Impact of the Report

An example of the possible impact of the PSAC Report is the interest expressed in the findings of this Panel by the officials of the Indonesian government and the scientists of that nation. Dr. Sarwono Prawirohardjo, who heads the Indonesian Institute of Sciences, asked the National Academy of Sciences to arrange a panel of experts from the United States and other nations to come to Indonesia to study its food problems with the possibility of developing solutions along the lines that were outlined in the PSAC World Food Report. A 27-man delegation did conduct a Workshop of Food, headquartered in Djarkarta, Indonesia, during the period May 27-June 1, 1968. We had the able assistance and cooperation of approximately 80 Indonesian scientists and technical workers from the various islands of that nation. A report was developed following the intensive study of problems of a nation experiencing a major food crisis. Major recommendations of that study have already been incorporated into the revised agricultural program presently being developed for immediate implementation in Indonesia. Indications are that other nations may be interested in similar intensive studies utilizing this PSAC Report as a basis for these investigations. This is rewarding to those who gave so generously of their valuable and limited time to the study. Many biological scientists may be called to participate in increased efforts in technical assistance. This must happen if hungry people are to be fed and if nations are to experience the forward thrust in economic development needed to bring some equitable balance in living conditions.

The Green Revolution
Anonymous

Until quite recently it seemed that the hopes for rapid economic progress in many of the poorest countries were doomed by very slow growth of the huge agricultural sector. Agricultural production barely kept pace with population growth, and in many areas fell behind.

The less developed countries, which had been net *exporters* of 14 million tons of cereal grains each year in the 1930s, became net *importers* of 10 million tons per year in the 1960s. Imports rose even further in 1966 and 1967 after the monsoon rains in South Asia had failed two years in succession. There was widespread prediction of imminent famine.

These dire forecasts could not take into account the progress that has now been made in seed research, in irrigation practices, in extension work, and in agricultural education. They also ignored the impact price incentives could have on agricultural practices and production. As controls over production were relaxed and prices for farm products were made remunerative, farmers proved willing to adopt a new technology with amazing rapidity.

The situation today is radically different from earlier pessimistic forecasts. A good part of the developing world is now experiencing a major breakthrough in food production, widely characterized as the Green Revolution.

In 1968-69, India's food output was about 8 million tons larger than the previous record of 89 million tons in 1964-65. Pakistan increased its wheat production by 50 per cent in two years. Ceylon's rice production has gone up by 34 per cent between 1966 and 1968, and in the Philippines two bumper rice harvests seem to have ended half a century of dependence on imported rice. The most dramatic advances have been concentrated in Asia.

These dramatic increases are well illustrated by experience in India and Pakistan. Indian agriculture employs 70 per cent of the population and contributes 46 per cent of the national product. Yet, before the mid-1960's, agriculture received only about 15 per cent of the Indian public development expenditures and was not regarded as a potential growth sector.

Little attention was paid to price incentives for farmers, to the provision of such agricultural inputs as fertilizer, improved seeds, and farm machinery, or to problems of farm credit.

The droughts which struck large parts of India in 1965 and 1966 made it necessary to raise the level of food imports to 10 million tons per year. This highlighted the necessity of a drastic revision of agricultural policy.

Almost simultaneously, new high-yield varieties of wheat and rice became available. These "dwarf" varieties had been developed in Mexico and the Philippines under the sponsorship of the Ford and Rockefeller Foundations. They permit profitable application of up to three or four times as much fertilizer as traditional varieties, which combined with irrigation and pesticides makes possible a doubling or tripling of yields.

These improved strains were quickly adopted in India, investment in agriculture was expanded, and support prices were raised. Together, these steps represented substantial new incentives, leading to improved yields.

In West Pakistan, lack of drainage had by the 1950s raised the groundwater level, which put large areas of farmland out of production by increasing salinity in the soil. However, the threat of disaster was turned into a blessing.

The solution to the salinization problem was the sinking of deep tubewells to lower the water level by intensive pumping, which permitted the saline soil to be reclaimed. The tubewells also produced large supplies of irrigation water.

Originally, it was assumed that massive public programmes would be re-required, but individual farmers adopted this innovation spontaneously because of the tremendous use in yields made possible by the increased groundwater supply.

The surge in private tubewell development became one of the most spectacular elements of agricultural modernization in West Pakistan. In 1959-60, about 1,300 of these wells were installed by private farmers. By 1963-64, the annual rate of installation had accelerated to 6,600, and by 1967-68, to about 9,500.

Although, from a technical point of view, these wells were not very efficient, they were extremely profitable. Private and public tubewell irrigation accounted for nearly half of the expanded agricultural output during the period 1959-60 to 1965-66.

As in India, the introduction of the new varieties was swift. The acreage under the new

wheat seeds rose from 200,000 acres in 1966-67 to 2.3 million acres the following year.

The new rice variety called IR-8 was planted on 9,000 acres in 1967, which produced the seed for almost 900,000 acres in 1968. Fertilizer supplies, largely financed by foreign credits, were rapidly expanded, and dramatic increases in grain production did in fact occur.

New Advances in Seed Research and Irrigation

The Green Revolution has been a matter of both new technology and new policy. Although it is too early to say how deep and how rapid the impact will be and whether similar breakthroughs will be repeated in other parts of the world, prospects for growth obviously look far brighter.

Above all, it has been demonstrated that the peasant farmer, contrary to many expectations, is not hopelessly fettered by custom and tradition and that he is not insensitive to costs and prices. Given the reasonably secure expectation of large returns, he is likely to respond. The lessons of this experience for development policy extend beyond the sphere of agriculture.

In the fertilizer field, which is essential to the new agricultural technology, progress has been striking. There have been substantial cost reductions in the postwar period and further improvement is in prospect. The capital cost of new nitrogen plants is almost 50 per cent lower than for "old" (pre-1963) plants. Consumption has risen rapidly.

In India, fertilizer supplies were increased by almost 80 per cent in 1966-67 over the previous year, and by another 50 per cent the following year. Domestic fertilizer production more than doubled between 1965 and 1968; imports of fertilizer and fertilizer raw materials now equal over one-fifth of India's total export earnings.

In Pakistan, fertilizer use has doubled every two years since 1960. Increases in other developing countries have been equally dramatic. Compared to the average for the five years ending in 1956-57, total consumption of nitrogenous and phosphatic fertilizer in 1967-68 was five times as great and potash use had increased sixfold.

The growth of several non-food crops, though not as dramatic as that in wheat and rice, has also been substantial. Between 1953 and 1967, cotton production in Asia and Africa rose by 40 per cent and this has been the basis for rapid growth of the textile industry in these areas.

Coffee production in Africa has almost tripled from 393,000 tons to 1,145,000 tons in the same period so that Africa now produces about 30 per cent of the world's coffee. This rapid growth of coffee exports from Africa has created marketing problems, but it does illustrate the potential of tropical agriculture for diversification and higher productivity.

If the Green Revolution signals a major breakthrough in food grain production, it also brings with it an array of new problems.

For one thing, continued heavy expenditures on agricultural research are necessary, as one seed variety is likely to last only for a few years and must be replaced by new varieties as new diseases evolve.

Moreover, accelerated agricultural extension and massive investments in irrigation and fertilizer production are needed. Increasing production also raises the demand for better marketing and distribution facilities and for more farm credit.

It will also be difficult to maintain a set of incentives for farmers which is adequate to elicit the necessary production, stimulate the continued adoption of new technology, and support diversification into other crops.

Increased tax revenue will be needed, but to tax agricultural income directly is difficult in most developing countries for the good reasons that most farmers are very poor, that such a tax would be politically explosive, and that the cost of collecting the tax might well exceed the yield.

However, the new technology is raising some rural incomes sharply. If large increases in income are to arise in agriculture, some of the increased revenue must come from these incomes. Agricultural taxation and the general division of the fruits of increased agricultural productivity among urban consumers, rural producers, and landowners will present thorny policy issues which have grave political implications and will also affect future development.

Areas untouched by the Green Revolution, such as most of Africa and Latin America, face a more difficult task in stimulating technological change in the countryside. Many of them still seriously neglect rural development. For all countries it is important to achieve new technical breakthroughs in crops other than foodgrains, especially in exportable ones, not only to increase the food supply but also to improve its quality.

Land reform and consolidation of fragmented holdings will be needed in many developing countries not only to accelerate technological change and stimulate production in the long run, but also to generate rural employment.

History teaches us that land reform is seldom a tidy affair and is always time-consuming. However, most governments now have at their disposal the means to minimize the short-run disruptions and conflicts arising from a programme of structural change in land ownership.

Population and Panaceas
A Technological Perspective
Paul R. Ehrlich and John P. Holdren

Today more than one billion human beings are either undernourished or malnourished, and the human population is growing at a rate of 2% per year. The existing and impending crises in human nutrition and living conditions are well-documented but not widely understood. In particular, there is a tendency among the public, nurtured on Sunday-supplement conceptions of technology, to believe that science has the situation well in hand—that farming the sea and the tropics, irrigating the deserts, and generating cheap nuclear power in abundance hold the key to swift and certain solution of the problem. To espouse this belief is to misjudge the present severity of the situation, the disparate time scales on which technological progress and population growth operate, and the vast complexity of the problems beyond mere food production posed by population pressures. Unfortunately, scientists and engineers have themselves often added to the confusion by failing to distinguish between that which is merely theoretically feasible, and that which is economically and logistically practical.

As we will show here, man's present technology is inadequate to the task of maintaining the world's burgeoning billions, even under the most optimistic assumptions. Furthermore, technology is likely to remain inadequate until such time as the population growth rate is drastically reduced. This is not to assert that present efforts to "revolutionize" tropical agriculture, increase yields of fisheries, desalt water for irrigation, exploit new power sources, and implement related projects are not worthwhile. They may be. They could also easily produce the ultimate disaster for mankind if they are not applied with careful atten-

tion to their effects on the ecological systems necessary for our survival (Woodwell, 1967; Cole, 1968). And even if such projects are initiated with unprecedented levels of staffing and expenditures, without population control they are doomed to fall far short. No effort to expand the carrying capacity of the Earth can keep pace with unbridled population growth.

To support these contentions, we summarize briefly the present lopsided balance sheet in the population/food accounting. We then examine the logistics, economics, and possible consequences of some technological schemes which have been proposed to help restore the balance, or, more ambitiously, to permit the maintenance of human populations much larger than today's. The most pertinent aspects of the balance are:

1) The world population reached 3.5 billion in mid-1968, with an annual increment of approximately 70 million people (itself increasing) and a doubling time on the order of 35 years (Population Reference Bureau, 1968).

2) Of this number of people, at least one-half billion are undernourished (deficient in calories or, more succinctly, slowly starving), and approximately an additional billion are malnourished (deficient in particular nutrients, mostly protein) (Borgstrom, 1965: Sukhatme, 1966). Estimates of the number actually perishing annually from starvation begin at 4 million and go up (Ehrlich, 1968) and depend in part on official definitions of starvation which conceal the true magnitude of hunger's contribution to the death rate (Lelyveld, 1968).

3) Merely to maintain present inadequate nutrition levels, the food requirements of Asia, Africa, and Latin America will, conserva-

tively, increase by 26% in the 10-year period measured from 1965 to 1975 (Paddock and Paddock, 1967). World food production must double in the period 1965-2000 to stay even; it must triple if nutrition is to be brought up to minimum requirements.

Food Production

That there is insufficient additional, good quality agricultural land available in the world to meet these needs is so well documented (Borgstrom, 1965) that we will not belabor the point here. What hope there is must rest with increasing yields on land presently cultivated, bringing marginal land into production, more efficiently exploiting the sea, and bringing less conventional methods of food production to fruition. In all these areas, science and technology play a dominant role. While space does not permit even a cursory look at all the proposals on these topics which have been advanced in recent years, a few representative examples illustrate our points.

CONVENTIONAL AGRICULTURE

Probably the most widely recommended means of increasing agricultural yields is through the more intensive use of fertilizers. Their production is straightforward, and a good deal is known about their effective application, although, as with many technologies we consider here, the environmental consequences of heavy fertilizer use are ill understood and potentially dangerous[1] (Wadleigh, 1968). But even ignoring such problems, we find staggering difficulties barring the implementation of fertilizer technology on the scale required. In this regard, the accomplishments of countries such as Japan and the Netherlands are often cited as offering hope to the underdeveloped world. Some perspective on this point is afforded by noting that if India were to apply fertilizer at the per capita level employed by the Netherlands, her fertilizer needs would be nearly half the present world output (United Nations, 1968).

On a more realistic plane, we note that although the goal for nitrogen fertilizer production in 1971 under India's fourth 5-year plan is 2.4 million metric tons (Anonymous, 1968a), Raymond Ewell (who has served as fertilizer production adviser to the Indian government for the past 12 years) suggests that less than 1.1 million metric tons is a more probable figure for that date.[2] Ewell cites poor plant maintenance, raw materials shortages, and power and transportation breakdowns as contributing to continued low production by existing Indian plants. Moreover, even when fertilizer is available, increases in productivity do not necessarily follow. In parts of the underdeveloped world lack of farm credit is limiting fertilizer distribution; elsewhere internal transportation systems are inadequate to the task. Nor can the problem of educating farmers on the advantages and techniques of fertilizer use be ignored. A recent study (Parikh et al., 1968) of the Intensive Agriculture District Program in the Surat district of Gujarat, India (in which scientific fertilizer use was to have been a major ingredient) notes that "on the whole, the performance of adjoining districts which have similar climate but did not enjoy relative preference of input supply was as good as, if not better than, the programme district . . . . A particularly disheartening feature is that the farm production plans, as yet, do not carry any educative value and have largely failed to convince farmers to use improved practices in their proper combinations."

As a second example of a panacea in the realm of conventional agriculture, mention

1. Barry Commoner, address to 135th Meeting of the AAAS, Dallas, Texas (28 December 1968).
2. Raymond Ewell, private communication (1 December 1968).

must be given to the development of new high-yield or high-protein strains of food crops. That such strains have the potential of making a major contribution to the food supply of the world is beyond doubt, but this potential is limited in contrast to the potential for population growth, and will be realized too slowly to have anything but a small impact on the immediate crisis. There are major difficulties impeding the widespread use of new high-yield grain varieties. Typically, the new grains require high fertilizer inputs to realize their full potential, and thus are subject to all the difficulties mentioned above. Some other problems were identified in a recent address by Lester R. Brown, administrator of the International Agriculture Development Service: the limited amount of irrigated land suitable for the new varieties, the fact that a farmer's willingness to innovate fluctuates with the market prices (which may be driven down by high-yield crops), and the possibility of tieups at market facilities inadequate for handling increased yields.[3]

Perhaps even more important, the new grain varieties are being rushed into production without adequate field testing, so that we are unsure of how resistant they will be to the attacks of insects and plant diseases. William Paddock has presented a plant pathologist's view of the crash programs to shift to new varieties (Paddock, 1967). He describes India's dramatic program of planting improved Mexican wheat, and continues: "Such a rapid switch to a new variety is clearly understandable in a country that tottered on the brink of famine. Yet with such limited testing, one wonders what unknown pathogens await a climatic change which will give the environmental conditions needed for their growth." Introduction of the new varieties creates enlarged monocultures of plants with essentially unknown levels of resistance to disaster. Clearly,

one of the prices that is paid for higher yield is a higher risk of widespread catastrophe. And the risks are far from local: since the new varieties require more "input" of pesticides (with all their deleterious ecological side effects), these crops may ultimately contribute to the defeat of other environment-related panaceas, such as extracting larger amounts of food from the sea.

A final problem must be mentioned in connection with these strains of food crops. In general, the hungriest people in the world are also those with the most conservative food habits. Even rather minor changes, such as that from a rice variety in which the cooked grains stick together to one in which the grains fall apart, may make new foods unacceptable. It seems to be an unhappy fact of human existence that people would rather starve than eat a nutritious substance which they do not recognize as food.[4]

Beyond the economic, ecological, and sociological problems already mentioned in connection with high-yield agriculture, there is the overall problem of time. We need time to breed the desired characteristics of yield and hardiness into a vast array of new strains (a tedious process indeed), time to convince farmers that it is necessary that they change their time-honored ways of cultivation, and time to convince hungry people to change the staples of their diet. The Paddocks give 20 years as the "rule of thumb" for a new technique or plant variety to progress from conception to substantial impact on farming (Paddock and Paddock, 1967). They write: "It is true that a *massive* research attack on the

3. Lester R. Brown, address to the Second International Conference on the War on Hunger, Washington, D.C. (February 1968).
4. For a more detailed discussion of the psychological problems in persuading people to change their dietary habits, see McKenzie, 1968.

problem could bring some striking results in less than 20 years. But I do not find such an attack remotely contemplated in the thinking of those officials capable of initiating it." Promising as high-yield agriculture may be, the funds, the personnel, the ecological expertise, and the necessary years are unfortunately not at our disposal. Fulfillment of the promise will come too late for many of the world's starving millions, if it comes at all.

Bringing More Land Under Cultivation

The most frequently mentioned means of bringing new land into agricultural production are farming the tropics and irrigating arid and semi-arid regions. The former, although widely discussed in optimistic terms, has been tried for years with incredibly poor results, and even recent experiments have not been encouraging. One essential difficulty is the unsuitability of tropical soils for supporting typical foodstuffs instead of jungles (McNeil, 1964; Paddock and Paddock 1964). Also, "the tropics" are a biologically more diverse area than the temperate zones, so that farming technology developed for one area will all too often prove useless in others. We shall see that irrigating the deserts, while more promising, has serious limitations in terms of scale, cost, and lead time.

The feasible approaches to irrigation of arid lands appear to be limited to large-scale water projects involving dams and transport in canals, and desalination of ocean and brackish water. Supplies of usable ground water are already badly depleted in most areas where they are accessible, and natural recharge is low enough in most arid regions that such supplies do not offer a long-term solution in any case. Some recent statistics will give perspective to the discussion of water projects and desalting which follows. In 1966, the United States was

using about 300 billion gallons of water per day, of which 135 billion gallons were consumed by agriculture and 165 billion gallons by municipal and industrial users (Sporn, 1966). The bulk of the agricultural water cost the farmer from 5 to 10 cents/1000 gal.; the highest price paid for agricultural water was 15 cents/1000 gal. For small industrial and municipal supplies, prices as high as 50 to 70 cents/1000 gal. were prevalent in the U.S. arid regions, and some communities in the Southwest were paying on the order of $1.00/1000 gal. for "project" water. The extremely high cost of the latter stems largely from transportation costs, which have been estimated at 5 to 15 cents/1000 gal. per 100 miles (International Atomic Energy Agency, 1964).

We now examine briefly the implications of such numbers in considering the irrigation of the deserts. The most ambitious water project yet conceived in this country is the North American Water and Power Alliance, which proposes to distribute water from the great rivers of Canada to thirsty locations all over the United States. Formidable political problems aside (some based on the certainty that in the face of expanding populations, demands for water will eventually arise at the source), this project would involve the expenditure of $100 billion in construction costs over a 20-year completion period. At the end of this time, the yield to the United States would be 69 million acre feet of water annually (Kelly, 1966), or 63 billion gallons per day. If past experience with massive water projects is any guide, these figures are overoptimistic; but if we assume they are not, it is instructive to note that this monumental undertaking would provide for an increase of only 21% in the water consumption of the United States, during a period in which the population is expected to increase by between 25 and 43% (U.S. Dept. of Commerce, 1966). To assess the possible contribution to the *world* food situation, we assume that all this water could be devoted to agriculture, although extrapolation of present consumption patterns indicates that only about one-half would be. Then using the rather optimistic figure of 500 gallons per day to grow the food to feed one person, we find that this project could feed 126 million additional people. Since this is less than 8% of the projected world population growth during the construction period (say 1970 to 1990), it should be clear that even the most massive water projects can make but a token contribution to the solution of the world food problem in the long term. And in the crucial short term —the years preceding 1980—*no* additional people will be fed by projects still on the drawing board today.

In summary, the cost is staggering, the scale insufficient, and the lead time too long. Nor need we resort to such speculation about the future for proof of the failure of technological "solutions" in the absence of population control. The highly touted and very expensive Aswan Dam project, now nearing completion, will ultimately supply food (at the present miserable diet level) for less than Egypt's population growth during the time of construction (Borgstrom, 1965; Cole, 1968). Furthermore, its effect on the fertility of the Nile Delta may be disastrous, and, as with all water projects of this nature, silting of the reservoir will destroy the gains in the long term (perhaps in 100 years).

Desalting for irrigation suffers somewhat similar limitations. The desalting plants operational in the world today produce water at individual rates of 7.5 million gal./day and less, at a cost of 75 cents/1000 gal. and up, the cost increasing as the plant size decreases (Bender, 1969). The most optimistic firm proposal which anyone seems to have made for desalting with present or soon-to-be availa-

ble technology is a 150 million gal. per day nuclear-powered installation studied by the Bechtel Corp. for the Los Angeles Metropolitan Water District. Bechtel's early figures indicated that water from this complex would be available at the site for 27-28 cents/1000 gal. (Galstann and Currier, 1967). However, skepticism regarding the economic assumptions leading to these figures (Milliman, 1966) has since proven justified—the project was shelved after spiralling construction cost estimates indicated an actual water cost of 40-50 cents/1000 gal. Use of even the original figures, however, bears out our contention that the *most* optimistic assumptions do not alter the verdict that technology is losing the food/population battle. For 28 cents/1000 gal. is still approximately twice the cost which farmers have hitherto been willing or able to pay for irrigation water. If the Bechtel plant had been intended to supply agricultural needs, which it was not, one would have had to add to an already unacceptable price the very substantial cost of transporting the water inland.

Significantly, studies have shown that the economies of scale in the distillation process are essentially exhausted by a 150 million gal. per day plant (International Atomic Energy Agency, 1964). Hence, merely increasing desalting capacity further will not substantially lower the cost of the water. On purely economic grounds, then, it is unlikely that desalting will play a major role in food production by conventional agriculture in the short term.[5] Technological "break-throughs" will presumably improve this outlook with the passage of time, but world population growth will not wait.

5. An identical conclusion was reached in a recent study (Clawson et al., 1969) in which the foregoing points and numerous other aspects of desalting were treated in far more detail than was possible here.

Desalting becomes more promising if the high cost of the water can be offset by increased agricultural yields per gallon and, perhaps, use of a single nuclear installation to provide power for both the desalting and profitable on-site industrial processes. This prospect has been investigated in a thorough and well-documented study headed by E. S. Mason (Oak Ridge National Laboratory, 1968). The result is a set of preliminary figures and recommendations regarding nuclear-powered "agro-industrial complexes" for arid and semi-arid regions, in which desalted water and fertilizer would be produced for use on an adjacent, highly efficient farm. In underdeveloped countries incapable of using the full excess power output of the reactor, this energy would be consumed in on-site production of industrial materials for sale on the world market. Both near-term (10 years hence) and far-term (20 years hence) technologies are considered, as are various mixes of farm and industrial products. The representative near-term case for which a detailed cost breakdown is given involved a seaside facility with a desalting capacity of 1 billion gal./day, a farm size of 320,000 acres, and an industrial electric power consumption of 1585 Mw. The initial investment for this complex is estimated at $1.8 billion, and annual operating costs at $236 million. If both the food and the industrial materials produced were sold (as opposed to giving the food, at least, to those in need who could not pay),[6] the estimated profit for such a complex, before subtracting financing costs, would be 14.6%.

The authors of the study are commendably cautious in outlining the assumptions and uncertainties upon which these figures rest. The key assumption is that 200 gal./day of water will grow the 2500 calories required to feed one person. Water/calorie ratios of this order or less have been achieved by the top 20% of

farmers specializing in such crops as wheat, potatoes, and tomatoes; but more water is required for needed protein-rich crops such as peanuts and soybeans. The authors identify the uncertainty that crops usually raised separately can be grown together in tight rotation on the same piece of land. Problems of water storage between periods of peak irrigation demand, optimal patterns of crop rotation, and seasonal acreage variations are also mentioned. These "ifs" and assumptions, and those associated with the other technologies involved, are unfortunately often omitted when the results of such painstaking studies are summarized for more popular consumption (Anonymous, 1968b, 1968c). The result is the perpetuation of the public's tendency to confuse feasible and available, to see panaceas where scientists in the field concerned see only potential, realizable with massive infusions of time and money.

It is instructive, nevertheless, to examine the impact on the world food problem which the Oak Ridge complexes might have if construction were to begin today, and if all the assumptions about technology 10 years hence were valid *now*. At the industrial-agricultural mix pertinent to the sample case described above, the food produced would be adequate for just under 3 million people. This means that 23 such plants per year, at a cost of $41 billion, would have to be put in operation merely to keep pace with world population growth, to say nothing of improving the substandard diets of between one and two billion members of the present population. (Fertilizer

6. Confusing statements often are made about the possibility that food supply will outrun food demand in the future. In these statements, "demand" is used in the economic sense, and in this context many millions of starving people may generate no demand whatsoever. Indeed, one concern of those engaged in increasing food production is to find ways of increasing demand.

production beyond that required for the on-site farm is of course a contribution in the latter regard, but the substantial additional costs of transporting it to where it is needed must then be accounted for.) Since approximately 5 years from the start of construction would be required to put such a complex into operation, we should commence work on at least 125 units post-haste, and begin at least 25 per year thereafter. If the technology *were* available now, the investment in construction over the next 5 years, prior to operation of the first plants, would be $315 billion—about 20 times the total U.S. foreign aid expenditure during the past 5 years. By the time the technology *is* available the bill will be much higher, if famine has not "solved" the problem for us.

This example again illustrates that scale, time, and cost are all working against technology in the short term. And if population growth is not decelerated, the increasing severity of population-related crises will surely neutralize the technological improvements of the middle and long terms.

OTHER FOOD PANACEAS

"Food from the sea" is the most prevalent "answer" to the world food shortage in the view of the general public. This is not surprising, since estimates of the theoretical fisheries productivity of the sea run up to some 50-100 times current yields (Schmitt, 1965; Christy and Scott, 1965). Many practical and economic difficulties, however, make it clear that such a figure will never be reached, and that it will not even be approached in the foreseeable future. In 1966, the annual fisheries harvest was some 57 million metric tons (United Nations, 1968). A careful analysis (Meseck, 1961) indicates that this might be increased to a world production of 70 million metric tons by 1980. If this gain were realized, it would repre-

sent (assuming no violent change in population growth patterns) a small per capita *loss* in fisheries yield.

Both the short- and long-term outlooks for taking food from the sea are clouded by the problems of overexploitation, pollution (which is generally ignored by those calculating potential yields), and economics. Solving these problems will require more than technological legerdemain; it will also require unprecedented changes in human behavior, especially in the area of international cooperation. The unlikelihood that such cooperation will come about is reflected in the recent news (Anonymous, 1968d) that Norway has dropped out of the whaling industry because overfishing has depleted the stock below the level at which it may economically be harvested. In that industry, international controls were tried—and failed. The sea is, unfortunately, a "commons" (Hardin, 1968), and the resultant management problems exacerbate the biological and technical problems of greatly increasing our "take." One suspects that the return per dollar poured into the sea will be much less than the corresponding return from the land for many years, and the return from the land has already been found wanting.

Synthetic foods, protein culture with petroleum, saline agriculture, and weather modification all may hold promise for the future, but all are at present expensive and available only on an extremely limited scale. The research to improve this situation will also be expensive, and, of course, time-consuming. In the absence of funding, it will not occur at all, a fact which occasionally eludes the public and the Congress.

Domestic and Industrial Water Supplies

The world has water problems, even exclusive of the situation in agriculture. Although total precipitation should in theory be adequate in quantity for several further doublings of population, serious shortages arising from problems of quality, irregularity, and distribution already plague much of the world. Underdeveloped countries will find the water needs of industrialization staggering: 240,000 gal. of water are required to produce a ton of newsprint; 650,000 gal. to produce a ton of steel (International Atomic Energy Agency, 1964). Since maximum acceptable water costs for domestic and industrial use are higher than for agriculture, those who can afford it are or soon will be using desalination (40-100 + cents/1000 gal.) and used-water renovation (54-57 cents/1000 gal. [Ennis, 1967]). Those who cannot afford it are faced with allocating existing supplies between industry and agriculture, and as we have seen, they must choose the latter. In this circumstance, the standard of living remains pitifully low. Technology's only present answer is massive externally-financed complexes of the sort considered above, and we have already suggested there the improbability that we are prepared to pay the bill rung up by present population growth.

The widespread use of desalted water by those who *can* afford it brings up another problem only rarely mentioned to date, the disposal of the salts. The product of the distillation processes in present use is a hot brine with salt concentration several times that of seawater. Both the temperature and the salinity of this effluent will prove fatal to local marine life if it is simply exhausted to the ocean. The most optimistic statement we have seen on this problem is that *"smaller plants* (our emphasis) at seaside locations may return the concentrated brine to the ocean if proper attention is paid to the design of the outfall, and to the effect on the local marine ecology" (McIlhenny, 1966). The same writer identifies the major economic uncertainties connected

with extracting the salts for sale (to do so is straightforward, but often not profitable). Nor can one simply evaporate the brine and leave the residue in a pile—the 150 million gal./day plant mentioned above would produce brine bearing 90 million lb. of salts daily (based on figures by Parker, 1966). This amount of salt would cover over 15 acres to a depth of one foot. Thus, every year a plant of the billion gallon per day, agro-industrial complex size, would produce a pile of salt over 52 ft. deep and covering a square mile. The high winds typical of coastal deserts would seriously aggravate the associated soil contamination problem.

Energy

Man's problems with energy supply are more subtle than those with food and water: we are not yet running out of energy, but we are being forced to use it faster than is probably healthy. The rapacious depletion of our fossil fuels is already forcing us to consider more expensive mining techniques to gain access to lower-grade deposits, such as the oil shales, and even the status of our high-grade uranium ore reserves is not clearcut (Anonymous, 1968e).

A widely held misconception in this connection is that nuclear power is "dirt cheap," and as such represents a panacea for developed and underdeveloped nations alike. To the contrary, the largest nuclear-generating stations now in operation are just competitive with or marginally superior to modern coal-fired plants of comparable size (where coal is not scarce); at best, both produce power for on the order of 4-5 mills (tenths of a cent) per kilowatt-hour. Smaller nuclear units remain less economical than their fossil-fueled counterparts. Underdeveloped countries can rarely use the power of the larger plants. Simply speaking, there are not enough industries, appliances, and light bulbs to absorb the output, and the cost of industrialization and modernization exceeds the cost of the power required to sustain it by orders of magnitude, regardless of the source of the power. (For example, one study noted that the capital requirement to consume the output of a 70,000 kilowatt plant —about $1.2 million worth of electricity per year at 40% utilization and 5 mills/kwh—is $111 million per year if the power is consumed by metals industries, $270 million per year for petroleum product industries [E. A. Mason, 1957].) Hence, at least at present, only those underdeveloped countries which are short of fossil fuels or inexpensive means to transport them are in particular need of nuclear power.

Prospects for major reductions in the cost of nuclear power in the future hinge on the long-awaited breeder reactor and the still further distant thermonuclear reactor. In neither case is the time scale or the ultimate cost of energy a matter of any certainty. The breeder reactor, which converts more nonfissile uranium (^{238}U) or thorium to fissionable material than it consumes as fuel for itself, effectively extends our nuclear fuel supply by a factor of approximately 400 (Cloud, 1968). It is not expected to become competitive economically with conventional reactors until the 1980's (Bump, 1967). Reductions in the unit energy cost beyond this date are not guaranteed, due both to the probable continued high capital cost of breeder reactors, and to increasing costs for the ore which the breeders will convert to fuel. In the latter regard, we mention that although crushing granite for its few parts per million of uranium and thorium is possible in theory, the problems and cost of doing so are far from resolved.[7] It is too soon to predict the costs associated with a fusion reactor (few who work in the field will predict whether such a device will work at all within the next 15-20 years). One guess puts the unit energy cost at something over half that for a

coal or fission power station of comparable size (Mills, 1967), but this is pure speculation. Quite possibly the major benefit of controlled fusion will again be to extend the energy supply rather than to cheapen it.

A second misconception about nuclear power is that it can reduce our dependence on fossil fuels to zero as soon as that becomes necessary or desirable. In fact, nuclear power plants contribute only to the electrical portion of the energy budget; and in 1960 in the United States, for example, electrical energy comprised only 19% of the total energy consumed (Sporn, 1963). The degree to which nuclear fuels can postpone the exhaustion of our coal and oil depends on the extent to which that 19% is enlarged. The task is far from a trivial one, and will involve transitions to electric or fuel-cell powered transportation, electric heating, and electrically powered industries. It will be extremely expensive.

Nuclear energy, then, is a panacea neither for us nor for the underdeveloped world. It relieves, but does not remove, the pressure on fossil fuel supplies; it provides reasonably-priced power where these fuels are not abundant; it has substantial (but expensive) potential in intelligent applications such as that suggested in the Oak Ridge study discussed above; and it shares the propensity of fast-growing technology to unpleasant side effects (Novick, 1969). We mention in the last connection that, while nuclear power stations do not produce conventional air pollutants, their radioactive waste problems may in the long run prove a poor trade. Although the AEC seems to have made a good case for solidification and storage in salt mines of the bulk of the radioactive fission products (Blanko et al., 1967), a number of radioactive isotopes are

7. A general discussion of extracting metals from common rock is given by Cloud, 1968.

released to the air, and in some areas such isotopes have already turned up in potentially harmful concentrations (Curtis and Hogan, 1969). Projected order of magnitude increases in nuclear power generation will seriously aggravate this situation. Although it has frequently been stated that the eventual advent of fusion reactors will free us from such difficulties, at least one authority, F. L. Parker, takes a more cautious view. He contends that losses of radioactive tritium from fusion power plants may prove even more hazardous than the analogous problems of fission reactors (Parker, 1968).

A more easily evaluated problem is the tremendous quantity of waste heat generated at nuclear installations (to say nothing of the usable power output, which, as with power from whatever source, must also ultimately be dissipated as heat). Both have potentially disastrous effects on the local and world ecological and climatological balance. There is no simple solution to this problem, for, in general, "cooling" only moves heat; it does not *remove* it from the environment viewed as a whole. Moreover, the Second Law of Thermodynamics puts a ceiling on the efficiency with which we can do even this much, i.e., concentrate and transport heat. In effect, the Second Law condemns us to aggravate the total problem by generating still *more* heat in any machinery we devise for local cooling (consider, for example, refrigerators and air conditioners).

The only heat which actually leaves the whole system, the Earth, is that which can be radiated back into space. This amount steadily is being diminished as combustion of hydrocarbon fuels increases the atmospheric percentage of CO_2 which has strong absorption bands in the infrared spectrum of the outbound heat energy. (Hubbert, 1962, puts the increase in the CO_2 content of the atmosphere at 10% since 1900). There is, of course, a com-

peting effect in the Earth's energy balance, which is the increased reflectivity of the upper atmosphere to incoming sunlight due to other forms of air pollution. It has been estimated, ignoring both these effects, that man risks drastic (and perhaps catastrophic) climatological change if the amount of heat he dissipates in the environment on a global scale reaches 1% of the incident solar energy at the Earth's surface (Rose and Clark, 1961). At the present 5% rate of increase in world energy consumption,[8] this level will be reached in less than a century, and in the immediate future the direct contribution of man's power consumption will create serious local problems. If we may safely rule out circumvention of the Second Law or the divorce of energy requirements from population size, this suggests that, whatever science and technology may accomplish, population growth must be stopped.

Transportation

We would be remiss in our offer of a technological perspective on population problems without some mention of the difficulties associated with transporting large quantities of food, material, or people across the face of the Earth. While our grain exports have not begun to satisfy the hunger of the underdeveloped world, they already have taxed our ability to transport food in bulk over large distances. The total amount of goods of *all* kinds loaded at U.S. ports for external trade was 158 million metric tons in 1965 (United Nations, 1968). This is coincidentally the approximate amount of grain which would have been required to make up the dietary shortages of the underdeveloped world in the same year (Sukhatme,

8. The rate of growth of world energy consumption fluctuates strongly about some mean on a time scale of only a few years, and the figures are not known with great accuracy in any case. A discussion of predicting the mean and a defense of the figure of 5% are given in Dueron et al., 1957.

1966). Thus, if the United States *had* such an amount of grain to ship, it could be handled only by displacing the entirety of our export trade. In a similar vein, the gross weight of the fertilizer, in excess of present consumption, required in the underdeveloped world to feed the additional population there in 1980 will amount to approximately the same figure—150 million metric tons (Sukhatme, 1966). Assuming that a substantial fraction of this fertilizer, should it be available at all, will have to be shipped about, we had best start building freighters! These problems, and the even more discouraging one of internal transportation in the hungry countries, coupled with the complexities of international finance and marketing which have hobbled even present aid programs, complete a dismal picture of the prospects for "external" solutions to ballooning food requirements in much of the world.

Those who envision migration as a solution to problems of food, land, and water distribution not only ignore the fact that the world has no promising place to put more people, they simply have not looked at the numbers of the transportation game. Neglecting the fact that migration and relocation costs would probably amount to a minimum of several thousand dollars per person, we find, for example, that the entire long-range jet transport fleet of the United States (about 600 planes [Molloy, 1968] with an average capacity of 150), averaging two round trips per week, could transport only about 9 million people per year from India to the United States. This amounts to about 75% of that country's annual population *growth* (Population Reference Bureau, 1968). Ocean liners and transports, while larger, are less numerous and much slower, and over long distances could not do as well. Does anyone believe, then, that we are going to compensate for the world's population growth by sending the excess to the planets? If

there were a place to go on Earth, financially and logistically we could not send our surplus there.

Conclusion

We have not attempted to be comprehensive in our treatment of population pressures and the prospects of coping with them technologically; rather, we hope simply to have given enough illustrations to make plausible our contention that technology, without population control, cannot meet the challenge. It may be argued that we have shown only that any one technological scheme taken individually is insufficient to the task at hand, whereas *all* such schemes applied in parallel might well be enough. We would reply that neither the commitment nor the resources to implement them all exists, and indeed that many may prove mutually exclusive (e.g., harvesting algae may diminish fish production).

Certainly, an optimum combination of efforts exists in theory, but we assert that no organized attempt to find it is being made, and that our examination of its probable eventual constituents permits little hope that even the optimum will suffice. Indeed, after a far more thorough survey of the prospects than we have attempted here, the President's Science Advisory Committee Panel on the world food supply concluded (PSAC, 1967): "The solution of the problem that will exist after about 1985 *demands* that programs of population control be initiated now." We most emphatically agree, noting that "now" was 2 years ago!

Of the problems arising out of population growth in the short, middle, and long terms, we have emphasized the first group. For mankind must pass the first hurdles—food and water for the next 20 years—to be granted the privilege of confronting such dilemmas as the exhaustion of mineral resources and physical space later.[9] Furthermore, we have not conveyed the extent of our concern for the environmental deterioration which has accompanied the population explosion, and for the catastrophic ecological consequences which would attend many of the proposed technological "solutions" to the population/food crisis. Nor have we treated the point that "development" of the rest of the world to the standards of the West probably would be lethal ecologically (Ehrlich and Ehrlich, 1970). For even if such grim prospects are ignored, it is abundantly clear that in terms of cost, lead time, and implementation on the scale required, technology without population control will be too little and too late.

What hope there is lies not, of course, in abandoning attempts at technological solutions; on the contrary, they must be pursued at unprecedented levels, with unprecedented judgment, and above all with unprecedented attention to their ecological consequences. We need dramatic programs now to find ways of ameliorating the food crisis—to buy time for humanity until the inevitable delay accompanying population control efforts has passed. But it cannot be emphasized enough that if the population control measures are *not* initiated immediately and effectively, all the technology man can bring to bear will not fend off the misery to come.[10] Therefore, confronted as we are with limited resources of time and money, we must consider carefully what fraction of our effort should be applied to the cure of the disease itself instead of to the temporary relief of the symptoms. We should ask, for example, how many vasectomies could be performed by a program funded with the 1.8 billion dollars required to build a single nuclear agro-industrial complex, and what the relative impact on the problem would be in both the short and long terms.

The decision for population control will be opposed by growth-minded economists and businessmen, by nationalistic statesmen, by zealous religious leaders, and by the myopic and well-fed of every description. It is therefore incumbent on all who sense the limitations of technology and the fragility of the environmental balance to make themselves heard above the hollow, optimistic chorus—to convince society and its leaders that there is no alternative but the cessation of our irresponsible, all-demanding, and all-consuming population growth.

ACKNOWLEDGMENTS

We thank the following individuals for reading and commenting on the manuscript: J. H. Brownell (Stanford University); P. A. Cantor (Aerojet General Corp.); P. E. Cloud (University of California, Santa Barbara); D. J. Eckstrom (Stanford University); R. Ewell (State University of New York at Buffalo); J. L. Fisher (Resources for the Future, Inc.); J. A. Hendrickson, Jr. (Stanford University); J. H. Hessel (Stanford University); R. W. Holm (Stanford University); S. C. McIntosh, Jr., (Stanford Univeristy); K. E. F. Watt (University of California, Davis). This work was supported in part by a grant from the Ford Foundation.

9. Since the first draft of this article was written, the authors have seen the manuscript of a timely and pertinent forthcoming book, *Resources and Man,* written under the auspices of the National Academy of Sciences and edited by Preston E. Cloud. The book reinforces many of our own conclusions in such areas as agriculture and fisheries and, in addition, treats both short- and long-term prospects in such areas as mineral resources and fossil fuels in great detail.

10. This conclusion has also been reached within the specific context of aid to underdeveloped countries in a Ph.D. thesis by Douglas Daetz: "Energy Utilization and Aid Effectiveness in Nonmechanized Agriculture: A Computer Simulation of a Socioeconomic System" (University of California, Berkeley, May 1968).

REFERENCES

Anonymous. 1968a. India aims to remedy fertilizer shortage. *Chem. Eng. News,* 46 (November 25): 29.

———— 1968b. Scientists Studying Nuclear-Powered Agro-Industrial Complexes to Give Food and Jobs to Millions. *New York Times,* March 10, p. 74.

———— 1968c. Food from the atom. *Technol. Rev.,* January, p. 55.

———— 1968d. Norway-The end of the big blubber. *Time,* November 29, p. 98.

———— 1968c. Nuclear fuel cycle. *Nucl. News,* January, p. 30.

Bender, R. J. 1969. Why water desalting will expand. *Power,* 113 (August): 171.

Blanko, R. E., J. O. Blomeke, and J. T. Roberts. 1967. Solving the waste disposal problem. *Nucleonics,* 25: 58.

Borgstrom, Georg. 1965. *The Hungry Planet.* Collier-Macmillan, New York.

Bump, T. R. 1967. A third generation of breeder reactors. *Sci. Amer.,* May, p. 25.

Christy, F. C., Jr., and A. Scott. 1965. *The Commonwealth in Ocean Fisheries.* Johns Hopkins Press, Baltimore.

Clawson, M., H. L. Landsberg, and L. T. Alexander. 1969. Desalted seawater for agriculture: Is it economic? *Science,* 164: 1141.

Cloud, P. R. 1968. Realities of mineral distribution. *Texas Quart.,* Summer, p. 103.

Cole, LaMont C. 1968. Can the world be saved? *BioScience,* 18: 679.

Curtis, R., and E. Hogan. 1969. *Perils of the Peaceful Atom.* Doubleday, New York. p. 135, 150-152.

Ennis, C. E. 1967. Desalted water as a competitive commodity. *Chem. Eng. Progr.,* 63: (1): 64.

Ehrlich, P. R. 1968. *The Population Bomb.* Sierra Club/Ballantine, New York.

Ehrlich, P. R., and Anne H. Ehrlich. 1970. *Population, Resources, and Environment.* W. H. Freeman, San Francisco (In press).

Galstann, L. S., and E. L. Currier. 1967. The Metropolitan Water District desalting project. *Chem. Eng. Progr.,* 63, (1): 64.

Gúeron, J., J. A. Lane, I. R. Maxwell, and J. R. Menke. 1957. *The Economics of Nuclear Power. Progress in Nuclear Energy.* McGraw-Hill Book Co., New York. Series VIII. p. 23.

Hardin, G. 1968. The tragedy of the commons. *Science.* 162: 1243

Hubbert, M. K. 1962. Energy resources, A report to the Committee on Natural Resources. National Research Council Report 1000-D. National Academy of Sciences.

International Atomic Energy Agency. 1964. Desalination of water using conventional and nuclear energy. Technical Report 24, Vienna.

Kelly, R. P. 1966. North American water and power alliance. In: *Water Production Using Nuclear Energy,* R. G. Post and R. L. Seale (eds.). University of Arizona Press, Tucson, p. 29.

Lelyveld, D. 1968. Can India survive Calcutta? *New York Times Magazine,* October 13, p. 58.

Mason, E. A. 1957. Economic growth and energy consumption. In: *The Economics of Nuclear Power. Progress in Nuclear Energy,* Series VIII, J. Gúeron et al. (eds.) McGraw-Hill Book Co., New York, p. 56.

McIlhenny, W. F. 1966. Problems and potentials of concentrated brines. In: *Water Production Using Nuclear Energy,* R. G. Post and R. L. Seale (eds.) University of Arizona Press, Tucson, p. 187.

McKenzie, John. 1968. Nutrition and the soft sell. *New Sci.,* 40: 423.

McNeil, Mary. 1964. Lateritic soils. *Sci. Amer.,* November, p.99.

Meseck, G. 1961. Importance of fish production and utilization in the food economy. Paper R11.3, presented at FAO Conference on Fish in Nutrition, Rome.

Milliman, J. W. 1966. Economics of water production using nuclear energy. In: *Water Production Using Nuclear Energy.* R. G. Post and R. L. Seale (eds.) University of Arizona Press, Tucson, p. 49.

Mills, R. G. 1967. Some engineering problems of thermonuclear fusion. *Nucl. Fusion.* 7: 223.

Molloy, J. F., Jr. 1968. The $12-billion financing problem of U.S. airlines. *Astronautics and Aeronautics,* October, p. 76.

Novick, S. 1969. *The Careless Atom.* Houghton Mifflin, Boston.

Oak Ridge National Laboratory. 1968. Nuclear energy centers, industrial and agro-industrial complexes, Summary Report. ORNL-4291, July.

Paddock, William. 1967. Phytopathology and a hungry world. *Ann. Rev. Phytopathol.,* 5: 375.

Paddock, William, and Paul Paddock. 1964. *Hungry Nations.* Little, Brown & Co., Boston.

———— 1967. *Famine 1975!* Little, Brown & Co., Boston.

Parikh, G., S. Saxena, and M. Maharaja. 1968. Agricultural extension and IADP, a study of Surat. *Econ. Polit. Weekly,* August 24, p. 1307.

Parker, F. L. 1968. Radioactive wastes from fusion reactors. *Science,* 159: 83.

Parker, H. M. 1966. Environmental factors relating to large water plants. In: *Water Production Using Nuclear Energy,* R. G. Post and R. L. Seale (eds.). University of Arizona Press, Tucson, p. 209.

Population Reference Bureau. 1968. Population Reference Bureau Data Sheet. Pop. Ref. Bureau, Washington, D.C.

PSAC. 1967. *The World Food Problem.* Report of the President's Science Advisory Committee. Vols. 1-3. U.S. Govt. Printing Office, Washington, D.C.

Rose, D. J., and M. Clark, Jr. 1961. *Plasma and Controlled Fusion.* M.I.T. Press, Cambridge, Mass., p. 3.

Schmitt, W. R. 1965. The planetary food potential. *Ann. N. Y. Acad. Sci.,* 118: 645.

Sporn, Philip. 1963. *Energy for Man.* Macmillan, New York.

———— 1966. *Fresh Water from Saline Waters.* Pergamon Press, New York.

Sukhatme, P. V. 1966. The world's food supplies. *Roy. Stat. Soc. J.,* 129A: 222.

United Nations. 1968. *United Nations Statistical Yearbook for 1967.* Statistical Office of the U.N., New York.

U. S. Department of Commerce. 1966. *Statistical Abstract of the U.S.* U.S. Govt. Printing Office, Washington, D.C.

Wadleigh, C. H. 1968. Wastes in relation to agriculture and industry. USDA Miscellaneous Publication No. 1065. March.

Woodwell, George M. 1967. Toxic substances and ecological cycles. *Sci. Amer.,* March, p. 24.

UPI

Riders of the
Apocalypse

Manifestations of our uncontrolled population growth and unrestrained expenditure of resources are identifiable in an apparently unending list of social and environmental ills. The pollution of air and soil, of lakes, rivers and oceans, the alienation of the young, the deterioration of the cities, the rising use of sedatives, the increase of degenerative diseases of obscure origin in our own species and the demise of other species are all outriders of disaster. It is in the nature of ecology that inter-relationships should be many and subtle, that the true urgency of the situation should be obscure and debatable, and that no single ultimate enemy should loom clearly among the mass of woes. Thus, while this anthology has led inexorably to a conclusion of catastrophe, no particular environmental armageddon has been specified and the intensity and scope of catastrophe remains unclear.

It may be that we still have the option of choosing among our poisons and plotting a course that will give the highest probability of survival for ourselves as individuals, for civilization (at least the best of it), for man as a species, or at least for the biosphere.

Past history and present realities dictate that warfare is almost certain to have a role to play. Rapid change is an inherent part of the situation and change always invites political unrest and "the pursuit of politics by other means." Failure of efforts at economic development, and a static or increasing gap between the developed and underdeveloped countries, famine, competition for vanishing resources, and over-crowding all predispose toward military activity. The military-industrial complexes, and the economics of armaments races, also make significant contributions to the high probability of armed conflict.

Large scale warfare, then will become more and more probable. What are the ecological significances of warfare, and the choices open as to the nature of warfare?

Warfare is traditionally an excuse for total ecological irresponsibility. It is conceived of as the supreme human emergency, and lack of concern for ecological consequences (as well as ignorance of ecology) is a hallmark of the defence departments, the strategic planners, and the military mind in general. Lack of concern for ecological effects of war may have been realistic in the past, but in the contexts of modern nuclear and chemical warfare, and of our precarious ecological position, this attitude is unrealistic. Nevertheless, it persists. Our "think-tanks" make elaborate studies of industrial and economic damage, and produce contingency plans for various degrees of *human* over and underskill, but are oddly unsolicitous on behalf of the biosphere.

Major nations are armed for nuclear conflict, and increasing numbers of nations are in position to so arm themselves. Despite the apparent absence of thorough studies on the *ecological* consequences of large scale nuclear warfare, some generalizations are possible. First, the "balance of terror" approach has already produced considerable environmental contamination (as well as nuclear stockpiles of absurd size). Present stockpiles are so large that if a major power were to explode its weapons on enemy targets, even without any retaliation, biospheric contamination would result in enormous environmental damage, if not complete sterilization of the planet. Second, fission products are not equivalent to background radiation, since many of the radioactive elements unique to nuclear explosions are biologically active and are concentrated along food chains. Third, nuclear warfare will inevitably bring famine and epidemic in its aftermath, since it involves maximum destruction of transportation networks and manufacturing facilities on which the present human population depends. These considerations suggest that a nuclear conflagration is the worst alternative, ecologically. Despite this, the folly of our society is such that there are now 20,000 or more nuclear warheads poised on the launching pads of our planet and it is highly probable that the apocalyptic horsemen will ride by the light of nuclear fireballs.

Chemical warfare, ecologically, has some of the same disadvantages as nuclear warfare. It involves a very high degree of environmental pollution, with unexplored ecological consequences. Modern chemical agents are likely to be related to some of the persistent pesticides and thus their use in the absence of knowledge of their ecological effects is particularly idiotic. On the other hand, there is a serious question as to whether the use of chemical agents would be any more destructive then the conglomeration of synthetic chemicals of all sorts which we are already pouring into the environment in ever increasing quantities.

Biological warfare appears to be a rather different matter. The biosphere has, after all, been dealing with biological renegades for 3.5 billion years. The mechanisms for destruction of biological agents, and resistance to them on various levels, are beautifully elaborated in every species. Biological warfare might involve little or no squandering of mineral resources, and would leave cities, transportation networks, and industrial plants intact for human survivors. Those survivors would, presumably, be individuals who had been in good physical condition and who had shown some form of immunity to the biological agents. In that restricted sense it would be a superior stock that had the opportunity to start anew and avoid past mistakes.

In an ecological context it is ironic that biological warfare should be renounced by governments, and be opposed with mindless emotion by "gut" liberals, while nuclear escalation continues unchecked. Unless war can be eliminated completely, biological warfare is certainly the ecological choice, or even, possibly, the only short route to salvation of both man and the biosphere.

But perhaps the "strategic planners" will be proven right and the nuclear stalemate will persist, while the establishment and its military-industrial complex will be able to limit warfare to the "ideal" variety. An "ideal" war is one in which a developed country can inflict casualties on a barefoot enemy while placating its own public with an economic boom and casualty lists low enough to dilute concern (casualties on both sides may be low enough to prove a stimulus, rather than a hindrance, to population growth). The "ideal" war is viewed as highly advantageous in an "economic" sense, involving a high rate of destruction and obsolesence of complex equipment, a high rate of resource exploitation, and a consequent high rate of employment, both in and out of the armed forces. The "ideal" war does provide an intensified form of oppression of the young, (who must, after all, do most of the dying) and operates to intensify their frustration, alienation and rebelliousness, but this is considered a minor price to pay for the economic return. Ideal wars probably also involve a high rate of pollution, but as far as is known there has been no calculation of Viet Nam's contribution to atmospheric and hydrospheric pollutants.

An important "side effect" of arms races and limited wars may be noted. Expenditures for "defense" consume a major part of national budgets, as well as vast amounts of manpower which could be put to other uses. NATO countries have, in recent years, maintained standing forces of at least 10 million men. In comparison with money and scientific effort expended for military research and development, the effort and funds devoted to study of ecological problems, and even the efforts devoted to applied pollution control, are trivial.

Among the four horsemen, war seems to be getting the popular vote at present. What of the others? Famine and Pestilence are the stuff of biological warfare, and are certain to follow nuclear or chemical warfare on a major scale. In association with conventional warfare, famine and pestilence have been limited in our recent experience, although the growing ecological crisis makes this "humane" accomplishment less likely in the future.

In view of the enormous gap in power, food, and other resources between the "have" and "have-not" nations, it seems unlikely that major famines will become worldwide killers which accomplish major reductions in the human population. It is unlikely, at least, that this could happen without large-scale wars.

What of disease? Can we anticipate major epidemics as primary agents? In recent years the spread of public health technology has been a strong defense against worldwide epidemics. There have been breaches in that defense, however, and asian flu demonstrates that it is still quite possible for major epidemics to spread with little hinderance. Protection against more virulent diseases has been purchased at a cost in terms of reduced natural immunities in protected popula-

UPI

tions. The worldwide transportation network now handles such masses of people that the spread of some new, highly contagious, and deadly virus could be almost instantaneous, once it reached a major center such as New York or London, and the increasing concentration of the population in urban centers has both positive and negative aspects in terms of control. Epidemic disease, then, does not appear out of the question. Under certain conditions it has some of the grim "advantages" of biological warfare.

Are there still other prospects? Ecological disasters, such as a catastrophic interference with the oxygen cycle, are apparently within the realm of possibility. Whether they have reached the level of probabilities we simply do not know, despite some disturbing pesticide "inside effects." As with most things of an ecological nature, the development of such a disaster would be ill defined, an illusive and complex shadow, slowly gaining substance. We may already have moved into such a shadow. If we have there may still be time for a "panic stop." Unlike the all-too-familiar automobile panic stop, however, a population panic stop takes a fair amount of time (say, 75 years) once the danger cannot be ignored by even the most myopic driver. The establishment that pilots "Spaceship Earth" today is pretty myopic about things ecological and if some ecological death does ride in the front rank of the apocalypse we probably won't even be wearing seat belts.

Can we, or will we, at least bargain for survival of the biosphere? From most disasters, the biosphere would, in its own good evolutionary time, be able to build anew. Could we, with some thought, leave a useful heritage? Why should we want to? While the question may have some heuristic value, my own opinion is that our altruism for continuance of the biosphere *with* man seems to be so low that the questions of at least leaving a biosphere that could regenerate *without* him is not likely to evoke much interest from "practical" men.

What We Must Do
John Platt

There is only one crisis in the world. It is the crisis of transformation. The trouble is that it is now coming upon us as a storm of crisis problems from every direction. But if we look quantitatively at the course of our changes in this century, we can see immediately why the problems are building up so rapidly at this time, and we will see that it has now become urgent for us to mobilize all our intelligence to solve these problems if we are to keep from killing ourselves in the next few years.

The essence of the matter is that the human race is on a steeply rising "S-curve" of change. We are undergoing a great historical transition to new levels of technological power all over the world. We all know about these changes, but we do not often stop to realize how large they are in orders of magnitude, or how rapid and enormous compared to all previous changes in history. In the last century, we have increased our speeds of communication by a factor of 10^7; our speeds of travel by 10^2; our speeds of data handling by 10^6; our energy resources by 10^3; our power of weapons by 10^6; our ability to control diseases by something like 10^2; and our rate of population growth to 10^3 times what it was a few thousand years ago.

Could anyone suppose that human relations around the world would not be affected to their very roots by such changes? Within the last 25 years, the Western world has moved into an age of jet planes, missiles and satellites, nuclear power and nuclear terror. We have acquired computers and automation, a service and leisure economy, superhighways, superagriculture, supermedicine, mass higher education, universal TV, oral contraceptives, environmental pollution, and urban crises. The rest of the world is also moving rapidly and may catch up with all these powers and problems within a very short time. It is hardly surprising that young people under 30, who have grown up familiar with these things from childhood, have developed very different expectations and concerns from the older generation that grew up in another world.

What many people do not realize is that many of these technological changes are now approaching certain natural limits. The "S-curve" is beginning to level off. We may never have faster communications or more TV or larger weapons or a higher level of danger than we have now. This means that if we could learn how to manage these new powers and problems in the next few years without killing ourselves by our obsolete structures and behavior, we might be able to create new and more effective social structures that would last for many generations. We might be able to move into that new world of abundance and diversity and well-being for all mankind which technology has now made possible.

The trouble is that we may not survive these next few years. The human race today is like a rocket on a launching pad. We have been building up to this moment of takeoff for a long time, and if we can get safely through the takeoff period, we may fly on a new and exciting course for a long time to come. But at this moment, as the powerful new engines are fired, their thrust and roar shakes and stresses every part of the ship and may cause the whole thing to blow up before we can steer it on its way. Our problem today is to harness and direct these tremendous new forces through this dangerous transition period to the new world instead of to destruction. But unless we can do this, the rapidly increasing strains and crises of the next decade may kill us all. They will make the last 20 years look like a peaceful interlude.

The "Half-Life" of Humanity

Several types of crisis may reach the point of explosion in the next 10 years: nuclear escalation, famine, participatory crises, racial crises, and what have been called the crises of adminstrative legitimacy. It is worth singling out two or three of these to see how imminent and dangerous they are, so that we can fully realize how very little time we have for preventing or controlling them.

Take the problem of nuclear war, for example. A few years ago, Leo Szilard estimated the "half-life" of the human race with respect to nuclear escalation as being between 10 and 20 years. His reasoning then is still valid now. As long as we continue to have no adequate stabilizing peace-keeping structures for the world, we continue to live under the daily threat not only of local wars but of nuclear escalation with overkill and megatonnage enough to destroy all life on earth. Every year or two there is a confrontation between powers—Korea, Laos, Berlin, Suez, Quemoy, Cuba, Vietnam, and the rest. MacArthur wanted to use nuclear weapons in Korea; and in the Cuban missile crisis, John Kennedy is said to have estimated the probability of a nuclear exchange as about 25 percent.

The danger is not so much that of the unexpected, such as a radar error or even a new nuclear dictator, as it is that our present systems will work exactly as planned!—from border testing, strategic gambles, threat and counter-threat, ... up to that "second-strike capability" that is already aimed, armed, and triggered to wipe out hundreds of millions of people in a 3-hour duel!

What is the probability of this in the average incident? 10 percent? 5 percent? There is no average incident. But it is easy to see that five or ten more such confrontations in this game of "nuclear roulette" might indeed give us only a 50-50 chance of living until 1980 or 1990. This is a shorter life expectancy than people have ever had in the world before. All our medical increases in length of life are meaningless, as long as our nuclear lifetime is so short.

Many agricultural experts also think that within the next decade the great famines will begin, with deaths that may reach 100 million people in densely populated countries like India and China. Some contradict this, claiming that the remarkable new grains and new agricultural methods introduced in the last 3 years in Southwest Asia may now be able to keep the food supply ahead of population growth. But others think that the re-education of farmers and consumers to use the new grains cannot proceed fast enough to make a difference.

But if famine does come, it is clear that it will be catastrophic. Besides the direct human suffering, it will further increase our international instabilities, with food riots, troops called out, governments falling, and international interventions that will change the whole political map of the world. It could make Vietnam look like a popgun.

In addition, the next decade is likely to see continued crises of legitimacy of all our overloaded administrations, from universities and unions to cities and national governments. Everywhere there is protest and refusal to accept the solutions handed down by some central elite. The student revolutions circle the globe. Suburbs protest as well as ghettoes, Right as well as Left. There are many new sources of collision and protest, but it is clear that the general problem is in large part structural rather than political. Our traditional methods of election and management no longer give administrations the skill and capacity they need to handle their complex new burdens and decisions. They become swollen, unresponsive—and repudiated. Every day now some distinguished administrator is pressured out of office by protesting constituents.

In spite of the violence of some of these confrontations, this may seem like a trivial problem compared to war or famine—until we realize the dangerous effects of these instabilities on the stability of the whole system. In a nuclear crisis or in any of our other crises today, administrators or negotiators may often work out some basis of agreement between conflicting groups or nations, only to find themselves rejected by their people on one or both sides, who are then left with no mechanism except to escalate their battles further.

What finally makes all of our crises still more dangerous is that they are now coming on top of each other. Most administrations are able to endure or even enjoy an occasional crisis, with everyone working late together and getting a new sense of importance and unity. What they are not prepared to deal with are multiple crises, a crisis of crises all at one time. This is what happened in New York City in 1968 when the Ocean Hill-Brownsville teacher and race strike was combined with a police strike, on top of a garbage strike, on top of a longshoremen's strike, all within a few days of each other.

When something like this happens, the staffs get jumpy with smoke and coffee and alcohol, the mediators become exhausted, and the administrators find themselves running two crises behind. Every problem may escalate because those involved no longer have time to think straight. What would have happened in the Cuban missile crisis if the East Coast power blackout had occurred by accident that same day? Or if the "hot line" between Washington and Moscow had gone dead? There might have been hours of misinterpretation, and some fatally different decisions.

I think this multiplication of domestic and international crises today will shorten that short half-life. In the continued absence of better ways of heading off these multiple crises, our half-life may no longer be 10 or 20 years, but more like 5 to 10 years, or less. We may have even less than a 50-50 chance of living until 1980.

This statement may seem uncertain and excessively dramatic. But is there any scientist who would make a much more optimistic estimate after considering all the different sources of danger and how they are increasing? The shortness of the time is due to the exponential and multiplying character of our problems and not to what particular numbers or guesses we put in. Anyone who feels more hopeful about getting past the nightmares of the 1970's has only to look beyond them to the monsters of pollution and populations rising up in the 1980's and 1990's. Whether we have 10 years or more like 20 or 30, unless we systematically find new large-scale solutions, we are in the gravest danger of destroying our society, our world, and ourselves in any of a number of different ways well before the end of this century. Many futurologists who have predicted what the world will be like in the year 2000 have neglected to tell us that.

Nevertheless the real reason for trying to make rational estimates of these deadlines is not because of their shock value but because they give us at least a rough idea of how much time we may have for finding and mounting some large-scale solutions. The time is short but, as we shall see, it is not too short to give us a chance that something can be done, if we begin immediately.

From this point, there is no place to go but up. Human predictions are always conditional. The future always depends on what we do and can be made worse or better by stupid or intelligent action. To change our earlier analogy, today we are like men coming out of a coal mine who suddenly begin to hear the rock rumbling, but who have also begun to see a little square of light at the end of the tunnel. Against this background, I am an optimist—in that I want to insist that there is a square of light and that it is worth trying to get to. I think what we must do is to start running as fast as possible toward that light, working to increase the probability of our survival through the next decade by some measurable amount.

For the light at the end of the tunnel is very bright indeed. If we can only devise new mechanisms to help us survive this round of terrible crises, we have a chance of moving into a new world of incredible potentialities for all mankind. But if we cannot get through this next decade, we may never reach it.

Needed: Wartime Research and Development

What can we do? I think that nothing less than the application of the full intelligence of our society is likely to be adequate. These problems will require the humane and constructive efforts of everyone involved. But I think they will also require something very similar to the mobilization of scientists for solving crisis problems in wartime. I believe we are going to need large numbers of scientists forming something like research teams or task forces for social research and development. We need full-time interdisciplinary teams combining men of different specialities, natural scientists, social scientists, doctors, engineers, teachers, lawyers, and many other trained and inventive minds, who can put together our stores of knowledge and powerful new ideas into improved technical methods, organizational designs, or "social inventions" that have a chance of being adopted soon enough and widely enough to be effective. Even a great mobilization of scientists may not be enough. There is no guarantee that these problems can be solved, or solved in time, no matter what we do. But for problems of this scale and urgency, this kind of focusing of our brains and knowledge may be the only chance we have.

Scientists, of course, are not the only ones who can make contributions. Millions of citizens, business and labor leaders, city and government officials, and workers in existing agencies, are already doing all they can to solve the problems. No scientific innovation will be effective without extensive advice and help from all these groups.

But it is the new science and technology that have made our problems so immense and intractable. Technology did not create human conflicts and inequities, but it has made them unendurable. And where science and technology have expanded the problems in this way, it may be only more scientific understanding and better technology that can carry us past them. The cure for the pollution of the rivers by detergents is the use of nonpolluting detergents. The cure for bad management designs is better management designs.

Also, in many of these areas, there are few people outside the research community who have the basic knowledge necessary for radically new solutions. In our great biological problems, it is the new ideas from cell biology and ecology that may be crucial. In our social-organizational problems, it may be the new theories of organization and management and behavior theory and game theory that offer the only hope. Scientific research and development groups of some kind may be the only effective mechanism by which many of these new ideas can be converted into practical invention and action.

The time scale on which such task forces would have to operate is very different from what is usual in science. In the past, most scientists have tended to work on something like a 30-year time scale, hoping that their careful studies would fit into some great intellectual synthesis that might be years away. Of course when they become politically concerned, they begin to work on something more like a 3-month time scale, collecting signatures or trying to persuade the government to start or stop some program.

But 30 years is too long, and 3 months is too short, to cope with the major crises that might destroy us in the next 10 years. Our urgent problems now are more like wartime problems, where we need to work as rapidly as is consistent with large-scale effectiveness. We need to think rather in terms of a 3-year time scale—or more broadly, a 1- to 5-year time scale. In World War II, the ten thousand scientists who were mobilized for war research knew they did not have 30 years, or even 10 years, to come up with answers. But they did have time for the new research, design, and construction that brought sonar and radar and atomic energy to operational effectiveness within 1 to 4 years. Today we need the same large-scale mobilization for innovation and action and the same sense of constructive urgency.

Priorities: A Crisis Intensity Chart

In any such enterprise, it is most important to be clear about which problems are the real priority problems. To get this straight, it is valuable to try to separate the different problem areas according to some measures of their magnitude and urgency. A possible classification of this kind is shown in Tables 1 and 2. In these tables, I have tried to rank a number of present or potential problems or crises, vertically, according to an estimate of their order of intensity of "seriousness," and horizontally, by a rough estimate of their time to reach climactic importance. Table 1 is such a classification for the United States for the next 1 to 5 years, the next 5 to 20 years, and the next 20 to 50 years. Table 2 is a similar classification for world problems and crises.

The successive rows indicate something like order-of-magnitude differences in the intensity of the crises, as estimated by a rough product of the size of population that might be hurt or affected, multiplied by some estimated average

TABLE 1

Classification of problems and crisis by estimated time and intensity (United States)

Grade	Estimated crisis intensity (number affected x degree of effect)		Estimated time to crisis*		
			1 to 5 years	5 to 20 years	20 to 50 years
1.		Total annihilation	Nuclear or RCBW escalation	Nuclear or RCBW escalation	†(Solved or dead)
2.	10^8	Great destruction or change (physical, biological, or political)	(Too soon)	Participatory democracy Ecological balance	Political theory and economic structure Population planning Patterns of living Education Communications Integrative philosophy
3.	10^7	Widespread almost unbearable tension	Administrative management Slums Participatory democracy Racial conflict	Pollution Poverty Law and justice	?
4.	10^6	Large-scale distress	Transportation Neighborhood ugliness Crime	Communications gap	?
5.	10^5	Tension producing responsive change	Cancer and heart Smoking and drugs Artificial organs Accidents Sonic boom Water supply Marine resources Privacy on computers	Educational inadequacy	?
6.		Other problems— important, but adequately researched	Military R & D New educational methods Mental illness Fusion power	Military R & D	
7.		Exaggerated dangers and hopes	Mind control Heart transplants Definition of death	Sperm banks Freezing bodies Unemployment from automation	Eugenics
8.		Noncrisis problems being "overstudied"	Man in space Most basic science		

*If no major effort is made at anticipatory solution.

effect in the disruption of their lives. Thus the first row corresponds to total or near-total annihilation; the second row, to great destruction or change affecting everybody; the third row, to a lower tension affecting a smaller part of the population or a smaller part of everyone's life, and so on.

Informed men might easily disagree about one row up or down in intensity, or one column left or right in the time scales, but these order-of-magnitude differences are already so great that it would be surprising to find much larger disagreements. Clearly, an important initial step in any serious problem study would be to refine such estimates.

In both tables, the one crisis that must be ranked at the top in total danger and imminence is, of course, the danger of large-scale or total annihilation by nuclear escalation or by radiological-chemical-biological-warfare (RCBW). This kind of crisis will continue through both the 1- to 5-year time period and the 5- to 20-years period as Crisis Number 1, unless and until we get a safer peace-keeping

TABLE 2

Classification of problems and crises by estimated time and intensity (World)

Grade	Estimated crisis intensity (number affected x degree of effect)		Estimated time to crisis*		
			1 to 5 years	5 to 20 years	20 to 50 years
1.	10^{10}	Total annihilation	Nuclear or RCBW escalation	Nuclear or RCBW escalation	†(Solved or dead)
2.	10^9	Great destruction or change (physical, biological, or political)	(Too soon)	Famines Ecological balance Development failures Local wars Rich-poor gap	Economic structure and political theory Population and ecological balance Patterns of living Universal education Communications-integration Management of world Integrative philosophy
3.	10^8	Widespread almost unbearable tension	Administrative management Need for participation Group and racial conflict Poverty-rising expectations Environmental degradation	Poverty Pollution Racial wars Political rigidity Strong dictatorships	?
4.	10^7	Large-scale distress	Transportation Diseases Loss of old cultures	Housing Education Independence of big powers Communications gap	?
5.	10^6	Tension producing responsive change	Regional organization Water supplies	?	?
6.		Other problems— important, but adequately researched	Technical development design Intelligent monetary design		
7.		Exaggerated dangers and hopes			Eugenics Melting of ice caps
8.		Noncrisis problems being "overstudied"	Man in space Most basic science		

*If no major effort is made at anticipatory solution.

arrangement. But in the 20- to 50-year column, following the reasoning already given, I think we must simply put a big "+" at this level, on the grounds that the peace-keeping stabilization problem will either be solved by that time or we will probably be dead.

At the second level, the 1- to 5-year period may not be a period of great destruction (except nuclear) in either the United States or the world. But the problems at this level are building up, and within the 5- to 20-year period, many scientists fear the destruction of our whole biological and ecological balance in the United States by mismanagement or pollution. Others fear political catastrophe within this period, as a result of participatory confrontations or backlash or even dictatorship, if our devisive social and structural problems are not solved before that time.

On a world scale in this period, famine and ecological catastrophe head the list of destructive problems. We will come back later to the items in the 20- to 50-year column.

The third level of crisis problems in the United States includes those that are already upon us: administrative management of communities and cities, slums, participatory democracy, and racial conflict. In the 5- to 20-year period, the problems of pollution and poverty or major failures of law and justice could escalate to this level of tension if they are not solved. The last column is left blank because secondary events and second-order effects will interfere seriously with any attempt to make longer-range predictions at these lower levels.

The items in the lower part of the tables are not intended to be exhaustive. Some are common headline problems which are included simply to show how they might rank quantitatively in this kind of comparison. Anyone concerned with any of them will find it a useful exercise to estimate for himself their order of seriousness, in terms of the number of people they actually affect and the average distress they cause. Transportation problems and neighborhood ugliness, for example, are listed as grade 4 problems in the United States because they depress the lives of tens of millions for 1 or 2 hours every day. Violent crime may affect a corresponding number every year or two. These evils are not negligible, and they are worth the efforts of enormous numbers of people to cure them and to keep them cured—but on the other hand, they will not destroy our society.

The grade 5 crises are those where the hue and cry has been raised and where responsive changes of some kind are already under way. Cancer goes here, along with problems like auto safety and an adequate water supply. This is not to say that we have solved the problem of cancer, but rather that good people are working on it and are making as much progress as we could expect from anyone. (At this level of social intensity, it should be kept in mind that there are also positive opportunities for research, such as the automation of chemical biochemistry or the invention of new channels of personal communication, which might affect the 20-year future as greatly as the new drugs and solid state devices of 20 years ago have begun to affect the present.)

Where The Scientists Are

Below grade 5, three less quantitative categories are listed, where the scientists begin to outnumber the problems. Grade 6 consists of problems that many people believe to be important but that are adequately researched at the present time. Military R & D belongs in this category. Our huge military establishment creates many social problems, both of national priority and international stability, but even in its own terms, war research, which engrosses hundreds of thousands of scientists and engineers, is being taken care of generously. Likewise, fusion power is being studied at the $100-million level, though even if we had it tomorrow, it would scarcely change our rates of application of nuclear energy in generating more electric power for the world.

Grade 7 contains the exaggerated problems which are being talked about or worked on out of all proportion to their true importance, such as heart transplants, which can never affect more than a few thousands of people out of the billions in the world. It is sad to note that the symposia on "social implications of science" at many national scientific meetings are often on the problems at grade 7.

In the last category, grade 8, are two subjects which I am sorry to say I must call "overstudied," at least with respect to the real crisis problems today. The Man in Space flights to the moon and back are the most beautiful technical achievements of man, but they are not urgent except for national display, and they absorb tens of thousands of our most ingenious technical brains.

And in the "overstudied" list I have begun to think we must now put most of our basic sciences. This is a hard conclusion, because all of science is so important in the long run and because it is still so small compared, say, to advertising or the tobacco industry. But basic scientific thinking is a scarce resource. In a national emergency, we would suddenly find that a host of our scientific problems could be postponed for several years in favor of more urgent research. Should not our total human emergency make the same claims? Long-range science is useless unless we survive to use it. Tens of thousands of our best trained minds may now be needed for something more important than "science as usual."

The arrows at level 2 in the tables are intended to indicate that problems may escalate to a higher level of crisis in the next time period if they are not solved. The arrows toward level 2 in the last columns of both tables show the escalation of all our problems upward to some general reconstruction in the 20- to 50-year time period, if we survive. Probably no human institution will continue unchanged for another 50 years, because they will all be changed by the crises if they are not changed in advance to prevent them. There will surely be widespread rearrangements in all our ways of life everywhere, from our patterns of society to our whole philosophy of man. Will they be more humane, or less? Will the world come to resemble a diverse and open humanist democracy? Or Orwell's *1984?* Or a postnuclear desert with its scientists hanged? It is our acts of commitment and leadership in the next few months and years that will decide.

Mobilizing Scientists

It is a unique experience for us to have peacetime problems or technical problems which are not industrial problems, on such a scale. We do not know quite where to start, and there is no mechanism yet for generating ideas systematically or paying teams to turn them into successful solutions.

But the comparison with wartime research and development may not be inappropriate. Perhaps the antisubmarine warfare work or the atomic energy project of the 1940's provide the closest parallels to what we must do in terms of the novelty, scale, and urgency of the problems, the initiative needed, and the kind of large success that has to be achieved. In the antisubmarine campaign, Blackett assembled a few scientists and other ingenious minds in his "back room," and within a few months they had worked out the "operations analysis" that made an order-of-magnitude difference in the success of the campaign. In the atomic energy work, scientists started off with extracurricular research, formed a central committee to channel their secret communications and then studied the possible solutions for some time before they went to the government, for large-scale support for the great development laboratories and production plants.

Fortunately, work on our crisis problems today would not require secrecy. Our great problems today are all beginning to be world problems, and scientists from many countries would have important insights to contribute.

Probably the first step in crisis studies now should be the organization of intense technical discussion and education groups in every laboratory. Promising lines of interest could then lead to the setting up of part-time or full-time studies and teams and co-ordinating committees. Administrators and boards of directors might find active crisis research important to their own organizations in many cases. Several foundations and federal agencies already have in-house research and make outside grants in many of these crisis areas, and they would be important initial sources of support.

But the step that will probably be required in a short time is the creation of whole new centers, perhaps comparable to Los Alamos or the RAND Corporation, where interdisciplinary groups can be assembled to work full-time on solutions to these crisis problems. Many different kinds of centers will eventually be necessary, including research centers, development centers, training centers, and even production centers for new sociotechnical inventions. The problems of our time—the $100-billion food problem or the $100-billion arms control problem—are no smaller than World War II in scale and importance, and it would be absurd to think that a few academic research teams or a few agency laboratories could do the job.

Social Inventions Needed

The thing that discourages many scientists

—even social scientists—from thinking in these research-and-development terms is their failure to realize that there are such things as social inventions and that they can have large-scale efforts in a surprisingly short time. A recent study with Karl Deutsch has examined some 40 of the great achievements in social science in this century, to see where they were made and by whom and how long they took to become effective. They include developments such as the following: Keynesian economics; Opinion polls and statistical sampling; Input-output economics; Operations analysis; Information theory and feedback theory; Theory of games and economic behavior; Operant conditioning and programmed learning; Planned programming and budgeting (PPB); Non-zero-sum game theory.

Many of these have made remarkable differences within just a few years in our ability to handle social problems or management problems. The opinion poll became a national necessity within a single election period. The theory of games, published in 1946, had become an important component of American strategic thinking by RAND and the Defense Department by 1953, in spite of the limitation of the theory at that time to zero-sum games, with their dangerous bluffing and "brinksmanship." Today, within less than a decade, the PPB management technique is sweeping through every large organization.

This list is particularly interesting because it shows how much can be done outside official government agencies when inventive men put their brains together. Most of the achievements were the work of teams of two or more men, almost all of them located in intellectual centers such as Princeton or the two Cambridges.

The list might be extended by adding commercial social inventions with rapid and widespread effects, like credit cards. And sociotechnical inventions, like computers and automation or like oral contraceptives, which were in widespread use within 10 years after they were developed. In addition, there are political innovations like the New Deal, which made great changes in our economic life within 4 years, and the pay-as-you-go income tax, which transformed federal taxing power within 2 years.

On the international scene, the Peace Corps, the "hot line," the Test-Ban Treaty, the Antarctic Treaty, and the Nonproliferation Treaty were all implemented within 2 to 10 years after their initial proposal. These are only small contributions, a tiny patchwork part of the basic international stabilization system that is needed, but they show that the time to adopt new structural designs may be surprisingly short. Our clichés about "social lag" are very misleading. Over half of the major social innovations since 1940 were adopted or had widespread social effects within less than 12 years—a time as short as, or shorter than, the average time for adoption of technological innovations.

Areas for Task Forces

Is it possible to create more of these social inventions systematically to deal with our present crisis problems? I think it is. It may be worth listing a few specific areas where new task forces might start.

Peace-keeping mechanisms and feedback stabilization. Our various nuclear treaties are a beginning. But how about a technical group that sits down and thinks about the whole range of possible and impossible stabilization and peace-keeping mechanisms? Stabilization feedback-design might be a complex modern counterpart of the "checks and balances" used in designing the constitutional structure of the United States 200 years ago. With our new knowledge today about feedbacks, group behavior, and game theory, it ought to be possible to design more complex and even more successful structures.

Some peace-keeping mechanisms that might be hard to adopt today could still be worked out and tested and publicized, awaiting a more favorable moment. Sometimes the very existence of new possibilities can change the atmosphere. Sometimes, in a crisis, men may finally be willing to try out new ways and may find some previously prepared plan of enormous help.

Biotechnology. Humanity must feed and care for the children who are already in the world, even while we try to level off the further population explosion that makes this so difficult. Some novel proposals, such as food from coal, or genetic copying of champion animals, or still simpler contraceptive methods, could possibly have large-scale effects on human welfare within 10 to 15 years. New chemical, statistical, and management methods for measuring and maintaining the ecological balance could be of very great importance.

Game theory. As we have seen, zero-sum game theory has not been too academic to be used for national strategy and policy analysis. Unfortunately, in zero-sum games, what I win, you lose, and what you win, I lose. This may be the way poker works but it is not the way the world works. We are collectively in a non-zero-sum game in which we will all lose together in nuclear holocaust or race conflict or economic nationalism, or all win together in survival and prosperity. Some of the many variations of non-zero-sum game theory, applied to group conflict and cooperation, might show us profitable new approaches to replace our sterile and dangerous confrontation strategies.

Psychological and social theories. Many teams are needed to explore in detail and in practice how the powerful new ideas of behavior theory and the new ideas of responsive living might be used to improve family life or community and management structures. New ideas of information handling and management theory need to be turned into practical recipes for reducing the daily frustrations of small businesses, schools, hospitals, churches, and town meetings. New economic inventions are needed, such as urban development corporations. A deeper systems analysis is urgently needed to see if there is not some practical way to separate full employment from inflation. Inflation pinches the poor, increases labor-management disputes, and multiplies all our domestic conflicts and our sense of despair.

Social indicators. We need new social indicators, like the cost-of-living index, for measuring a thousand social goods and evils. Good indicators can have great "multiplier effects" in helping to maximize our welfare and minimize our ills. Engineers and physical scientists working with social scientists might come up with ingenious new methods of measuring many of these important but elusive parameters.

Channels of effectiveness. Detailed case studies of the reasons for success or failure of various social inventions could also have a large multiplier effect. Handbooks showing what channels or methods are now most effective for different small-scale and large-scale social problems would be of immense value.

The list could go on and on. In fact, each study group will have its own pet projects. Why not? Society is at least as complex, as say, an automobile with its several thousand parts. It will probably require as many research-and-development teams as the auto industry in order to explore all the inventions it needs to solve its problems. But it is clear that there are many areas of great potential crying out for brilliant minds and brilliant teams to get to work on them.

For a Science of Survival

This is an enormous program. But there is

nothing impossible about mounting and financing it, if we, as concerned men, go into it with commitment and leadership. Yes, there will be a need for money and power to overcome organizational difficulties and vested interests. But it is worth remembering that the only real source or power in the world is the gap between what is and what might be. Why else do men work and save and plan? If there is some future increase in human satisfaction that we can point to and realistically anticipate, men will be willing to pay something for it and invest in it in the hope of that return. In economics, they pay with money; in politics, with their votes and time and sometimes with their jail sentences and their lives.

Social change, peaceful or turbulent, is powered by "what might be." This means that for peaceful change, to get over some impossible barrier of unresponsiveness or complexity or group conflict, what is needed is an inventive man or group—a "social entrepreneur"—who can connect the pieces and show how to turn the advantage of "what might be" into some present advantage for every participating party. To get toll roads, when highways were hopeless, a legislative-corporation mechanism was invented that turned the future need into present profits for construction workers and bondholders and continuing profitability for the state and all the drivers.

This principle of broad-payoff anticipatory design has guided many successful social plans. Regular task forces using systems analysis to find payoffs over the barriers might give us such successful solutions much more often. The new world that could lie ahead, with its blocks and malfunctions removed, would be fantastically wealthy. It seems almost certain that there must be many systematic ways for intelligence to convert that large payoff into the profitable solution of our present problems.

The only possible conclusion is to call to

action. Who will commit himself to this kind of search for more ingenious and fundamental solutions? Who will begin to assemble the research teams and the funds? Who will begin to create those full-time interdisciplinary centers that will be necessary for testing detailed designs and turning them into effective applications?

The task is clear. The task is huge. The time is horribly short. In the past, we have had science for intellectual pleasure, and science for the control of nature. We have had science for war. But today, the whole human experiment may hang on the question of how fast we now press the development of science for survival.

History's Greatest Dead End
Frank Blackaby

The past twenty years have produced a record of virtually total failure in attempts to introduce any effective form of arms control, let alone disarmament. Clear evidence of this failure lies in world-wide trends in military expenditure. According to figures recently compiled by the Stockholm International Peace Research Institute, the rate at which world military spending has been rising over the past two decades gives some idea of what may happen if attempts to curb the arms race are as unsuccessful between now and 1990 as they have been between 1950 and 1970.

The two basic figures concerning the world war industry—to use Professor Kenneth Boulding's apt term—are those needed for a first appraisal of any world-wide industry: the share it takes of the world's output and its trend rate of growth.

Over the past twenty years, military spending has taken close to 8 per cent of the world's output. Only during the two World Wars did the proportion exceed this percentage. Even in 1913, after the naval arms race, the military's share in world output was probably not more than 3 per cent. The present high figure is not simply a result of the immense expense of the gleaming weaponry needed to support strategic confrontation between the United States and the Soviet Union. There have never before been standing armies of millions of men continuing under arms for decades. Throughout the 1960's, NATO and Warsaw Pact countries combined had more than ten million men in their armed forces.

Although it is not easy to give relevance to this close to 8 per cent figure, a report from the United States Arms Control and Disarmament Agency does provide a potent comparison: "Global military expenditures are equivalent to the total annual income produced by one billion people living in Latin America, South Asia, and the Near East. They are greater by 40 per cent than world-wide expenditures on education by all levels of government, and more than three times world-wide expenditures on public health."

(Although this is a valuable report, it nonetheless has its limitations in that it was prepared by a government agency of one of the major world powers. The report's text discusses the rise in world military expenditures between 1965 and 1967 without once mentioning the word Vietnam. The increase in United States military spending in Vietnam accounted for two-thirds of the rise in world military expenditure between those years.)

The growth rate in the world's military expenditure over the past twenty years has been more on the order of 6 per cent a year, in real terms—that is, omitting the effect of inflation. By going back fifty years instead of twenty, the rate of rise is probably about 5 per cent a year. A 5 per cent a year growth rate means a doubling every fifteen years; a trebling every twenty-three years; and a quintupling every thirty-three years. It means that the world is now devoting to the military a quantity of resources roughly equal to the world's whole output at the turn of the century. This figure of 5 per cent is the long-term trend rate. Year-to-year movements have been irregular. World military expenditure has tended to move up in jumps, and then level off for a while. For example, it rose some 30 per cent between 1965 and 1968, and since 1968 has leveled off, as United States spending in Vietnam fluctuated.

There is no automatic economic brake to prevent world military expenditure from continuing to show a rising trend of around 5 per cent a year, in real terms, indefinitely. It cannot, of course, indefinitely rise faster than world output; but world output itself has been increasing at around 5 per cent a year during the past decade.

So all the military leaders have to do, in order to continue doubling the quantity of resources they have at their disposal every fifteen years, is to ensure that military expenditure maintains its present share of world output. That, indeed, often tends to be their line in budget arguments. In many countries a conventional percentage of national output has been established for military use and has become hallowed by custom. If the figure falls below the established norm, the chiefs of staff

Long and Short Term Trends in the Volume of World Military Expenditure

	Average per cent change per year					Based on constant price figures
	Long-term trend[a] 1949-68	Short-term trend 1965-68	Year-to-year changes			Budgeted change in 1969	Size of military expenditure in 1968 US$bn, current prices and exchange-rates
			1965-66	1966-67	1967-68		
United States	+ 7.7	+12.0	+19.2	+15.4	+ 2.0	− 0.6	79.5
Other NATO	+ 5.3	+ 1.6	+ 0.9	+ 4.7	− 0.9	+ 0.4	24.4
Total NATO	+ 7.1	+ 9.3	+14.1	+12.8	+ 1.3	− 0.4	104.0
USSR	+ 4.0	+ 9.3	+ 4.7	+ 8.0	+15.5	+ 5.9	18.6[b] or 39.8[c]
Other Warsaw Pact	+ 7.1	+10.9	+ 7.2	+ 7.4	+18.5	+12.8	13.4[b] or 6.3[c]
Total Warsaw Pact	+ 4.1	+ 9.5	+ 5.0	+ 7.9	+15.9	+ 6.8	32.0[b] or 46.1[c]
Other European	+ 5.2	+ 2.0	+ 3.1	− 0.3	+ 3.2	+ 0.3	2.5
Middle East	+12.8	+19.9	+ 8.2	+32.5	+20.0	...	2.7
South Asia	+ 5.2	− 2.5	+ 1.9	−11.7	+ 3.0	...	1.9
Far East (excl. China)	+ 6.5	+ 8.4	+ 0.7	+10.3	+14.8	...	4.0
Oceania	+ 7.9	+17.7	+18.9	+18.2	+16.1	...	1.4
Africa	...	+ 7.6	+11.5	[+ 1.4]	[+10.0]	...	1.2
Central America	+ 3.2	+ 5.2	+ 9.7	+ 5.1	+ 1.1	...	0.5
South America	+ 2.7	+ 3.7	− 9.3	+12.8	[+ 8.8]	...	2.1
World[d]	+ 5.9	+ 8.9	+10.2	+10.7	+ 5.8	...	159.3[b] or 173.4[c]

[a]1957-68 for "Other Warsaw Pact" and Far East, excl. China; 1949-67 for Central and South America.

[b]At official basic exchange rates.

[c]At Benoit-Lubell estimated defense purchasing-power-parity exchange-rates. See reference section, page 198.

[d]Including an estimate for China of $7 bn in 1968.

SOURCE: *SIPRI Yearbook of World Armaments and Disarmament*

complain they are not getting their fair share. Military expenditure—or defense, as it is almost always called in national budgets—tends to be treated in budget debates as if it were a common good, like public health or education, which one naturally wants to have more of as one grows richer.

In fact, a constant share of the world's product for military purposes is a recipe for the indefinite continuation of the world's arms race at the rate of the last two decades. If world military expenditure's present share in world output is maintained, and if world output continues to grow as fast as it did during the last decade—and neither of these assumptions can be dismissed as implausible—then by the early years of the next century the world will be devoting to military purposes a quantity of resources equal to the whole of its present output. It is hard to present a picture of the sort of advanced weaponry that might then be around. In 1945, anyone who had forecast the array of weapons employed in the present strategic confrontation would have been dismissed as wholly fanciful, and deeply pessimistic, too. It is awesome how far and how fast the world has come in weaponry in the past twenty-five years. Clearly, if the arms race goes on, anti-anti-missile-missile-missiles will stop being a joke and will become a reality somewhere along the way. Armies may be deployed along the sea bottom. Surface-to-air missiles and medium-range surface-to-surface missiles may then be thickly spread all around the world.

Behind the mountainous quantity of military expenditure in the last twenty years there has been a wide variety of different forms of arms competition. At one end of the spectrum, there has been the strategic arms competition between the two major world powers; at the other end, the proliferation of conventional weapons in the developing countries.

The strategic arms race between the United States and the Soviet Union has always been an arms race, never a plateau. The "strategic plateau" was a strategist's concept that never at any time approached reality. It was logical, in one sense, that at some point both sides might tacitly decide that they had enough lethal power, and stop increasing the number of nuclear warheads and elaborating the methods of delivery. But this situation never occurred. In fact, in spite of all the intellectual expertise devoted to strategic thinking in the United States, the strategic arms race has been a fairly unsophisticated process—in the decision-making, of course; not in the weaponry itself.

A considerable part of the "action-reaction phenomenon" of this race can be traced now on the United States' side, with the help of a growing body of slightly penitent testimony from some of those experts who had positions of responsibility then. On the Soviets' side, there is no testimony, penitent or otherwise; the rationale of their strategic decisions can only be guessed at from what is known of their arms procurement and deployment.

Two episodes illustrate this phenomenon. When the Kennedy administration came into power at the beginning of 1961, it was already politically committed to a high forward estimate of likely Soviet missile construction. This commitment was used as the justification for a program of expansion of both land-based and sea-based missile forces that swelled the number of U. S. strategic missiles from 160 in 1961 to 1,350 in 1965. This program went inexorably ahead, with an immense amount of technical improvement in the weaponry along the way, despite steadily increasing evidence that the Soviet Union was in fact doing comparatively little to build up its missile forces. By 1965, the United States had a better than 4-to-1 missile superiority in numbers alone, and a significant qualitative edge—particularly in submarine-launched missiles—as well. Faced with statistics such as these, the Soviet Union, not surprisingly, began to accelerate its missile installations the following year. This Soviet move is now being used as the rationale for another major step by the United States: the construction of an anti-ballistic missile system.

To take another example: In a statement of position at the beginning of 1967, Secretary of Defense Robert McNamara said, "We must, for the time being, plan our forces on the assumption that the Soviet Union will have deployed some sort of ABM system around their major cities by the early 1970's." Consequently, the Pentagon began to insist on the need to double the number of existing targetable warheads—then around 4,000—in order to allow for the ones which would be shot down by anti-ballistic missiles. The program for developing MIRVs was pushed ahead. Now it is admitted that the Soviet ABM system consists of some sixty launchers around Moscow and no more: that is, sixty additional warheads would be needed, not 4,000. But the MIRV program still goes ahead.

The pattern is not a complex one. First, the potential threat is overestimated; defense planners are, in fact, generally required to plan on the basis of a higher-than-expected threat. Then a program is started to meet this threat. The threat proves to have been overestimated, but the program—already well under way—is not changed. And, likely as not, it provokes as a reaction the sort of increase in weapons by the other side that was originally feared. Belatedly, the overestimated threat is made real.

There is one particular characteristic of this strategic arms race that is important to an understanding of its nature; it applies to some extent to the race in tactical weapons, too. It is that a large part of the competition is conducted in the research and development establishments on either side. Weapons research and development is a process that proceeds quite happily under its own impetus, without any need for direct reference to what the other side is doing. Actual deployment tends to be justified, on the United States' side at any rate, by some quantified statement, however inaccurate, of the potential threat. Research and development decisions do not need any such justification; work on new weapons can be initiated and carried on for five or six years, more or less independently of the enemy. Large research establishments have to have something to do. Once the research on a new weapon system has been successfully completed, there develops strong pressure for deployment. Nobody is keen to see his invention lying around unused.

This partially self-propagating character of the present-day strategic arms race is presumably one explanation for the oddity that the competition has gone merrily on for nearly eight years now through a period when tension between the two great powers has been slackening. By all the standard criteria, tension is much lower now than it was before 1962. The weaponry nonetheless has continued to build up. So much for the view that arms races are the results of increasing tension and, if tension is reduced, the arms race will wither away by itself. For this particular arms race, the facts lend no support to this view.

In the developing nations, military expenditure is tiny, compared with that of the major powers. But, in fact, it has been rising faster in these countries, over the past decade, than in the rest of the world. Most of these countries do not produce weapons themselves, except for small arms. All their heavier weapons are imported from the big powers. Here, then, is another important facet of the world arms race since World War II—the substantial and increasing flow of major weapons from the producing countries to the non-producers.

The result has been a "horizontal proliferation" of fairly sophisticated conventional weapons. For example, in 1955 no developing countries had supersonic military planes; by 1968, twenty-nine such countries had them. In 1957, no developing country had long-range surface-to-air missiles; by 1968, these weapons had spread to eighteen such countries. All this weaponry was supplied by the big powers. The United States and the Soviet Union were the prime suppliers, mainly for political reasons. Britain and France combined supplied about a fifth of the total; their motives were mainly economic.

The United States was first with a policy of supplying arms to these countries, for a variety of reasons. It had a world-wide system of military bases, and arms supplies were one way of keeping a host country friendly. It also had an active policy of "containing communism." Most of the United States' weapons were supplied to countries designated "forward defense areas," that is, countries on the fringe of what was then called the "communist bloc." Some supplies were also sent to areas labeled "free world-orientated areas." The whole policy fitted in with a picture of the world in which the forces of light everywhere needed to be strengthened to resist attacks from the forces of darkness.

The Soviet Union did not begin to show an interest in non-communist developing countries until 1955. Thereafter, it found an increasing number of clients, and in recent years the quantity of major weapons it has been sending to developing countries has been roughly the same as that of the United States. Taking the whole of the 1960s, the United States and the Soviet Union between them have sent about $6-billion worth of major weapons—that does not include considerable quantities of small arms—to developing countries. The Soviet Union has in general sold the weapons; and the United States' supplies, which were mainly gifts in the 'fifties, were mainly sales by the end of the 'sixties.

Weapons have not always stayed where they were first sent. There have been several notorious incidents where they have moved on to unintended destinations, such as when ninety fighter planes sent to Iran ended up in Pakistan. Nor have they always had the political effects intended. Indeed, it is interesting to judge this arms trade from the point of view of a committed cold-war warrior on either side: Did it achieve the objectives set for it? A dispassionate cold-war warrior must judge that a great deal of it was waste. The United States' supply of arms to Pakistan, for example, hardly served to strengthen South Asia against the communist bloc. The Soviet Union supplies to Indonesia did not do much to aid the communist cause there. Both sides must sometimes question what returns they are getting from their entanglement as arms suppliers to the Middle East conflict.

Does it matter that these sophisticated conventional weapons are springing up all over the globe? A country that invests resources in them will expect some return on its investment. It will be tempted to look for military solutions to its problems. The people in power —in a hypothetical country—will be inclined to ask what is the point of having spent large sums of money on Saladin armored cars and MIG 17s, if it cannot use them to deal with the incursions on its eastern border.

There is another consideration. Modern fighter planes, as the Six-Day War has shown, are admirable weapons for a first strike. A large number of countries must have noted the lesson of that war: If you are going to do battle with your neighbor, the thing to do is to wipe out his air force before he wipes out yours.

It is not easy to think of any effective ways of setting some kind of international curb on arms sales. It is hard enough, to begin with, to envisage the United States, the Soviet Union, Britain, and France agreeing on any joint rules. Even if they did, these rules would

be subject to bitter complaint from nonpro-
ducing countries, who would say, quite
rightly, that the limitations were just one more
discrimination practiced by the rich countries
against the poor. The strength of this feeling
among the developing countries is shown by
the fate of a recent Danish proposal at the
United Nations. It was a proposal—phrased in
the mildest possible terms—inviting the Secretary General to inquire of member countries
whether, in their view, it was practicable to set
up any kind of register or record of arms sales.
The opposition to the proposal from the developing countries was so strong that the
proposal was dropped without being put to
vote. It is arguable that, if anything is to be
done about this trade at all, the supplying
powers will just have to put up with the charge
of imperialism.

One-twelfth of world output for the military—for preparations for mutual slaughter—
is too much. This is really the grossest of the
absurdities in the way man has organized his
life on this planet; it deserves to be moved up
to a higher place on the list of world concerns.

What about the argument that this mass of
war material is a symptom and not a cause,
and that the important thing is to get at the
causes? This is too simple a view. The existence of a large military machine is not neutral; it predisposes toward military solutions.
In the early 'sixties, the United States built up
a military capacity for swift and distant intervention; the weaponry was there at hand, and
it was used in Vietnam. Since the Soviet Union
had the tanks and the men, and could occupy
the whole of Czechoslovakia in a few hours,
this opportunity appeared as a possible solution—and it was in fact the one that was
chosen. The bigger the military establishment,
the more thinkable the use of force, and the
more likely that in various contingencies some
military action will be high up on the list of
possible courses.

**The spread of long-range surface-to-air missiles
among third world countries**

*The first "X'd"-year indicates the year when a country first
acquired long-range surface-to-air missiles[a]*

	1958	59	60	61	62	63	64	65	66	67	68
China, P.R.	X	X	X	X	X	X	X	X	X	X	X
Greece		X	X	X	X	X	X	X	X	X	X
Taiwan		X	X	X	X	X	X	X	X	X	X
Turkey		X	X	X	X	X	X	X	X	X	X
Korea, South			X	X	X	X	X	X	X	X	X
Cuba				X	X	X	X	X	X	X	X
Indonesia				X	X	X	X	X	X	X	X
United Arab Rep.						X	X	X	X	X	X
Iraq						X	X	X	X	X	X
Israel						X	X	X	X	X	X
Australia							X	X	X	X	X
India							X	X	X	X	X
Iran							X	X	X	X	X
Korea, North									X	X	X
Viet-Nam, North									X	X	X
Saudi Arabia									X	X	X
Algeria									X	X	X
Syria										X	X
Thailand											X

In the same way, it is not the whole truth that successful arms–control treaties are simply the result of periods of reduced international tension. Such treaties contribute to a reduction of that tension. In their effect on weapons, they have so far been marginal. In their effect on mutual trust, they have been a good deal more important. There are three arms control treaties now in force affecting both the United States and the Soviet Union—the demilitarization of Antarctica, the ban on nuclear weapons in outer space, and the partial test ban. Perhaps the most significant thing about them is that neither side has ever accused the other of breaking them. This is the more surprising in that both sides have technically broken the partial test ban. The treaty forbids any action that results in radioactive material crossing international borders; radioactive material from both the United States and the Soviet Union has crossed international borders after underground nuclear tests vented. Yet, neither side has made any open representations to the other about it.

At this time when it is possible that the logjam may be beginning to move, it would be useful if there were a more active world public opinion agitating about the need for arms control and massive reductions in military expenditure. Public concern—certainly in Europe—has tended to move on to other things, such as environmental pollution or the world population explosion. The world arms race is an old subject, and an uncomfortable one. Yet it remains the world's greatest waste of resources and its most pressing danger.

Eco-Catastrophe!
Paul R. Ehrlich

I

The end of the ocean came late in the summer of 1979, and it came even more rapidly than the biologists had expected. There had been signs for more than a decade, commencing with the discovery in 1968 that DDT slows down photosynthesis in marine plant life. It was announced in a short paper in the technical journal, *Science*, but to ecologists it smacked of doomsday. They knew that all life in the sea depends on photosynthesis, the chemical process by which green plants bind the sun's energy and make it available to living things. And they knew that DDT and similar chlorinated hydrocarbons had polluted the entire surface of the earth, including the sea.

But that was only the first of many signs. There had been the final gasp of the whaling industry in 1973, and the end of the Peruvian anchovy fishery in 1975. Indeed, a score of other fisheries had disappeared quietly from over-exploitation and various eco-catastrophes by 1977. The term "eco-catastrophe" was coined by a California ecologist in 1969 to describe the most spectacular of man's attacks on the systems which sustain his life. He drew his inspiration from the Santa Barbara offshore oil disaster of that year, and from the news which spread among naturalists that virtually all of the Golden State's seashore bird life was doomed because of chlorinated hydrocarbon interference with its reproduction. Eco-catastrophes in the sea became increasingly common in the early 1970's. Mysterious "blooms" of previously rare micro-organisms began to appear in offshore waters. Red tides —killer outbreaks of a minute single-celled plant—returned to the Florida Gulf coast and were sometimes accompanied by tides of other exotic hues.

It was clear by 1975 that the entire ecology of the ocean was changing. A few types of phytoplankton were becoming resistant to chlorinated hydrocarbons and were gaining the upper hand. Changes in the phytoplankton community led inevitably to changes in the community of zooplankton, the tiny animals which eat the phytoplankton. These changes were passed on up the chains of life in the ocean to the herring, plaice, cod, and tuna. As diversity of life in the ocean diminished, its stability also decreased.

Other changes had taken place by 1975. Most ocean fishes that returned to fresh water to breed, like the salmon, had become extinct, their breeding streams so dammed up and polluted that their powerful homing instinct only resulted in suicide. Many fishes and shellfishes that bred in restricted areas along the coasts followed them as onshore pollution escalated.

By 1977 the annual yield of fish from the sea was down to 30 million metric tons, less than one-half of the per capita catch of a decade earlier. This helped malnutrition to escalate sharply in a world where an estimated 50 million people per year were already dying of starvation. The United Nations attempted to get all chlorinated hydrocarbon insecticides banned on a worldwide basis, but the move was defeated by the United States. This opposition was generated primarily by the American petrochemical industry, operating hand in glove with its subsidiary, the United States Department of Agriculture. Together they persuaded the government to oppose the U.N. move—which was not difficult since most Americans believed that Russia and China were more in need of fish products than was the United States. The United Nations also attempted to get fishing nations to adopt strict and enforced catch limits to preserve dwindling stocks. This move was blocked by Russia, who, with the most modern electronic equipment, was in the best position to glean what was left in the sea. It was, curiously, on the very day in 1977 when the Soviet Union announced its refusal that another ominous article appeared in *Science.* It announced that incident solar radiation had been so reduced by worldwide air pollution that serious effects on the world's vegetation could be expected.

II

Apparently it was a combination of ecosystem destabilization, sunlight reduction, and a rapid escalation in chlorinated hydrocarbon pollution from massive Thanodrin applications which triggered the ultimate catastrophe. Seventeen huge Soviet-financed Thanodrin plants were operating in underdeveloped countries by 1978. They had been part of a massive Russian "aid offensive" designed to fill the gap caused by the collapse of America's ballyhooed "Green Revolution."

It became apparent in the early '70s that the "Green Revolution" was more talk than substance. Distribution of high yield "miracle" grain seeds had caused temporary local spurts in agricultural production. Simultaneously, excellent weather had produced record harvests. The combination permitted bureaucrats, especially in the United States Department of Agriculture and the Agency for International Development (AID), to reverse their previous pessimism and indulge in an outburst of optimistic propaganda about staving off famine. They raved about the approaching transformation of agriculture in the underdeveloped countries (UDCs). The reason for the propaganda reversal was never made clear. Most historians agree that a combination of utter ignorance of ecology, a desire to justify past errors, and pressure from agro-industry (which was eager to sell pesticides, fertilizers, and farm machinery to the UDCs and agencies helping the UDCs) was behind the campaign. Whatever the motivation, the results were clear. Many concerned people, lacking the expertise to see through the Green Revolution drivel, relaxed. The population-food crisis was "solved."

But reality was not long in showing itself. Local famine persisted in northern India even after good weather brought an end to the ghastly Bihar famine of the mid-'60s. East Pakistan was next, followed by a resurgence of general famine in northern India. Other foci of famine rapidly developed in Indonesia, the Philippines, Malawi, the Congo, Egypt, Colombia, Ecuador, Honduras, the Dominican Republic, and Mexico.

Everywhere hard realities destroyed the illusion of the Green Revolution. Yields dropped as the progressive farmers who had first accepted the new seeds found that their higher yields brought lower prices—effective demand (hunger plus cash) was not sufficient in poor countries to keep prices up. Less progressive farmers, observing this, refused to make the extra effort required to cultivate the "miracle" grains. Transport systems proved inadequate to bring the necessary fertilizer to the fields where the new and extremely fertilizer-sensitive grains were grown. The same systems were also inadequate to move products to markets. Fertilizer plants were not built fast enough, and most of the underdeveloped countries could not scrape together funds to purchase supplies, even on concessional terms. Finally, the inevitable happened, and pests began to reduce yields in even the most carefully cultivated fields. Among the first were the famous "miracle rats" which invaded Philippine "miracle rice" fields early in 1969. They were quickly followed by many insects and viruses, thriving on the relatively pest-susceptible new grains, encouraged by the vast and dense plantings, and rapidly acquiring resistance to the chemicals used against them. As chaos spread until even the most obtuse agricultur-

ists and economists realized that the Green Revolution had turned brown, the Russians stepped in.

In retrospect it seems incredible that the Russians, with the American mistakes known to them, could launch an even more incompetent program of aid to the underdeveloped world. Indeed, in the early 1970's there were cynics in the United States who claimed that outdoing the stupidity of American foreign aid would be physically impossible. Those critics were, however, obviously unaware that the Russians had been busily destroying their own environment for many years. The virtual disappearance of sturgeon from Russian rivers caused a great shortage of caviar by 1970. A standard joke among Russian scientists at that time was that they had created an artificial caviar which was indistinguishable from the real thing—except by taste. At any rate the Soviet Union, observing with interest the progressive deterioration of relations between the UDCs and the United States, came up with a solution. It had recently developed what it claimed was the ideal insecticide, a highly lethal chlorinated hydrocarbon complexed with a special agent for penetrating the external skeletal armor of insects. Announcing that the new pesticide, called Thanodrin, would truly produce a Green Revolution, the Soviets entered into negotiations with various UDCs for the construction of massive Thanodrin factories. The USSR would bear all the costs; all it wanted in return were certain trade and military concessions.

It is interesting now, with the perspective of years, to examine in some detail the reasons why the UDCs welcomed the Thanodrin plan with such open arms. Government officials in these countries ignored the protests of their own scientists that Thanodrin would not solve the problems which plagued them. The governments now knew that the basic cause of their problems was overpopulation, and that these problems had been exacerbated by the dullness, daydreaming, and cupidity endemic to all governments. They knew that only population control and limited development aimed primarily at agriculture could have spared them the horrors they now faced. They knew it, but they were not about to admit it. How much easier it was simply to accuse the Americans of failing to give them proper aid; how much simpler to accept the Russian panacea.

And then there was the general worsening of relations between the United States and the UDCs. Many things had contributed to this. The situation in America in the first half of the 1970's deserves our close scrutiny. Being more dependent on imports for raw materials than the Soviet Union, the United States had, in the early 1970's, adopted more and more heavy-handed policies in order to insure continuing supplies. Military adventures in Asia and Latin America had further lessened the international credibility of the United States as a great defender of freedom—an image which had begun to deteriorate rapidly during the pointless and fruitless Viet-Nam conflict. At home, acceptance of the carefully manufactured image lessened dramatically, as even the more romantic and chauvinistic citizens began to understand the role of the military and the industrial system in what John Kenneth Galbraith had aptly named "The New Industrial State."

At home in the USA the early '70s were traumatic times. Racial violence grew and the habitability of the cities diminished, as nothing substantial was done to ameliorate either racial inequities or urban blight. Welfare rolls grew as automation and general technological progress forced more and more people into the category of "unemployable." Simultaneously a taxpayers' revolt occurred. Although there was not enough money to build the schools, roads, water systems, sewage systems, jails,

hospitals, urban transit lines, and all the other amenities needed to support a burgeoning population, Americans refused to tax themselves more heavily. Starting in Youngstown, Ohio in 1969 and followed more closely by Richmond, California, community after community was forced to close its schools or curtail educational operations for lack of funds. Water supplies, already marginal in quality and quantity in many places by 1970, deteriorated quickly. Water rationing occurred in 1723 municipalities in the summer of 1974, and hepatitis and epidemic dysentery rates climbed about 500 per cent between 1970-1974.

III

Air pollution continued to be the most obvious manifestation of environmental deterioration. It was, by 1972, quite literally in the eyes of all Americans. The year 1973 saw not only the New York and Los Angeles smog disasters, but also the publication of the Surgeon General's massive report on air pollution and health. The public had been partially prepared for the worst by the publicity given to the U.N. pollution conference held in 1972. Deaths in the late '60s caused by smog were well known to scientists, but the public had ignored them because they mostly involved the early demise of the old and sick rather than people dropping dead on the freeways. But suddenly our citizens were faced with nearly 200,000 corpses and massive documentation that they could be the next to die from respiratory disease. They were not ready for that scale of disaster. After all, the U.N. conference had not predicted that accumulated air pollution would make the planet uninhabitable until almost 1990. The population was terrorized as TV screens became filled with scenes of horror from the disaster areas. Especially vivid was NBC's coverage of hundreds of unattended people choking out their lives outside of New York's hospitals. Terms like nitrogen oxide, acute bronchitis and cardiac arrest began to have real meaning for most Americans.

The ultimate horror was the announcement that chlorinated hydrocarbons were now a major constituent of air pollution in all American cities. Autopsies of smog disaster victims revealed an average chlorinated hydrocarbon load in fatty tissues equivalent to 26 parts per million of DDT. In October, 1973, the Department of Health, Education and Welfare announced studies which showed unequivocally that increasing death rates from hypertension, cirrhosis of the liver, liver cancer and a series of other diseases had resulted from the chlorinated hydrocarbon load. They estimated that Americans born since 1946 (when DDT usage began) now had a life expectancy of only 49 years, and predicted that if current patterns continued, this expectancy would read 42 years by 1980, when it might level out. Plunging insurance stocks triggered a stock market panic. The president of Velsicol, Inc., a major pesticide producer, went on television to "publicly eat a teaspoonful of DDT" (it was really powdered milk) and announce that HEW had been infiltrated by Communists. Other giants of the petro-chemical industry, attempting to dispute the indisputable evidence, launched a massive pressure campaign on Congress to force HEW to "get out of agriculture's business." They were aided by the agro-chemical journals, which had decades of experience in misleading the public about the benefits and dangers of pesticides. But by now the public realized that it had been duped. The Nobel Prize for medicine and physiology was given to Drs. J. L. Radomski and W. B. Deichmann, who in the late 1960's had pioneered in the documentation of the long-term lethal effects of chlorinated hydrocarbons. A Presidential Commission with unimpeachable credentials directly accused the agro-chemical complex of "condemning many millions of Americans to an early death." The year 1973 was the year in which Americans finally came to understand the direct threat to their existence posed by environmental deterioration.

And 1973 was also the year in which most people finally comprehended the indirect threat. Even the president of Union Oil Company and several other industrialists publicly stated their concern over the reduction of bird populations which had resulted from pollution by DDT and other chlorinated hydrocarbons. Insect populations boomed because they were resistant to most pesticides and had been freed, by the incompetent use of those pesticides, from most of their natural enemies. Rodents swarmed over crops, multiplying rapidly in the absence of predatory birds. The effect of pests on the wheat crop was especially disastrous in the summer of 1973, since that was also the year of the great drought. Most of us can remember the shock which greeted the announcement by atmospheric physicists that the shift of the jet stream which had caused the drought was probably permanent. It signalled the birth of the Midwestern desert. Man's air-polluting activities had by then caused gross changes in climatic patterns. The news, of course, played hell with commodity and stock markets. Food prices skyrocketed, as savings were poured into hoarded canned goods. Official assurances that food supplies would remain ample fell on deaf ears, and even the government showed signs of nervousness when California migrant field workers went out on strike again in protest against the continued use of pesticides by growers. The strike burgeoned into farm burning and riots. The workers, calling themselves "The Walking Dead," demanded immediate compensation for their shortened lives, and crash research programs to attempt to lengthen them.

It was in the same speech in which President Edward Kennedy, after much delay, finally declared a national emergency and called out the National Guard to harvest California's crops, that the first mention of population control was made. Kennedy pointed out that the United States would no longer be able to offer any food aid to other nations and was likely to suffer food shortages herself. He suggested that, in view of the manifest failure of the Green Revolution, the only hope of the UDCs lay in population control. His statement, you will recall, created an uproar in the underdeveloped countries. Newspaper editorials accused the United States of wishing to prevent small countries from becoming large nations and thus threatening American hegemony. Politicians asserted that President Kennedy was a "creature of the giant drug combine" that wished to shove its pills down every woman's throat.

Among Americans, religious opposition to population control was very slight. Industry in general also backed the idea. Increasing poverty in the UDCs was both destroying markets and threatening supplies of raw materials. The seriousness of the raw material situation had been brought home during the Congressional Hard Resources hearings in 1971. The exposure of the ignorance of the cornucopian economists had been quite a spectacle—a spectacle brought into virtually every American's home in living color. Few would forget the distinguished geologist from the University of California who suggested that economists be legally required to learn at least the most elementary facts of geology. Fewer still would forget that an equally distinguished Harvard economist added that they might be required to learn some economics, too. The overall message was clear: America's resource situation was bad and bound to get worse. The hearings had led up to a bill requiring the Departments of State, Interior, and Commerce to set up a joint resource procurement council with the express purpose of "insuring that proper consideration of American resource needs be an integral part of American foreign policy."

Suddenly the United States discovered that it had a national consensus: population control was the only possible salvation of the underdeveloped world. But that same consensus led to heated debate. How could the UDCs be persuaded to limit their populations, and should not the United States lead the way by limiting its own? Members of the intellectual community wanted America to set an example. They pointed out that the United States was in the midst of a new baby boom: her birth rate, well over 20 per thousand per year, and her growth rate of over one per cent per annum were among the very highest of the developed countries. They detailed the deterioration of the American physical and psychic environments, the growing health threats, the impending food shortages, and the insufficiency of funds for desperately needed public works. They contended that the nation was clearly unable or unwilling to properly care for the people it already had. What possible reason could there be, they queried, for adding any more? Besides, who would listen to requests by the United States for population control when that nation did not control her own profligate reproduction?

Those who opposed population controls for the U.S. were equally vociferous. The military-industrial complex, with its all-too-human mixture of ignorance and avarice, still saw strength and prosperity in numbers. Baby-food magnates, already worried by the growing nitrate pollution of their products, saw their market disappearing. Steel manufacturers saw a decrease in aggregate demand and slippage for that holy of holies, the Gross National Product. And military men saw, in the growing population-food-environment crisis, a serious threat to their carefully nurtured Cold War. In the end, of course, economic arguments held sway, and the "inalienable right of every American couple to determine the size of its family," a freedom invented for the occasion in the early '70s, was not compromised.

The population control bill, which was passed by Congress early in 1974, was quite a document, nevertheless. On the domestic front, it authorized an increase from 100 to 150 million dollars in funds for "family planning" activities. This was made possible by a general feeling in the country that the growing army on welfare needed family planning. But the gist of the bill was a series of measures designed to impress the need for population control on the UDCs. All American aid to countries with overpopulation problems was required by law to consist in part of population control assistance. In order to receive any assistance each nation was required not only to accept the population control aid, but also to match it according to formula. "Overpopulation" itself was defined by a form based on U.N. statistics, and the UDCs were required not only to accept aid, but also to show progress in reducing birth rates. Every five years the status of the aid program for each nation was to be re-evaluated.

The reaction to the announcement of this program dwarfed the response to President Kennedy's speech. A coalition of UDCs attempted to get the U.N. General Assembly to condemn the United States as a "genetic aggressor." Most damaging of all to the American cause was the famous "25 Indians and a dog" speech by Mr. Shankarnarayan, Indian Ambassador to the U.N. Shankarnarayan pointed out that for several decades the United States, with less than six per cent of the people of the world, had consumed roughly 50 per cent of the raw materials used every year. He described vividly America's contribution to worldwide environmental deterioration, and he scathingly denounced the miserly record of United States foreign aid as "unworthy of a fourth-rate power, let alone the most powerful nation on earth."

It was the climax of his speech, however, which most historians claim once and for all destroyed the image of the United States. Shankarnarayan informed the assembly that the average American family dog was fed more animal protein per week than the average Indian got in a month. "How do you justify taking fish from protein-starved Peruvians and feeding them to your animals?" he asked. "I contend," he concluded, "that the birth of an American baby is a greater disaster for the world than that of 25 Indian babies." When the applause had died away, Mr. Sorensen, the American representative, made a speech which said essentially that "other countries look after their own self-interest, too." When the vote came, the United States was condemned.

IV

This condemnation set the tone of U.S.-UDC relations at the time the Russian Thanodrin proposal was made. The proposal seemed to offer the masses in the UDCs an opportunity to save themselves and humiliate the United States at the same time; and in human affairs, as we all know, biological realities could never interfere with such an opportunity. The scientists were silenced, the politicians said yes, the Thanodrin plants were built, and the results were what any beginning ecology student could have predicted. At first, Thanodrin seemed to offer excellent control of many pests. True, there was a rash of human fatalities from improper use of the lethal chemical, but, as Russian technical advisors were prone to note, these were more than compensated for by increased yields. Thanodrin use skyrocketed throughout the underdeveloped world. The Mikoyan design group developed a dependable, cheap agricultural aircraft which the Soviets donated to the effort in large numbers. MIG sprayers became even more common in UDCs than MIG interceptors.

Then the troubles began. Insect strains with cuticles resistant to Thanodrin penetration began to appear. And as streams, rivers, fish culture ponds and onshore waters became rich in Thanodrin, more fisheries began to disappear. Bird populations were decimated. The sequence of events was standard for broadcast use of a synthetic pesticide: great success at first, followed by removal of natural enemies and development of resistance by the pest. Populations of crop-eating insects in areas treated with Thanodrin made steady comebacks and soon became more abundant than ever. Yields plunged, while farmers in their desperation increased the Thanodrin dose and shortened the time between treatments. Death from Thanodrin poisoning became common. The first violent incident occurred in the Canete Valley of Peru, where farmers had suffered a similar chlorinated hydrocarbon disaster in the mid- '50s. A Russian advisor serving as an agricultural pilot was assaulted and killed by a mob of enraged farmers in January, 1978. Trouble spread rapidly during 1978, especially after the word got out that two years earlier Russia herself had banned the use of Thanodrin at home because of its serious effects on ecological systems. Suddenly Russia, and not the United States, was the *bête noir* in the UDCs. "Thanodrin parties" became epidemic, with farmers, in their ignorance, dumping carloads of Thanodrin concentrate into the sea. Russian advisors fled, and four of the Thanodrin plants were leveled to the ground. Destruction of the plants in Rio and Calcutta led to hundreds of thousands of gallons of Thanodrin concentrate being dumped directly into the sea.

Mr. Shankarnarayan again rose to address the U.N., but this time it was Mr. Potemkin, representative of the Soviet Union, who was on the hot seat. Mr. Potemkin heard his nation described as the greatest mass killer of all time as Shankarnarayan predicted at least 30 million deaths from crop failures due to overdependence on Thanodrin. Russia was accused of "chemical aggression," and the General Assembly, after a weak reply by Potemkin, passed a vote of censure.

It was in January, 1979, that huge blooms of a previously unknown variety of diatom were reported off the coast of Peru. The blooms were accompanied by a massive die-off of sea life and of the pathetic remainder of the birds which had once feasted on the anchovies of the area. Almost immediately another huge bloom was reported in the Indian Ocean, centering around the Seychelles, and then a third in the South Atlantic off the African coast. Both of these were accompanied by spectacular die-offs of marine animals. Even more ominous were growing reports of fish and bird kills at oceanic points where there were no spectacular blooms. Biologists were soon able to explain the phenomena: the diatom had evolved an enzyme which broke down Thanodrin; that enzyme also produced a breakdown product which interfered with the transmission of nerve impulses, and was therefore lethal to animals. Unfortunately, the biologists could suggest no way of repressing the poisonous diatom bloom in time. By September, 1979, all important animal life in the sea was extinct. Large areas of coastline had to be evacuated, as windrows of dead fish created a monumental stench.

But stench was the least of man's problems. Japan and China were faced with almost instant starvation from a total loss of the seafood on which they were so dependent. Both blamed Russia for their situation and demanded immediate mass shipments of food. Russia had none to send. On October 13, Chinese armies attacked Russia on a broad front. . . .

V

A pretty grim scenario. Unfortunately, we're a long way into it already. Everything mentioned as happening before 1970 has actually occurred; much of the rest is based on projections of trends already appearing. Evidence that pesticides have long-term lethal effects on human beings has started to accumulate, and recently Robert Finch, Secretary of the Department of Health, Education and Welfare expressed his extreme apprehension about the pesticide situation. Simultaneously the petrochemical industry continues its unconscionable poison-peddling. For instance, Shell Chemical has been carrying on a high-pressure campaign to sell the insecticide Azodrin to farmers as a killer of cotton pests. They continue their program even though they know that Azodrin is not only ineffective, but often *increases* the pest density. They've covered themselves nicely in an advertisement which states, "Even if an overpowering migration [sic] develops, the flexibility of Azodrin lets you regain control fast. Just increase the dosage according to label recommendations." It's a great game—get people to apply the poison and kill the natural enemies of the pests. Then blame the increased pests on "migration" and sell even more pesticide!

Right now fisheries are being wiped out by over-exploitation made easy by modern electronic equipment. The companies producing the equipment know this. They even boast in advertising that only their equipment will keep fishermen in business until the final kill. Profits must obviously be maximized in the short run. Indeed, Western society is in the process of completing the rape and murder of the planet for economic gain. And, sadly, most of the rest of the world is eager for the opportunity to emulate our behavior. But the underdeveloped peoples will be denied that opportunity—the days of plunder are drawing inexorably to a close.

Most of the people who are going to die in the greatest cataclysm in the history of man have already been born. More than three and a half billion people already populate our moribund globe, and about half of them are hungry. Some 10 to 20 million will starve to death *this year.* In spite of this, the population of the earth will increase by 70 million souls in 1969. For mankind has artificially lowered the death rate of the human population, while in general birth rates have remained high. With the input side of the population system in high gear and the output side slowed down, our fragile planet has filled with people at an incredible rate. It took several million years for the population to reach a total of two billion people in 1930, while a *second two billion will have been added by 1975!* By that time some experts feel that food shortages will have escalated the present level of world hunger and starvation into famines of unbelievable proportions. Other experts, more optimistic, think the ultimate food-population collision will not occur until the decade of the 1980's. Of course more massive famine may be avoided if other events cause a prior rise in the human death rate.

Both worldwide plague and thermonuclear war are made more probable as population growth continues. These, along with famine, may make up the trio of potential "death rate solutions" to the population problem—solutions in which the birth rate-death rate imbalance is redressed by a rise in the death rate rather than by a lowering of the birth rate. Make no mistake about it, *the imblance will be redressed.* The shape of the population growth curve is one familiar to the biologist. It is the outbreak part of an outbreak-crash sequence. A population grows rapidly in the presence of abundant resources, finally runs out of food or some other necessity, and crashes to a low level or extinction. Man is not only running out of food, he is also destroying the life support systems of the Spaceship Earth. The situation was recently summarized very succinctly: "It is the top of the ninth inning. Man, always a threat at the plate, has been hitting Nature hard. It is important to remember, however, that NATURE BATS LAST."

UPI

Alternatives

The recuperative powers of the biosphere are great, and our ability to predict in ecology is small. Though the outlook is grim, those alternatives that are less likely to lead us deeper into the morass deserve exploration.

Visions of a way out for man and the biosphere must be as radical as the problem. There is no precedent for effective reduction of armaments, or an effective system of international security, or world population control, but neither is there any precedent for world overpopulation supported and driven by today's technology and today's economics.

Alternatives must be sought in cognizance that the carrying capacity of the planet is limited and has probably been exceeded several fold. We must recognize that growth-promoting value systems and growth-promoting social institutions have suddenly become suicidal and intolerable. We will have to find new outlets for human drives that have been nurtured, biologically and culturally, for as long as man has been man.

Much of what we will have to discard has been fundamental to our past. Are any of the varied forms of marriage and the family which mankind has known still appropriate? Many may not be since they have functioned to sustain growth of population. Technological solutions which simply substitute one pollutant of unknown significance for another which has become intolerable are no longer acceptable. We must now look beyond the law of supply and demand and consider the environmental cost when we reach for a low grade ore, or a less accessible resource. We have sought, and failed to find, an effective system of international security and arms control. Now we *must* find them, and find them in association with worldwide conservation and allocation of diminishing resources, and worldwide control of population and pollution. Even some of our most cherished and ancient irrationalities are going to have to be discarded. Can we go on burying our dead in bronze and concrete, sowing the land above them with large rocks? Hardly. Human biomass will have to be returned to its source.

Loyalty to our dead, and loyalty to nations, will have to be discarded and replaced with loyalty to life and loyalty to Earth.

The carrying capacity of our planet, measured in numbers of individuals, will vary with the amount of regimentation and standard of living we accept. We can, with sufficient environmental bludgeoning and genetic surgery, adapt mankind to a low common denominator of docile and apathetic human poultry in a gleaming planet-wide chicken factory, and support a large population. Political freedom, and an opportunity for each human being who is born to realize his highest innate potentialities, are compatible only with a much lower population. Our present moral codes do not equip us for such choices, and our political systems do not provide the means to effect them. Our courts of law do not even, in general, yet recognize the right of the individual to a decent environment.

The first practical steps towards whatever alternatives may be open must include an inventory, in detail, of our present ecological realities. The gloomy intuitions of ecologists, and the shiny fantasies of the technomaniacs will have to be replaced with hard information on the status and trends of the biosphere, gathered, collated, and monitored by a single central organization. An international organization capable of acting on that information will have to be created.

While each of us searches, personally, for new value systems and new life styles, society will have to develop and test economic theories for the hoped-for equilibrium condition, as well as the period of transition. The developed and underdeveloped countries alike will have to adopt unilateral reduction of military expenditures and weaponry, as well as effective austerity programs to conserve and redistribute resources and diminish pollution.

Most essential, among all these things, is the development of a worldwide, comprehensive plan for the control and adjustment of population. Without that step no alternatives are possible.

The initiatives must come largely from those who have yet to be bought off by the status quo. The inertia of the worldwide establishment, the age and conservatism of its leaders, the irrelevance of its concerns, largely immobilize our present system of getting things done. Much will depend on whether the alienated and restless young who are still willing to try can escape from the equally irrelevant political radicalisms of the past and present and come up with an all encompassing ecological radicalism for the future.

The Unanswerable Questions
W. H. Ferry

The number of great issues to which no reasonable solution is possible appears to me to be growing almost daily. I am aware that mankind has gone on for a long time with a considerable backlog of unanswered questions, or, in any case, with questions that have been at best half-answered or have been resolved in the muffled tones of compromise. But it seems that today's backlog is different in important ways. Yesterday's questions tended to be local —for example, World War I only in a limited way had global dimensions. Today's questions, on the other hand, are nearly universal in their significance. Unless we have been misled, we shall have to call the next international conflict not World War III but Space War I.

There is the added sense that half-answers to today's questions cannot be satisfactory—or at least not satisfactory for very long. Mr. Micawber has been our proconsul for too many years. I think he must be retired instantly.

Reference to Mr. Micawber brings me to the first, the largest, and the most obvious of my questions: *Will civilization survive?* Man is now providing himself with the instruments of annihilation at a cost of almost two hundred billion dollars a year. Such is the fever pitch that the figure is probably now nearing the quarter-trillion-dollar mark. Most of this lunacy is taking place in the best-educated, best-informed, and most affluent nations in the world. No comparable figures for works of non-annihilation have been assembled, as far as I know. If one puts together expenditures for the United Nations and its satellites, plus the output for such items as aid to developing nations, and adds desultory efforts like disarmament commissions, world constitutions, and the like, a figure of around five to ten billion dollars can perhaps be reached for the works of survival, or about two to five per cent of the wealth expended on armaments. It must be agreed, I think, that there is little market for peace and an apparently insatiable one for the works of war. There arises the unhappy vision of mankind surviving only in nests of missiles and biological warfare installations, chattering across the barricades about peace "on our terms," in perpetual hock to fear. Will civilization survive? I deem the question unanswerable.

The second question is equally obvious: *If civilization survives, will man live in a world worth living in?* Since I am an American, let

me localize this second question: I'll make it *Will democracy survive?* I have nothing complicated in mind here—by democracy I mean self-government.

The evidence is now plain that, in the minds of many Americans today, someone else in some other place is making all the decisions. There are many reasons for this increasing sense of political remoteness. All are of surface validity—for example, the sometimes incomprehensible interconnections of the industrial, internationalized state. But understanding these reasons does not change the citizen's feeling that politics is becoming form without substance. The notorious military-industrial complex has become an almost palpable presence in our lives. The citizen is talked *to* unceasingly—by the media, by politicians and self-seekers, but he has no way to talk back. The residents of Santa Barbara, where I live, looking across streaked beaches to a swiftly increasing grove of oil platforms in their channel, are discovering that the amenities of their city depend on such unintelligible factors as the world price of oil, the gaps and holes in tax laws, and quiet agreements between magnates of industry and magnates of government thousands of miles away. Political scientists are in honest dispute as to the extent of democratic practice—not, mind you, democratic rhetoric —but all seem agreed that democratic practice is dwindling or is so changing in character as to call for a different name. Will democracy survive? Unanswerable again.

All unanswerable questions are not of this cosmic character. There are many homely examples. Consider, for example, whether we can, if we wish to do so, actually control the conditions of our daily lives. Posit a community of fifty thousand people, and suppose it to be in a setting pleasing to the inhabitants. Suppose further that the great majority of these inhabitants feel that their city is quite

large enough and that they have no desire for an increase in population. Further suppose that, turning their backs on our alleged progress, they do not want any more industry, any more invading freeways. You may think these unlikely suppositions, but there is always a chance that some community will come to its senses and try to have something to say about its own development or non-development.

Let us suppose a few more things: that this community opts for home-owned businesses, for complete public control of the land within its boundaries, for privacy, and for freedom from the high decibels of crashing sound that have become commonplace in urban life, and also for a minimum of automobiles—say, one to a family. Supposing all such goals, could they be achieved? There is another unanswerable question. No one can confidently assert that such a community is impossible because it has never been tried. Nor can anyone say that it is possible if only the requisite determination can be achieved.

Now this is no frivolous set of suppositions. In my judgment there is an intense yearning among Americans for an environment that is manageable and small enough for citizens to lend their weight in the balance of decisions. The country is dotted with towns and small cities desperately seeking to resist a nearby megalopolis. The impulse is most clearly seen in the so-called outer suburbs. Are these communities at the mercy of the urban suction-pump or can they do something about it? Aside from these outer suburbs there are those more distant enclaves which still retain some sylvan charm. Trying to control one's own destiny took on a modern dimension when the suburbs finally cut themselves off from their parent cities and said, "We want a life different from yours." Of course most suburbanites never really meant this. What they really meant was that they wanted a quiet bedroom, better schools, and nothing but white faces in the neighborhood stores. The near-in suburbs have all but given up the struggle for real distinctiveness, but the effort goes on in the outer rings of megalopolis. And there are many communities, large and small, remote from the great centers, which are now watching their powers of self-government being eroded by distant forces that they cannot comprehend, much less regulate.

All this is only to say that many obstacles lie across the path of a community such as I have supposed. I pass over certain obstacles that will occur to everyone, such as the drive for super-efficient government, the difficulty of getting the citizenry to consider and debate significant issues, the indignation of neighboring communities about actions that injure them, and the subsequent reprisals of these neighbors. The less apparent obstacles are just as numerous. There is, for example, the homogenizing effect of television. Any effort to maintain a distinctive community in this country has the tube to contend with. Television decrees our national styles in attitude, dress, consumption, and who can say what else. Thus, to use a kitchen example, television is mainly responsible for the elimination of local cheese producers, so that we now have a choice only of rubbery samenesses on most of our supermarket cheese counters. Can a community, however determined it might be, prevail against such pressures?

The federal Constitution is, however, the chief obstruction, for it might well decree that our imaginary city could not ration births, halt the immigrant, deny the trespassing free-way builder, or keep a second car out of the garage. Nonetheless, I do not consider this a conclusive objection, for the Constitution can always be changed. Indeed, it should be changed if it gets in the way of the human development of the nation. I would not care to argue that the Constitution ought to be amended or rewritten merely to meet the desires of my suppositious community. Perhaps a better way of stating the issue, then, is to ask whether the Constitution is facilitating undesirable development, so that under its permissive gaze we stand in danger of losing our souls as well as our water supplies, our clean air, and our arable land. Here, I believe, there is another unanswerable question: *Do we have the political wit or will to rewrite the Constitution to reflect what we have learned about ourselves and technical society in the past generation or two?*

To conclude the instance of the small city I have imagined, it would undertake to control its population. It would limit entry. It would work out a transportation policy based on the reasonable needs of the community. And so on and on. All such matters would require a communal dedication and enterprise in political action far beyond our current imagination. Combinations of persuasion and incentive and restraint, of which there is no modern example, would be called for.

Once again such a program will be dismissed as repugnant and regressive. Yet a good deal of the commotion today may be interpreted as efforts in that very direction. The traffic boss of New York City recently warned that private vehicles may soon have to be barred from the streets of that suffocating city. Los Angeles, a special example of demented mobility, is moving through smog in a similar direction, though no decisive action may be expected there until a few thousand citizens gasp to death. Garrett Hardin, of the University of California at Santa Barbara, proposes a tax in Los Angeles of a dollar a gallon on gasoline. The New York traffic boss and Professor Hardin and anyone like them trying to do what is necessary will, however, find themselves up against a constitutional challenge if they try to put their plans into effect.

As a final instance, consider some recent despairing words of Daniel Moynihan, President Nixon's adviser on the cities. Moynihan, ordinarily one of the most ebullient of men, said:

"For reasons no one understands, government has been impotent in trying to solve urban problems. The inability of government to bring about urban change is a fundamental problem of government today. Our delivery systems are not in fact working."

I am trying to point out that the matters which have to be dealt with by a small city determined to keep its singular character are similar to those faced by every metropolis. But of course for most large cities it is already too late to think of finding human solutions. For these cities the big question is survival itself. Not long ago, *Newsweek* asked:

"Can the suburban 'white noose' around the cities be broken by genuine integration? . . . Is there any real way to end the cycle of welfare dependency? Will better-paid, more mobile police cut the spiraling crime rate? Do decentralization and community control really promise better education? . . . Is big-city life still worth living?"

The difference, it is clear, is one of degree. If the questions asked about the small city are unanswerable, all the more unanswerable are the questions being asked about the great cities.

We see in these issues some of what is agitating our young people. They have no care for a civilization held in thrall by questions that are at once awesome and unanswerable. Their plea for participation can be read as an appeal for intelligibility and manageability. A good deal of adult resentment of the young (which as often as not seems to verge on outright hatred) may arise from a subconscious realization that our children are stirring things up in favor of a *human* civilization, not a technical order, and in favor of restoring the kind of relations among people that adults themselves somehow let slip from their grasp. So there may be a heavy dose of envy mixed into the adult demand that the young be repressed and that they conform to a style of existence that their elders, in occasional fits of candor, acknowledge to be aimless and excruciatingly boring. It is a fair guess that adults do not want their children participating in any significant way in determining the conditions of their own lives if they themselves are not capable of doing so. In Kenneth Millar's phrase, the conservatism of despair holds them back.

Race is the next entry. I find some difficulty in phrasing the question correctly. The following will have to do: *Can we resolve the race issue, with freedom and justice for all?* This seems as unanswerable a query as any. On the one side we see millions of people getting to their feet, beginning to stand upright for the first time, experiencing a sense of selfhood and dignity that had been squashed throughout American history. Their talk is loud, angry, confused, and direct. Blacks are economically near-broke but spiritually and politically affluent—and I would like to come down hard on that word "politically." They are saying, we know what is best for ourselves, so please get that white hand off our shoulder.

On the other side there is a vast white majority that is frightened, with no genuine understanding of what the blacks are clamoring for in their campaigns for autonomy and for a culture based on black history. This white majority on the whole wishes its black colonial subjects well, and has even passed laws in evidence of its benign outlook. It hopes that the blacks succeed in improving their lot, but only if the improvement does not cost much and only if the blacks do not threaten white superiority or white peace of mind. The white majority is also heavily armed, officially and unofficially. These arms are intended for use any time the minority goes beyond the demilitarized zone that now by tacit consent separates blacktown and whitetown. Finally, the white majority says to the black minority: "We know better than you do what is best for you, and we'll see that you get it, even if you find it painful." The argument resembles the viewpoint of those who think we can best resolve the Vietnam dilemma by killing off all the Vietnamese.

I deny that there is an answer to the race question; I deny that we—that is, the conscientious members of the white majority—are really working at it and have achieved fair success already. I reject this whole line of argument, for on examination what we see is a series of compromises, or concessions, or half measures usually arrived at as a means of damping down explosive situations. Meanwhile the principal institutions of our racist society remain intact and unchanged, the towers of privilege are unshaken. Bargains and compromises, I am aware, are said to be the essence of American political life. But what, in this situation, has such an explanation to do with freedom and justice, let us say for the urban or rural blacks chained in their respective slums? I am reminded of one recent political bargain in Washington, under the terms of which starving black children are to be allowed—as an experiment—free food stamps for a while.

I conclude that race is a question without an answer in the United States, another problem to which there is no solution. There may well be a resolution of the matter, but it will in no sense be a genuine answer. I refer to the possibility of a civil war, which I regard as likely. But of course this resolution will not be a civil war but a pogrom, incited by those millions of whites to whom blacks today are almost as frightening a presence as Nat Turner and his band were to Southern plantation-owners. Such a resolution will necessarily have

only a temporary effect in subduing blacks, will enormously increase the freight of white guilt, and will only compound the unanswerability of the question with which this article opened.

My last question is this: *Can Americans acquire the new consciousness required to give purpose and zest to their common life, and to redeem the nation's proclaimed ideals?* It may be thought that there is a preliminary query; namely, whether such a new consciousness is needed. Based on my view of the state of the nation, the answer has to be roaringly affirmative. A superficial casting-up of accounts establishes the United States as blessed beyond history's wildest intimations of success: a constitutional regime of two hundred years' stability, a plentiful domain moated by great oceans and friendly (or at least non-hostile) neighbors, propelled by an economic engine that year after year breaks records in all departments from G.N.P. to the amounts spent on non-economic follies like arms and space, and constantly rising achievement in all the conventional indices from childhood mortality to literacy, to books and concert tickets sold, to salaries paid rock singers and football players. Such cataloguings of the American Way go on and on. They are all valid enough.

Yet this is not the central story of the United States today, only a small part of it. Such statistics are merely the backdrop of success against which the ironies stand forth plainly. Something has gone wrong. Erich Fromm says that Americans are "anxious, lonely, and bored." Wealth and accomplishment are not nearly enough. There is a hollowness at the center of the national life. Americans are cynical about politics and politicians and think that government is a matter of logrolling or of buying and selling favors. They are frightened by the prospects of abundance and leisure whether this leisure is for them or for their neighbors. Miracle-a-day science and technology fascinate them and at the same time bring forebodings as they sense their incapacity to deal with the moral issues generated in the laboratories. For example: What to do about the geneticists' alarming promise that we shall soon be able to determine the chief physical and mental characteristics of the population? What to do about our possession of enough poison to kill the people of the world ten times over, and enough deadly bugs and hydrogen bombs to eliminate any life that might manage to survive around the edges? What to do about the advent of a society in which relatively few people will have to work as the machine slaves take over?

We are distressed to discover that we have so misused nature as to put our grandchildren, if not our children, in danger of suffocation through lack of oxygen, or of being buried alive in their own garbage, or being carried off by lead poisoning. There is a halting recognition that as we pave our farmlands we may be doing away with croplands we shall one day need.

In the thoughtless and catastrophic attack on the environment, from the bottom of the sea to the outer envelope of the atmosphere, there may just be a new consciousness forming. Belatedly as always, the mass media are beginning to give a little alarmed attention to the rapid deterioration of our physical world. So at last we may be getting sensible and learning to fear the right things—not the Communists but our own suicidal practices. But I do not think that we will spend the billions needed to redeem our environment, nor, indeed, that we have the political machinery to carry on the great clean-up even if we were willing to tax ourselves for this purpose.

New orders of magnitude appear daily, and human beings become more dwarfed and insignificant. Size has always been a major criterion in the American consciousness—the bigger the better—but we are beginning to understand that size in the end is the death of individuality and distinctiveness. The illicit claims of bigness are apparent in the condition of our cities. They continue to grow, becoming less and less livable, each itself an increasing cluster of unanswerable questions. Citizens quiver in the drumhead tautness lying across metropolitan centers. The idea of community is itself the most conspicuous fatality of urban civilization. There can never have been a more sour joke than placing the label Fun City on New York. Our greatest metropolis is gritty with hatred and threatened with disintegration.

Like the earlier list, this catalogue of American ills could be extended indefinitely. Problems multiply at an unprecedented rate. I know that I shall be charged with having drawn an overblown picture. But it is the picture I see. There is little ease or serenity and little laughter among us. An uncommon malignity is beginning to mark our disputes. The legacy of America, as matters now stand, is not a life that is nasty, brutish, and short, but one that is apprehensive, surfeited, and long.

So I return to the question, Can Americans acquire a new consciousness? I believe this to be unanswerable. Americans must! cry the young. We will not, say the old. We cannot, intone the political pragmatists. It must be evident that I come down on the side of the young. The old consciousness is worn out. It has served its purpose and deserves respectful burial. But having cast my vote for the new consciousness, I have not answered the question whether we can achieve it, I have only declared my hope for its election. Nor is it within the scope of this article to discuss the *content* of the new consciousness. It is a separate matter—youth's dream of revolutionizing American morality, in John Updike's words. I do not mind, however, indicating the key to my understanding of this phrase. It takes in the word we hear more and more often from our young friends, the word "radicalization."

Radicalization is the process of looking afresh at the institutions and procedures of the society, beginning at their roots and working onward, taking nothing for granted or as being of continuing value except these indispensable things: community, self-rule, the integrity of the human person, and the necessity of living in harmony with the natural order. In this sense our protesting young people must be deemed not revolutionaries but counter-revolutionaries. They wish to undo or trip up the technical-industrial revolution and get back to pre-revolutionary ideas like political participation and personal responsibility.

I take it that the new consciousness is aimed at doing something about all the conditions that cripple these essential aims. I wish its proponents well, for I cannot imagine how it will all end otherwise. It does not, I suppose, matter much whether a great civilization ends in a horrible bang or an ignoble whimper. The apocalyptic current in me has slowed down. Nor does my morose outlook incline me to suicide—men just don't seem to jump off the bridge often for big reasons, they usually do so for little ones.

Anyway, I have no taste for such long-range forecasts; the focus of this article is on the ambiance of life, the tone of hatred and suspicion and cynicism that we are coming to accept as part of what citizenship in post-industrial society means. Perhaps we shall never know when civilization, in the ordinary and honored sense, actually ends and refined barbarism takes over.

I realize that the mood of this essay has been unrelieved melancholy. This is partly because the topic is not a cheery one. It is mainly because no answers occur to me; and because I sense strongly that we have pushed our luck very near the breaking point. It is hard for me to accept that the human annals contain chapters as bleak and unpromising as the present one. There has surely never been so much morbid speculation amid plenty and apparent progress. Our hardest and most high-minded labors seem inevitably Sisyphean. Those who take comfort in the solidity of our institutions may do so; these institutions no longer feel solid to me. Those who derive confidence from our "tradition" are, I think, falsely led, for our tradition of justice, equality, fraternity, mutual respect is mostly a tissue of affectionate memory, not reality functioning in our lives except here and there along the margins. I do not know what to say to those asserting that God is working in history. That statement only recalls to me the frightening graffito reported from a San Francisco lavatory: "God has cancer." Our heroes are in outer space, on the basketball courts, or in the grave. Martin Luther King, Jr., is a hero, now that he is safely dead. President Eric Walker of the University of Pennsylvania has said that salvation is to be found in a rebirth of "practical know-how," and engineering on a gigantic scale. This strikes me as hair-of-the-dog advice, and about as effective.

There is one bright spot and one only on my screen. I see wisps of hope in the young, in the blacks and browns, in the militants of all colors. They are the partisans of the new consciousness. But I do not want to idealize them. They are disorganized, rude, contemptuous. Sometimes they are even a tenth as violent as those beating on them. But they are thinking of the right things, thinking of how to save our environment from human predators, thinking of how to retain their manhood as the structures of post-industrial society close in, thinking of how to convert accomplishment into human ends, thinking of how to achieve dignity. They are thinking of Malcolm X's wish that he might "save America from a grave, even fatal catastrophe." Their hope and mine rises as we consider that their efforts helped to drive a war-committed President from office, and as we see that genuine gains have been made in the independence of the black spirit.

Yet it must be conceded that these are miniscule breaches in the mighty wall of self-satisfaction and self-righteousness that is the chief defense of an over-endowed society. In my mind the question is whether this small band of young people and blacks and idealists can, by demonstration, exhortation, confrontation, example, or raid, bring this nation to its senses before it undergoes the final great lesson of catastrophe, domestic or international.

And this is the least answerable question of all.

Criteria For An Optimum Human Environment
Hugh H. Iltis, Orie L. Loucks, and Peter Andrews

Almost every current issue of the major science journals contains evidence of an overwhelming interest in one urgent question: Shall a single species of animal, man, be permitted to dominate the earth so that life, as we know it, is threatened? The uniformity of the theme is significant but if there is consensus, it is only as to the need for concern. Each discipline looks differently at the problem of what to do about man's imminent potential to modify the earth through environmental control. Proposals to study ways of directing present trends in population, space and resource relationships toward an "optimum" for man are so diverse as to bewilder both scientists and the national granting agencies.

Arrogance Toward Nature

It is no thirst for argument that compels us to add a further view. Rather it is the sad recognition of major deficiencies in policies guiding support of research on the restoration of the quality of our environment. Many of us find the present situation so desperate that even short-term treatments of the symptoms look attractive. We rapidly lose sight of man's recent origins, probably on the high African plains and the natural environment that shaped him. Part of the scientific community also accepts what Lynn White has called our Judeo-Christian arrogance toward nature, and is gambling that our superior technology will deliver the necessary food, clean water and fresh air. But are these the only necessities? Few research proposals effectively ask whether man has other than these basic needs, or whether there is a limit to the artificiality of the environment that he can tolerate.

In addition, we wish to examine which disciplines have the responsibility to initiate and carry out the research needed to reveal the limits of man's tolerance to environmental modification and control. We are especially concerned that there is, on the one hand, an unfortunate conviction that social criteria for environmental quality can have no innate biological basis—that they are only conventions. Yet, on the other hand, there is increasing evidence suggesting that mental health and the emotional stability of populations may be profoundly influenced by frustrating aspects of an urban, biologically artificial environment.

There have been numerous proposals for large-scale inter-disciplinary studies of our environment and of the future of man, but such studies must have sufficient breadth to treat conflicting views and to seek to reconcile them. We know of no proposal that would combine the research capabilities of a group studying environmental design with those of a group examining the psychological and mental health responses of man to natural landscapes. The annual mass migration of city man into natural landscapes which provide diversity is a matter of concern to the social scientist, whose research will only be fully satisfactory when joined with studies that quantify the landscape quality, the psychology of individual human response, and the evolutionary basis of man's possible genetic adaptations to nature. The following summary of recent work may provide a basis for scientists in all areas to seek and support even greater breadth in our studies of present and future environments for man.

"Web of Life"

Two major theses are sufficiently well established to provide the positive foundation of our argument. First, we believe the inter-dependency of organisms, popularly known as the "web of life," is essential to maintaining life and a natural environment as we know it. The suffocation of aquatic life in water systems, and the spread of pollutants in the air and on the land, make it clear that the "web of life" for many major ecosystems is seriously threatened. The abrupt extinction of otherwise incidental organisms, or their depletion to the point of no return, threatens permanently to impair our fresh water systems and coastlines, as well as the vegetation of urban regions.

Second, man's recent evolution is now well enough understood for it to play a major part in elucidating the total relation of man to his natural environment. The major selection stresses operating on man's physical evolution

have also had some meaning for the development of social structures. These must be considered together with the immense potential of learned adaptations over the entire geologic period of this physical evolution. Unfortunately, scientists, like most of us moderns, are city dwellers dependent on social conventions, and so have become progressively more and more isolated from the landscape where man developed, and where the benchmarks pointing to man's survival may now be found. They, of all men, must recognize that drastic environmental manipulations by modern man must be examined as part of a continuing evolutionary sequence.

The immediacy of problems relating to environmental control is so startling that the threat of a frightening and unwanted future is another point of departure for our views. At the present rate of advance in technology and agriculture, with an unabated expansion of population, it will be only a few years until all of life, even in the atmosphere and the oceans, will be under the conscious dictates of man. While this general result must be accepted by all of us as inevitable, the methods leading to its control offer some flexibility. It is among these that we must weigh and reweigh the cost-benefit ratios, not only for the next 25 or 50 years, but for the next 25,000 years or more. The increasing scope of the threat to man's existence within this controlled environment demands radically new criteria for judging "benefits to man" and "optimum environments."

It would be perverse not to acknowledge the immense debt of modern man to technological development. In mastering his environment, man has been permitted a cultural explosion and attendant intricate civilization made possible by the very inventiveness of modern agriculture, an inventiveness which must not falter if the world is to feed even its present population. Agricultural technology of the nineteenth and twentieth centuries, from Liebig and the gasoline engine to hybrid corn, weed killers and pesticides, has broken an exploitative barrier leading to greatly increased production and prosperity in favored regions of the world. But this very success has imposed upon man an even greater responsibility for managing all of his physical and biotic environment to his best and sustained advantage.

The view also has been expressed recently that the "balance of nature," upset by massive use of non-disintegrating detergents and pesticides, will be restored by "new engineering." Such a view is necessarily based on the assumption that it is only an engineering problem to provide "an environment [for man] relatively free from unwanted man-produced stress." But when the engineering is successful, the very success dissipates our abilities to see the human being as part of a complex biological balance. The more successful technology and agriculture become, the more difficult it is to ask pertinent questions and to expect sensible answers on the long-range stability of the system we build.

The Right Questions?

Inspired by recent success, some chemical and agricultural authorities still hold firmly that we can feed the world by using suitable means to increase productivity, and there is a conviction that we can and must bend all of nature to our human will. But if open space were known to be as important to man as is food, would we not find ways to assure both? Who among us has such confidence in modern science and technology that he is satisfied we know enough, or that we are even asking the right questions, to ensure our survival beyond the current technological assault upon our en-

vironment. The optimism of post-World War II days that man can solve his problems—the faith in science that we of Western culture learn almost as infants—appears more and more unfounded.

To answer "what does man now need?" we must ask "where has he come from?" and "what evidence is there of continuing genetic ties to surroundings similar to those of his past?"

Theodosius Dobzhansky and others have stressed that man is indeed unique, but we cannot overlook the fact that the uniqueness does not separate him from animals. Man is the product of over a hundred million years of evolution among mammals, over 45 million years among primates, and over 15 million years among apes. While his morphology has been essentially human for about two million years, the most refined neurological and physical attributes are perhaps but a few hundred thousand years old.

Selection and Adaptation

G. G. Simpson notes that those among our primate ancestors with faulty senses, who misjudged distances when jumping for a tree branch or who didn't hear the approach of predators, died. Only those with the agility and alertness that permitted survival in ruthless nature lived to contribute to our present-day gene pool. Such selection pressure continued with little modification until the rise of effective medical treatment and social reforms during the last five generations. In the modern artificial environment it is easy to forget the implications of selection and adaptation. George Schaller points out in "The Year of the Gorilla" that the gorilla behaves in the zoo as a dangerous and erratic brute. But in his natural environment in the tropical forests of

Africa, he is shy, mild, alert and well-coordinated. Neither gorilla nor man can be fully investigated without considering the environments to which he is adapted.

Unique as we may think we are, it seems likely that we are genetically programmed to a natural habitat of clean air and a varied green landscape, like any other mammal. To be relaxed and feel healthy usually means simply allowing our bodies to react as evolution has equipped them to do for 100 million years. Physically and genetically we appear best adapted to a tropical savanna, but as a civilized animal we adapt culturally to cities and towns. For scores of centuries in the temperate zones we have tried to imitate in our houses not only the climate, but the setting of our evolutionary past: warm humid air, green plants, and even animal companions. Today those of us who can afford it may even build a greenhouse or swimming pool next to our living room, buy a place in the country, or at least take our children vacationing at the seashore. The specific physiological reactions to natural beauty and diversity, to the shapes and color of nature, especially to green, to the motions and sounds of other animals, we do not comprehend and are reluctant to include in studies of environmental quality. Yet it is evident that nature in our daily lives must be thought of, not as a luxury to be made available if possible, but as part of our inherent indispensable biological need. It must be included in studies of resource policies for man.

Dependence on Nature

Studies in anthropology, psychology, ethology and environmental design have obvious implications for our attempts to structure a biologically sound human environment. Unfortunately, these results frequently are masked by the specifics of the studies themselves. Except for some pioneer work by Konrad Lorenz followed up at several symposia in Europe, nothing has been done to systematize these studies or extend their implications to modern social and economic planning. For example, Robert Ardrey's popular work, "The Territorial Imperative," explores territoriality as a basic animal attribute, and tries to extend it to man. But his evidence is somewhat limited, and we have no clear conception of what the thwarting of this instinct does to decrease human happiness. The more extensive studies on the nature of aggression explore the genetic roots of animal conflicts, roots that were slowly developed by natural selection over millions of generations. These studies suggest that the sources of drive, achievement, and even of conflict within the family and war among men are likely to be related to primitive animal responses as well as to culture.

Evidence exists that man is genetically adapted to a nomadic hunting life, living in small family groups and having only rare contact with larger groups. As such he led a precarious day-to-day existence, with strong selective removal due to competition with other animals, including other groups of humans. Such was the population structure to which man was ecologically restricted and adapted until as recently as 500 generations ago. Unless there has since been a shift in the major causes of human mortality before the breeding age (and except for resistance to specific diseases there is no such evidence), this period is far too short for any significant changes to have occurred in man's genetic makeup.

Studies of neuro-physiological responses to many characteristics of the environment are also an essential part of investigating genetic dependence on natural as opposed to artificial environment. The rapidly expanding work on electroencephalography in relation to stimuli is providing evidence of a need for frequent change in the environment for at least short periods, or, more specifically, for qualities of diversity in it. There is reason to believe that the electrical rhythms in the brain are highly responsive to changes in surroundings when these take the full attention of the subject. The rise of mechanisms for maintaining constant attention to the surroundings can be seen clearly as a product of long-term selection pressures in a "hunter and hunted" environment. Conversely, a monotonous environment produces wave patterns contributing to fatigue. One wonders what the stimuli of brick and asphalt jungles, or the monotony of corn fields, do to the nervous system. Biotic as well as cultural diversity, from the neurological point of view, may well be fundamental to the general health that figures prominently in the discussions of environmental quality.

Results with Patients

The interesting results of Maxwell Weismann in taking chronically hospitalized mental patients camping are also worth noting. Hiking through the woods was the most cherished activity. Some 35 of the 90 patients were returned to their communities within three months after the two-week camping experience. Other studies have shown similar results. Many considerations are involved, but it seems possible that in a person whose cultural load has twisted normal functioning into bizarre reactions, his innate genetic drives still continue to function. Responses attuned to natural adaptations would require no conscious effort. An equally plausible interpretation of Weismann's results is that the direct stimuli of the out-of-doors, of nature alone, produces a response toward the more normal. A definitive investigation of the bases for these responses is

needed as guidance to urban planners and public health specialists.

These examples are concerned with the negative effects which many see as resulting from the unnatural qualities of man's present, mostly urban, environment. Aldous Huxley ventures a further opinion as he considers the abnormal adaptation of those hopeless victims of mental illness who appear most normal: "These millions of abnormally normal people, living without fuss in a society to which, if they were fully human beings, they ought not to be adjusted, still cherish 'the illusion of individuality,' but in fact they have been to a great extent de-individualized. Their conformity is developing into something like uniformity. But uniformity and freedom are incompatible. Uniformity and mental health are incompatible as well. . . . Man is not made to be an automaton, and if he becomes one, the basis for mental health is lost."

Clearly, a program of research could tell us more about man's subtle genetic dependence on the environment of his evolution. But of one thing we can be sure: only from study of human behavior in its evolutionary context can we investigate the influence of the environment on the life and fate of modern man. Even now we can see the bases by which to judge quality in our environment, if we are to maintain some semblance of one which is biologically optimum for humans.

We do not plead for a return to nature, but for re-examination of how to use science and technology to create environments for human living. While sociological betterment of the environment can do much to relieve poverty and misery, the argument that an expanding economy and increased material wealth alone would produce a Utopia is now substantially discounted. Instead, a natural concern for the quality of life in our affluent society is evident.

But few economists or scientists have tried to identify the major elements of the quality we seek, and no one at all has attempted to use evolutionary principles in the search for quality. Solutions to the problems raised by attempts to evaluate quality will not be found before there is tentative agreement on the bases for judging an optimum human environment. A large body of evidence from studies in evolution, medicine, psychology, sociology, and anthropology suggests clearly that *such an environment will be a compromise between one in which humans have maximum contact with the properties of the environment to which they are innately adapted, and a more urban environment in which learned adaptations and social conventions are relied upon to overcome primitive needs.*

Our option to choose a balance between these two extremes runs out very soon. Awareness of the urgency to do something is national, and initial responses may be noted in several well-established but relatively narrow scientific disciplines. There has been the recent revival of eugenics. A balanced view has been proposed by Leonard Ornstein (*Bulletin,* June 1967), who agrees with others that positive improvements in man's genetic make-up must wait until we are vastly more knowledgeable. He recommends control of degenerating effects from uncontrolled mutation (in the absence of high selection) until more positive measures can be taken.

An "Impossible" Challenge

More extreme views have been expressed that man could be changed genetically to fit any future, but the means to do this and the moral justification of the aims sought are still far from being resolved. Many support the so-called evolutionary and technological optimists who, unlike their forefathers of little more than a generation ago, believe man can be changed radically when the time comes. They show a faith that science has proved its ability to draw on an expanding technology to do the impossible. The technologically impossible seems to have been accomplished time and time again during the past two or three generations, and may happen again. But some important scientific objectives have not been achieved, and we are likely to become more aware of the failures of science, of the truly impossible, as the irreversible disruptions of highly complex biological systems become more evident.

We suggest that the alternative to genetic modification of man is to select a course where the objectives only verge on the impossible. Let us regard the study and documentation of criteria for an environmental optimum as the "impossible" challenge for science and technology in the next two decades! Although considerable research in biology, sociology, and environmental design is already directed to this objective, there are several other types of study required that we outline briefly, simply to indicate the scope of the challenge.

First, a thorough examination must be undertaken of the extent to which man's evolutionary heritage dominates his activity both as an individual and in groups. The survival advantage of certain group activities has clearly figured in his evolutionary success and adaptive culture. Although cultural adaptation now dominates the biological in the evolution of man, his basic animal nature has not changed. Research leading to adequate understanding of the need to meet innate genetic demands lies in the field of biology, and more specifically in a combination of genetics, physical anthropology and ethology.

Second, we need to understand more of how cultural adaptations and social conventions of man permit him to succeed in an artificial environment. Cultural adaptation is the basis of his success as a gregarious social animal, and it will continue to be the basis by which he modifies evolutionarily imposed adaptations. Medical studies suggest there may be a limit to the magnitude of cultural adaptations, and that for some people this is nearly reached. Studies in sociology, cultural anthropology and psychology are all necessary to such research, in combination with environmental design and quantitative analysis of diversity in the native landscape.

Third, relationships between the health of individuals, both mental and physical, and the properties of the environment in which they live should be a fundamental area of research. It is easy to forget that we should expect as much genetic variability in the capacity of individuals to adjust to artificial environments as we find in the physical characteristics of man. Some portions of the population should be expected to have a greater inherent commitment to the natural environment, and will react strongly if deprived of it. Others may be much more neutral. Studies of the population as a whole must take into account the variability in reaction, and must therefore consider population genetics as well as psychiatry and environmental design.

Fourth, environmental qualities should be programmed so as to optimize for the maximal expression of evolutionary (i.e., human) capabilities at the weakest link in the ontogenic development of human needs. While there are many critical periods during our life, we believe the ties to natural environments to be most vital during youth. We have abundant evidence on our campuses and in our cities that the dislodgement of youth presents one— if not the most—serious obstacle to successful adoption of more complex social structures. The dislodgement of man in an artificial environment will vary throughout his ontogeny. Even the small child or infant cannot be expected to be indifferent to changes in the gross

characteristics of his own community, as he cannot within his own family.

Young men and women accept many of the modern social conventions, but retain the highly questioning mind that once led to new and better ways to hunt and forage. By early middle age, man's physical and mental agility has changed and he becomes a stronger adherent to the social conventions that make his own society possible. During the rise of modern man on the high African plains, and continuing into modern primitive societies, each community was very much dependent on its young men. They contributed to hunting and community protection through their strength and agility, commodities for which there is declining demand in modern society. Survival in the primitive groups was to some degree dependent on the willingness of youth to innovate and take risks, and this has become a fixed adaptation, requiring outlets of expression.

Over 30 years ago, sociologist W. F. Ogburn suggested that society in the future would require "prolonging infancy to, say, thirty or forty years or even longer." Is not our 20-year educational sequence a poorly-veiled attempt to do just that? From an evolutionary point of view will not this dislodgement of youth present the most serious obstacle to successful adoption of more complex social structures? We are compelled to acknowledge that our over-all technological environment for youth has not compensated for the loss of challenges of the hunt and the freedom of the Veldt. The disruptions on our campuses and in the cities indicate the need to plan environmental optima for this weakest link in the human need for expression of evolutionary capabilities.

Finally, systems ecology is developing the capacity for considering all of the relationships and their interactions simultaneously. The notion of fully describing the optimum for any organism may seem presumptuous. It requires measurement of every type of response, particularly behavioral responses, and their statement as a series of component equations. Synthesis in the form of a complex model permits mathematical examination of an optimum for the system as a whole. Until recently it seemed more reasonable to study such optimization for important resources such as fisheries, but the capability is available and relevant to the study of the environmental optimum of man, and its application must now be pursued vigorously.

These five approaches to the study of human environment provide an objective base for investigating the environmental optimum for man. We cannot close this discussion, however, without pointing out that the final decision, both as to the choice of the optimum and its implementations, is an ethical one. There is an optimum for the sick, and another for the well; there is an optimum for the maladjusted, and another for the well-adjusted. But in treating the problems of the poor and minority groups, in our preoccupation with their immediate relief, we may continue to overlook the ways in which cultural demands of the modern, sub-optimum environment go far beyond the capacity of learned adaptations.

A Compromise?

Considering our scientific effort to learn the functions and structure of the human body, and of the physical environment around us, the limited knowledge of man's relationships to his environment is appalling. Because of the very success of our scientific establishment we are faced with population densities and environmental contaminants that have left us no alternative but to undertake control of the environment itself. In this undertaking let us understand the need to choose a humane compromise—a balance between the evolutionary

demands we cannot deny except with great emotional and physical misery, and the fruits of an unbelievably varied civilization we are loath to give up.

Yet are we even considering such a compromise? With rare exceptions are we not continuing to destroy much that remains of man's natural environment with little thought for the profit of the remote future? In the conflict between preservationists and industrialists (or agriculturists) the latter have had it their way, standing as they do for "progress" and "modern living." While the balance between these conflicts is slowly changing, preservationists continue to be regarded as sentimentalists rather than realists.

Theodosius Dobzhansky says that "the preponderance of cultural over biological evolution will continue to increase in the foreseeable future." We could not wish this to be otherwise; adaptation to the environment by culture is more rapid and efficient than biological adaptation. But social structures cannot continue indefinitely to become more complex and further removed from evolutionary forces. At some stage a compromise must be reached with man's innate evolutionary adaptability.

Need for Continuing Study

We believe that the evidence of man's need for nature, particularly its diversity, is sufficient to justify a determined effort by the scientific community to obtain definitive answers to the questions we have posed. The techniques for studying the problems are to be found in separate disciplines, and there is a sufficient measure of willingness among scientists to undertake the new approaches. But the first steps will be faltering and financial support will be slow in coming.

Now that buttercups are rare, at least symbolically, and springs often silent, why study them? Have there not already been several generations for whom the fields and woods are nearly a closed book? We could encourage the book to close forever, and we might succeed, but in so doing we might fail disastrously. The desire to see and smell and know has not yet been suppressed and enthusiasm for natural history continues to bring vitality to millions. Let us recognize that we are a product of evolution, without apology for the close affinities with our primate forebears. We need only prepare consciously to make a compromise between our cultural and our genetic heritage by striking a balance of social structures with maintenance of natural environments. Most important, we must discover the mechanisms of environmental influence on man. There is no other satisfactory approach to an optimum environment.

Ecumenopolis, World-City of Tomorrow
Constantinos A. Doxiadis

Why the Crises in Our Cities

About thirty-five years ago, when one talked of cities, the only questions raised were, as a rule, questions of the aesthetics of buildings: whether a particular house or monument was beautiful or ugly. Later, when the world began to suffer severely from the poor state of communications, all one heard on every hand was about the crisis in urban communications, and more particularly about too many motorcars. Later still, social problems arose in certain countries, and people began to view the urban crisis from that particular aspect. In some countries, the problems were specifically racial, as in the United States, where the situation is more delicate than elsewhere. And so the urban crisis then took on the appearance of a social crisis.

Sometimes, also, the urban crisis was poorly understood, because each person tended to regard it from his own particular point of view. Certain people, in fact, referred to it as a crisis in small-scale organic unity, that of the family; others saw it as resulting from the disappearance of small neighborhoods and towns, and yet others as something inherent in large cities and big centers of population. Each one, in fact, saw only a single factor of the general crisis in space.

Actually, the crisis is nothing other than that of the entire system. This is an essential principle we must understand: it is the crisis of a system we commonly refer to as the city, but which it would be more accurate to call the human settlement.

The crisis is a general one. We shall understand this if we consider the city from a rational point of view. To do so, we must try to view it under three aspects.

The Five Elements

First we must try to understand of what it is composed. Five elements enter into its composition (Fig. 1). The first of these is nature, the soil which gave it birth. By not taking this simple fact into account, our efforts have been doomed to failure. We have, in fact, polluted the atmosphere and the waters, destroyed beautiful country-sides, exhausted the natural resources, killed off the animals, insects and plants—in a word, annihilated the city's natural setting.

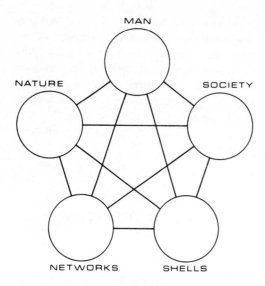

Figure 1.

The second element is man, who is frequently left out of the calculations. All we need to do is look at a city from the air to realize that the most important place is allotted to motor-cars, and man occupies only the second place; that the city is covered with colossal buildings. We have so completely failed to recognize man's existence as to prevent our children from using the streets. In this way man is left exposed to various forms of psychosis and neurosis which are far more serious than the accidents which stain our streets with blood, for they are the diseases of a man who is no longer free to wander about and grows up, for the first time in history, with the feeling of being more at ease in the heart of nature than in the city.

Thirdly, society, painstakingly created by man, produces, in turn, human settlements. We are incapable of creating a society. All we can create are enormous masses of people incapable of performing their normal functions. Men are more and more separated in space

from each other; the necessary contacts between them are lacking. The women, left alone in the suburbs without a second car at their disposal become "nervy," and the men who have to drive for several hours back and forth also become tense and nervous. It is curious to observe that when men are scattered over vast tracts of territory they lose contact with the small surface points of unity—the family, the neighbors. In this way we create a city with nothing human about it.

The fourth element is buildings in general, what the architectural trade refers to as "shells". Technically, there has certainly been progress. But does this progress serve man's true interests? This cannot be proved. On the other hand, we cannot help thinking that these great blocks isolate men and turn them into cave-dwellers. And the idea of wandering through car-choked streets admiring the beauty of some of the buildings as you go along is, strictly speaking, inconceivable.

The fifth and last element is the networks: highways, railways, water-supply systems, electricity and telecommunications. All these become more and more technically perfect every day. If they are underground systems, as in the case of water-supply, electricity and telephone, they cannot possibly inconvenience men from the technical or aesthetic point of view. If, on the other hand, they are on the surface, such as certain electrical and telephone systems, they can be disadvantageous, at any rate aesthetically. But it is in the road system that failure has been most marked. Motorways have the effect of breaking up a city's continuity and preventing it from functioning normally.

We therefore reach the conclusion that the failure of the system is due to the destruction of its elements and of the relationships between them. Because man cannot find joy in his home, because his habitation cannot offer him a better life, the entire city system suffers. Because the city expands rapidly, thanks to the motor-car and other centrifugal forces, it destroys the surrounding country-side and the entire system goes slowly from bad to worse.

The Five Points of View

We can also regard this system in other ways, from the scientific point of view, for example; or we can regard it as an economic, social or political system or, again, as a technological, civilizing or aesthetic one. From whichever angle we look at it, we see that the city, the human settlement, the system, instead of getting better is only getting worse. Everything goes to show that the results obtained bear no relation to the progress being made nowadays in all the spheres of human activity.

The problem becomes even more complicated when, looking at these five elements from different angles and from the five basic points of view—economic, social, political, technological, and cultural—you arrive at a great number of combinations, which show just how many demands will have to be satisfied and how difficult it will be to meet them all. Figure 2 just shows a simple combination of these variables.

These problems become even more complex if we admit that what we understand by the word "city" is only an extremely simplified term for a phenomenon of infinitely greater complexity. The significant feature of the way in which our life in space is organized is no longer the city *per se,* but rather a system of human settlements. In antiquity, at the time of the city-states, we could perhaps have claimed that the system was composed of a city and some villages. Such city-states varied neither in size nor in the relationships between them; these latter were few in number and often much more hostile than pacific.

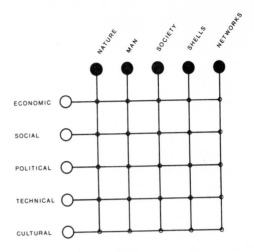

Figure 2.

In maintaining that the same is true today —which unfortunately is what we do—we make a profound mistake, for the city systems in which we live are not limited by cities' natural boundaries. No modern city could survive if its exits were to be closed. If its conduits were to be cut off, too, it would cease to exist, having neither water nor electricity. Moreover, even if it were allowed to maintain relations with the surrounding villages, the city, as we know it, would still be unable to live. Its system of communications is very extended; it imports goods from a great distance and exports likewise to remote countries, and as it functions at an intense rhythm it ends by absorbing the neighboring villages. A village which has radio and television is no longer a village; it has been incorporated into the city network. It is ever receiving orders, and little by little begins to obey them.

Hence the following conclusion: we should not hold the concept of a dissociated "city"— a concept which is, moreover, impossible to define—but envisage it as a system made up of many different units.

The Fifteen Space-Units

If we closely and systematically analyze our living space, we shall discover that we live in fifteen different space units of increasingly greater dimensions (Fig. 3).

The first of these, and the smallest, is that of man himself—it is precisely the space occupied by the human body with all its limbs extended; the second is the room; the third, the dwelling; the fourth, the dwelling group; the fifth, the small neighborhood. Leaping upward, we come to the eighth unit, the traditional town of 30,000 to 50,000 inhabitants; then to the tenth, comprising the metropolis with around two million inhabitants; the eleventh, the conurbation with several million inhabitants, and the twelfth, a new type of urban concentration going by the name of "megalopolis," like the one stretching along the east coast of the United States or like those to be found in the region of the American and Canadian Great Lakes, or in the Netherlands, or along the banks of the Rhine, or yet again in the area stretching between Tokyo and Osaka which is known as Tokaido.

Finally, we come to the fourteenth and fifteenth units, the urban continent and Ecumenopolis, the universal city. These constitute a world system that we cannot, of course, actually see because it remains to be created, but which we should be able to visualize were we able to record the total movements of aircraft in the sky, those of trains and motor vehicles on the earth, and the torrents of news circulating by telephone, telegraph and television. These fifteen spatial units govern that total urban system which I call the city.

If we now consider this system with its fifteen units, and if we combine them with the five elements and five points of view so as to form a complex of forces exerting their influence on the city, we shall then realize that we are talking of billions and trillions of aspects and problems in an enormously complicated system.

We might naturally ask ourselves if this is the first time that the city has fallen sick. Obviously it is not. A city is, generally speaking, always sick. Sometimes it is the dwellings that are unsuitable for habitation and become slums; in other instances service installations are faulty or lacking, as in the past, when there were no town mains of any kind. In certain cases, it is the people themselves who are abnormal so that the community does not function properly. Diseases peculiar to cities, just as human diseases, have always existed. Sometimes remedies could be found for them, sometimes not. But all these diseases differed from our own in that they were restricted to a single element, a single aspect.

Nowadays, the disease of the city is also that of the whole system. It forms part of that system's very existence and increases as the system expands. We can say today that every city is, by definition, sick and that it is always moving towards a crisis, because no single one of its elements is immune from disease.

Intensification of the Crisis

If we have succeeded in understanding the exact nature of the crisis and the reasons for it, we shall also be able to realize easily enough that it can only get worse and worse as time

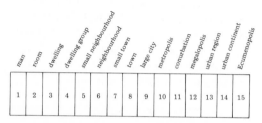

Figure 3. The Fifteen Space Units.

goes on, if things are allowed to remain as they are. It is because we have been unable to grasp this state of affairs that we find ourselves incapable of dealing with it rationally and have no assurance that we will be able to avert the worst.

Since the cause of the crisis in the system is essentially the latter's size, obviously as the size grows, the problems arising out of it will grow in the same proportion. Assuming that the world population at the end of the century will be double the present population of 3,500 million, (and more likely it will be something more), that means it will have reached 7,000 million. In the generation immediately following, that is to say, by the year 2030, the population will probably have quadrupled, but even if this were not so, it is bound to be very much higher than 7,000 million. If this rate of increase continues, we shall have reached a figure of over 20,000 million by the end of the twenty-first century. Such a population increase will call for a corresponding increase in the units which compose the main elements of the city.

But population is only one aspect of the question. A city is not composed only of human elements; there are others besides, such as buildings and mains of various kinds, which complicate the problem. To be able to deal with questions of buildings and mains, we have to know something about the economy. We know, of course, that the economic potential of the population is expanding. We may therefore expect *per capita* income to rise steeply, to at least double the present one in the course of the generation. This means that the gross revenue of a medium-sized town will have quadrupled.

As a first approximation, as the population increases, the need for surface space increases proportionately. However, since incomes go up, people demand more space for their dwell-

ings and service networks. They also have more cars at their disposal and they insist on more room for them, too. And with rising prosperity, the mileage covered by each car constantly increases, so that new motorways have to be built. We can, therefore, say that it is not just the city itself which needs more room, but also every individual in it. This explains how it is that in a good many urban centers, over the past forty years, the surface area has multiplied twice or even three times. We are led to face this additional problem: that the demand for urban space is bound to increase at a faster rate than the population and, in certain cases, will greatly outstrip the growth of the economy.

We now see that, where we have a population with an increased economic potential and therefore insisting on more living space, our whole system of human settlements is bound to become much more complex than the actual growth in population would seem to justify. Likewise, the whole body of problems increases at a faster rate than does either the world population or the urban population.

Escapist Solutions

Most people are not yet aware of the major problems. But there are some who understand them because they think and they make calculations, even though these may not always be very accurate. Such people endeavor to find solutions. At present these are, in fact, but escapist solutions. I propose to enumerate the main ones in the order of their appearance.

In the first category we can place solutions based on various myths. These are, for example, the myth of optimism, which foresees the solution of all problems in emigration to other planets or in the abolishing of motor-cars; the myth founded upon imaginary concepts, such as the assertion that people are living today in high-density urban agglomerations, whereas the medium-sized town of today is in fact less densely populated than it was a generation ago; the myth that our problems would be solved by increasing the height of buildings, although, on the contrary, these enormous constructions create new problems without at all solving the human ones.

There is a whole series of other more unrealistic utopias, those which are based on some dream of reconstruction which is entirely divorced from logic. According to these, cities are no longer necessary.

There is still another form of utopia based on the application of the same abortive solutions to various problems which were tried at the end of the nineteenth century; these led to a host of utopian groups which founded utopian communities. This is a typical escapist solution.

Then there are escapist-motivated solutions which take the form of ideal cities and technological utopias. These advocate the creation of parking lots on the flat roofs of dwellings, or the construction of buildings shaped like huge metal tanks capable of moving from place to place.

The most dangerous escapist solutions are those which advocate a return to small towns. Actually, many of us have been born and raised in such towns and we still see them in our mind's eye and dream of going back to them. This form of utopia takes on different aspects, such as the ideal little towns like those imagined by Skinner with his *Walden Two,* or like those described in Aldous Huxley's last book, *Island,* in which there is a small island where people live in little towns.

Still more dangerous is the theory recommending as the ideal solution the establishment of new satellite towns outside the large cities. Yet do we not now possess the evidence of experience, showing that the satellite towns established sixty years ago have for the past thirty years been urban sectors and that the same fate has befallen those established thirty years ago?

The Path of the Future

We have now reached the point where we must decide on the future road to follow. The question is: are we now capable of examining systematically the various practicable alternatives for the future which are open to us? I believe we are.

First among the roads we can follow is that of research into basic causes, and the first among these to be studied must be the world population increase. Even if a decision on birth control could be adopted immediately, two generations would go by before we could convince the inhabitants of remote villages in India or South America to apply it. This means that in all probability we will reach the figure of 7,000 million and then 12,000 million before the population increase can be arrested.

Given this increase in world population, it is permissible to ask whether the urban population must necessarily rise too. Could we not arrange to keep this population in the countryside? That is something quite out of the question. Man's belief in the freedom of the individual (which all peoples are coming to hold as firmly as their belief in human progress) makes it impossible for us to intervene directly in a man's decisions. We cannot say: 'Live in the villages, even if you are not needed for rural work to produce food, which can today be produced by a reduced number of people.'

The general increase in farm productivity will sustain the swelling of urban populations and the decrease of the rural population. The result will be that, in a world population of 7,000 million in the year 2000, 5,000 million will be city dwellers. Consequently, the urban population will not simply have doubled, as is sometimes naively thought, but will have quadrupled. And when the world population reaches 12,000 million, the lowest leveling-off point, 10,000 million people will be city-dwellers, six or seven times the present number.

Let us now consider the third road open to us. Given the fact of the increase in the urban population, might it not be possible to restrict the growth of our cities by directing the surplus population to new towns? We should give this solution serious thought. We could, indeed, build new towns to absorb that population if the necessary funds were available; for such towns call for a bigger capital investment per head for fewer services, especially during the first few years, the first decades, the first generations, until they reach the size of our present-day cities.

But why do we want the new towns to reach the size of today's cities? The answer is that only towns of a certain size can give men a greater number of options. Some people will say that even a town of modest size can have a theatre and a hospital, and, indeed, towns of 50,000, 80,000 or 250,000 inhabitants (this last being the fashionable figure just now) will be attractive to a certain proportion of the population.

The answer to this argument is simple. There was a time when a man was presumably content in a village of 700 inhabitants, which could support a primary school. What people forget was that such a village could well have insisted on having a secondary school, a vocational school, and a university for its children.

Some will still argue that small towns like this can even meet the cost of maintaining a theatre. But where is the man who will be content with just one theatre, on the model of the small cities of ancient Greece, where the theatre was only obliged to open its doors during important festivals? Might he not well prefer a city with five, ten or twenty theatres, where he could choose between various productions? And why would he be content with a hospital with 200 beds if his disease calls for the attention of a number of specialists, which only university centers are able to provide? In fact, we cannot logically conceive of a city of fixed size which can satisfy our needs. The larger a city the more the needs it serves, which is why people are increasingly attracted to cities.

Should we enlarge our present cities or should we build new ones? The answer is that, theoretically, we could build new ones, provided we make them at least as large as the present ones. In view, however, of the fact that the exodus is from small cities to large ones and that, whatever the size of the new cities we wish to build, their actual construction is bound to take a considerable number of years —one, two or possibly three generations—the population flow will obviously be away from them. Consequently they will prove a failure, for we cannot possibly force these generations to accept dictated solutions to problems which involve their being told where they are to live.

So the answer to this difficult question is as follows: evolutionary and constructive forces will inevitably lead to the growth of cities of the present type in response to the need to satisfy all the requirements of the inhabitants. This does not mean that a certain number of cities of a new type will not be built, but the building of them will be difficult, and will only affect a small proportion of the population.

Thus we are led to the following conclusion: the most probable, logical and practical solution among the three we have been discussing is the progressive expansion of the present type of city as a result of the massive influx of an ever-increasing population.

Towards Ecumenopolis

Our present cities are developing into increasingly complex systems. Starting with the city which develops in concentric circles (see Fig. 4), we finally reach the one which is strung out along the main artery linking it up with the nearest town, port or coast. We thus pass naturally from the city we are familiar with to a system of cities linked together, forming an urban complex with great numbers of inhabitants (Fig. 5).

This brings us at last to the following conclusions: under the growing pressures of these various forces—economic, biological, demographic, etc.—we are gradually creating a bigger and bigger system of settlements, a system

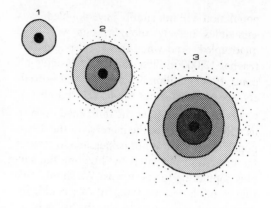

Figure 4.

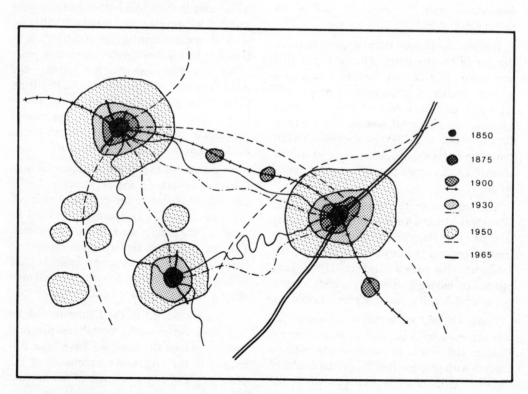

1850

1875

1900

1930

1950

1965

Figure 5.

which, left to develop blindly, can only worsen daily and eventually bring us to catastrophe. This system will very quickly assume world proportions. From the megalopolis we shall pass to cities extending over continents, and thence to Ecumenopolis, or world city. Its advent is inevitable. Any strict analysis reveals that there is nothing logical, rational or practical we can do to avoid it.

Ecumenopolis will be shaped by the forces engendered in cities of the present type as they attract huge populations in the future; by great systems for transportation, the inevitable magnet for industry and other activities; and by forces of an aesthetic nature, such as the attraction exerted by the seaboard on increasing numbers of people (Fig. 6).

People will want to enjoy aesthetic pleasures at home; they will want to be able to build their houses overlooking an attractive valley, or along the coast with beaches opening before them, even though the crowded city center is some distance away. At the same time, we must bear in mind the attraction of the vast plains, where water abounds, where the climate is mild.

All the above allows us gradually to form an idea of what the universal city will look like. As a result of research conducted by the Athens Centre of Ekistics we can already imagine to some degree how it will appear within a century or a century and a half (Fig. 7).

Ecumenopolis, this world-city that will englobe the whole of humanity, will be a frightening conurbation, but, as we made clear earlier, we have no evidences that enable us to conclude that a better sort of city can be created. Once we are convinced that this city is inevitable, we can only form one conclusion: if it is built on today's lines, according to present-day trends, it will be a city doomed to destruction, that which Lewis Mumford re-

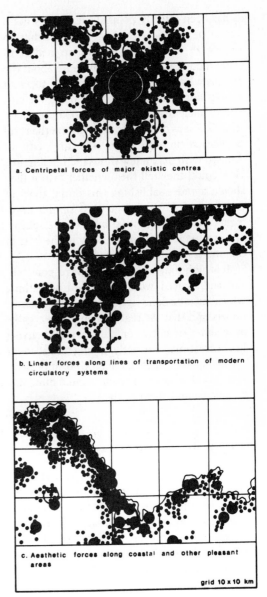

a. Centripetal forces of major ekistic centres

b. Linear forces along lines of transportation of modern circulatory systems

c. Aesthetic forces along coastal and other pleasant areas

grid 10 x 10 km

Figure 6.

ferred to some time ago as a necropolis—city of the dead.

There is still, of course, another road that could be followed: to avoid it altogether. But, as has been already pointed out, this is not a logical or practical solution. I would like to emphasize here that we have no reason to claim today that we know any more about the reasons why it would be a good thing to avoid establishing a universal city than did the citizens of ancient Athens, about 3,000 years ago, when Theseus decided to concentrate the rural population in one town, the tiny city of Athens, with only a few thousand inhabitants. How could they tell then whether or not they should avoid establishing this initial town in the plain of Athens? Doubtless similar arguments were used then in favor of the scheme and against it as are used today in the case of the universal city.

It is high time we accepted our responsibilities and started working toward something which must be done right. To do this, we must understand that the real challenge does not lie in whether or not to create the world-city; it lies in creating it correctly, taking into account the human factor, so that man who, at present, sees his values disintegrating around him, may be able to find them again.

The Outlines of Ecumenopolis

In the preceding sections of this study, I have tried above all to clarify the problem so it can be fully understood, for understanding is an essential condition for success. If we fail today to deal with this problem, it will be because we have not understood it.

In Athens we are presently applying ourselves to the collection of data on human settlements. This systematized knowledge forms an organized discipline which is becoming a science, called "ekistics"—the science of human settlements.

We must realize that ekistics cannot be restricted merely to understanding the problem; it must also lead to solutions for tomorrow.

How is it that man in the past possessed the necessary strength, imagination and courage to build permanent settlements when he was still a hunter? How is it that he then went on to build villages, towns, industrial cities and metropolises? Why shouldn't we today have the necessary courage to conceive and build the world-city? To do it, we need, in addition to science, technology and art. Thus, we shall have to make ekistics a science, a technology and an art, all in one.

If we set to work in this way, we shall come to realize that we really can create Ecumenopolis. It is of no special importance to us to know exactly what the size of the city will be. For it will not make much difference to us that, going in certain directions, we would pass through hundreds of miles of urban centers. What will really matter is to know that, after a journey of ten or twenty minutes or of one to five hours, as the case may be, we can be certain of finding the country-side.

When we see the problem in this light, we shall understand that size and shape are of no special concern; what matters is a proper balance between elements. We shall then affirm this conclusion: that nature must be converted into a gigantic network with tentacles penetrating deeply into all parts of the universal city so as to reach every residential area—a system of woodlands transformed into parks, intersected by avenues and gardens, within easy reach of our homes (Fig. 8).

The size of a city should not worry us if we know that we can control the atmosphere and keep it unpolluted. There are small towns where the atmosphere is contaminated by a great number of cars and large towns where the atmosphere is clean. What is important is to ensure that pollution from industrial plants and vehicular traffic is under control. Then we will be able to breathe better air in a large city than we could in a small town without proper pollution control.

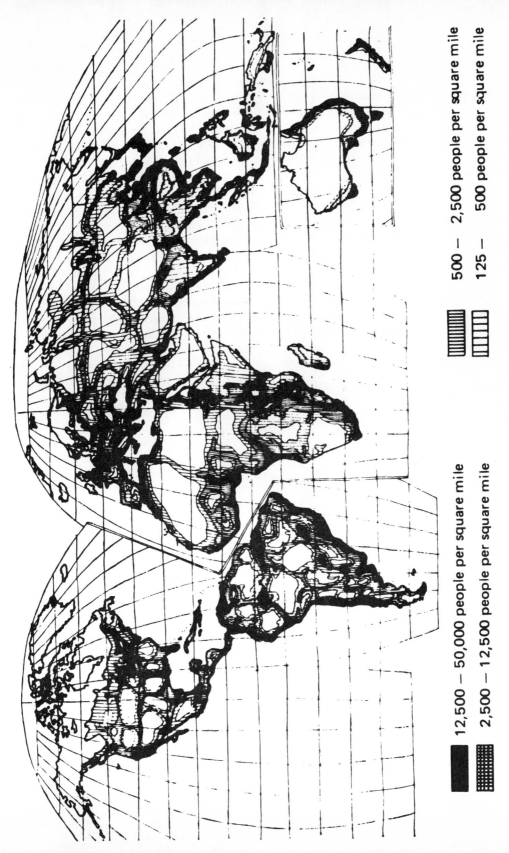

12,500 – 50,000 people per square mile

2,500 – 12,500 people per square mile

500 – 2,500 people per square mile

125 – 500 people per square mile

Figure 7.

Proceeding in this way enables us to understand, little by little, how the elements of nature will penetrate the city. A similar approach can establish the nature of the various networks. Transportation problems have nothing whatever to do with the size of a city. They are the result of lack of organization and because we have not yet learned that men and machines cannot exist on the same footing.

Urban transportation will function properly when it is placed beneath the surface, like arteries in the human body. As soon as we grasp this fact, we can take the first steps in the right direction—indeed, we have already done so. At one time, water was carried in surface conduits; the same was true—and still is in many places—of sewage drains; overhead electric and telephone wires are still a common sight.

The day will come when all such installations will be below the surface. Goods will be transported through underground tubes. Some of these already exist. In Canada, important networks of the kind are under construction for the transport of industrial products.

In the near future, no one will object to using underground roadways, just as no one made objections when the subways began transporting people at speeds of 15-20 miles an hour in London, Paris and New York, considerably faster than they had been used to in their horse-drawn cabs. Before very long, we shall be travelling underneath our cities at speeds of 60-200 miles an hour, spending perhaps between 5 and 10 minutes below ground, instead of driving for hours on roads, constantly irritated by the stop and go of traffic lights.

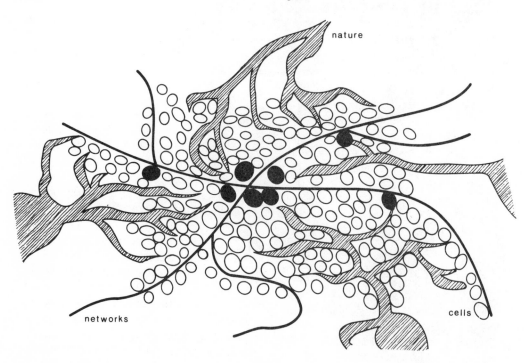

Figure 8.

These new transportation networks will be much more satisfactory and will make it possible to have cities spread over very much wider areas while being much better organized.

When this programme has been carried out, we will then attain the solution which is of prime importance to us: freeing the surface of the earth for man to enjoy and to use for the development of his artistic gifts. In a word, the earth's surface will be used in harmony with man's way of life. We harken back thus to the time, thousands of years ago, when man was both a researcher and a guinea-pig in the vast laboratory of life and did the experiments which enabled him to build throughout the world those beautiful cities which we still admire: ancient Athens, Florence, the old Paris and old London, as well as Williamsburg in the United States. An intrinsic worth attached to these cities because they had been built on a human scale, to man's own measurements.

A careful study of the cities of the past shows that they never exceeded more than about 1 mile in length or about 1.5 square miles in area, and included no more than 50,000 inhabitants, when they were at their most successful. Cities which were much bigger were, in fact, the capitals of large empires and were never able to retain their organization for very long. They often deteriorated into anarchy, as in the case of Rome and Byzantium. If they hoped to sustain an organized and integral life of their own, they had to be carefully planned from the start, like Peking and Changan (modern Sian), two ancient Chinese capitals.

Generally, then, any cities which exceeded the usual, the reasonable maxima were doomed to fail, were short-lived, and offer us no solution. Those that do offer a solution were the small cities of 30,000 to 50,000 inhabitants, covering an area of about 1.5 square miles. If we examine their structure rationally,

we shall realize that we must return to something similar if we want to organize our life properly.

We thus reach what seems to be a paradox: on the one hand, the inevitable huge Ecumenopolis; on the other, the absolute need for man to live in small cities. But this only appears paradoxical. For following such reasonings to their logical conclusion, we arrive at a gigantic city of superhuman dimensions, made up of small units.

Thus, our thought processes have led us to the construction of huge cities composed of small towns, of vast urban complexes served by underground transportation systems, leaving the surface of the ground free, at man's disposal, and supplied with every human amenity. The conclusion we reach then is that Ecumenopolis, the world-city, will be made up of cells of 30,000 to 50,000 people.

In practice, we are already beginning to build such cities. Islamabad, the new capital of Pakistan, intended for 2.5 million inhabitants, has been laid out in this way. Before such cities are even finished, life there is already in full swing and anyone can go and study them in operation. It is very important to keep on studying them until they've been brought to perfection. Then by applying similar principles, it will be possible to transform some of the older cities.

Indeed, these principles have been applied to certain limited areas or even to certain cities which are planning the organization of the entire system on this sound basis. One example is Philadelphia (United States), which is now engaged in a gradual urban renewal project which will house 10,000 families on ground once covered with slums. There is also the immense urban region of Detroit, now in course of development and planned to accommodate by the year 2000 more than 10 million inhabitants enjoying maximum amenities.

So, after examining the nature of the crisis of the cities and the various escapist solutions, we have gradually synthesized a solution. This solution enables us to envisage the development of human settlements in a practical manner and to build in such a way as to offer men a much happier form of existence by combining the advantages of the small towns of old—which were certainly considerable from the point of view of a humane way of life—with those of the large cities which alone are capable of enlarging our freedoms and chances of development.

Four Changes
Anonymous

Population

THE CONDITION

Position: Man is but a part of the fabric of life—dependent on the whole fabric for his very existence. As the most highly developed tool-using animal, he must recognize that the unknown evolutionary destinies of other life forms are to be respected, and act as gentle steward of the earth's community of being.

Situation: There are now too many human beings, and the problem is growing rapidly worse. It is potentially disastrous not only for the human race but for most other life forms.

Goal: The goal would be half of the present world population, or less.

ACTION

Social/political: First, a massive effort to convince the governments and leaders of the world that the problem is severe. And that all talk about raising food-production—well intentioned as it is—simply puts off the only real solution: reduce population. Demand immediate participation by all countries in programs to legalize abortion, encourage vasectomy and sterilization (provided by free clinics)—free insertion of intrauterine loops—try to correct traditional cultural attitudes that tend to force women into child-bearing—remove income tax deductions for more than two children above a specified income level, and scale it so that lower income families are forced to be careful too—or pay families to limit their number. Take vigorous stand against the policy of the right-wing in the Catholic hierarchy and any other institutions that exercise an irresponsible social force in regard to this question; oppose and correct simple-minded boosterism that equates population growth with continuing prosperity. Work ceaselessly to have all political questions be seen in the light of this prime problem.

The community: Explore other social structures and marriage forms, such as group marriage and polyandrous marriage, which provide family life but may produce less children. Share the pleasures of raising children widely, so that all need not directly reproduce to enter into this basic human experience. We must hope that no woman would give birth to more than one child, during this period of crisis. Adopt children. Let reverence for life and reverence for the feminine mean also a reverence for other species, and future human lives, most of which are threatened.

Our own heads: "I am a child of all life, and all living beings are my brothers and sisters, my children and grandchildren. And there is a child within me waiting to be brought to birth, the baby of a new and wiser self." Love, love-making, a man and woman together, seen as the vehicle of mutual realization, where the creation of new series and a new world of being is as important as reproducing our kind.

Pollution

THE CONDITION

Position: Pollution is of two types. One sort results from an excess of some fairly ordinary substance—smoke, or solid waste—which cannot be absorbed or transmitted rapidly enough to offset its introduction into the environment, thus causing changes the great cycle is not prepared for. (All organisms have wastes and by-products, and these are indeed part of the total biosphere: energy is passed along the line and refracted in various ways, "the rainbow body." This is cycling, not pollution.) The other sort is powerful modern chemicals and poisons, products of recent technology, which the biosphere is totally unprepared for. Such is DDT and similar chlorinated hydrocarbons—nuclear testing fall-out and nuclear waste—poison gas, germ and virus storage and leakage by the military; and chemicals which are put into food, whose long-range effects on human beings have not been properly tested.

Situation: The human race in the last century has allowed its production and scattering of wastes, by-products, and various chemicals to become excessive. Pollution is directly harming life on the planet: which is to say, ruining the environment for humanity itself. We are fouling our air and water, and living in noise and filth that no "animal" would tolerate, while advertising and politicians try and tell us we've never had it so good. The dependence of the modern governments on this kind of untruth leads to shameful mind-pollution: mass media and most school education.

Goal: Clean air, clean clear-running rivers, the presence of Pelican and Osprey and Gray Whale in our lives; salmon and trout in our streams; unmuddled language and good dreams.

ACTION

Social/political: Effective international legislation banning DDT and related poisons —with no fooling around. The collusion of certain scientists with the pesticide industry and agri-business in trying to block this legislation must be brought out in the open. Strong penalties for water and air pollution by industries—"Pollution is somebody's profit." Phase out the internal combustion engine and fossil fuel use in general—more research into nonpolluting energy sources; solar energy; the tides. No more kidding the public about atomic waste disposal: it's impossible to do it safely, and nuclear-power generated electricity cannot be seriously planned for as it stands now. Stop all germ and chemical warfare research and experimentation; work toward a hopefully safe disposal of the present staggering and stupid stockpiles of H-Bombs, cobalt gunk, germ and poison tanks and cans. Laws and sanctions against wasteful use of paper, etc. which adds to the solid waste of cities— develop methods of re-cycling solid urban wastes. Re-cycling should be the basic principle behind all waste-disposal thinking. Thus, all bottles should be re-usable; old cans should make more cans; old newspapers back into newsprint again. Stronger controls and research on chemicals in foods. A shift toward a more varied and sensitive type of agriculture (more small-scale and subsistence farming) would eliminate much of the call for blanket use of pesticides.

The community: DDT and such: don't use them. Air pollution: use less cars. Cars pollute the air, and one or two people riding lonely in a huge car is an insult to intelligence and the Earth. Share rides, legalize hitch-hiking, and build hitch-hiker waiting stations along the highways. Also—a step toward the new world—walk more; look for the best routes through beautiful countryside for long-distance walking trips: San Francisco to Los Angeles down the Coast Range, for example. Learn how to use your own manure as fertilizer if you're in the country—as the Far East has done for centuries. There's a way, and it's safe. Solid waste: boycott bulky wasteful Sunday papers which use up trees. It's all just advertising anyway, which is artificially inducing more industry consumption. Refuse paper bags at the store. Organize Park and Street clean-up festivals. Don't work in any way for or with an industry which pollutes, and don't be drafted into the military. Don't waste. (A monk and an old master were once walking in the mountains. They noticed a little hut upstream. The monk said, "A wise hermit must live there"—the master said, "That's no wise hermit, you see that lettuce leaf floating down the stream, he's a Waster." Just then an old man came running down the hill with his beard flying and caught the floating lettuce leaf.) Carry your own jug to the winery and have it filled from the barrel.

Our own heads: Part of the trouble with talking about DDT is that the use of it is not just a practical device, it's almost an establishment religion. There is something in Western culture that wants to totally wipe out creepy-crawlies, and feels repugnance for toadstools and snakes. This is fear of one's own deepest natural inner-self wilderness areas, and the answer is, relax. Relax around bugs, snakes, and your own hairy dreams. Again, farmers can and should share their crops with a certain percentage of buglife as "paying their dues." Thoreau says: "How then can the harvest fail? Shall I not rejoice also at the abundance of the weeds whose seeds are the granary of the birds? It matters little comparatively whether

the fields fill the farmer's barns. The true husbandman will cease from anxiety, as the squirrels manifest no concern whether the woods will bear chestnuts this year or not, and finish his labor with every day, relinquish all claim to the produce of his fields, and sacrificing in his mind not only his first but his last fruits also." In the realm of thought, inner experience, consciousness, as in the outward realm of interconnection, there is a difference between balanced cycle, and the excess which cannot be handled. When the balance is right, the mind recycles from highest illuminations to the stillness of dreamless sleep; the alchemical "transmutation."

Consumption

THE CONDITION

Position: Everything that lives eats food, and is food in turn. This complicated animal, man, rests on a vast and delicate pyramid of energy-transformations. To grossly use more than you need, to destroy, is biologically unsound. Most of the production and consumption of modern societies is not necessary or conducive to spiritual and cultural growth, let alone survival; and is behind much greed and envy, age-old causes of social and international discord.

Situation: Man's careless use of "resources" and his total dependence on certain substances such as fossil fuels (which are being exhausted, slowly but certainly) are having harmful effects on all the other members of the life-network. The complexity of modern technology renders whole populations vulnerable to the deadly consequences of the loss of any one key resource. Instead of independence we have over-dependence on life-giving substances such as water, which we squander. Many species of animals and birds have become extinct in the service of fashion fads—

or fertilizer—or industrial oil—the soil is being used up; in fact, mankind has become a locust-like blight on the planet that will leave a bare cupboard for its own children—all the while in a kind of Addict's Dream of affluence, comfort, eternal progress—using the great achievements of science to produce software and swill.

Goal: Balance, harmony, humility, growth which is a mutual growth with Redwood and Quail (would you want your child to grow up without ever hearing a wild bird?) —to be a good member of the great community of living creatures. True affluence is not needing anything.

ACTION

Social/political: It must be demonstrated ceaselessly that a continually "growing economy" is no longer healthy but a Cancer. And that the criminal waste which is allowed in the name of competition—especially that ultimate in wasteful needless competition, hot wars and cold wars with "communism" (or "capitalism")—must be halted totally with ferocious energy and decision. Economics must be seen as a small sub-branch of Ecology, and production/distribution/consumption handled by companies or unions with the same elegance and spareness one sees in nature. Soil banks: open space; phase out logging in most areas. "Lightweight dome and honeycomb structures in line with the architectural principles of nature." "We shouldn't use wood for housing because trees are too important." Protection for all predators and varmints: "Support your right to arm bears." Damn the International Whaling Commission which is selling out the last of our precious, wise whales: absolutely no further development of roads and concessions in National Parks and Wilderness Areas; build auto campgrounds in the least

desirable areas. Plan consumer boycotts in response to dishonest and unnecessary products. Radical Co-ops. Politically, blast both "Communist" and "Capitalist" myths of progress, and all crude notions of conquering or controlling nature.

The community: Sharing and creating. The inherent aptness of communal life— where large tools are owned jointly and used efficiently. The power of renunciation: If enough Americans refused to buy a new car for one given year it would permanently alter the American economy. Recycling clothes and equipment. Support handicrafts, gardening, home skills, midwifery, herbs—all the things that can make us independent, beautiful and whole. Learn to break the habit of unnecessary possessions—a monkey on everybody's back —but avoid a self-abnegating anti-joyous self-righteousness. Simplicity is light, carefree, neat and loving—not a self-punishing ascetic trip. (The great Chinese poet Tu Fu said "The ideas of a poet should be noble and simple.") Don't shoot a deer if you don't know how to use all the meat and preserve that which you can't eat, to tan the hide and use the leather— to use it all, with gratitude, right down to the sinew and hooves. Simplicity and mindfulness in diet is a starting point for many people.

Our own heads: It is hard to even begin to gauge how much a complication of possessions, the notions of "my and mine," stand between us and a true, clear, liberated way of seeing the world. To live lightly on the earth, to be aware and alive, to be free of egotism, to be in contact with plants and animals, starts with simple concrete acts. The inner principle is the insight that we are inter-dependent energy-fields of great potential wisdom and compassion—expressed in each person as a superb mind, a handsome and complex body, and the almost magical capacity of language. To these potentials and capacities, "owning things" can add nothing of authenticity. "Clad in the sky, with the earth for a pillow."

Transformation

Position: Everyone is the result of four forces: the conditions of this known-universe (matter/energy forms and ceaseless change); the biology of his species; his individual genetic heritage and the culture he's born into. Within this web of forces there are certain spaces and loops which allow total freedom and illumination. The gradual exploration of some of these spaces is "evolution" and, for human cultures, what "history" could be. We have it within our deepest powers not only to change our "selves" but to change our culture. If man is to remain on earth he must transorm the five-millenia-long urbanizing civilization tradition into a new ecologically-sensitive harmony-oriented wild-minded scientific/ spiritual culture. "Wildness is the state of complete awareness. That's why we need it."

Situation: Civilization, which has made us so successful a species, has overshot itself and now threatens us with its inertia. There is some evidence that civilized life isn't good for the human gene pool. To achieve the Changes we must change the very foundations of our society and our minds.

Goal: Nothing short of total transformation will do much good. What we envision is a planet on which the human population lives harmoniously and dynamically by employing a sophisticated and unobtrusive technology in a world environment which is "left natural." Specific points in this vision:

* A healthy and spare population of all races, much less in number than today.
* Cultural and individual pluralism, unified by a type of world tribal council. Division by natural and cultural boundaries rather than arbitrary political boundaries.

* A technology of communication, education, and quiet transportation, land-use being sensitive to the properties of each region. Allowing, thus, the Bison to return to much of the high plains. Careful but intensive agriculture in the great alluvial valleys; deserts left wild for those who would trot in them. Computer technicians who run the plant part of the year and walk along with the Elk in their migrations during the rest.
* A basic cultural outlook and social organization that inhibits power and property-seeking while encouraging exploration and challenge in things like music, meditation, mathematics, mountaineering, magic and all other ways of authentic being-in-the-world. Women totally free and equal. A new kind of family—responsible, but more festive and relaxed —is implicit.

ACTION

Social/political: It seems evident that there are throughout the world certain social and religious forces which have worked through history toward an ecologically and culturally enlightened state of affairs. Let these be encouraged: Gnostics, hip Marxists, Teilhard de Chardin Catholics, Druids, Taoists, Biologists, Witches, Yogins, Bhikkus, Quakers, Sufis, Tibetans, Zens, Shamans, Bushmen, American Indians, Polynesians, Anarchists, Alchemists . . . the list is long. All primitive cultures, all communal and ashram movements. Since it doesn't seem practical or even desirable to think that direct bloody force will achieve much, it would be best to consider this a continuing "revolution of consciousness" which will be won not by guns but by seizing the key images, myths, archetypes, eschatologies, and ecstasies so that life won't seem worth living unless one's on the transforming energy's side. By taking over "science and

technology" and releasing its real possibilities and powers in the service of this plant—which after all, produced us and it.

The community: New schools, new classes, walking in the woods and cleaning up the streets. Find psychological techniques for creating an awareness of "self" which includes the social and natural environment. "Consideration of what specific language forms—symbolic systems—and social institutions constitute obstacles to ecological awareness." Without falling into a facile interpretation of McLuhan, we can hope to use the media. Let no one be ignorant of the facts of biology and related disciplines; bring up our children as part of the wild-life. Some communities can establish themselves in back-water rural areas and flourish—others maintain themselves in urban centers, and the two types work together—a two-way flow of experience, people, money and home-grown vegetables. Ultimately cities will exist only as joyous tribal gatherings and fairs, to dissolve after a few weeks. Investigating new life-styles is our work, as is the exploration of Ways to explore our inner realms—with the known dangers of crashing that go with political-minded people where it helps, hoping to enlarge their vision, and with people of all varieties of politics or thought at whatever point they become aware

of environment urgencies. Master the archaic and the primitive as models of basic nature-related cultures—as well as the most imaginative extensions of science—and build a community where these two vectors cross.

Our own heads: is where it starts. Knowing that we are the first human beings in history to have all of man's culture and previous experience available to our study, and being free enough of the weight of traditional cultures to seek out a larger identity. —The first members of a civilized society since the early Neolithic to wish to look clearly into the eyes of the wild and see our self-hood, our family, there. We have these advantages to set off the obvious disadvantages of being as screwed up as we are—which gives us a fair chance to penetrate some of the riddles of ourselves and the universe, and to go beyond the idea of "man's survival" or "the survival of the biosphere" and to draw our strength from the realization that at the heart of things is some kind of serene and ecstatic process which is actually beyond qualities and certainly beyond birth-and-death. "No need to survive!" "In the fires that destroy the universe at the end of the kalpa, what survives?" — "The iron tree blooms in the void!"

Knowing that nothing need be done, is where we begin to move from.